Computer Communications and Networks

T0214575

The **Computer Communications and Networks** series is a range of textbooks, monographs and handbooks. It sets out to provide students, researchers and non-specialists alike with a sure grounding in current knowledge, together with comprehensible access to the latest developments in computer communications and networking.

Emphasis is placed on clear and explanatory styles that support a tutorial approach, so that even the most complex of topics is presented in a lucid and intelligible manner.

Also in this series:

Sid Katzen

The Quintessential PIC® Microcontroller

Second Edition

 Springer

Sid Katzen, BSc, MSc, DPhil, MIEE, MIEEE, CEng
School of Electrical and Mechanical Engineering, University of Ulster,
Northern Ireland

Series Editor
Professor A.J. Sammes, BSc, MPhil, PhD, FBCS, CEng
CISM Group, Cranfield University, RMCS, Shrivenham, Swindon SN6 8LA, UK

British Library Cataloguing in Publication Data
A catalogue record for this book is available from the British Library

Library of Congress Cataloging-in-Publication Data
Katzen, Sid.
    The quintessential PIC® microcontroller/Sid Katzen.—2nd ed.
    p. cm. — (Computer communications and networks)
    Includes bibliographical references and index.
    ISBN 1-85233-942-X (alk. paper)
    1. Programmable controllers. I. Title. II. Computer communications and networks.
TJ223.P76K38 2005
629.8′9—dc22   2005042504

The following are registered trademarks of Microchip Technology Incorporated in the United States of America and other countries: dsPIC, MPLAB, PIC, and PICSTART.

The following are trademarks of Microchip Technology Incorporated in the United States of America and other countries: ICSP, In-Circuit Serial Programming, and MPASM.

Computer Communications and Networks ISSN 1617-7975

ISBN-10: 1-85233-942-X
ISBN-13: 978-1-85233-942-5
First Edition ISBN: 1-85233-309-X
Springer Science+Business Media
springeronline.com

© Springer-Verlag London Limited 2005
First Edition published 2001

Typesetting: Output-ready electronic files provided by the author.
Printed and bound in the United States of America   (MVY)
9 8 7 6 5 4 3 2 1   Printed on acid-free paper

In memory of Eva Jones

Contents

Appendices

Preface to the Second Edition

A second edition of this book has given me the opportunity to respond to suggestions from both students and correspondents from around the world, from disparate regions ranging from Scotland to Hawaii. Since the time of the first edition written in the late 1990s, the Microchip PIC range has become the largest volume selling 8-bit MCU. The mid-range family used in the original edition has continued to expand vigorously, with some of the exemplars used becoming essentially obsolete. In addition, the enhanced-range 16-bit instruction line has been enlarged from virtually nothing to form a significant proportion of the family. At the same time, new introductions to the original low- (or base-) end architecture continue apace. Because of the close relationship between the low-, mid-, high- and enhanced-range lines, the focus of the new edition has stayed with the mid-range line up.

Virtually all diagrams have been modified, many extensively, and numerous additional new figures have been added. Throughout the text, special attention has been paid to clarify the basic concepts. In Part I, Chapter 3 has been extensively rewritten with this in mind and to better integrate with Chapters 4 and 5 in Part II, both of which bear only a superficial relation to the original text. Chapter 7, covering interrupt handling, has also been largely rewritten to elucidate a difficult topic. Part III not only has been revised to use current exemplars, but has been extended to cover additional peripherals such as the Analog Comparator and Voltage Reference modules. A new chapter covers the enhanced-range PIC18FXXX range.

With the exception of the first two and last chapters, all chapters have both fully worked examples and self-assessment questions. As an extension to this, an associated Web site at

`http://www.engj.ulst.ac.uk/sidk/quintessential`

has the following facilities:

- Solutions to self-assessment questions.
- Further self-assessment questions.
- Additional material.
- Source code for all examples and questions in the text.
- Pointers to development software and data sheets for devices used in the book.

- Errata.
- Feedback from readers.

The manuscript was typeset by the author on a variety of Microsoft® Windows™ PCs using a Y&Y implementation of LaTeX 2_ε and the Lucida Bright font family. Line drawings were created or modified with Autocad R13 and incorporated as encapsulated PostScript files. Photographs were taken by the author using several Olympus digital cameras—which are absolutely full of microcontrollers!

Hopefully, any gremlins have been exorcised, but if you find any or have any other suggestions, I will be happy to acknowledge such communications via the Web site.

Sid Katzen
University of Ulster at Jordanstown
July 2005

Preface to the First Edition

Microprocessors and their microcontroller derivatives are a widespread, if rather invisible, part of the infrastructure of our twenty-first-century electronic and communications society. In 1998, it was estimated[1] that hidden in every home there were about 100 microcontrollers and microprocessors: in the singing birthday card, washing machine, microwave oven, television controller, telephone, personal computer and so on. About 20 more lurked in the average family car, for example, monitoring in-tire radio pressure sensors and displaying critical data through a control area network (CAN).

Around 4 billion such devices are sold each year to implement the intelligence of these "smart" electronic devices, ranging from smart egg-timers through to aircraft management systems. The evolution of the microprocessor from the first Intel device introduced in 1971 has revolutionised the structure of society, effectively creating the second Industrial Revolution at the beginning of the twenty-first century. Although the microprocessor is better known for its role in powering the ubiquitous PC, where raw computing power is the goal, sales of microprocessors such as the Intel Pentium represent only around 2% of the total volume. The vast majority of sales are of low-cost microcontrollers embedded into a dedicated-function digital electronic device, such as the smart card. Here the emphasis is on the integration of the core processor with memory and input/output resources in the one chip. This integrated computing system is known as a *microcontroller*.

In seeking to write a book in this area, the overall objective was to get the reader up-to-speed in designing small embedded microcontroller-based systems, rather than using microcontrollers as a vehicle to illustrate computer architecture in the traditional sense. This will hopefully give the reader confidence that, even at such an introductory level, he/she can design, construct, and program a complete working embedded system.

Given the practical nature of this material, real-world hardware and software products are used throughout to illustrate the material. The microcontroller market is dominated by devices that operate on 8-bit data (although 4- and 16-bit examples are available) like early microprocessors and unlike the 64-bit Intel Pentium and Motorola Power PC

---

[1] *New Scientist*, vol. 59, no. 2141, 4 July 1998, p. 139.

"heavy brigade". In contrast, the essence of the microcontroller lies in its high system-integration/low-cost profile. Power can be increased by distributing processors throughout the system. Thus, for example, a robot arm may have a microcontroller for each joint implementing simple local processes and communicating with a more powerful processor making overall executive decisions.

In choosing a target architecture, acceptance in the industrial market, easy availability, and low-cost development software have made the Microchip family one of the most popular choices as the pedagogic vehicle in learning microprocessor/microcontroller technology at all levels of electronic engineering from grade school to university. In particular, the reduced instruction set, together with the relatively simple innovative architecture, reduces the learning curve. In addition to their industrial and educational roles, the PIC® MCU families are also the mainstay of hobbyist projects, as a leaf through any electronics magazine will show.

Microchip, Inc., is a relatively recent entrant to the microcontroller market with its family of Harvard architecture PIC devices introduced in 1989. By 1999, Microchip was the second largest producer of 8-bit units—behind only Motorola.

This book is split into three parts. Part I covers sufficient digital, logic and computer architecture to act as a foundation for the microcontroller engineering topics presented in the rest of the text. Inclusion of this material makes the text suitable for stand-alone usage, as it does not require a prerequisite digital systems module.

Part II looks mainly at the software aspects of the mid-range PIC microcontroller family, its instruction set, how to program it at assembly and high-level **C** coding levels, and how the microcontroller handles subroutines and interrupts. Although the 14-bit PIC family is the exemplar, both architecture and software are comparable to both the 12- and 16-bit ranges.

Part III moves on to the hardware aspects of interfacing and interrupt handling, with the integration of the hardware and software being a constant theme throughout. Parallel and serial input/output, timing, analog, and EEPROM data-handling techniques are covered. A practical build and program case study integrates the previous material into a working system, as well as illustrating simple testing strategies.

Sid Katzen
University of Ulster at Jordanstown
December 2000

PART I

The Fundamentals

This book is about microcontrollers (MCUs). These are digital engines modeled after the architecture of a stored-program computer and integrated onto a single very largescale integrated circuit together with support circuitry, memories and peripheral interface devices. Although the microcontroller is often confused with its better-known cousin, the microprocessor, in its role as the driving force of the ubiquitous personal computer, the vast majority of both microprocessors and microcontrollers are embedded into a variety of other digital components. The first microprocessors in the early 1970s were marketed as an alternative way of implementing digital circuitry. Here the task would be determined by a series of instructions encoded as binary code groups in read-only memory. This is more flexible than the alternative approach of wiring hardware integrated circuits in the appropriate manner. The microcontroller is simply the embodiment of this original role of the integrated computer.

We will look at embedded microcontrollers in a general digital processing context in Parts II and III. Here our objective is to lay the foundation for this material. We will be covering:

- Digital code patterns.
- Binary arithmetic.
- Digital circuitry.
- Computer architecture and programming.

This will by no means be a comprehensive review of the subject, but there are many other excellent texts in this area[1] which will launch you into greater depths.

---

[1]Such as S.J. Cahill's *Digital and Microprocessor Engineering*, 2nd ed., Prentice Hall, Englewood Cliffs, NJ, 1993.

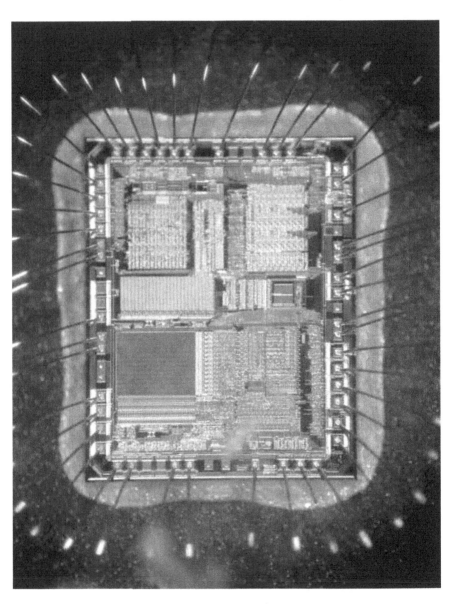

Peeking into the silicon.

CHAPTER 1

Digital Representation

To a computer or microcontroller, the world is seen in terms of patterns of digits. The **decimal** (or denary) system represents quantities in terms of the ten digits $0, \cdots, 9$. Together with the judicious use of the symbols $+$, $-$ and . any quantity in the range $\pm\infty$ can be depicted. Indeed non-numeric concepts can be encoded using numeric digits. For example the American Standard Code for Information Interchange (ASCII) defines the alphabetic (alpha) characters A as 65, B = 66, $\cdots$, Z = 90 and a = 97, b = 98, $\cdots$, z = 122, etc. Thus the string "Microcontroller" could be encoded as "77, 105, 99, 114, 111, 99, 111, 110, 116, 114, 111, 108, 108, 101, 114". Provided you know the context—that is, what is a pure quantity and what is text—just about any symbol can be coded as numeric digits.[1]

Electronic circuits are not very good at storing and processing a multitude of different values. It is true that the first American digital computer, the Electronic Numerical Integrator And Calculator (ENIAC) in 1946 did its arithmetic in decimal form,[2] but all computers since then have handled data in **binary** (base 2) form. The decimal (base 10) system is really only convenient for humans, in that we have ten fingers.[3] Thus, in this chapter we will solely look at the properties of binary digits, their groupings and processing. After reading it you will:

- Understand why a binary data representation is the preferred base for digital circuitry.
- Know how a quantity can be depicted in natural binary, hexadecimal and binary coded decimal.
- Be able to apply the rules of addition and subtraction for natural binary quantities.
- Know how to multiply by shifting left.
- Know how to divide by shifting right and propagating the sign bit.
- Understand the Boolean operations of NOT, AND, OR and XOR.

The information technology revolution is based on the manipulation, computation and transmission of digitized information. This informa-

---

[1] Of course, there are lots of digital encoding standards; for instance, the 6-dot Braille code for the visually impaired.

[2] As did Babbage's mechanical computer of a century earlier.

[3] And ten toes, but base-20 systems are rare though not unknown.

tion is virtually universally represented as aggregrates of *binary digits* (**bits**).[4] Most of this processing is effected using microprocessors[5] and microcontrollers, and it is sobering to reflect that there is more computing power in a singing birthday card than existed on the entire planet in 1950!

Binary is the universal choice for data representation, as an electronic switch is just about the easiest device that can be implemented using a transistor. Such 2-state switches are very small; they change state very quickly and consume little power. Furthermore, as there are only two states to distinguish between, a binary depiction is likely to be resistant to the effects of noise. The upshot of this is that both the packing density on a silicon chip and switching rate can be very high. Although a switch on its own does not represent much computing power, 5 million switches changing at 100 million times a second manage to present at least a façade of intelligence!

The two states of a bit are conventionally designated **logic 0** and **logic 1**, or just 0 and 1. A bit may be represented by two states of any number of physical quantities; for instance, electric current or voltage, light, or pneumatic pressure. Most microcontrollers use 0 V (or ground) for state 0 and 3 – 5 V for state 1, but this is not universal. For instance, the RS232 serial port on your computer uses nominally +12 V for state 0 and −12 V for state 1.

A single bit on its own can only represent two states. By dealing with groups of bits, rather more complex entities can be coded. For example, the standard alphanumeric characters can be coded using 7-bit groups of digits, as listed in Table 1.1. Thus the ASCII code for "Microcontroller" becomes:

1001101 1101001 1100011 1110010 1101111 1100011 1101111 1101110
1110100 1110010 1101111 1101100 1101100 1100101 1110010

Unicode is an extension of ASCII and with its 16-bit code groups is able to represent characters from many languages and mathematical symbols.

The ASCII code is **unweighted**, as the individual bits do not signify a particular quantity; only the overall pattern has any significance. Other examples are the die code on gaming dice and the 7-segment code of Fig. 6.8 on page 161. Here we will deal with **natural binary** weighted codes, where the position of a bit within the number field determines its

---

[4]The binary base is not a new fangled idea invented for digital computers; many cultures have used base 2 numeration in the past. The Harappān civilization existed more than 4000 years ago in the Indus River basin. Found in the ruins of the Harappān city of Mohenjo-Daro, in the beadmakers' quarter, was a set of stone pebble weights. These were in ratios that doubled in the pattern, 1,1,2,4,8,16,..., with the base weight of around 25 g ($\approx$ 1 oz). Thus bead weights were expressed by digits which represented powers of 2; that is, in binary.

[5]*Microprocessors* and *microcontrollers* are closely related (see Fig. 3.8 on page 62) and so we will use the terms here interchangeably.

Table 1.1: 7-bit ASCII characters.

| MS nybble — | h'0' | h'1' | h'2' | h'3' | h'4' | h'5' | h'6' | h'7' |
LS nybble	b'000'	b'001'	b'010'	b'011'	b'100'	b'101'	b'110'	b'111'
h'0' b'0000'	NUL	DLE	SP	0	@	P	`	p
h'1' b'0001'	SOH	XON	!	1	A	Q	a	q
h'2' b'0010'	STX	DC2	"	2	B	R	b	r
h'3' b'0011'	ETX	XOFF	#	3	C	S	c	s
h'4' b'0100'	EOT	DC4	$	4	D	T	d	t
h'5' b'0101'	ENQ	NAK	%	5	E	U	e	u
h'6' b'0110'	ACK	SYN	&	6	F	V	f	v
h'7' b'0111'	BEL	ETB	'	7	G	W	g	w
h'8' b'1000'	BS	CAN	(	8	H	X	h	x
h'9' b'1001'	HT	EM	)	9	I	Y	i	y
h'A' b'1010'	LF	SUB	*	:	J	Z	j	z
h'B' b'1011'	VT	ESC	+	;	K	[	k	{
h'C' b'1100'	FF	FS	,	<	L	\	l	|
h'D' b'1101'	CR	GS	-	=	M	]	m	}
h'E' b'1110'	SO	RS	.	>	N	^	n	~
h'F' b1111'	SI	US	/	?	O	_	o	DEL

value or weight. In an integer binary number the rightmost digit is worth $2^0 = 1$, the next left column $2^1 = 2$, and so on to the nth column which is worth 2^{n-1}. For instance, the decimal number 1998 is represented as:

$$10^3 \ 10^2 \ 10^1 \ 10^0$$
$$1 \quad 9 \quad 9 \quad 8$$

i.e., $1 \times 10^3 + 9 \times 10^2 + 9 \times 10^1 + 8 \times 10^0$, or just 1998. In **natural binary** the same quantity is:

$$2^{10} \ 2^9 \ 2^8 \ 2^7 \ 2^6 \ 2^5 \ 2^4 \ 2^3 \ 2^2 \ 2^1 \ 2^0$$
$$1 \ \ 1 \ 1 \ 1 \ 1 \ 0 \ 0 \ 1 \ 1 \ 1 \ 0$$

i.e., $1 \times 2^{10} + 1 \times 2^9 + 1 \times 2^8 + 1 \times 2^7 + 1 \times 2^6 + 0 \times 2^5 + 0 \times 2^4 + 1 \times 2^3 + 1 \times 2^2 + 1 \times 2^1 + 0 \times 2^0$, or b'11111001110'.[6] Fractional numbers may equally well be represented by columns to the right of the binary point using negative powers of 2. Thus b'1101.11' is equivalent to 13.75. As can be seen from this example, binary numbers are rather longer than their

---

[6]The b'· · ·' notation is not universal; for example, $(1111011110)_2$ is an alternative. If the base is unambiguous then the base indicator may be omitted.

decimal equivalent—on average, a little over three times longer. Nevertheless, 2-way switches are considerably simpler than 10-way devices, so the binary representation is preferable.

An n-digit binary number can represent up to 2^n patterns. Most computers store and process groups of bits. For instance, the first microprocessor, the Intel 4004, handled its data four bits (a **nybble**) at a time. Many current processors cope with blocks of 8 bits (a **byte**), 16 bits (a **word**), 32 bits (a **long-word**) or 64-bits (a **quad-word**). Some of these groupings are shown in Table 1.2. The names illustrated are somewhat de facto, and variations are sometimes encountered.

As in the decimal number system, large binary numbers are often expressed using the prefixes k (kilo), M (mega) and G (giga). A binary kilo is $2^{10} = 1024$; for instance, 64 kbyte of memory. In an analogous way, a binary mega is $2^{20} = 1,048,576$; thus a 1.44 Mbyte (or MB) floppy disk. Similarly a 20-Gbyte (or GB) hard disk has a storage capacity of $20 \times 2^{30} = 21,474,836,480$ bytes. The former representation is certainly preferable.

Table 1.2: Some common bit groupings.

Bit	(1 bit)	0 – 1
(0 – 1)		
Nybble	(4 bits)	0 – 15
(0000 – 1111)		
Byte	(8 bits)	0 – 255
(0000 0000 – 1111 1111)		
Word	(16 bits)	0 – 65,535
(0000 0000 0000 0000 – 1111 1111 1111 1111)		
Long-word	(32 bits)	0 – 4,294,967,295
(0000 0000 0000 0000 0000 0000 0000 0000 – 1111 1111 1111 1111 1111 1111 1111 1111)		

Long binary numbers are not very human friendly. In Table 1.2, binary numbers were zoned into fields of four digits to improve readability. Thus the address of a data unit stored in memory might be b'1000 1100 0001 0100 0000 1010'. If each group of four can be given its own symbol, 0,···,9 and A,···,F, as shown in Table 1.3, then the address becomes h'8C140A';[7] a rather more manageable characterization. This code is called **hexadecimal**, as there are 16 symbols. Hexadecimal (base-16) numbers are a viable number base in their own right, rather than just being a convenient binary representation. Each column is worth $16^0, 16^1, 16^2, \ldots, 16^n$ in the normal way.[8]

Binary coded decimal (BCD) is a hybrid binary/decimal code extensively used at the input/output ports of a digital system (see Example 11.5

---

[7]Other representations for the hexadecimal base are 8C140Ah and 0x8C140A.

[8]Many scientific calculators, including that in the Accessories group under Microsoft's Windows, can do hexadecimal (and binary) arithmetic.

Table 1.3: Different ways of representing the quantities decimal 0,...,20.

Decimal	Natural binary	Hexadecimal	Binary coded decimal
00	00000	00	0000 0000
01	00001	01	0000 0001
02	00010	02	0000 0010
03	00011	03	0000 0011
04	00100	04	0000 0100
05	00101	05	0000 0101
06	00110	06	0000 0110
07	00111	07	0000 0111
08	01000	08	0000 1000
09	01001	09	0000 1001
10	01010	0A	0001 0000
11	01011	0B	0001 0001
12	01100	0C	0001 0010
13	01101	0D	0001 0011
14	01110	0E	0001 0100
15	01111	0F	0001 0101
16	10000	10	0001 0110
17	10001	11	0001 0111
18	10010	12	0001 1000
19	10011	13	0001 1001
20	10100	14	0010 0000

on page 325). Here each decimal digit is individually replaced by its 4-bit binary equivalent. Thus 1998 is coded as $(0001\ 1001\ 1001\ 1000)_{BCD}$. This is very different from the equivalent natural binary code, even if it is represented by 0s and 1s. As might be expected, arithmetic in such a hybrid system is difficult, and BCD is normally converted to natural binary at the system input, and processing is done in natural binary before being converted back (see Program 5.7 on page 138).

The rules of arithmetic are the same in natural binary[9] as they are in the more familiar base 10 system, or indeed in any base-n radix scheme. The simplest of these is **addition**, which is a shorthand way of totaling quantities, as compared to the more primitive counting or incrementation process. Thus $2 + 4 = 6$ is rather more efficient than $2 + 1 = 3; 3 + 1 = 4; 4 + 1 = 5; 5 + 1 = 6$. However, it does involve memorizing the rules of addition.[10] In decimal this involves 45 rules, assuming that order is irrelevant; from $0 + 0 = 0$ to $9 + 9 = 18$. Binary addition is much simpler as it is covered by only three rules:

---

[9]Sometimes called 8-4-2-1 code after the weightings of the first four lowest columns.
[10]Which you had to do way back in the mists of time in primary/elementary school!

$$0 + 0 \quad = \; 0$$

$$\left.\begin{array}{l} 0 + 1 \\ 1 + 0 \end{array}\right\} \; = \; 1$$

$$1 + 1 \quad = \; 10 \quad (0 \text{ carry } 1)$$

Based on these rules, the least significant bit (LSB) is totalled first, passing a **carry** if necessary to the next left column. The process ends with the most significant bit (MSB) column, its carry being the new MSD of the sum. For example:

1		1	
0 1		2 6 3 1	
0 0 1		8 4 2 6 8 4 2 1	
96	Augend	1100000	Augend
+ 37	Addend	+ 0100101	Addend
1 1	Carries	1 1	Carries
133	Sum	10000101	Sum

(a) Decimal   *(b) Binary*

Just as addition implements an up count, **subtraction** corresponds to a down count, where units are removed from the total. Thus $8 - 5 = 3$ is the equivalent of $8 - 1 = 7; 7 - 1 = 6; 6 - 1 = 5; 5 - 1 = 4; 4 - 1 = 3$.

The technique of decimal subtraction you are familiar with applies the subtraction rules commencing from LSB and working through to the MSB. In any given column where a larger quantity is to be taken away from a smaller quantity, a unit digit is **borrowed** from the next higher column and given back after the subtraction is completed. Based on this borrow principle, the subtraction rules are given by:

$$0 - 0 = 0$$
$$^{1}0 - 1 = 1 \quad \text{Borrowing 1 from the higher column}$$
$$1 - 0 = 1$$
$$1 - 1 = 0$$

For example:

1		6 3 1	
0 1		4 2 6 8 4 2 1	
96	Minuend	1100000	Minuend
- 37	Subtrahend	- 0100101	Subtrahend
1	Borrows	1 1 1 1 1 1	Borrows
59	Difference	0111011	Difference

(a) Decimal   *(b) Binary*

Although this familiar method works well, there are several problems implementing it in digital circuitry.

- How can we deal with situations where the subtrahend is larger than the minuend?
- How can we distinguish between positive and negative quantities?
- Can a digital system's adder circuits be coerced into subtracting?

To illustrate these points, consider the following example:

37	Minuend	0100101	Minuend
- 96	Subtrahend	-1100000	Subtrahend
41	Difference (-59)	1000101	Difference (-0111011)

(a) Decimal   *(b) Binary*

Normally when we know that the minuend is greater than the subtrahend, the two operands are interchanged and a minus sign is appended to the outcome; that is − (subtrahend − minuend). If we do not swap, as in (a) above, then the outcome appears to be incorrect. In fact, 41 is correct, in that this is the difference between 59 (the correct outcome) and 100; 41 is described as the **10's complement** of 59. Furthermore, the fact that a borrow digit was generated from the MSD indicates that the difference is negative, and therefore will be in this 10's complement form. Converting from 10's complement decimal numbers to the "normal" magnitude form is simply a matter of inverting each digit and then adding one to the outcome. A decimal digit is inverted by computing its difference from 9. Thus the 10's complement of 3941 is −6059:

$$\overline{3941} \mapsto 6058; +1 = -6059$$

However, there is no reason why negative numbers should not remain in this complement form just because we are not familiar with this type of notation.

The complement method of negative quantity representation of course applies to binary numbers. Here the ease of inversion ($0 \rightarrow 1; 1 \rightarrow 0$) makes this technique particularly attractive. Thus in our example above:

$$\overline{1000111} \mapsto 0111000; +1 = -0111001$$

Again, negative numbers should remain in a **2's complement** form.[11] This complement process is reversible. Thus:

complement $\Longleftrightarrow$ normal

Signed decimal numeration has the luxury of using the symbols + and − to denote positive and negative quantities. A 2-state system is stuck with 1s and 0s. However, looking at the last example gives us a clue about how to proceed. A negative outcome gives a borrow back out to the highest column. Thus we can use this MSD as a **sign bit**, with 0 for + and 1 for −. This gives b'1,1000101' for −59 and b'0,0111011' for +59. Although for clarity the sign bit has been highlighted above using a comma delimiter, the advantage of this system is that it can be treated in all arithmetic processes in the same way as any other ordinary bit. Doing this, the outcome will give the correct sign:

---

[11]If you enter a negative decimal number in the Microsoft Windows calculator and change base to Binary, the number will be displayed in 2's complement form.

```
0,1100000  (+96)              0,0100101  (+37)
1,1011011  (-37)              1,0100000  (-96)
                                   
0,0111011  (+59)              1,1000101  (-59)
```

(a) Minuend less than subtrahend   *(b) Minuend greater than subtrahend*

From this example we see that if negative numbers are in a signed 2's complement form, then we no longer have the requirement to implement hardware subtractors, as adding a negative number is equivalent to subtracting a positive number. Thus $A - B = A + (-B)$. Furthermore, once numbers are in this form, the outcome of any subsequent processing will always remain 2's complement signed throughout.

There are two difficulties associated with signed 2's complement arithmetic. The first of these is **overflow**. It is possible that adding two positive or two negative numbers will cause overflow into the sign bit; for instance:

```
0,1000  (+8)                  1,1000  (-8)
0,1011  (+11)                 1,0101  (-11)
                                   
1,0011  (-13!!!)              0,1101  (+13!!!)
```

(a) Sum of two +ve numbers gives -ve   *(b) Sum of two -ve numbers gives +ve*

In (a) the outcome of $(+8) + (+11)$ is -13! The 2^4 numerical digit has overflowed into the sign position (actually, $10011b = 19$ is the correct outcome). Example (b) shows a similar problem for the addition of two signed negative numbers. Overflow can only happen if both operands have the *same* sign bits. Detection is then a matter of determining this situation with an outcome that differs. See Fig. 1.5 for a logic circuit to implement this overflow condition.

The final problem concerns arithmetic on signed operands with different sized fields. For instance:

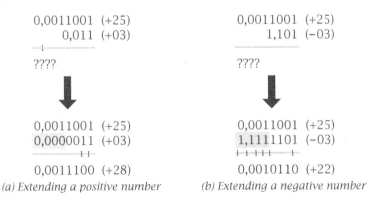

```
0,0011001  (+25)              0,0011001  (+25)
    0,011  (+03)                  1,101  (-03)
                                   
????                          ????
```

```
0,0011001  (+25)              0,0011001  (+25)
0,0000011  (+03)              1,1111101  (-03)
                                   
0,0011100  (+28)              0,0010110  (+22)
```

(a) Extending a positive number   *(b) Extending a negative number*

Both the examples involve adding an 8-bit to a 16-bit operand. Where the former is positive, the data may be increased to 16 bits by padding with 0s. The situation is slightly less intuitive where negative data requires extension. Here the prescription is to extend the data by padding out with 1s. In the general case the rule is simply to pad out data by propagating the sign bit left. This technique is known as **sign extension**.

Multiplication by the nth power of two is simply implemented by shifting the data left n places. Thus $00110(6) << 01100(12) << 11000(24)$ multiplies 5 by 2^2, where the $<<$ operator is used to denote shifting left. The process works for signed numbers as well:

0,00000011 (3)	1,11111101 (−3)	0,00000110 (3 × 2)
<<	<<	+ 0,00011000 (3 × 8)
0,00000110 (6)	1,11111010 (−6)	0,00011110 (3 × 10 = 30)
<<	<<	
0,00001100 (12)	1,11110100 (−12)	
<<	<<	
0,00011000 (24)	1,11101000 (−24)	

   (a) +3 × 8 = +24   *(b) − 3 × 8 = − 24*   *(c) +3 × 10 = 30*

Should the sign bit change polarity, then a magnitude bit has overflowed. Some computers/microprocessors have an **Arithmetic Shift Left** operation that signals this situation, as opposed to the standard **Logic Shift Left** process used in unsigned number shifts.

Multiplication by nonpowers of 2 can be implemented by a combination of shifting and adding. Thus, as shown in (c) above, 3×10 is implemented as $(3 \times 8) + (3 \times 2) = (3 \times 10)$ or $(3 << 3) + (3 << 1)$.

In a similar fashion, division by powers of 2 is implemented by shifting right n places. Thus $1100(12) >> 0110(6) >> 0011(3) >> 0001.1(1.5)$. This process also works for signed numbers:

0,1111.000 (+15)	1,0001.000 (−15)	0001.1
>>	>>	1010⌐1111.0
0,0111.100 (+7.5)	1,1000.100 (−7.5)	−1010
>>	>>	0101
0,0011.110 (+3.75)	1,1100.010 (−3.75)	−101.0
>>	>>	000.0
0,0001.111 (+1.875)	1,11110.001 (−1.875)	

   (a) +15/8 = 1.875   *(b) −15/8 = −1.875*   *(c) 15/10 = 1.5*

Notice that rather than always shifting in 0s, the sign bit should be propagated in from the left. Thus positive numbers shift in 0s and negative numbers shift in 1s. This is known as **arithmetic shift right** as opposed to **logic shift right** which always shifts in 0s.

Division by nonpowers of 2 is illustrated in (c) above. This shows the familiar long division process used in decimal division. This is an analagous process to the shift and add technique for multiplication, using a combination of shifting and subtracting.

Arithmetic is not the only way to manipulate binary patterns. George Boole[12] in the mid-nineteenth century developed an algebra dealing with symbolic processing of logic propositions. This **Boolean algebra** deals with variables which can be true or false. In the 1930s it was realized that this mathematical system could equally well be used to analyze switching networks and thus binary logic systems. Here we will confine ourselves to looking at the fundamental logic operations of this switching algebra.

The inversion or **NOT** operation is represented by overscoring. Thus $f = \overline{A}$ states that the variable f is the inverse of A; that is if $A = 0$ then $f = 1$ and if $A = 1$ then $f = 0$. In Fig. 1.1(a) this transfer characteristic is presented in the form of a **truth table**. By definition, inverting twice returns a variable to its original state; thus $\overline{\overline{f}} = f$.[13]

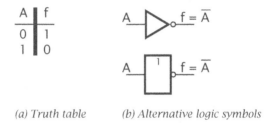

A	f
0	1
1	0

$f = \overline{A}$

$f = \overline{A}$

(a) Truth table   (b) Alternative logic symbols

Fig. 1.1 The NOT operation.

Logic function implementations are normally represented in an abstract manner rather than as a detailed circuit diagram. The **NOT gate** is symbolized as shown in Fig. 1.1(b). The circle *always* represents inversion in a logic diagram, and is often used in conjunction with other logic elements, such as in Fig. 1.2(c).

The **AND operator** gives an *all or nothing* function. The outcome will only be true when *every* one of the n inputs are true. In Fig. 1.2 two input variables are shown, and the output is symbolized as $f = B \cdot A$, where · is the Boolean AND operator. The number of inputs is not limited to two, and in general $f = A(0) \cdot A(1) \cdot A(2) \cdots A(n)$. The AND operator is

---

[12]The first professor of mathematics at Queen's College, Cork.

[13]In days of yore when logic circuits were built out of discrete devices, such as diodes, resistors and transistors, problems arising from sneak current paths were rife. In one such laboratory experiment the output lamp was rather dim, and the lecturer in charge suggested that two NOTs in series would in a suspect line would not disturb the logic but would block off the unwanted current leak. On returning sometime later, the students complained that the remedy had had no effect. On investigation the lecturer discovered two knots in the offending wire—obviously not tied tightly enough!

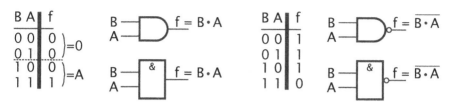

B	A	f		
0	0	0	)=0	
0	1	0		
1	0	0	)=A	
1	1	1		

(a) Truth table   (b) Alternative logic symbols (c) NAND

Fig. 1.2 The AND function.

sometimes called a logic product, as ANDing (cf. multiplying) any bit with logic 0 always yields a 0 output.

If we consider B as a control input and A as a stream of data, then consideration of the truth table shows that the output follows the data stream when B = 1 and is always 0 when B = 0. Thus the circuit can be considered to be acting as a valve, gating the data through on command. The term **gate** is generally applied to any logic circuit implementing a fundamental Boolean operator.

Most practical AND gate implementations have an inverting output. The logic of such implementations is NOT AND, or NAND for short, and is symbolized as shown in Fig. 1.2(c).

The **inclusive-OR operator** gives an *anything* function. Here the outcome is true when *any* input or inputs are true (hence the ≥ 1 label in the logic symbol). In Fig. 1.3 two inputs are shown, but any number of variables may be ORed together. ORing is sometimes referred to as a logic sum, and the + used as the mathematical operator; thus f = B + A. In an analogous manner to the AND gate detecting all ones, the OR gate can be used to detect all zeros. This is illustrated in Fig. 2.20 on page 35 where an 8-bit zero outcome brings the output of the NOR gate to 1. Inclusive-ORing any bit with a logic 1 always yields a 1 output.

Considering B as a control input and A as data (or vice versa), then from Fig. 1.3(a) we see that the data is gated through when B is 0 and inhibited (always 1) when B is 1. This is a little like the inverse of the AND function. In fact, the OR function can be expressed in terms of AND

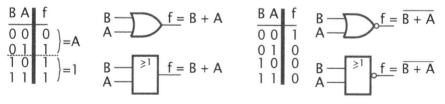

B	A	f		
0	0	0	)=A	
0	1	1		
1	0	1	)=1	
1	1	1		

(a) Truth table   (b) Alternative logic symbols (c) NOR

Fig. 1.3 The inclusive-OR operation.

using the duality relationship $\overline{A + B} = \overline{B} \cdot \overline{A}$. This states that the NOR function can be implemented by inverting all inputs into an AND gate.

The three fundamental Boolean operators are AND, OR and NOT. There is one more operation commonly available as an electronic gate; the **eXclusive-OR operator (XOR)**. The XOR function is true if *only one* input is true (hence the =1 label in the logic symbol). Unlike the inclusive-OR, the situation where both inputs are true gives a false outcome.

If we consider B is a control input and A as data (they are fully interchangeable) then:

- When B = 0 then f = A; that is, the output follows the data input.
- When B = 1 then f = $\overline{A}$; that is, the output is the inverse of the data input.

Thus an XOR gate can be used as a programmable inverter.

Another useful property considers the XOR function as a logic differentiator. The XOR truth table shows that the gate gives a true output if the two inputs differ. Alternatively, the XNOR truth table of Fig. 1.4(c) shows a true output when the two inputs are the same. Thus an XNOR gate can be considered to be a 1-bit equality detector. The equality of two n-bit words can be tested by ANDing an array of XNOR gates (see Fig. 2.7 on page 23), each generating the function $\overline{B_k \oplus A_k}$; that is:

$$f_{B=A} = \sum_{k=0}^{n-1} \overline{B_k \oplus A_k}$$

(a) Truth table   (b) Alternative logic symbols   (c) ENOR

Fig. 1.4 The XOR operation.

As a simple example of the use of the XOR/XNOR gates, consider the problem of detecting sign overflow (see page 10). This occurs if both the sign bits of word B and word A are the same $(\overline{S_B \oplus S_A})$ AND the sign bit of the outcome word C is not the same as either of these sign bits, say $S_B \oplus S_C$. The logic diagram for this detector is shown in Fig. 1.5 and implements the Boolean function:

$$(\overline{S_B \oplus S_A}) \cdot (S_B \oplus S_C)$$

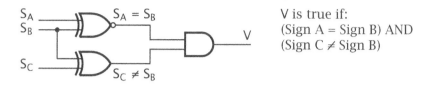

Fig. 1.5 Detecting sign overflow.

Finally, the XOR function can be considered as detecting when the number of true inputs are odd. By cascading $n + 1$ XOR gates, the overall **parity** function is true if the n-bit word has an odd number of ones. Some measure of error protection can be obtained by adding an additional bit to each word, so that overall the number of bits is odd. This oddness can be checked at the receiver and any deviation indicates corruption.

CHAPTER 2
Logic Circuitry

We have noted that digital processing is all about transmission, manipulation and storage of binary word patterns. Here we will extend the concepts introduced in the last chapter as a lead into the architecture of the computer and microcontroller. We will look at some relevant logic functions, their commercial implementations and some practical considerations.

After reading this chapter you will:

- Understand the properties and use of active pull-up, open-collector and 3-state output structures.
- Appreciate the logic structure and function of the natural decoder.
- See how a MSI implementation of an array of XNOR gates can compare two words for equality.
- Understand how a 1-bit adder can be constructed from gates, and can be extended to deal with the addition of two n-bit words.
- Appreciate how the function of an ALU is so important to a programmable system.
- Be aware of the structure and utility of a read-only memory (ROM).
- Understand how two cross-coupled gates can implement a R S latch.
- Appreciate the difference between a D latch and a D flip flop.
- Understand how an array of D flip flops or latches can implement a register.
- See how a serial cascade of D flip flops can perform a shifting function.
- Understand how a D flip flop can act as a frequency divide by two, and how a cascade of these can implement a binary count.
- See how an ALU/PIPO register can implement an accumulator processor unit.
- Appreciate the function of a RAM.

The first integrated circuits, available at the end of the 1960s, were mainly NAND, NOR and NOT gates. The most popular family of logic functions was, and still is, the 74 series Transistor Transistor Logic (TTL) introduced by Texas Instruments and soon copied by all the major semiconductor manufacturers.

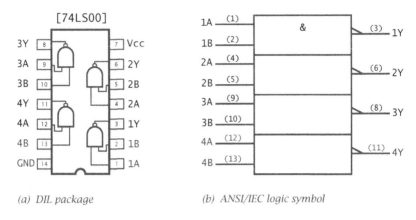

(a) DIL package   (b) ANSI/IEC logic symbol

Fig. 2.1 The 74LS00 quad 2-I/P NAND package.

The 74LS00[1] comprises four 2-input NAND gates in a 14-pin package. The integrated circuit (IC) is powered with a 5 ± 0.25 V supply between V_{CC}[2] (usually about 5 V) and GND. The logic outputs are 2.4 – 5 V for the High state and 0 – 0.4 V for the Low state. Most IC logic families require a 5 V supply, but 3 V versions are available, and some CMOS implementations can operate with a range of supplies between 3 V and 15 V.

The 74LS00 IC is shown in Fig. 2.1(a) in its dual in-line (DIL) package. Strictly it should be described as a positive-logic quad 2-I/P NAND, as the electrical equivalent for the two logic levels 0 and 1 are Low (L is around ground potential) and High (H is around V_{CC}, usually about 5 V). If the relationship $0 \leadsto H$; $1 \leadsto L$ is used (negative logic) then the 74LS00 is actually a quad 2-I/P NOR gate. The ANSI/IEC[3] logic symbol of Fig. 2.1(b) denotes a Low electrical potential by using the polarity $\searrow$ symbol. The ANSI/IEC NAND symbol shown is thus based on the *real* electrical operation of the circuit. In this case the logic coincides with a positive-logic NAND function. The & operator shown in the top block is assumed applicable to the three lower gates.

The output structure of a 74LS00 NAND gate is **active pull-up**. Here both the High and Low states are generated by connection via a low-resistance switch to V_{CC} or GND, respectively. In Fig. 2.2(a) these switches

---

[1]The LS stands for "low-power schottky transistor." There are very many other versions, such as ALS (advanced LS), AS (advanced schottky) and HC (high-speed complementary metal-oxide transistor, CMOS). These family variants differ in speed and power consumption, but for a given number designation have the same logic function and pinout.

[2]For historical reasons the positive supply on logic ICs are usually designated as V_{CC}; the C referring to a bipolar's transistor collector supply. Similarly field-effect circuitry sometimes use the designation V_{DD} for drain voltage. The zero reference pin is normally designated as the ground point (GND), but sometimes the V_{EE} (for emitter) or V_{SS} (for source) label is employed.

[3]The American National Standards Institution/International Electrotechnical Commission.

are shown for simplicity as metallic contacts, but they are of course transistor derived.

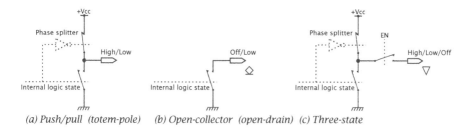

(a) Push/pull (totem-pole)   (b) Open-collector (open-drain)   (c) Three-state

Fig. 2.2 Output structures.

Logic circuits, such as the 74LS00, change output state in around 10 nanoseconds.[4] To be able to do this, the capacitance of any interconnecting conductors and other logic circuits' inputs must be rapidly discharged. Mainly for this reason, active pull-up (sometimes called totem-pole) outputs are used by most logic circuits. There are certain circumstances where alternative output structures have some advantages. The **open-collector** (or open-drain) configuration of Fig. 2.2(b) provides a "hard" Low state, but the High state is in fact an open circuit. The High-state voltage can be generated by connecting an external resistor to either V_{CC} or indeed to a different power rail. Nonorthodox devices, such as relays, lamps or light-emitting diodes, can replace this pull-up resistor. The output transistor is often rated with a higher than usual current and/or voltage rating for such purposes.

The application of most interest to us here is illustrated in Fig. 2.3. Here four open-collector gates share a *single* pull-up resistor. Note the use of the ◇ symbol to denote an open-collector output. Assume that there are four peripheral devices, any of which may wish to attract the attention of the processor, e.g., computer or microcontroller. If this processor has only one Attention pin, then the four Signal lines must be **wire-ORed** together as shown. With all Signal lines inactive (logic 0) the outputs of all buffer NOT gates are off (state H), and the party line is pulled up to +V by RL. If *any* Signal line is activated (logic 1), as in Sig_1, then the output of the corresponding buffer gate goes hard Low. This pulls the party line Low, irrespective of the state of the other signal lines, and thus interrupts the processor.

The **three-state** structure of Fig. 2.2(c) has the properties of both preceding output structures. When enabled, the two logic states are represented in the usual way by high and low voltages. When disabled, the

---

[4]A nanosecond is 10^{-9} s, so 100,000,000 transitions each second are possible.

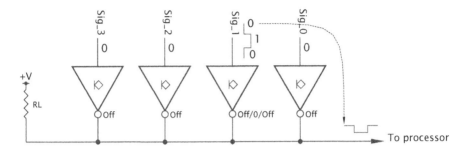

Fig. 2.3 Open-collector buffers driving a party line.

output is open circuit irrespective of the activities of the internal logic circuitry and any change in input state. A logic output with this three-state is indicated by the $\triangledown$ symbol.

As an example of the use of this structure, consider the situation depicted in Fig. 2.4. Here a master controller wishes to read one of several devices, all connected to this master over a set of party lines. As this data highway or **data bus** is a common resource, only the selected device can be allowed access to the bus at any one time. The access has to be withdrawn immediately after the data has been read, so that another device can use the resource. As shown in the diagram, each "thing" connected to the bus outputs is designated by the $\triangledown$ symbol. When selected, *only* the active logic levels will drive the bus lines. The 74LS244 octal ($\times$8) 3-state (sometimes called tristate or TRIS) buffer has high-current outputs (designated by the $\triangleright$ symbol) specifically designed to charge/discharge the capacitance associated with long bus lines.

Integrated circuits with a complexity of up to 12 gates are categorized as small-scale integration (SSI). Gate counts upwards to 100 on a single IC are medium-scale integration (MSI); up to 1000 are known as large-scale

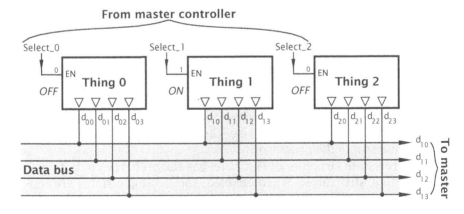

Fig. 2.4 Sharing a bus.

integration (LSI) and over this, very large scale integration (VLSI). Memory chips and microcontrollers are examples of this latter category.

The NAND gate networks shown in Fig. 2.5 are typical MSI-complexity ICs. Remembering that the output of a NAND gate is logic 0 only when *all* its inputs are logic 1 (see Fig. 1.2(c) on page 13) then we see that for any combination of the *select* inputs B A (2^1 2^0) in Fig. 2.5(a) only *one* gate will go to logic 0. Thus output $\overline{Y_2}$ will be activated when B A = 10. The associated truth table shows the circuit *decodes* the binary address B A so that address n selects output $\overline{Y_n}$. The 74LS139 is described as a dual 2- to 4-line **natural decoder**. Dual because there are two such circuits in the one chip. The symbol X/Y denotes converting code X (natural binary) to code Y (unary – one of n). The enabling input $\overline{G}$ is connected to all gates in parallel. Thus the decoder function only operates if $\overline{G}$ is Low (logic 0). If $\overline{G}$ is High, then irrespective of the state of B A (the X entries in the truth table denote a "don't care" situation) all outputs remain deselected—logic 1. An example of the use of the 74LS139 is given in Fig. 2.25 on page 40.

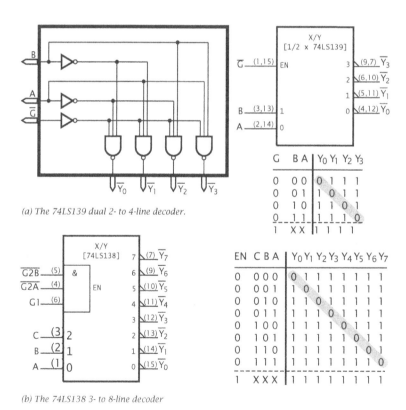

(a) The 74LS139 dual 2- to 4-line decoder.

(b) The 74LS138 3- to 8-line decoder

Fig. 2.5 The 74LS138 and 74LS139 MSI natural decoders.

The 74LS138 of Fig. 2.5(b) is similar, but implements a 3- to 8-line decoder function. The state of the three address lines C B A (2^2 2^1 2^0) n selects only one of the eight outputs $\overline{Y_n}$. The 74LS138 has three Gate inputs which generate an internal enabling signal $\overline{G2B} \cdot \overline{G2A} \cdot G1$. Only if both $\overline{G2A}$ and $\overline{G2B}$ are Low and G1 is High will the device be enabled. The 74LS138 is used in Fig 11.12 on page 316 to decode microcontroller port lines to enable several devices to communicate to the one port.

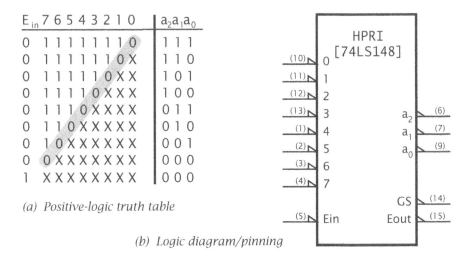

E_{in} 7 6 5 4 3 2 1 0	$a_2 a_1 a_0$
0 1 1 1 1 1 1 1 0	1 1 1
0 1 1 1 1 1 1 0 X	1 1 0
0 1 1 1 1 1 0 X X	1 0 1
0 1 1 1 1 0 X X X	1 0 0
0 1 1 1 0 X X X X	0 1 1
0 1 1 0 X X X X X	0 1 0
0 1 0 X X X X X X	0 0 1
0 0 X X X X X X X	0 0 0
1 X X X X X X X X	0 0 0

(a) Positive-logic truth table

(b) Logic diagram/pinning

Fig. 2.6 The 74LS148 highest-priority encoder.

The **priority encoder** illustrated in Fig. 2.6 is a sort of reverse decoder. Bringing one of the eight input lines Low results in the active-Low three-bit binary equivalent appearing at the output. Thus if $\overline{5}$ is Low, then $\overline{a_2}\,\overline{a_1}\,\overline{a_0}$ = 010 (active Low 101).

If more than one input line is active, then the output code reflects the highest. Thus if both $\overline{5}$ and $\overline{3}$ are Low, the output code is still 010. Hence the label HPRI for Highest PRIority. The device is enabled when Enable_In $(\overline{E_{in}})$ is Low. Enable_Out $(\overline{E_{out}})$ and Group_Strobe $(\overline{GS})$ are used to cascade 74LS148s to expand the number of lines.

A large class of ICs implement arithmetic operations. The gate array illustrated in Fig. 2.7 detects when the 8-bit byte P7,...,P0 is identical to the byte Q7,...,Q0. Eight XNOR gates each give a logic 1 when its two input bits Pn, Qn are identical, as described on page 14. Only if *all* 8-bit pairs are the same, will the output NAND gate go Low. The 74LS688 **equality comparator** also has a direct input $\overline{G}$ into this NAND gate, acting as an overall enabling signal.

The ANSI/IEC logic symbol, shown in Fig. 2.7(b), uses the COMP label to denote the arithmetic comparator function. The output is prefixed

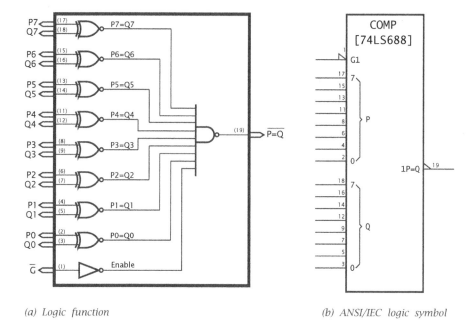

(a) Logic function

(b) ANSI/IEC logic symbol

Fig. 2.7 The 74LS688 octal equality detector.

with the numeral 1, indicating that its operation P=Q is dependent on any input qualifying the same numeral; that is G1. Thus the active-Low enabling input G1 gates the active-Low output, 1P=Q.

One of the first functions beyond simple gates to be integrated into a single IC was that of addition. The truth table of Fig. 2.8(a) shows the sum (S) and carry-out (C_1) resulting from the addition of the two bits A and B and any carry-in (C_0). For instance, row 6 states that adding two 1s with a carry-in of 0 gives a sum of 0 and a carry-out of 1 ($1 + 1 + 0 = {}^1 0$). To implement this row we need to detect the pattern 1 1 0; that is, $A \cdot B \cdot \overline{C_0}$; which is gate 6 in the logic diagram. Thus we have by ORing all applicable patterns together for each output:

$$S \; = \; (\overline{A} \cdot \overline{B} \cdot C_0) + (\overline{A} \cdot B \cdot \overline{C_0}) + (A \cdot \overline{B} \cdot \overline{C_0}) + (A \cdot B \cdot C_0)$$
$$C_1 \; = \; (\overline{A} \cdot B \cdot C_0) + (A \cdot \overline{B} \cdot C_0) + (A \cdot B \cdot \overline{C_0}) + (A \cdot B \cdot C_0)$$

Using such a circuit for *each* column of a binary addition, with the carry-out from column $k - 1$ feeding the carry-in of column k means that the addition of any two n-bit words can be implemented simultaneously.

As shown in Fig. 2.8(b), the 74LS283 adds two 4-bit nybbles in 25 ns. In practice the final carry-Out C_4 is generated using additional circuitry to avoid the delays inherent on the carries rippling through each stage from the least to the most significant digit. n 74LS283s can be cascaded to implement addition for words of $4 \times n$ width. Thus two 74LS283s

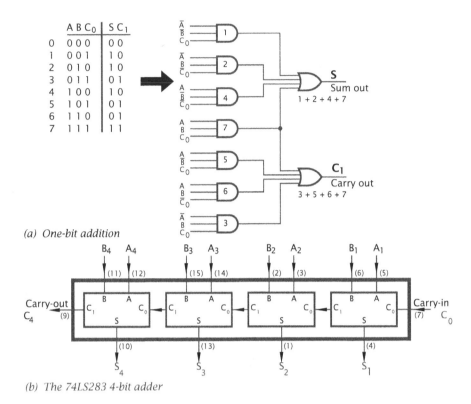

(a) One-bit addition

(b) The 74LS283 4-bit adder

Fig. 2.8 Addition.

perform a 16-bit addition in 45 ns, the extra time being accounted for by the carry propagation between the two units.

Adders can, of course, be coaxed into subtraction by inverting the minuend and adding one, that is 2's complementation. An adder/subtractor circuit could be constructed by feeding the minuend word through an array of XOR gates acting as programmable inverters (see page 14). The mode line $\overline{\text{Add}}$/Sub in Fig. 2.9 that controls these inverters also feeds the carry-In, effectively adding one when in the Subtract mode.

Extending this line of argument leads to the **arithmetic logic unit** (**ALU**). An ALU is a circuit which can undertake a selection of arithmetic and logic processes on input data as controlled by mode inputs. The 74LS382 in Fig. 2.10 processes two 4-bit operands in eight ways, as controlled by the three Mode Select bits $S_2 S_1 S_0$ and tabulated in Fig. 2.10(a). Besides addition and subtraction, the logic operations of AND, OR and XOR are supported. The 74LS382 even generates the 2's complement overflow function (see page 10).

As we shall see, the ALU is at the heart of the computer and microcontroller architectures. By feeding the Mode Select inputs with a series

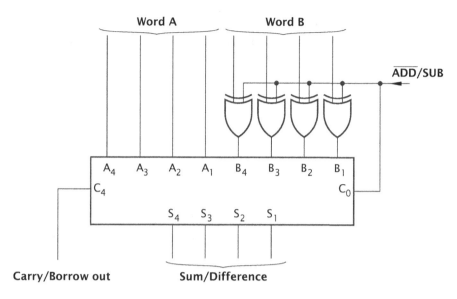

Fig. 2.9 Implementing a programmable adder/subtractor.

of binary words, a program of operations can be performed by the ALU. Such **operation codes** are stored in an external memory, and are accessed sequentially by the computer's control circuits.

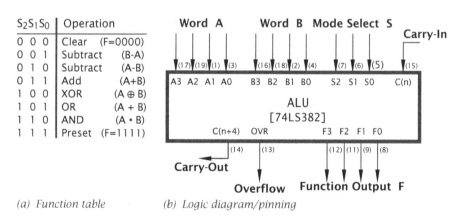

$S_2 S_1 S_0$	Operation	
0 0 0	Clear	(F=0000)
0 0 1	Subtract	(B-A)
0 1 0	Subtract	(A-B)
0 1 1	Add	(A+B)
1 0 0	XOR	(A ⊕ B)
1 0 1	OR	(A + B)
1 1 0	AND	(A • B)
1 1 1	Preset	(F=1111)

(a) Function table

(b) Logic diagram/pinning

Fig. 2.10 The 74LS382 ALU.

Sequences of program operation codes are normally stored in some kind of LSI read-only memory. Consider the architecture illustrated in Fig. 2.11. This is essentially a 3- to 8-line decoder driving an 8 × 2 array of diodes. The 3-bit address selects only row n for each input combination n. If a diode is connected to this row, then it conducts and brings the appropriate column Low. The inverting 3-state output buffer consequently gives a High for each connected diode and Low where the link is broken. The pattern of diode links then defines the output code for each input. For illustrative purposes, the structure has been programmed to implement the 1-bit full adder of Fig. 2.8(a), but *any* two functions of three variables can be generated.

The diode matrix look-up table shown here is known as a **read-only memory** (ROM), as its "memory" is in the diode pattern, which is programmed in when the device is manufactured. Early devices, which were typically decoder/32×8 matrices, usually came in user-programmable versions in which the links were implemented with fusible links. By using a high voltage, a selection of diodes could be taken out of contact. Such devices are called **programmable ROMs (PROMs)**.

Fuses are messy when implementing the larger sizes of VLSI PROMs necessary to store computer programs. For instance, the small 27C64

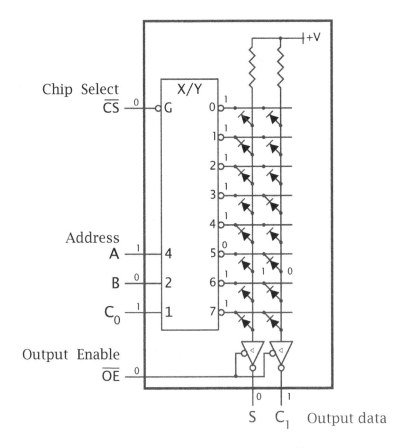

Fig. 2.11 A ROM-implemented 1-bit adder.

PROM shown in Fig. 2.12 has the equivalent of 65,536 fuse/diode pairs, and this is a relatively small device capable of storing 8192 bytes of memory. The 27C64 uses the electrical charge on the floating gate of a metal-oxide field-effect transistor (MOSFET) as the programmable link, with another MOSFET to replace the diode. Charge can be tunneled onto this isolated gate by, again, using a high voltage. Once on the gate, the electric field keeps the link MOSFET conducting. This charge takes many decades to leak away, but this can be dramatically reduced to about 20 minutes by exposure to intense ultraviolet radiation. For this reason the 27C64 is known as an **erasable PROM (EPROM)**. When an EPROM is designed for reusability, a quartz window is integrated into the package, as shown in Fig. 2.12 and on page 2. Programming is normally done externally with special equipment, known as PROM programmers, or colloquially as PROM blasters. Versions without windows are referred to as one-time programmable (OTP) ROMs, as they cannot easily be erased once programmed. They are, however, much cheaper to produce and are thus suitable for small- to medium-scale production runs.

Figure 2.13 shows a simplified representation of such a floating-gate MOSFET link. The cross-point device is a metal-oxide enhancement n-channel field-effect transistor TR1, rather than a diode. This MOSFET has its gate G1 connected to the X line and its source S1 to the Y line. If

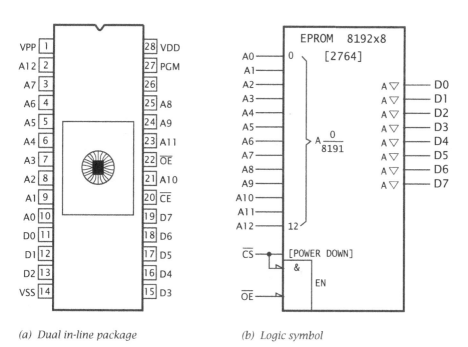

(a) Dual in-line package   (b) Logic symbol

Fig. 2.12 The 2764 erasable PROM (EPROM).

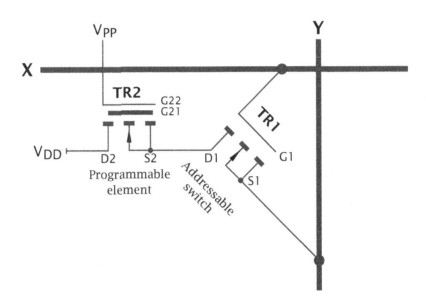

Fig. 2.13 Floating-gate MOSFET link.

its drain D1 is connected to the positive supply and the X line is selected (positive), then the Y line too becomes positive (positive-logic 1) as TR1 is conducting (switch is on). However, if TR1 is disconnected from V_{DD} then it does not conduct and the output on the Y line is logic 0. Transistor TR2 is in series with V_{DD} and thus acts as the programmable element. Transistor TR2 has an extra unconnected gate buried in the silicon dioxide insulation layer. Normally there is no charge on this gate and TR2 is off. If the programming voltage V_{PP} is pulsed high to typically 20–25 V, negative charges tunnel across the extremely thin insulation surrounding the buried gate. This turns TR2 on permanently and thus connects TR1 to its supply. This shows up as a logic 1 on the Y line when selected by the internal memory decoder.

This charge remains more or less permanently on the buried gate until it is exposed to ultraviolet light. The high-energy light photons knock electrons (negative charges) out of the buried (floating) gate[5] effectively discharging in around 20 minutes and wiping out all stored information.

There are PROM structures which can be erased electrically, often in situ in the circuit. These are known variously as electrically-erasable PROMs (EEPROMs) or flash memories. In the former case a large negative pulse at V_{PP} causes the captured electrons on the buried gate to tunnel back out. Generally the negative voltage is generated on the chip, which saves having to provide an additional external supply. The **flash** vari-

---

[5]This is called the Einstein effect. Einstein was awarded his Nobel prize for this discovery and not for his theories of relativity, as these were considered too revolutionary!

ant of EEPROM relies on hot electron injection rather than tunneling to charge the floating gate. The geometry of the cell is approximately half the size of a conventional EEPROM cell which increases the memory density. Programming voltages are also somewhat lower. An example of a commercial EEPROM memory is given in Fig. 12.26 on page 393.

Most modern EPROMs/EEPROMs are fairly fast, taking around 150 ns to access and read. Programming is slow, at perhaps 10 ms per word, but this is an infrequent activity. Flash EEPROMs program around 1000 times faster, in around 10 μs per cell.

All the circuits shown thus far are categorized as **combinational logic**. They have no memory in the sense that the output depends only on the present input, and not the sequence of events leading up to that input. Logic circuits, such as latches, counters, registers and read/write memories are described as **sequential logic**. Their output not only depends on the current input, but the sequence of prior inputs.

R S	Q
0 0	Q (no change)
0 1	1 (set)
1 0	0 (reset)

(a) Defining RS latch truth table

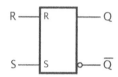

(b) Logic symbol with true/complement outputs

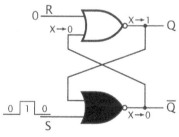

(c) Setting the latch

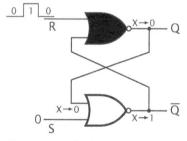

(d) Resetting the latch

Fig. 2.14 The R S latch.

Consider a typical doorbell pushswitch. When you press such a switch the bell rings, and it stops as soon as you release it. This switch has no memory.

Compare this with a standard light switch. Set the switch and the light comes on. Moreover, it remains on when you remove the stimulus (usually your finger!). To turn the light off you must reset the switch.

Again it remains off when the input is taken away. This type of switch is known as a **bistable**, as it has two stable states. Effectively it is a 1-bit memory cell, that can store either an on or off state indefinitely.

A read/write memory, such as the 6264 device of Fig. 2.26, implements each bistable cell using two cross-coupled transistors. Here we are not concerned with this microscopic view. Instead, consider the two cross-coupled NOR gates of Fig. 2.14. Remembering from Fig. 1.3(c) on page 13 that any logic 1 into a NOR gate will always give a logic 0 output irrespective of the state of the other inputs, allows us to analyse the circuit:

- If the S input goes to 1, then output $\overline{Q}$ goes to 0. Both inputs to the top gate are now 0 and thus output Q goes to 1. If the S input now goes back to 0, then the lower gate remains 0 (as the Q feedback is 1) and the top gate output also remains unaltered. Thus the latch is *set* by pulsing the S input.
- If the R input goes to 1, then output Q goes to 0. Both inputs to the bottom gate are now 0 and thus output $\overline{Q}$ goes to 1. If the R input now goes back to 0, then the upper gate remains 0 (as the $\overline{Q}$ feedback is 1) and the bottom gate output also remains unaltered. Thus the latch is *reset* by pulsing the R input.

In the normal course of events—that is assuming that the R and S inputs are not both active at the same time[6] — then the two outputs are always complements of each other, as indicated by the logic symbol of Fig. 2.14(b).

There are many bistable implementations. For example, replacing the NOR gates by NAND gives a $\overline{R}\,\overline{S}$ latch, where the inputs are active on a logic 0. The circuit illustrated in Fig. 2.15 shows such a latch used to debounce a mechanical switch. Manual switches are frequently used as inputs to logic circuits. However, most metallic contacts will bounce off the destination contact many times over a period of several tens of milliseconds before settling. For instance, using a mechanical switch to interrupt a computer/microcontroller will give entirely unpredictable results.

In Fig. 2.15, when the switch is moved up and hits the contact the latch move into its Set state. When the contact is broken, the latch remains unchanged, provided that the switch does not bounce all the way back to the lower contact. The state will remain Set no matter how many bounces occur. By symmetry, the latch will in this state when the switch is moved to the bottom contact, and remain in this Reset state on subsequent bounces.

---

[6]If they were, then both Q and $\overline{Q}$ would go to 0. On relaxing the inputs, the latch would end up in one of its stable states, depending on the relaxation sequence. The response of a latch to a simultaneous Set and Reset input signal is not part of the latch definition, shown in Fig. 2.14(a), but depends on its implementation. For instance, trying to turn a light switch on and off together could end in splitting it in two!

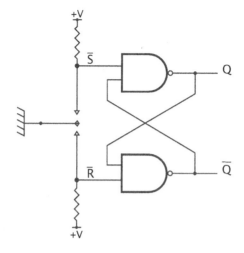

Fig. 2.15 Using a $\overline{R}\,\overline{S}$ latch to debounce a switch.

The **D latch** is an extension to the R S latch, where the output follows the D (Data) input when the C (Control) input is active (logic 1 in our example) and freezes when C is inactive. The D latch can be considered to be a 1-bit memory cell where the datum is retained at its value at the end of the sample pulse.

In Fig. 2.16(b) the dependency of the Data input with its Control signal is shown by the symbols C1 and 1D. The 1 prefix to D shows that it depends on any signal with a 1 suffix, in this case the C input. That is, C1, clocks in the 1D data.

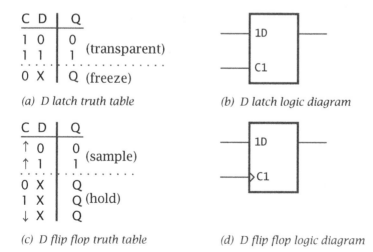

C	D	Q	
1	0	0	(transparent)
1	1	1	
0	X	Q	(freeze)

(a) D latch truth table

(b) D latch logic diagram

C	D	Q	
↑	0	0	(sample)
↑	1	1	
0	X	Q	
1	X	Q	(hold)
↓	X	Q	

(c) D flip flop truth table

(d) D flip flop logic diagram

Fig. 2.16 The D latch and flip flop.

A flip flop is also a 1-bit memory cell, but the datum is only sampled on an *edge* of the control (known here as the Clock) input. The **D flip flop** described in Fig. 2.16(c) is triggered on a _/‾ (as illustrated in the truth table as ↑), but ‾_ clocked flip flops are common. The edge-triggered activity is denoted as > on a logic diagram, as shown in Fig. 2.16(d).

The 74LS74 shown in Fig. 2.17 has two D flip flops in the one SSI circuit. Each flip flop has an overriding Reset (R̄) and Set (S̄) input, which are asynchronous, that is, not controlled by the Clock input. MSI functions include arrays of four, six and eight flip flops all sampling simultaneously with a common Clock input.

The 74LS377 shown in Fig. 2.18 consists of eight D flip flops all clocked by the same single Clock input C, which is gated by input Ḡ. Thus the 8-bit data $8D, ..., 1D$ is clocked in on the _/‾ of C if Ḡ is Low. In the ANSI/ISO logic diagram shown in Fig. 2.18(b), this dependency is indicated as $G1 \rightarrow 1C2 \rightarrow 2D$, which states that Ḡ enables the Clock input, which in turn acts on the Data inputs.

Arrays of D flip flops are known as **registers**, that is, read/write memories that hold a single word. The 74LS377 is technically known as a parallel-in parallel-out (PIPO) register, as data is entered in parallel (that is, all in one go) and is available to read at one go. D latch arrays are also available, such as the 74LS373 octal PIPO register shown in Fig. 2.19, in which the eight D flip flops are replaced by D latches. In addition, the latch outputs have a 3-state capability. This is useful if data is to be captured and later put onto a common data bus to be read subsequently as desired by a computer.

A pertinent example of the use of a PIPO register is shown in Fig. 2.20. Here an 8-bit ALU is coupled with an 8-bit PIPO register, accepting as its input the ALU output, and in turn feeding one input word back to the ALU. This register accumulates the outcome of a series of operations, and is

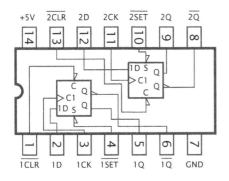

(a) Logic function

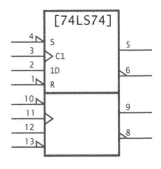

(b) ANSI/IEC logic symbol

Fig. 2.17 The 74LS74 dual D flip flop.

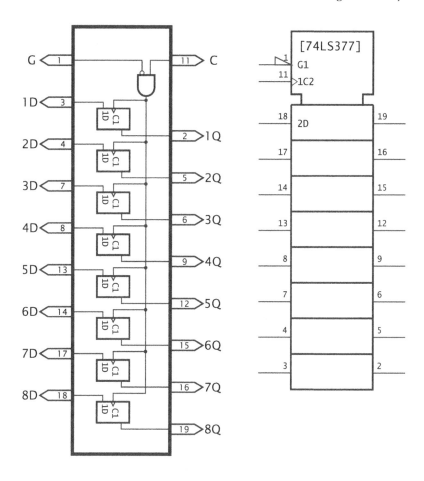

(a) Logic function   (b) ANSI/IEC logic symbol

Fig. 2.18 The 74LS377 octal D flip flop array.

sometimes called an **Accumulator** or **Working register**. To describe the operation of this circuit, consider the problem of adding two words A and B. The sequence of operations, assuming the ALU is implemented by cascading two 74LS382s might be:

1. Program step.
   - Mode = 000 (Clear).
   - Pulsing Execute loads the ALU output (0000 0000) into the register.
   - Data out is zero (0000 0000).
2. Program step.
   - Fetch word A down to the ALU input.
   - Mode = 011 (Add).

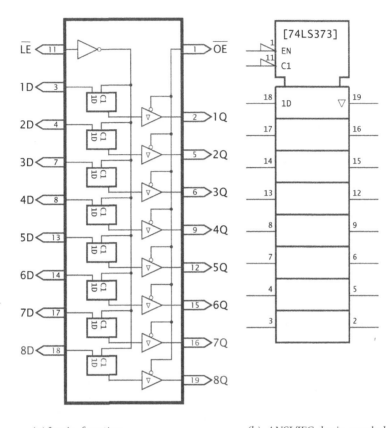

(a) Logic function   (b) ANSI/IEC logic symbol

Fig. 2.19 The 74LS373 octal D latch array.

- Pulse ⎍ Execute to load the ALU output (word A + zero) into the register.
- Data out is word A.
3. Program step.
  - Fetch word B down to the ALU input.
  - Mode = 011 (Add).
  - ⎍ Execute to load the ALU output (word B + word A) into the register.
  - Data out is word B plus word A.

The sequence of operation codes, that is 000 – 100 – 100 constitutes the program. In practice each instruction would also contain the address (where relevant) in memory of the data to be processed; in this case the locations of word A and word B.

Each outcome of a process will have associated properties. For instance, it may be zero or have a carry-out. Such properties may be significant in the future progress of the program. In the diagram, two D flip flops, clocked by Execute, are used to grab this status information. In this situation the flip flops are usually known as **flags** (or sometimes semaphores). Thus we have **Z** (Zero) and **C** (Carry from bit 7) flags, which form a Code Condition or Status register.

As we will see in the next chapter, the ALU/Working register processor is the heart of digital computing engines. In complex systems, such as a computer or microcontroller, the detail of a diagram like Fig. 2.20 is not necessary and will hide the underlying system process from the observer. Figure 2.21 shows the same process at a higher level of abstraction. For instance, the various multiple wire data connections or **buses** are shown as a single thick path; the actual details are unimportant. The number of connections in a path is not shown, but if important, is usually indicated by a diagonal tick, thus .

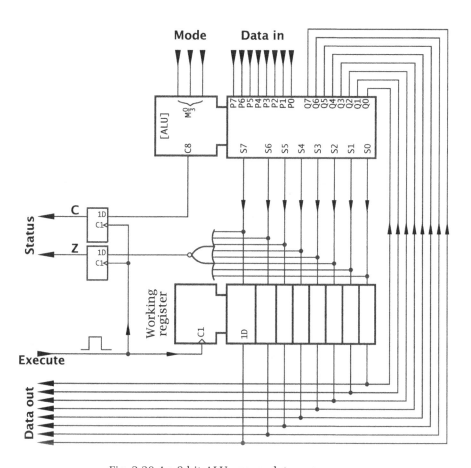

Fig. 2.20 An 8-bit ALU-accumulator processor.

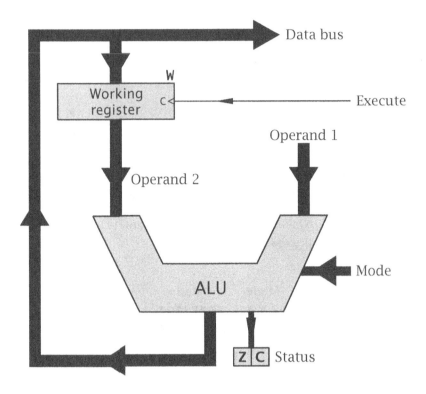

Fig. 2.21 A system look at our ALU-accumulator processor.

The ALU, with its distinctive shape, is at the center of our system. Its two data inputs, or operands, are processed according to the Mode input. Operand 1 comes from outside our system, whilst Operand 2 is connected from the Working register. In a computer, the Mode input codes normally come from the program memory, whilst Operand 1 is obtained from the data memory.

The ALU output can be either latched back into the Working register when sampled by the Execute signal, or it can be fed outside into a data memory via the bus. This enhancement is shown in Fig. 3.2 on page 46.

There are various other forms of register. The 4-bit **shift register** of Fig. 2.22(a) is an example of a serial-in serial-out (SISO) structure. In this instance the data held in the nth D flip flop is presented to the input of the $(n+1)$th stage. On receipt of a clock pulse (or shift pulse in this context), this data moves into this $(n + 1)$th flip flop, i.e., effectively moving from stage n to stage $n + 1$. As all flip flops are clocked simultaneously, the entire word moves once to the right on each shift pulse.

In the example of Fig. 2.22 a 4-bit external data nybble is fed into the leftmost stage bit-by-bit as synchronized by the clock. After four shift pulses the serial 4-bit word is held in the register. To get it out again,

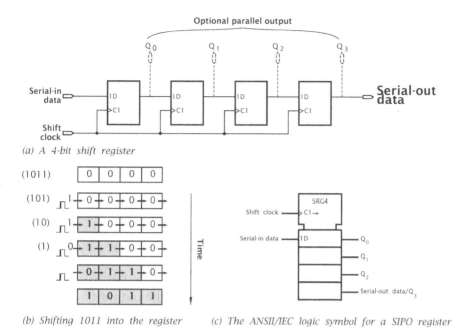

(a) A 4-bit shift register

(b) Shifting 1011 into the register   (c) The ANSII/IEC logic symbol for a SIPO register

Fig. 2.22 The SISO shift register.

four further shifts move the word bit-by-bit out of the shift register; this is SISO. If the individual flip flops are accessible then the data can be accessed at one go, that is, serial-in parallel-out.

The logic diagram of Fig. 2.22(b) uses the → symbol prefixed by the clock input to indicate the shift action; i.e., C1 →. SRG4 indicates a Shift ReGister 4-stage architecture. An example of an 8-stage shift register is given in Fig. 12.2 on page 333.

Other architectures include parallel-in serial-out which is useful for parallel to serial conversion. Counting registers (counters) increment or decrement on each clock pulse, according to a binary sequence. Typically an n-bit counter can perform a count of 2^n states. Some can also be loaded in parallel and thus act as a store.

Consider the negative-edge triggered D flip flop shown in Fig. 2.23 where its $\bar{Q}$ output is connected back to the 1D input. On each ⌐_ at the Clock input C1 the data at the 1D input will be latched in to appear at the Q output. As it is the complement of this output that is fed back to the input, then the next time the flip flop is clocked the *opposite* logic state will be latched in. This constant alternation is called *toggling* and is depicted on the diagram by T. The output waveform resulting from a constant-frequency input pulse train is half this frequency. This waveform is a precision squarewave, provided that the input frequency

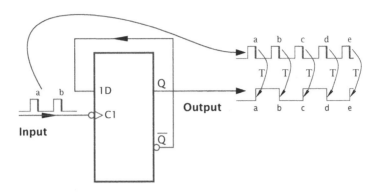

Fig. 2.23 The T flip flop.

remains constant. This **T flip flop** is sometimes known as a binary or a divide-by-2.

T flip flops can, of course, be cascaded, as shown in Fig. 2.24(a). Here four ⌐⌐ triggered flip flops are chained, with the output of binary n clocking binary $n+1$. Thus if the input Count frequency was 8 kHz, then Q_A would be a 4 kHz square waveform and similarly Q_B would measure in at 2 kHz, Q_C at 1 kHz, Q_D at 500 Hz.

The waveform Q_A of Fig. 2.24(b) was derived in the same manner as in Fig. 2.23. Q_B is toggled on each ⌐⌐ of Q_A and likewise for the subsequent outputs. Marking a High as logic 1 and a Low as logic 0 gives the 2^4 (16) positive-logic binary patterns as time advances, with the count rolling over back to state 0 on a continual basis. Each pattern remains in the register until the next event clocks the chain; an event being defined in our example as a ⌐⌐ at Count. Examining the sequence shows it to be a natural 8-4-2-1 binary up count, incrementing from b'0000' to b'1111'. In fact, the circuit is a modulo-16 **binary counter**. A modulo-n count is the sequence taking only the first n numbers into account.[7]

In theory there is no limit to the number of stages that can be cascaded. Thus using eight T flip flops would give a modulo-256 (2^8) counter. In practice there is a small propagation delay through each stage and this limits the ultimate frequency. For instance, the 74LS74 dual D flip flop of Fig. 2.17 has a maximum propagation from an event at its Clock input to output of 25 ns. The maximum toggling frequency for a single stage, such as in Fig. 2.23, is given as 25 MHz. An 8-stage counter thus has a maximum ripple-through time of 200 ns. If such a **ripple counter** were clocked at the resulting 5 MHz ($\frac{1}{200\,ns}$) then no sooner would one particular code pattern stabilize then the next one would begin to appear. This is only really a problem if the various states of the counter are to be de-

---

[7]Mathematically any number can be converted to its modulo-n equivalent by dividing by n. The remainder, or modulus, will be a number from 0 to $n-1$.

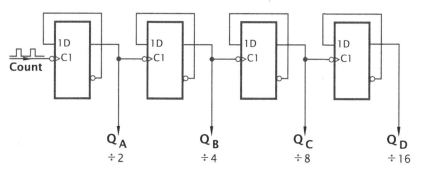

(a) Cascading toggle flip flops

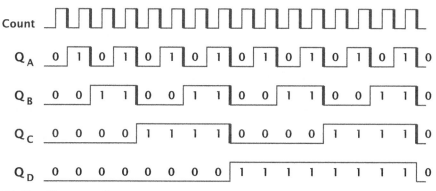

(b) Resulting waveforms

Fig. 2.24 A modulo-16 ripple counter.

coded and used to control other logic. The decoding logic, such as shown in Fig. 2.25, may inadvertently respond to these short transient states and cause havoc. In such cases more sophisticated synchronous counter configurations are more applicable where the flip flops are clocked simultaneously and steered by the appropriate logic configuration to count in the desired sequence.

The circuit illustrated here implements an up count. If the complement $\bar{Q}$ lines are used as the outputs, but with the clocking arrangements remaining the same, then the count sequence will decrement, that is a down count. Likewise, if _⌐ triggered flip flops, such as the 74LS74 dual flip flop (see Fig. 2.25), are used as the storage element, then the count will be down. It is easily possible to use some simple logic to combine the two functions to produce a programmable up/down counter. It is also feasible to provide logic to load the flip flop array in parallel with any number and then count up or down from that point. Such an arrangement can be thought of as a parallel-in counting register.

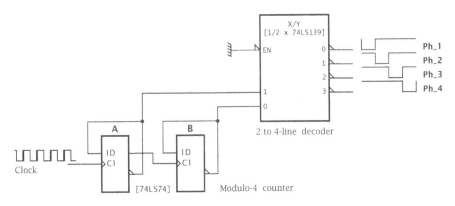

Fig. 2.25 Generating timing waveforms.

In addition to the more obvious uses of a counter register to add up the number of events, such as cans of peas coming along a conveyor belt, there are other uses. One of these is to time a sequence of operations. In Fig. 2.25 a modulo-4 counter is used to address one section of a 74LS139 2- to 4-line decoder; see Fig. 2.5(a). This detects each of the four states of the counter, and the outcome is four time-separated outputs that can be used to sequence, say, the operation of a computer's control section logic. As a practical point, the complement $\overline{Q}$ flip flop outputs have been used to address the decoder to compensate for the _/‾ triggered action that would normally give a down count. Larger counters with the appropriate decoding circuitry can be used to generate fairly sophisticated sequences of control operations.

The term register is commonly applied to a read/write memory that can store a single binary word, typically 4 – 64 bits. Larger memories can be constructed by grouping n such registers and selecting one of n. Such a structure is sometimes known as a register file. For example, the 74LS670 is a 4×4 register file with a separate 4-bit data input and data output and separate 2-bit address. This means that any register can be read at any time, independently of any concurrent writing process.

Larger read/write memories are customarily known as **read/write random-access memories**, or **RAMs** for short. The term random-access indicates that any memory word may be selected with the same access time, irrespective of its position in the memory matrix.[8] This contrasts with a magnetic tape memory, where the reel must be wound to the sector in question—and if this is at the end of the tape....

For our example, Fig. 2.26 shows the 6264 RAM. This has a matrix of 65,536 (2^{16}) bistables organized as an array of 8192 (2^{13}) words of

---

[8]Strictly speaking, ROMs should also be described as random access, but custom and practice have reserved the term for read/write memories.

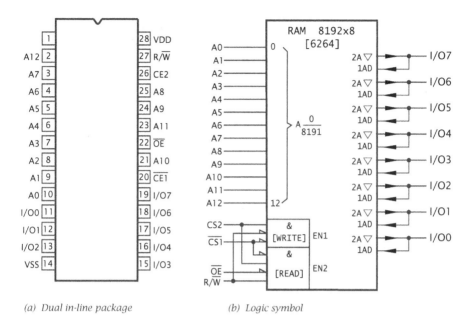

(a) Dual in-line package   (b) Logic symbol

Fig. 2.26 The 6264 8196 × 8 RAM.

8 bits. Word n is accessed by placing the binary pattern of n on the 13-bit Address pins A12,...,A0.

When in the Read mode (Read/$\overline{\text{Write}}$ = 1), word n will appear at the eight data outputs (I/O7,...,I/O0) as determined by the state n of the address bits. The A symbol at the input/outputs (as was the case in Fig. 2.12) indicates this addressability. In order to enable the 3-state output buffers, the Output Enable input must be Low.

The addressed word is written into if R/$\overline{\text{W}}$ is Low. The data to be written into word n is applied by the outside controller to the eight I/O pins. This bidirectional traffic is a feature of computer buses.

In both cases, the RAM chip as a whole is enabled when $\overline{\text{CS1}}$ is Low and CS2 is High. Depending on the version of the 6264, this access from enabling takes around 100 – 150 ns. There is no upper limit to how long the data can be held, provided power is maintained. For this reason, the 6264 is described as static (SRAM). Rather than using a transistor pair bistable to implement each bit of storage, data can be stored as charge on the gate-source capacitance of a single field-effect transistor. Such charge leaks away in a few milliseconds, so needs to be refreshed on a regular basis. Dynamic RAMs (DRAMs) are cheaper to fabricate than SRAM equivalents and obtainable in larger capacities. They are usually found where very large memories are to be implemented, such as found in a personal computer. In such situations, the expense of refresh circuitry is more than amortized by the reduction in cost of the memory devices.

Both types of read/write memories are volatile, that is, they do not retain their contents if power is removed. Some SRAMs can support existing data at a very low holding current and lower than normal power supply voltage. Thus a backup battery can be used in such circumstances to keep the contents intact for many months.

CHAPTER 3

Stored Program Processing

In Chapter 2 we designed a simple computing engine based on an arithmetic logic unit (ALU) paired with a parallel-in parallel-out register. The ALU did the number crunching and the Working register held one of the operands and also stored any outcome. In our example, shown on page 33, we added three numbers together, with the result accumulating in the Working register. If the ALU's mode code is set up before each step, then we can potentially make the computing engine carry out any task that can be described as a sequence of arithmetic and logic operations. This set of command codes (e.g., Add, Subtract, AND,...) can be stored in digital memory, as can the various operands fed to the ALU and likewise any outcomes. These codes constitute both the **software** of the programmable machine and the various operands or **data**. By *fetching* these **instructions** down one at a time, we can *execute* the system's **program**. This structure, together with its associated data paths, decoders and logic circuitry is known as a digital **computer**.

As we will see, microcontroller architecture is modeled on that of a computer. With this in mind, this chapter looks at the architecture and operating rhythm of the computer structure. Although this computer is strictly hypothetical, it has been very much "designed" with our book's target microcontroller in mind.

After reading this chapter you will have an understanding of:

- The von Neumann computer structure and recognise its weakness.
- The Harvard architecture with its parallel fetch and execute units, and separate memory spaces;
- The relationship between a digital computer, microprocessor and a microcontroller.
- The structure of a Program store and its interaction with the Program Counter and Pipeline.
- The binary anatomy of typical program instructions.
- The function and structure of a Data store.

Historically the electronic digital computer that we know today was an indirect outcome of the Second World War. Several experimental computers were designed, and some actually functioned in that period.[1] These computing machines were either special-purpose structures, mainly designed to do a single task on various data, or else needed to be partly rewired to change their behavior.

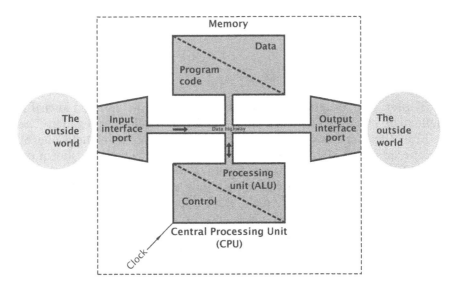

Fig. 3.1 An elementary von Neumann computer; Address bus not shown.

Given the feasibility of building such computing structures, a major breakthrough by a team of engineers working with von Neumann[2] was to recognize that the program could be stored in memory along with any data. The advantage of this approach is flexibility. To alter the software the new program bit patterns are simply loaded into the appropriate area of memory. In essence, the von Neumann architecture, shown in Fig. 3.1, comprises a **Central Processing Unit** (CPU), a memory and a common connecting bus (or highway) carrying data back and forth. In practice the CPU must also communicate with the environment outside the com-

---

[1] A prime example was the British Colossus which spent several years breaking Enigma codes. See the book's website for more historical and technical details of these early machines.

[2] Von Neumann was a Hungarian mathematician working for the American Manhattan nuclear weapons program during the Second World War. After the war he became a consultant for the Moore School of Electrical Engineering at the University of Pennsylvania's EDVAC computer project, for which he was to employ his new concept that the program was to be stored in memory along with its data. He published his ideas in 1946 and EDVAC became operational in 1951. Ironically, a somewhat lower key project at Manchester University, UK made use of this approach and the Mark 1 executed its first stored program in June 1948! This was closely followed by Cambridge University's EDSAC which ran its program in May 1949, almost two years ahead of EDVAC.

puter. For this purpose, data to and from suitable interface ports are also funneled through this *single* **data bus**.

The great advantage of the von Neumann architecture is simplicity, and because of this the majority of general-purpose computers are modeled after this concept. However, the use of a common bus means that only one thing can happen at at time. Thus an execution transaction between the CPU and the **Data store** cannot occur at the same time that an instruction is being fetched from the **Program store**. This phenomena is sometimes known as the von Neumann bottleneck.

In the first decade after the war, Harvard University designed and implemented the Mark 1 through Mark 4 series of computers, which used a variation of this structure where the program memory was completely separate from the data memory. In the original Mark 1 and Mark 2 machines the program was read from a punched paper tape reader. This strategy was more efficient then the von Neumann (or, as it is sometimes known, Princeton) architecture, since code could be fetched from program memory concurrently with activity between the CPU and the data memory or input/output transactions. However, such machines were more complex and expensive, and with 1950s technology never became widely accepted after loosing out in a Department of Defence competition to build a computer to monitor the far-flung radar stations in the continental United States. With the evolution of complex integrated circuits this **Harvard** architecture has made a reappearance.

Figure 3.2 shows the two physically distinct buses used to carry information to the CPU from these disjoint memories. Each memory has its own address bus and thus there is no interaction between a Program store cell's address and a Data store cell's address. The two memories are said to lie in *separate memory spaces*. The Data store is sometimes known as the **File store**, with a cell at location n being described as **File n**.

Let us look at these elements in a little more detail.

The Central Processing Unit
The CPU consists of the ALU/Working register together with the associated control logic. Under the management of the control circuitry, program instructions are fetched from memory, decoded and executed. Data resulting from, or used by, the program is also accessed from memory. This fetch-and-execute cycle constitutes the operating rhythm of the computer and continues indefinitely, as long as the system is active.

Memory
All computing structures use memory to hold both program code and data. Random access memories are characterised by the *contents* they hold in a set of cells and the location or *address* of each cell. In the case of von Neumann type architectures these are held in a single memory space, whereas in Harvard structures they are located in completely separate memory spaces. That is, the addresses of one type of memory do not

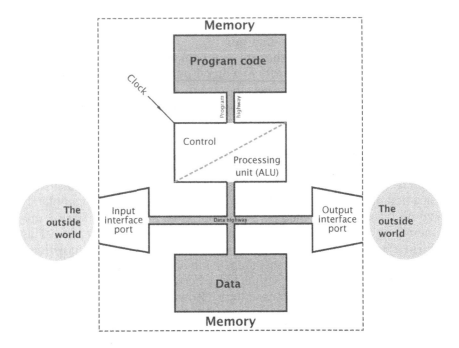

Fig. 3.2 An elementary Harvard architecture computer; Address bus not shown.

relate in any way to the addresses of the other memory. In all cases the data stored in a memory is transported to the CPU via a data bus. The CPU sends the address code of the cell it wishes to communicate with via an address bus. In Harvard structures, there will be separate data and address buses for each memory space; see Fig. 3.4. In a random access memory the time to read from or write to any addressed cell is independent of where in the memory address space the cell is located.

Most computers have large backup memories, usually magnetic or optical disk-based, in which case access time depends on the cell's physical position in the memory rather than being random access. Apart from this sequential access problem, such media are normally too slow to act as the main memory and are used for backup storage of large arrays of data (e.g., student exam records) or programs that must be loaded (or swapped) into main memory before execution.

Program Memory
The Program store holds the bit patterns which define the program or software. The word is a play on the term hardware; as such patterns do not correspond to any physical rearrangement of the circuitry. Memory

holding software should ideally be as fast as the CPU, and normally use semiconductor technologies, such as that described in the last chapter.[3]

Data Memory

The Data store holds data being processed by the program. Again, this memory is normally as fast as the CPU. The processor may also locate special-purpose registers in this memory space; for instance, input/output ports.

The Interface Ports

To be of any use, a computer must be able to interact with its environment. Although conventionally one thinks of a keyboard and screen, any of a range of physical devices may be read and controlled. Thus the flow of fuel injected into a cylinder together with engine speed may be used to alter the instant of spark ignition in the combustion chamber of an internal combustion engine.

Data Highway

All the elements of the von Neumann computer are wired together with the one *common* data highway, or bus (see Fig. 2.4 on page 20 for a definition of a bus). With the CPU acting as the master controller, all information flow is back and forward along these shared wires. A Harvard computer has a separate data bus for the Program store allowing the instruction codes to be fetched in parallel with activity on the Data store's data bus. Other buses carry addresses to the various memories and control/status information; see Fig. 3.4.

Our target microcontroller is a Harvard computing engine and so we will concentrate on this structure from now on. Based on the CPU of Fig. 2.20 on page 35, we can add program and data memory with some control and decoding logic to give us the elementary Harvard computer of Fig. 3.3. The shaded portion of the diagram is the original circuit of Fig. 2.21 on page 36.

By extending the data bus out to the Data store, we can both source operand 1 from this memory and also optionally put the outcome back there. The address of this operand is part of the instruction fetched from the Program store and decoded by the control circuitry. This Control Unit also extracts the mode bits for the ALU, which depends on the current instruction. The outcome from the ALU can either be loaded into the Working register (the control unit pulses W) or back into the File in the Data store where the operand originated (the Control Unit pulses F). Again, this destination information is part of the instruction code.

---

[3] This wasn't always so; the earliest practical large high-speed program memories used miniature ferrite cores (donuts) that could be magnetized in any one of two directions. Core memories were in use from the 1950s to the early 1970s, and program memory is sometimes still referred to as core.

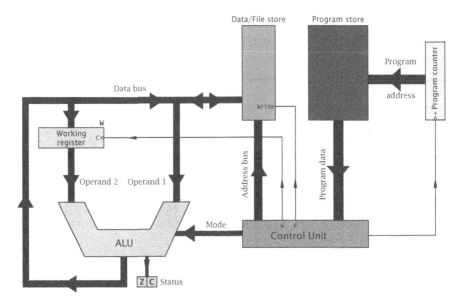

Fig. 3.3 A system look at a rudimentary Harvard computer.

Instructions are normally located as sequential code words in the Program store. A binary up counter (see Fig. 2.24 on page 39) is used to address each instruction word in turn. If we assume that this **Program Counter** is zeroed when the computer is reset, then the first instruction is located at address h'000' in the Program store, the second at h'001' and so on; see Fig. 3.4. The Control Unit simply increments the counter after each instruction has been fetched. By parallel loading a new address into the program counter, overriding this incrementation, the program can be forced to jump to another routine.

The fetch instruction down/decode it/execute sequence, the so-called **fetch-and-execute cycle**, is fundamental to the understanding of the operation of the computer. To illustrate this operating rhythm we are going to look at a simple program that takes a variable called NUM_1, then adds four to it and finally assigns the resultant value to the variable called NUM_2. In the high-level language **C** this may be written as:[4]

```
NUM_2 = NUM_1 + 4;
```

A rather more detailed close-up of our computer, which I have named BASIC (for Basic All-purpose Stored Instruction Computer) is shown in Fig. 3.4. This shows the CPU and memories, together with the two data buses and corresponding address buses.

---

[4]If you are more familiar with PASCAL or Modula-2, then this program statement would be expressed as the NUM_2 := NUM_1 + 4 statement.

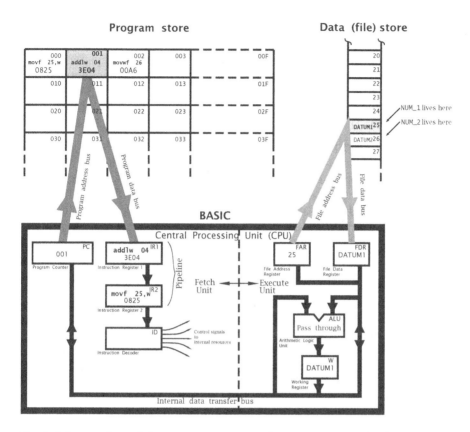

Fig. 3.4 A snapshot of the CPU executing the first instruction whilst simultaneously fetching the second instruction. All addresses/data are in hexadecimal.

The CPU can broadly be partitioned into two sectors. The leftmost circuitry deals with *fetching* the instruction codes and sequentially presenting them to the Instruction decoder. The rightmost sector *executes* each instruction, as controlled by this Instruction decoder ID.

Looking first at the fetch process:

Program Counter
Instructions are normally stored sequentially in program memory, and the PC is the counter register that keeps track of the current instruction word. This up-counter is sometimes called (perhaps more sensibly) an Instruction Pointer.

As the PC is connected to the execution unit — via an internal data bus — the ALU can be used to manipulate this register and disrupt the orderly execution sequence. In this way various Goto and Skip to another part of the program operations can be implemented.

Pipeline

Two instruction registers hold instruction codes from the Program store. At the top, instruction word n is latched into Instruction Register 1 (IR1) and held for processing during the next cycle. This enables instruction $n - 1$ at the bottom of the Pipeline (Instruction Register 2, IR2) to be executed at the same time as instruction n is being fetched into the top of the Pipeline. The Pipeline operation is illustrated in Fig 3.7.

Instruction Decoder

The ID is the "brains" of the CPU, deciphering the instruction word in IR2 and sending out the appropriate sequence of signals to the execution unit as necessary to locate the operand in the Data store (if any) and to configure the ALU to its appropriate mode. In the diagram, the instruction shown is movf h'25',w (MOVe contents of File h'25' to the Working register).

The execution sector deals with accesses to the Data store and configuring the ALU. Execution circuitry is controlled from the Instruction decoder, which is in turn commanded by instruction word $n - 1$ in IR2.

All number crunching in the execute unit is done eight bits at a time, and all the registers and Data store likewise hold data in byte-sized chunks. Because of this, the computer would usually be described as an 8-bit processor.

File Address Register

When the CPU wishes to access a cell (or File) in the Data store, it places the File address in the FAR. This directly addresses the memory via the File address bus. As shown in the diagram, File h'25' is being read from the Data store and the resulting datum is latched into the CPU's File Data register.

File Data Register

This is a bidirectional register which either:
- Holds the contents of an addressed File if the CPU is executing a **read cycle**. This is the case for instruction 1 (movf h'25',w) that MOVes (copies or reads from) a datum from File h'25' into the Working register.
- Holds the datum that a CPU wishes to send out (write to) to an addressed File. This **write cycle** is implemented for the movwf h'26' instruction that moves (writes) out the contents of the Working register to File h'26'.

Arithmetic Logic Unit

The ALU carries out an arithmetic or logic operation as commanded by its mode code (see Fig. 2.10 on page 25) which is extracted from the instruction code by the Instruction decoder.

Status Register

This holds the **Z** and **C** flags, which are set respectively when the outcome of the operation is zero and when a carry-out is generated after an addition operation.

Working Register

W is the ALU's Working register, generally holding one of an instruction's operands, either source or destination. For instance, addwf h'20',w ADDs the contents of the Working register to the contents of File h'20' and places the sum back in W. Some computers call this a data or accumulator register.

In addition to the CPU, our BASIC computer has two stores to hold the program code and data.

Program Store

Each location (or cell) in the Program store holds one instruction which is coded as a 14-bit word. In Fig. 3.4 each one of these cells has an address, which originates from the Program Counter via the Program store's address bus. In the diagram the contents of the PC are h'001' (or b'0000000000001'), and this enables the contents of cell h'001' to be placed on the Program store's data bus and hence read into the top of the Pipeline. In the illustrated case this is h'3E04' (or b'11111000000100'), which is the machine code for the instruction addlw 04. This will eventually be interpreted by the Instruction decoder as a command to add the constant four to the Working register.

Data Store

Each cell (or File) in the Data store holds one byte (eight bits) of data. The File address is generated by the execute unit via the File Address Register (FAR) and the Data store's address bus. The contents of the addressed File is either read into the File Data Register (FDR) or written from it.

The File address and data busses are completely separate from the the Program store counterparts and so processes can proceed on both stores at the same time. Also, Program and Data store addresses are not the same; e.g., the Program store address h'25' is completely different from the Data store address h'25'—that is File h'25'.

Now that we have our CPU with its Program and Data stores, we need to look in more detail at the program itself. There are three instructions in our illustrative software, and as we have already observed, the task is to copy the contents of a byte-sized variable located at the address we have called NUM_1 plus 4 into the location at the address we have called

NUM_2. We see from Fig. 3.4 that the variable named NUM_1 is simply a pseudonym for "the contents of File h'25'" and similarly NUM_2 is a symbolic representation for "the contents of File h'26'."

movf

The instruction MOVe File copies the contents of the specified File, usually down to the Working register, or sometimes just back on top of itself; see page 121. Thus movf NUM_1,w loads the byte out in data memory at location File h'25' into the Working register. This will set the Z flag if the contents of the specified File are all zero—b'0000000'. Otherwise it will be zeroed.

addlw

The ADD Literal to Working register instruction adds a byte-sized literal (constant) to the Working register. Thus addlw 04 adds four to the byte in the Working register and overwrites it with the outcome. The C flag is set if a carry-out is generated and the Z flag is set if the outcome is zero.

movwf

The MOV Working register to File instruction copies the contents of the Working register to the specified File in the Data store.   Thus movwf NUM_2 stores the byte in the Working register in File h'26'. None of the flags are altered by this instruction.

In our description of the instructions we have used mnemonics, such as addlw. Of course the actual digital logic decoding these instructions only operate with binary patterns. Mnemonics are simply just a symbolic aide-mémoire for the programmer. Although it is unlikely he/she will ever program in **machine code**, the binary structure of instructions are logical and a working knowledge of this will be useful in understanding the foibles and limitations of the instruction set and the real hardware we will discuss in the next two chapters.

Here we will look at two categories of instructions:

File Direct | op-code | d | ffffff |

Instructions that specify the File address where their target operand is located use this type of addressing; e.g., movf h'25',w designates File h'25' as the target.

From Fig. 3.5 we see that the 14-bit instruction code is split into three zones.

- The leftmost six bits (bits 13 through 8) are known as the **operation code**, or **op-code** for short. Every instruction has a unique op-code, and it is this pattern that the decoding circuits use to define what type of instruction it is.

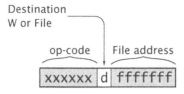

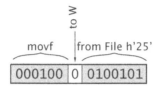

(a) *Machine-code structure.*

(b) *For instance,* movf h'25',w.

Fig. 3.5 Machine-code structure for Direct instructions.

- The middle bit (bit 7) labeled d defines the destination of the outcome. For instance, addwf h'30',w means "add the contents of the Working register to File h'30' and put the answer back in the Working register," whereas addwf h'30',f means "add the contents of the Working register to File h'30' and put the answer back in File h'30'." In the former case the destination is W and the d bit is 0, and in the latter case the destination is the File and the d bit is 1. We will look at this instruction in Chapter 5, page 99. In our symbolic representation of instructions, ,w symbolizes a destination bit of 0 whereas ,f means d = 1.
- The rightmost seven bits (bits 6 through 0) define the File address. Thus in our example File h'25' is b'0100101'. The fact that the address field is only seven bits wide means that only a *bank* of $2^7 = 128$ Files can be Directly addressed; that is File 00 through File h'7F'; see Fig. 4.7 on page 80.

Literal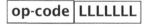

Instructions that deal with constants or **literals** are coded in a slightly different manner, as shown in Fig. 3.6. Again the upper 6-bit zone defines the instruction op-code. The lower 8-bit zone is the byte constant itself. The outcome is always in the Working register and so there is no need for a destination bit nor Data store address.

In our exemplar instruction addlw 04, the op-code is b'111110' and the literal is b'00000100'. The literal is limited to the range b'00000000' – b'11111111' (h'00' – h'FF', or decimal zero through 255), which makes

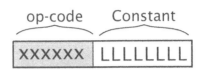

(a) Machine-code structure

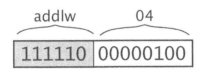

(b) For instance addlw 04.

Fig. 3.6 Machine-code structure for Literal instructions.

sense since the Working register, like all registers relating to the execution unit, is only eight bits wide.[5]

In addition to using symbolic instruction mnemonics, we have already seen that locations in the Data store can also be given names. Thus, in Fig. 3.4, NUM_1 is our name for "the contents of File h'25'" and NUM_2 names File h'26'. We thus can symbolise our program as:

NUM_2 = NUM_1 + 4;

Now as far as the computer is concerned, starting at location h'000' our program is:

```
00100000100101
11111000000100
00000010100110
```

Unless you are a CPU this is not much fun![6]
Using hexadecimal[7] is a little better.

```
0825
3E04
00A6
```

---

[5]One of the more frequent mistakes is to forget the 8-bit size restriction and try and use instructions, such as addlw d'500' in our program. That makes as much sense as trying to fill a liter (quart) bottle with the contents of a 4-liter (gallon) bucket!

[6]I know; I programmed this way back in the primitive middle 1970s.

[7]Remember that we are only using hexadecimal notation as a human convenience. If you took an electron microscope and looked inside these cells you would only "see" the binary patterns indicated.

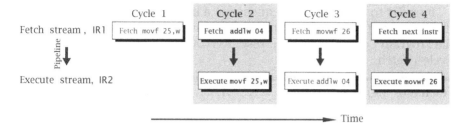

Fig. 3.7 Parallel fetch and execute streams.

but is still instantly forgettable. Furthermore, the CPU still only under-
stands binary, so you are likely to have to use a translator program running on, say a PC, to translate from hexadecimal to binary.

If you are going to use a computer as an aid to translate your program,
known as **source code**, to binary machine code, known as **object code**,
then it makes sense to go whole hog and express the program symbolically. Here the various instructions are represented by mnemonics and
variables' addresses are given names. Doing this our program becomes:

```
movf  NUM_1,w    ; Copy the variable NUM_1 to W
addlw 4          ; Add the literal constant 4 to it
movwf NUM_2      ; Copy NUM_1 + 4 into NUM_2
```

where the text after a semicolon is comment, which makes the program
easier to understand by the tame human programmer.

Code written in this symbolic manner is known as **assembly-level** programming. Chapter 8 is completely devoted to the syntax of assembly-
level language, and its translation to machine-executable binary.

In writing programs using assembly-level symbolic representation, it
is important to remember that each instruction has a one-to-one corre-
spondence to the underlying machine instructions and its binary code. In
Chapter 9 we will see that high-level languages lose that 1:1 relationship.

The core of computer operation is the rhythm of the **fetch-and-
execute cycle**. Here each instruction is successively fetched from the
Program store, interpreted and then executed. Because any execution
memory access will be on the Data store and as each store has its own
busses, then the fetch and execution processes can progress *in parallel*. Thus while instruction n is being fetched, instruction $n - 1$ is being
executed. In Fig. 3.4 the instruction codes for both the imminent and
current instructions are held in the two Instruction registers IR1 and IR2
respectively. Instructions are fetched into one end of this Pipeline and
"popped out," into the Instruction decoder at the other end. Figure 3.7
shows the timeline of our 3-instruction exemplar program, quantized in
instruction cycles. During each cycle, except for the first, both a fetch
and an execution is proceeding simultaneously.

In order to illustrate the sequence in a little more detail, let us trace through our specimen program. We assume that our computer (that is, the Program Counter) has been reset to h'000' and has just finished the Cycle 1 fetch.

Fetch (Fig. 3.4) .... Cycle 2

- Increment the Program Counter to point to instruction 2.
- Simultaneously move the instruction word 1 down the Pipeline (from Instruction register 1 to Instruction register 2).
- Program Counter (h'001') to the Program store's address bus.
- The instruction word 2 then appears on the Program store's data bus and is loaded into Instruction register 1.

Execute (Fig. 3.4) .... Cycle 2

- The operand address h'25' (i.e., NUM_1) moves to the File Address register and out onto the File address bus.
- The resulting datum at NUM_1 is read onto the Data store's data bus and loaded into the File Data register.
- The ALU is configured to the Pass Through mode, which feeds the datum through to the Working register.

Fetch .... Cycle 3

- Increment the Program Counter to point to instruction 3.
- Simultaneously move the instruction word 2 down the Pipeline (from Instruction register 1 to Instruction register 2).
- Program Counter (h'002') to the Program store's address bus.
- The instruction word 3 then appears on the Program store's data bus and is loaded into the Pipeline at Instruction register 1.

Execute .... Cycle 3

- The ALU is configured to the Add mode and the literal (which is part of instruction word 2) is added to the datum in W.
- The ALU output, NUM_1 + 4, is placed in W.

Fetch .... Cycle 4

- Increment the Program Counter to point to instruction 4.
- Simultaneously move instruction word 3 down the Pipeline to IR2.
- Program Counter (h'003') to the Program store's address bus.
- The instruction word 4 then appears on the Program store's data bus and is loaded into the Pipeline at IR1.

Execute .... Cycle 4

- The operand address (i.e., NUM_2) h'26' to the File Address register and out onto the File address bus.
- The ALU is configured to the Pass Through mode, which feeds the contents of W through to the File Data register and onto the Data store's data bus.
- The datum in the File Data register is written into the Data store at the address on the Data store's address bus and becomes the new datum in NUM_2.

Notice how the Program Counter is automatically advanced during each fetch cycle. This sequential advance will continue indefinitely unless an instruction to modify the PC occurs, such as goto h'200'. This would place the address h'200' into the PC, overwriting the normal incrementing process, and effectively causing the CPU to jump to whatever instruction was located at h'200'. Thereafter, the linear progression would continue.

Although our program doesn't do very much, it only takes around 1 μs to implement each instruction. A million unimpressive operations each second can amount to a great deal! In essence, all computers, no matter how intelligent they may appear, are executing relatively simple instructions very rapidly. The skill of course lies with the programmer in deciding what sequence of instructions and data structures are needed to implement the appropriate task!

Up to now we have referred specifically to computerlike structures. To finish the chapter we have to link the subject of this text to this material — that is, the microcontroller.

What exactly is a **microcontroller unit (MCU)**? In a nutshell, a microcontroller is a microprocessor unit (MPU) which is integrated with memory and input/output peripheral interface functions on the (usually) one integrated circuit. In essence it is a MPU with on-board system support circuitry. Thus we begin by investigating the origins of the MPU. From a historical perspective the story begins in 1968 when Robert Noyce (one of the inventors of the integrated circuit), Gordon Moore[8] and Andrew Grove left the Fairchild Corporation and founded their own company, which they called Intel.[9] Within three years, Intel had developed all the basic types of semiconductor memories used today — dynamic and static RAMs and EPROMs.

As a sideline, Intel also designed large-scale integrated circuits to customers' specifications. In 1970 they were approached by the Busicom corporation of Japan, and asked to manufacture a suitable chip set for

---

[8]Moore's law stated in 1965 (when ICs had around 50 transistors per chip) that the number of elements on a chip would double every 18 months. This was based on an extrapolation of growth from 1959 and this was subsequently revised to 2 years.

[9]Reputed to stand for INTELligence or INTegrated ELectronics.

a line of calculators. At that time calculators were a fast-evolving product and any LSI devices were likely to be superseded within a few years. This of course would reduce an LSI product's profitability and increase its cost. Engineer Ted Hoff—reputedly while on a topless beach in Tahiti—came up with a revolutionary way to tackle this project. Why not make a simple computer CPU on silicon? This could then be programmed to implement the calculator functions, and as time progressed these could be enhanced by developing this software. Besides giving the chip a longer and more profitable life, Intel was in the business of making memories—and computerlike architectures need lots of memory! Truly a brain wave. The Japanese company endorsed the Intel design for its simplicity and flexibility in late 1969, rather than the conventional implementation.

Federico Faggin joined Intel in spring 1970[10] and by the end of the year had produced working samples of the first chip set. This could only be sold to Busicom, but by the middle of 1971 they were in financial straits and in return for a payback of their $65,000 design costs, Intel was given the right to sell the chip set to anyone for non-calculator purposes. Intel was dubious about the market for this device, but went ahead and advertised the 4004 "Micro-Programmable Computer on a Chip" in the *Electronic News* of November 1971. The term **microprocessor unit** was not coined until 1972. The 4004 created a lot of interest as a means of introducing "intelligence" into electronic products.

The 4004 MPU featured a von Neumann architecture using a four-bit data bus, with direct addressing of 512 bytes of memory. Clocked at 108 kHz, it was implemented with a transistor count of 2300.[11] Within a year the 8-bit 200 kHz 8008 appeared, addressing 16 Kbytes and needing a 3500 transistor implementation. Four bits is satisfactory for the BCD digits used in calculators, but eight bits is more appropriate for intelligent data terminals (like cash registers) which need to handle a wide range of alphanumeric characters. The 8008 was replaced by the 8080[12] in 1974, and then the slightly modified 8085 in 1976. The 8085 is still the current Intel 8-bit device.

The MPU concept was such a hit that many other electronic manufacturers clambered onto the bandwagon. In addition, many designers set up shop on their own, such as Zilog. By 1976 there were 54 different MPUs either available or announced. For example, one of the most successful families was based on the 6800 introduced by Motorola.[13] The

---

[10]He was later to found Zilog (last word (Z) in Integrated LOGic) which became notable with the Z80 MPU, a rather superior Intel 8085.

[11]Compare with the Pentium Pro (also known as the P6 or 80686) at around 5.5 million!

[12]Designed by Masatoshi Shima, who went on to design the 8080-compatible Z80 for Zilog.

[13]Motorola was launched in the 1930s to manufacture motor car radios, hence the name "motor" and "ola", as in pianola. It has the largest share of the worldwide microcontroller market by value at the time of writing (2005).

Motorola 6800 had a clean and flexible von Neumann architecture, could be clocked at 2 MHz and address up to 64 Kbyte of memory. The 6802 (1977) even had 128 bytes of on-board memory and an internal clock oscillator. By 1979 the improved 6809 represented the last in the line of these 8-bit devices, competing mainly with the Intel 8085, Zilog Z80 and MOS Technology's 6502.

The MPU was not really designed to power conventional computers, but a small calculator company called MITS,[14] faced with bankruptcy, took a final desperate gamble in 1975 and decided to make and market a computer. This primitive machine, designed by Ed Roberts, was based on the 8080 MPU and interacted with the operator using front-panel toggle switches and lamps—no keyboard or VDU. The Altair[15] was advertised, and within a few weeks MITS had around 650 advance orders at about $400 each; going from $400,000 in the red to $250,000 in the black.

This first **personal computer** (**PC**) spawned a generation of computer hackers. Thus an unknown 19-year-old Harvard computer science student, Bill Gates, and a visiting friend, Paul Allen, in December 1975 noticed a picture of the Altair[16] on the front cover of *Popular Electronics* and decided to write software for this primordial PC. They called Ed Roberts with a bluff, telling him that they had just about finished a version of the BASIC programming language that would run on the Altair. Thus was the Microsoft® Corporation born.

In a parallel development, some two months later, 32 people in San Francisco set up the Home-brew Club, with initially one hard-to-get Altair between them. Two members were Steve Jobs and Steve Wozniak. As a club demonstration, they built a PC which they called the Apple.[17] By 1978 the Apple II made $700,000; in 1979 sales were $7 million, and then $48 million····.

The Apple II was based on the low-cost 6502 MPU which was produced by a company called MOS Technology. It was designed by Chuck Peddle, who was also responsible for the 6800 MPU, and had subsequently left Motorola. The 6502 bore an uncanny resemblance to the Motorola 6800 family and indeed Motorola sued to prevent the related 6501 MPU being sold, as it even had the same pinout as the 6800. The 6502 was one of the main players in PC hardware by the end of the 1970s, being the computing engine of the BBC series and Commodore PETs amongst many others.

What really powered up Apple II sales was the VisiCalc spreadsheet package. When the business community discovered that the PC was not just a toy, but could do "real" tasks, sales took off. The same thing

---

[14]Located next door to a massage parlor in New Mexico.

[15]After a planet in *Star Trek.*

[16]The picture was just a mock-up; they actually were not yet available; an early example of computer "vaporware"!

[17]Jobs was a fruitarian and had previously worked in an apple orchard.

happened to the IBM PC. Reluctantly introduced by IBM in 1981, the PC was powered by an Intel 8088 MPU clocked at 4.77 MHz together with 128 Kbyte of RAM, a twin 360 Kbyte disk drive and a monochrome text-only VDU. The operating system was Microsoft's® PC/MS-DOS version 1.0. The spreadsheet package was Lotus 1-2-3.

By the end of the 1970s the technology of silicon VLSI fabrication had progressed to the stage that several tens of thousands of transistors could be integrated on a single chip. Microprocessor designers were quick to exploit this capability in one of two ways. The better known of these was to increase the size of the ALU and buses/memory capacity. Intel was the first with the 29,000-transistor 8086, introduced in 1978 as a 16-bit version of the 8085 MPU.[18] It was designed to be compatible with its 8-bit predecessor in both hardware and software aspects. This was wise commercially, in order to keep the 8085's extensive customer base from looking at (better?) competitor products, but technically dubious. It was such previous experience that led IBM to use the 8088 version, which had a reduced 8-bit data bus and 20-bit address bus[19] to save board space.

In 1979 Motorola brought out its 16-bit offering called the 68000 and its 8-bit data bus version, the 68008 MPU. However, internally it was 32-bit, and this has provided compatibility right up to the 68060 introduced in 1995 and the ColdFire RISC device launched in 1997. With a much smaller 8-bit customer base to worry about, the 68000 MPU was an entirely new design and technically much in advance of its 80X86 rivals.

The 68000 was adopted by Apple for its Macintosh series of PCs. However, the Apple Mac only accounts for less than 5% of PC sales. Motorola MPUs have been much more successful in the embedded microprocessor market, the area of smart instrumentation from egg timers to aircraft management systems. Of course, this is just the area which MPUs were developed for in the first place, and the number, if not the proFile and value, of devices sold for this purpose exceeds those for computers by more than an order of magnitude.

In this applications area an MPU is "buried" in the application circuit together with memory and various input and output interface circuits. The MPU with its program acts as the controller of the system by virtue of the software in program memory. Over 3.5 billion microprocessor and related devices are sold each year for embedded control, making up over 90% of the MPU market.

The second way of using the additional integrated circuit complexity that became available by the end of the 1970s was to keep a relatively

---

[18]And the Intel 8086 architecture–based MPUs are by far the largest-selling MPU for computer-based circuitry.

[19]A 2^{20} address space is 1 Mbyte, and this is why for backwards compatibility MS-DOS was limited to 1 Mbyte of conventional memory, called real memory in a Microsoft® Windows environment.

simple CPU and use the extra silicon "real estate" to implement on-board memory and input/output interface. In doing so, simple embedded control systems on a single chip became possible and the overall chip count to implement a given function was thereby considerably reduced. The majority of control tasks require relatively little computing power, but the reduction in size (and therefore cost) is vital. A simple example of this is the intelligent smart card, which has a processor integrated into the card itself. Such microprocessor-based devices were called microcontrollers.[20] For instance, there are several hundred microcontrollers hidden in every home—in domestic appliances, entertainment units, PCs, communication devices, smart cards and in particular in the family's cars.

In terms of architecture, referring back to Figs. 3.1 and 3.2, the microprocessor is the central processor unit, whereas the microcontroller is the complete functioning computerlike system. As an example, consider the electronics of a car odometer monitoring system displaying total distance since manufacture and also a trip odometer. The main system input signal is a tachometer generating pulses on each rotation of the engine flywheel, which when totalled gives the number of engine revolutions—and the pulse-to-pulse duration could also give the road speed. Of course the actual road distance depends on the gearing of the engine, and thus we need to know which of the five gear ratios has been chosen by the driver at any time. This is shown as five lines G5,...,G1, (usually designated G[5:1]), originating from the gear box. One signal will be high for the appropriate forward gear, with reverse being ignored. Additional inputs are used to give a manufacturer's option of a mile or kilometer display, and a user input to reset the trip display to zero.

The display itself consists of seven 7-segment digits (see Fig. 6.8 on page 161) to indicate up to (optimistically) **999999.9**. As there are so many segments to control (49 in total), Fig. 3.8 shows the display data fed via a single digital line, shunted serially into a shift register; see Fig. 2.22 on page 37. A second line provides clock pulses for the register with 49 clock pulses being needed to refresh the display.[21]

The trip odometer display comprises four digits, which will record up to **999.9**. Similarly two output lines are used to feed and clock the shift register, and 28 clock pulses are needed to shift in a new 4-digit trip display.

The **resource budget** (list of subsystem functions) for this system is:

- An edge-triggered input for the tachometer pulse train, connected to a counter/timer to add up engine revolutions.
- Seven static digital input lines to interface to the gear ratio, mi/$\overline{\text{km}}$ option and trip reset.

---

[20]The term *microcomputer* was an alternative term but was easily confused with early personal computers and has dropped into disuse.

[21]Many displays have this shift register built in as a complete subsystem.

Microcontroller

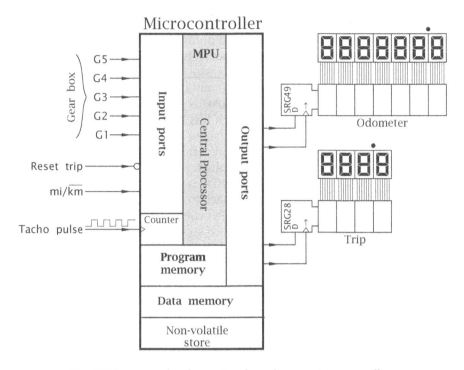

Fig. 3.8 An example of a system based on a microcontroller.

- Four output digital lines to clock the two shift registers and provide segment data.
- A microprocessor to do the calculations and to read/write to the input/output ports, respectively.
- Program memory, usually ROM of some kind.
- Data memory for temporary storage of program variables, usually static RAM.
- Non-volatile storage for physical variables, such as total distance and distance since trip reset.

This functionality could be implemented onto a single integrated circuit, and in this situation would be known as a microcontroller, that is, a microprocessor integrated with its support circuitry giving a complete microcomputer function. Of course the resource budget listed above is specific to our example. Although the core functions (microprocessor and memory) are common to a wide range of applications, the input/output (I/O) interface needs to be tailored to the task at hand. Some typical I/O functions are:

- I/O to interface to a serial bit stream of various synchronous and asynchronous protocols.

- Counter/timer functions to add up input events and to generate precision time-varying digital output signals.
- Analog-to-digital multiplex/conversion to be able to read and digitize analog inputs.
- Digital-to-analog conversion to output analog signals.
- Display ports to drive multidigit liquid crystal displays.

This alternative approach to using additional silicon resources led to the first MCUs in the late 1970s. For instance, the 35,000-transistor Motorola 6801, designed in response to a specific application from a car manufacturer, used the existing 6800 MPU as a core, with 2048 bytes of ROM program memory, 128 bytes of data RAM, 29 I/O lines and a 16-bit timer. With the viability of the MCU approach vindicated, a range of families, each based on a specific core but with individual family members having a different selection of I/O facilities, was introduced by the leading MPU manufacturers. For instance, the Motorola 68HC11 family (a development of the 6801 MCU) uses a slightly enhanced 6800 core. The 68HC12 and 68HC16 families use 16-bit cores but are designed to be upwardly compatible with the 8-bit 68HC11. It was quickly realized that many embedded applications did not even need the power of the (antique) 6800 core, and the 68HC05 family[22] had a severely reduced core and lower price. Actually, 4-bit MCUs, such as the Texas Instruments TMS1000 series outsold all other kinds of processor until the early 1990s (and are still going strong) and 8-bit MCUs, now the most popular, are likely to continue in this role for the foreseeable future. Indeed the Motorola 14500 processor even uses one bit!

All these MPUs and MCUs are based on the von Neumann architecture used by mainframe computers. The alternative Harvard architecture first reappeared in the Signetics 8X300 MPU, and this was adapted by General Instruments in the mid 1970s for use as a **Peripheral Interface Controller** (**PIC**) which was designed to be a programmable I/O port for their 16-bit CP1600 MPU. When General Instruments sold off their microelectronics division in 1988 to a startup company called Arizona Microchip Technology, this device resurfaced as a stand-alone microcontroller. This family of microcontroller devices is the subject of the rest of our book.

Examples

Example 3.1
A greenhouse controller is to monitor an analog signal from a soil moisture probe and if below a certain value turn on a water valve for 5 seconds and off for 5 seconds. The source of the water is a tank with a float and if

---

[22]The 68HC05 has found a niche as the computing engine of smart cards, where high-power computing is not a priority.

the level in the tank drops too low, a switch will be closed. In this event a buzzer is to be activated to sound the alarm.

Can you devise a system based on a microcontroller that will implement the system intelligence?

Solution

The solution given in Fig. 3.9 is based on the car odometer of Fig. 3.8. The only new peripheral device is the analog port which can read and digitize the analog output of the soil moisture transducer. This works on the principle that the resistance of the soil between the two electrodes depends on the moisture content. This forms a potential divider and thus a varying voltage at the junction with the fixed resistor. The MCU can digitize this analog voltage, giving an internal digital equivalent byte, which is then compared in software with a predetermined value. Alternatively, the input port can simply be an analog comparator, giving a digital on/off response if the input voltage exceeds a value which can be set by the program.

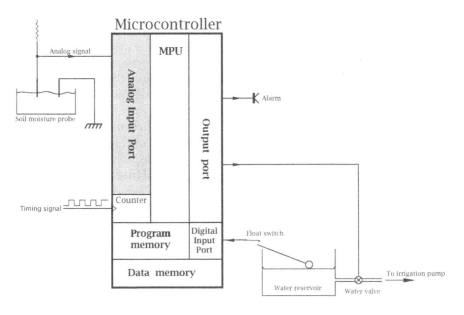

Fig. 3.9 A greenhouse environmental controller.

On the basis of this diagram we can list the resource budget.

- An input for an external oscillator, connected to a counter/timer to allow the MCU to calculate time. In practice the system clock can often be used by this internal timer to measure duration.

- A 1-input analog input line to measure the analog signal from the moisture detector.
- A 1-input digital line to check the level of the reservoir water tank.
- A 1-output digital line to open and close the water valve.
- A 1-output digital line to activate the buzzer alarm.
- A microprocessor to do the calculations and to read/write to the input/output ports, respectively.
- Program memory, usually ROM of some kind.
- Data memory for temporary storage of program variables, usually static RAM.

Assuming that the software does not take up all the MCU's processing time, extra inputs may be used to monitor other environmental signals, such as temperature and light, to give a more comprehensive climate control.

Example 3.2
The most difficult problem that is being solved by a programmer is to define the problem that is being solved. This is the logical thought process that humans are (quite) good at and machines are not. The mark of a good programmer is one who has this ability of problem solving. It is a developed skill, coupled with some talent, and a good understanding of the problem being solved.

To illustrate this process, devise a sequence of simple steps that a MCU-controlled robot must perform to cross a busy road at a pedestrian-controlled crossing.

Solution
1. Walk up to the pedestrian crossing and stop.
2. Look at the traffic light.
3. Is it green for your direction of travel—make a decision?
4. IF the light is red THEN go to step 2 ELSE continue.
5. Look to the right.
6. Are there still cars still passing by?
7. IF yes THEN go to step 5 ELSE continue.
8. Look to the left.
9. Are there cars still passing by (there shouldn't be any by now, but you never know!)?
10. IF yes THEN goto step 5 ELSE continue.
11. Proceed across the street—carefully!

An alternative visual representation is illustrated in Fig. 3.10. This **flow chart** uses boxes to show statements, diamonds for decisions and ovals for entry and exit points. Lines with arrows give action paths, with annotations at decision points. Although this is not much of an advantage in this relatively simple case, for more complex situations with multiple decisions and flow paths, the visual presentation may have the edge in documenting system behavior. Neither the task list or flow chart is much

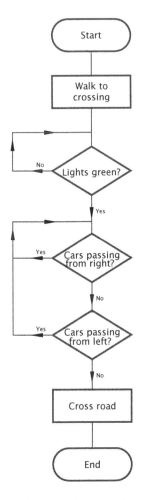

Fig. 3.10 A flow chart showing the robot how to cross the road.

use for very complex situations and here a hierarchy of descriptions starting with the more general and working down to the more particular must
be implemented.

Now, this example may seem childish at first glance, but this is exactly
what you should do every time you traverse a busy street with a pedestrian light-controlled crossing. This is also exactly what you need to tell
a MCU-controlled robot to cross a street at a pedestrian crossing. This
sequence of simple steps or instructions is called a **program**.. Taken as
a whole, the steps lead you across a busy road, which if a robot did it,
would seem very intelligent. It is not intelligence; people are intelligent.
The programmer that programmed these steps into the MCU would im-
part that intelligence to the robot.

Of course, the MCU-programmed robot would not know what to do when it got to the other side, since we did not tell it! In the case of the person, though, there has been some programming; it's called past experience!

Notice that the steps are numbered in the order they should be executed. The program counter, in this case the reader, starts with instruction 1 (the reset state) and ends with instruction 11. In a MCU the program counter automatically advances to the next step, after doing what the current step says, unless a *skip* or *goto* is encountered. A skip type of instruction directs the program counter to hop over the next instruction, usually based on some condition or outcome. A goto instruction directs the program counter to jump to any step in the program. Without these types of instructions, a program would be unable to make decisions or implement loops, where an action is repeated may times over; for instance, repetitively checking for a green light for as long as it takes.

Self-Assessment Questions

3.1 Given the three instructions discussed in this chapter, can you devise a way of incrementing and also decrementing the contents of the Working register of Fig. 3.4?

3.2 Devise a program to enable the MCU-controlled robot of Example 3.2 to fill a glass of water from a tap/faucet.

3.3 The BASIC computer of Fig. 3.4 can fetch an instruction at the same time as it can execute an instruction. Discuss what features enable it to do these tasks in parallel.

3.4 Design a task list to program a robot to go to the nearest ATM and withdraw a specified amount of money, request a statement and return. Your consideration should include a request to print a statement and also what to do if your account does not have sufficient funds!

3.5 The gear inputs to the microcontroller system shown in Fig. 3.8 require five pins on the integrated circuit. MCU packages often have a small pin count; see, for example, Fig 10.2 on page 273. Can you think of any way to reduce the pin count and do you think this is economically viable? *Hint:* See Fig. 2.6 on page 22.

3.6 In a similar attempt to reduce the pin count, can you think of a way to reduce the number of output pins driving the odometer and trip displays by one, and is this economically compatible?

PART II
The Software

In Part I we developed the concept of the Harvard architecture, ending up with our somewhat simplified BASIC computer. Although BASIC was entirely fictitious, it was designed with an eye toward the microcontroller that forms the basis for the rest of this book.

This part of the text looks mainly at the software aspects of our chosen microcontroller, the mid-range Microchip PIC® family. We will be covering:

- The internal structure of the mid-range PIC microcontrollers.
- The instruction set.
- Instruction addressing.
- Software development using the MPLAB® integrated development environment.
- The assembly translation process.
- Subroutines and modular program design.
- Interrupt handling.
- The high-level language **C** and compilation.

14-bit core instruction set.

14-bit op-code	mid-range instruction	Mnemonic	Dest W	Dest F	CCR Z	CCR D	CCR C	Operation summary
11 1110 LLLL LLLL	ADD Literal to W	addlw LL	√		√	√	√	w <- w + #LL
00 0111 dfff ffff	ADD W and F	addwf f,d	√	√	√	√	√	d <- w + f
11 1110 LLLL LLLL	AND Literal to W	andlw LL	√		√	•	•	w <- w · #LL
00 0101 dfff ffff	AND W to F	andwf f,d	√	√	√	•	•	d <- w · f
01 00nn nfff ffff	Bit Clear File bit n	bcf f,n		√	•	•	•	fn <- 0
01 01nn nfff ffff	Bit Set File bit n	bsf f,n		√	•	•	•	fn <- 1
01 10nn nfff ffff	Bit Test File bit n & Skip if Clear	btfsc f,n			•	•	•	pc++ IF fn == 0
01 11nn nfff ffff	Bit Test File bit n & Skip if Set	btfss f,n			•	•	•	pc++ IF fn == 1
10 0aaa aaaa aaaa	CALL (jump to) subroutine	call aaa			•	•	•	TOS <- pc, pc <- aaa
00 0001 1fff ffff	CLeaR File	clrf f		√	√	•	•	f <- 00
00 0001 0000 0011	CLeaR Working register	clrw	√		√	•	•	d <- 00
00 0000 0000 0100	CLeaR Watch Dog Timer	clrwdt			•	•	•	wdt <- 00
00 1001 dfff ffff	COMplement File	comf f,d	√	√	√	•	•	d <- f̄
00 0011 dfff ffff	DECrement File	decf f,d	√	√	√	•	•	d <- f-
00 1011 dfff ffff	DECrement File & Skip on Zero	decfsz f,d	√	√	•	•	•	d <- f-; pc++ IF == 0
10 1aaa aaaa aaaa	GOTO (jump to) aaa	goto aaa			•	•	•	pc <- aaa
00 1010 dfff ffff	INCrement File	incf f,d	√	√	√	•	•	d <- f++
00 1111 dfff ffff	INCrement File & Skip on Zero	incfsz f,d	√	√	•	•	•	d <- f++; pc++ IF == 0
11 1000 LLLL LLLL	Inclusive OR Literal to W	iorlw LL	√		√	•	•	w <- w + #LL
00 0100 dfff ffff	Inclusive OR W to F	iorwf f,d	√	√	√	•	•	d <- w + f
00 1000 dfff ffff	MOVe in File (load)	movf f,d	√	√	√	•	•	d <- f
11 0000 LLLL LLLL	MOVe Literal into W	movlw LL	√		•	•	•	w <- #LL
00 0000 1fff ffff	MOVe W out to File (store)	movwf f		√	•	•	•	f <- w
00 0000 0000 0000	No OPeration	nop			•	•	•	Do nothing
11 0100 LLLL LLLL	RETurn from subroutine; L in W	retlw	√		•	•	•	w <- #LL, pc <- TOS
00 0000 0000 1000	RETURN from subroutine	return			•	•	•	pc <- TOS
00 0000 0000 1001	RETurn From IntErrupt	retfie			•	•	•	GIE <- 1, pc <- TOS
00 1101 dfff ffff	Rotate Left File	rlf f,d	√	√	•	•	b7	
00 1100 dfff ffff	Rotate Right File	rrf f,d	√	√	•	•	b0	
00 0000 0110 0011	SLEEP mode on	sleep			•	•	•	wdt <- 0, Clock off
11 1100 LLLL LLLL	SUB W from Literal	sublw LL	√		√	√	√	w <- #LL - w
00 0010 dfff ffff	SUBtract W from F	subwf f,d	√	√	√	√	√	d <- f - w
00 1110 dfff ffff	SWAP File nybbles	swapf f,d	√	√	•	•	•	d <- f[7:4] <-> f[3:0]
11 1010 LLLL LLLL	eXclusive OR Literal to W	xorlw LL	√		√	•	•	w <- w ⊕ #LL
00 0110 dfff ffff	eXclusive OR W to F	xorwf f,d	√	√	√	•	•	d <- w ⊕ f

√ : Flag operates in the normal manner   • : Not affected   a... : Address
d : Destination; 0 = w, 1 = f   f... : File register   fn : File bit n
L... : Literal data   pc : Program Counter   w : Working register
wdt : Watch Dog Timer/prescaler   TOS : Top Of Stack   == : Equivalent to
pc++ : Jump over next instruction   ++ : Add one   − : Subtract one
GIE : Global Interrupt Enable mask   # : Constant number

CHAPTER 4

The PIC16F84 Microcontroller

Within a year of acquiring the intellectual rights to the General Instrument's Peripheral Interface Controller (PIC), as described on page 63, Microchip had developed the first of their range of Harvard architecture 8-bit microcontroller families. This low- (or base-) range PIC16C5XX family, and currently the PIC10FXXX and 12CXXX families, have a 33-instruction répertoire 12-bit Program store with parallel ports and an 8-bit timer/-counter. As in all subsequent PIC MCU families, the execution module processes all data as bytes to match the 8-bit Data store.

By 1992 the mid-range PIC16CXXX family appeared. This has a 14-bit Program store, which facilitates accessing of larger Data stores. Two instructions were added to the original low-range set. The base set of interface devices was extended, with functions such as 16-bit timers, A/D converters, and serial ports; together with an interrupt handling capability.

This mid-range family is the subject of the rest of this text, apart from Chapter 16, where the extended-range PIC18XXXX family introduced in 1999 is discussed. This has a 16-bit core with 42 additional instructions, many of which are more oriented towards high-level language compiler needs.

In this chapter we introduce the mid-range core from an architectural aspect. After completing this chapter you should: ·

- Understand the mid-range Harvard-based Microchip PIC microcontroller architecture;
- Appreciate the function, structure and memory map of the separate Program and Data stores;
- Appreciate the principle of banking in the Data store and its relationship to the RP0 control bit in the Status register;
- Be able to interpret the Status register bits that control memory paging and hold the **C**, **DC** and **Z** flags;
- Know how to manipulate the contents of the Program Counter in conjunction with the PCLATH special-purpose register;
- Recognize the interaction between the clock phases and the internal sequence of micro-operations;

- Be aware of the base peripheral functions, using the PIC16F84 as an exemplar.

From the point of view of software, all devices with the same core are identical. Indeed, there is considerable commonality across the entire range of PIC MCU cores. From the hardware perspective, the fetch and execute units differ only in minor respects; for instance, in memory size and reset circuitry.

Individual members of a family have a similar base set of peripheral interface ports and modules, but with a different mix of extended I/O facilities. For instance, the PIC16F73 supports a 5-channel 8-bit analog input port, the PIC16F874/7 has an 8-channel 10-bit analog input port and the PIC16F627/8 a 2-input 1-output analog comparator as well as an externally gatable 16-bit counter/timer. However, despite such variations, there is much commonality, with a range of similar modules being used across the mid- and extended-core families. We are not going to look at these modules until Part 3.

In this chapter we will be mainly concerned with the mid-range core. One of the initial introductions (1994) in this family was the 18-pin PIC16C83/4. This became popular, as it was the first PIC microcontroller to use an electrically erasable PROM (EEPROM) to implement the Program store; see Fig. 2.13 on page 28. This meant that it was not necessary to use an ultraviolet source to erase program code and reprogramming could be done in seconds. It also had a 16-byte integral non-volatile memory module for long-term storage of data. For several years this was unique in the PIC MCU range.

The PIC16C83/4 was superseded in 1996 by the PIC16F83/4, which uses Flash-EEPROM technology for the Program store and improved parametric data, such as a higher clock speed.

Although the PIC16F627/8 introduced in 2002 is now recommended for new 18-pin designs, the main difference lies in the extended-range peripheral modules, which will not be considered until Part 3. Thus for simplicity much of the material here will be based on the PIC16F84. However, where necessary we will refer to other family members.

The architecture of the PIC16F84 is shown in a simplified form in Fig. 4.1. The PIC16F83 is identical but with a 512-instruction Program store. Although initially this looks rather complex, it is little more than the architecture of our BASIC computer of Fig. 3.4 on page 49 but with interface ports connected to the internal Data store's data bus. You should review this material now as background to our discussion. In essence the PIC MCU family is based on a Harvard structure with its separate Program and Data stores, and with peripheral interface ports mapped onto the Data Store's address space. That is, the various ports appear to the software to be in the Data store. The miscellaneous status and control registers and even the Program Counter also appears to the software to be in the Data store.

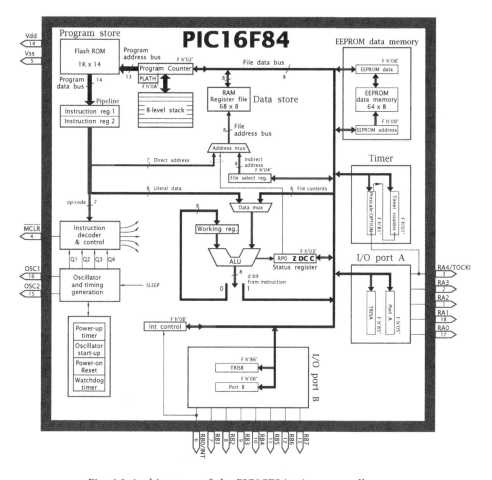

Fig. 4.1 Architecture of the PIC16F84 microcontroller.

Fetch Unit

The fetch unit, shown close-up in Fig. 4.2, is primarily concerned in fetching instructions down into the Pipeline from the Program store. The location of each instruction is maintained by the Program Counter. Each instruction is presented in turn to the decoding circuitry, which activates the appropriate logic in the execute unit in the correct sequence.

The Program Store

Central to the fetch unit is the Program store. Software in embedded systems is invariably fixed, in that on power-on the microcontroller is expected to perform its duty without having to load in its operating program. This means that the Program memory will normally be ROM of some kind. Most PIC MCUs use some sort of electrically programmable technology; in the case of F parts this is Flash EEPROM. For the PIC16F84, 1024 instructions can be stored, with each instruction comprising 14 binary bits; see Fig. 3.5 on page 53. In newer members of the mid-range

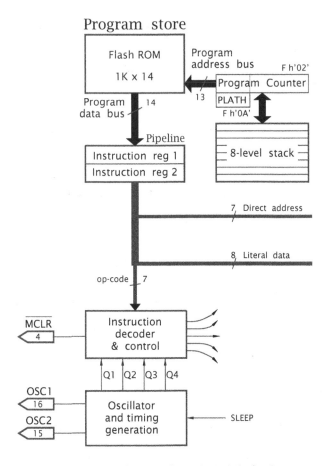

Fig. 4.2 A close-up look at the PIC16F84's fetch unit.

family, capacity varies from 1024 (e.g., PIC16F627) through to 8192 instructions (e.g., PIC16F876/7).

Program Counter

The Program Counter (PC) is used to address, or *point to*, the instruction being fetched at any instant of time. This 13-bit register normally increments after each fetch, effectively acting as a binary counter. However, as we will see in the following chapter, there are a few instructions, such as goto, that will cause execution of the program to jump to another part of the Program store. Thus the Program Counter's normal up count can be overridden. In addition the programmer has access to the PC via the Data store, as we see in Fig. 4.8.

Although the PC is shown as a 13-bit register in order to address $2^{13} = 8192$ instructions, in the PIC16F84 only the lower ten bits ($2^{10} = 1024$)

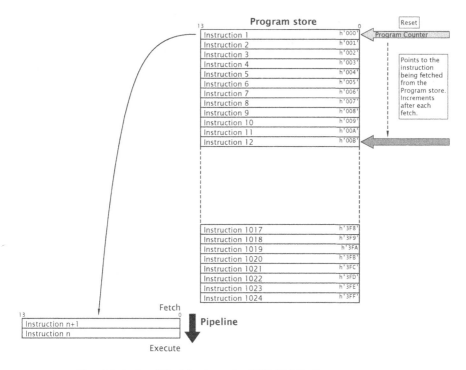

Fig. 4.3 A simplified look at the PIC16F84's Program store.

are actually connected. If the program attempts to go to any instruction above 1024, the address will wrap around back through address zero! In all cases the Program Counter will be cleared on a reset; that is, the first instruction is always in location h'000'. This address is known as the **reset vector**.

Pipeline

Two 14-bit registers implement the Pipeline. The top register holds the instruction that has just been fetched from the Program store at the location pointed to by the **PC**. The bottom register feeds the decoder circuits and is the instruction that is in the process of being executed. This allows an instruction to be fetched whilst at the same time the processor is executing the previously fetched instruction. This of course assumes that the instruction execution sequence is linear. For instructions that action a jump to another part of the Program store, the instruction sitting at the top of the Pipeline needs to be replaced by the far instruction. This process is known as *flushing* and adds an extra machine cycle to the execution time. As you can see from Fig. 4.4, normally an instruction takes one machine cycle to execute.

Instruction Decoder

The Instruction decoder uses logic circuitry to decode each field of the 14-bit instruction and gate the appropriate addresses and data to the correct execution unit's circuitry and configure the ALU.

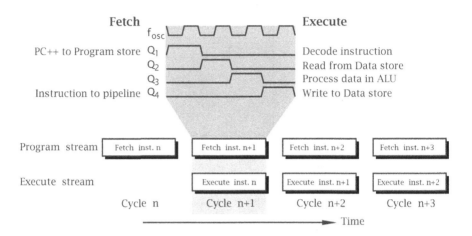

Fig. 4.4 Internal clock sequencing waveforms.

All PIC MCU families have an integral oscillator that generates the internal time-related sequences of micro-operations commanded by the Instruction decoder. The timing element is typically an external quartz crystal connected across the OSC1 and OSC2 pins (Fig. 4.2), and this determines the clock frequency f_{osc}. More details are given in Chapter 10. Most mid-range devices have an upper frequency of 20 MHz, but some early members were limited to 10 or even 4 MHz. There is no minimum frequency.

The oscillator is frequency divided as shown in Fig. 2.25 on page 40, to give four internal non-overlapping quadrature clocks. These four pulses are used as part of the decoding logic to activate internal processes in time-dependent sequences. A consequence is that an **instruction cycle** takes four external clock frequency f_{osc} periods to complete; see Fig. 4.4. Thus with a 4 MHz crystal, the instruction cycle rate is $f_{osc}/4$, or one million per second, corresponding to a period of 1 μs.

The clock-related sequence of operations in the fetch unit are:

Q_1: Increment the Program Counter and copy onto the Program store address bus.

Q_4: Read the instruction code off the Program store's data bus into Instruction register 1 and at the same time move the previous instruction down the Pipeline into Instruction register 2, where it is presented to the Instruction decoder.

Stack

Eight 13-bit registers are stacked below and are connected to the Program Counter. We will see in Chapter 6 that the Stack is used to hold past states of the Program Counter to "remember" the jumping-off point when a subroutine is called up.

Execute Unit

The 8-bit execute unit is responsible for reading a datum from the Data store or literal datum from the instruction and processing this byte as commanded by the Instruction decoder using the ALU. The outcome is placed either in the Working register or back in the Data store, overwriting the original datum.

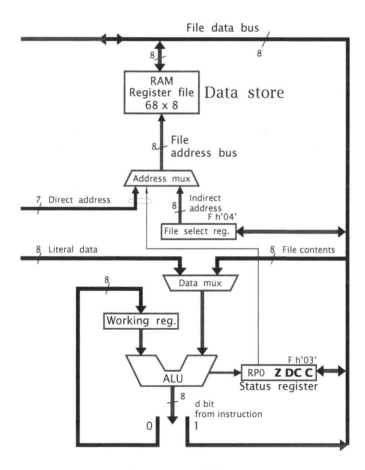

Fig. 4.5 A close-up view of the execute unit.

Arithmetic Logic Unit

Central to the execution unit is the ALU (see Fig. 2.10 on page 25) processing data from up to two sources. One of these is the 8-bit Working register. The other can be:

- A byte directly from a specified File in the Data store. For instance addwf h'20', f ADDs the contents of the Working register to the byte in File h'20'.
- A literal byte held as part of the instruction code; see Fig. 3.6 on page 54. For instance, addlw 5 ADDs the Literal 5 to the Working register.

The outcome in the former case can be directed either back into the Data store if the destination bit is 0 (see Fig. 3.5 on page 53) or into the Working register if this bit is 1, e.g., addwf h'20', w.

Status Register

Associated with the ALU is the Status register which holds three flag bits used to tell the software something about the outcome from an instruction. For instance, if there was a carry-out from an addition.

Carry Flag

Bit 0 of the Status register is the **C** flag. This primarily holds the carry out from the last addition operation. Subtraction operations activate this bit as the *complement* of the borrow out; see Example 4.2. For instance, $24 - 12 = 12\overline{B}1$ and $12 - 24 = 88\overline{B}0$. **C** also functions as an input/output bit for the Rotate instructions, as shown in Fig. 5.13 on page 127.

The label R/W ? in Fig. 4.6 indicates that this bit can be read from or written to and has an uncertain value on a Power-on reset; its value does not alter on any other type of reset.

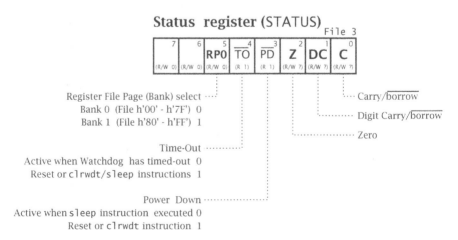

Fig. 4.6 The PIC16F84 Status register.

Digit Carry Flag

Bit 1 of the Status register is the **DC** flag. This operates in the same manner as the standard **C** flag but holds the carry out from the lower nybble to the upper nybble; that is, from bit 3 to bit 4. In the same manner **DC** holds the *complement* of the borrow out from bit 3 to bit 4.

Knowledge of the carry activity between the lower and upper halves of the byte is useful where binary coded decimal data is being manipulated; for instance, see Example 4.5. Here each nybble holds a 4-bit representation of the decimal digits 0,...,9 (see page 7) and the half carry then indicates carries between decimal decades.

Zero Flag

Bit 2 of the Status register is the **Z** flag. This is set whenever the outcome of the instruction is zero, otherwise it is cleared.

Unlike most MCUs, there are no instructions to specifically clear or set a flag, such as `sec` for SEt Carry.[1] However, as we shall see in Fig. 4.7, the Status register is accessible as File 3 in the Data store, and thus any instruction that can alter the contents of a File can potentially change the state of a flag. However, there is a problem in that many of these instructions inherently affect one or more flags (see for instance, Table 5.1 on page 109) as part of their execution logic and this *overrides* any change that would result from the outcome of the instruction's execution. For instance, trying to use the Clear instruction to zero the flags `clrf 3` (see Table 5.2 on page 112), actually sets the **Z** flag to 1 to indicate a zero outcome! The Bit Clear File and Bit Set File instructions (see Table 5.2) are recommended where an individual bit in the Status register needs to be altered, as these instructions do not inherently alter any flags. For instance, `bsf 3,0` (Set Bit 0 in File 3) is equivalent to `sec` and `bcf 3,2` (Clear Bit 2 in File 3) will clear the **Z** flag.

The function of the more specialized flags located in bit 3 and bit 4 will be described in later chapters. Briefly, $\overline{PD}$ (Power Down) is cleared when the `sleep` instruction is executed. The `sleep` instruction is used to disable the oscillator and place the MCU in a low current (typically $<1\,\mu A$) standby mode. $\overline{TO}$ (Time Out) is cleared when the Watchdog timer[2] times out. Both flags can only be read; they cannot be directly be altered by the program, and they are set to 1 on a Power-on reset.

All these bits are known as **flags**, or sometimes semaphores, as they signal an outcome of an instruction, such as a zero result. Bit 5 is rather different, as it is not passively altered by an event. Rather **RP0** is used by the programmer to change the state of the processor. To illustrate the function of this **switch**, we need to examine the PIC1684's Data store in some detail.

---

[1]For instance, the Motorola 6800/5/11 families.

[2]This optional counter, discussed in Chapter 13 will, if given enough time without encountering a `clrwdt` instruction, reset the device.

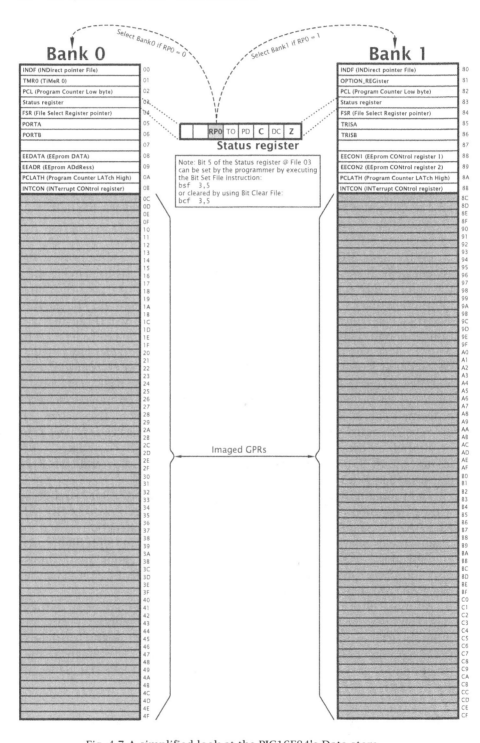

Fig. 4.7 A simplified look at the PIC16F84's Data store.

Figure 4.7 shows a simplified model of the PIC16F84's Data store. You can regard the Data store as a filing cabinet, which in this case has two drawers (banks). Inside each drawer are a number of folders or **Files**, each holding eight bits of data. The term **register** is often used interchangeably with File.

Actually there are two different types of File in our cabinet. Some of these Files are labeled and have special significance. These are known as **special-purpose registers**. SPRs are used to control and monitor the state of the microcontroller and its various peripheral devices. For instance, as we have seen in Fig. 4.6, File 3 is the Status register and File 6 is Parallel Port B connected to eight pins RB7, · · ·, RB0 (usually designated RB[7:0]), as shown in Fig. 4.1.

The remaining Files (shown shaded in the diagram) can be given relevant names by the programmer and used for general-purpose storage. The PIC16F84 has 68 of these **general-purpose registers** addressed from File h'0C' through File h'4F'. All mid-range PIC MCUs use the lower tranche of Files for the SPRs and upper addresses for the **GPRs**. However, newer family members, such as the PIC16F628, need more room in order to accommodate SPRs for the enhanced peripheral devices supported by such devices. In these cases memory up to File h'1F' is reserved for SPRs. In order to accommodate this, all programs in this book will assume that storage for variables is available from File h'20' upwards.

Most family members have more general-purpose storage than the PIC16F84, although even the largest mid-range devices have a maximum storage of 368 bytes. Even this amount of RAM storage is not very much, so programs have to make very efficient use of this limited capacity.[3]

Back in Fig. 3.5 on page 53 we noted that of the 14 bits comprising an instruction, seven bits are reserved for the *address* of the datum in the Data store. Seven bits gives $2^7 = 128$ addresses; that is, a page or **bank** of 128 Files. To get around this size restriction, we need extra bits to increase the range of addresses. The PIC16F84 uses an extra bit in the Status register called **RP0**—short for Register Page 0[4]. This bit gives an effective 8th address bit, allowing for a potential 256-File Data store. When RP0 is 0, as it is on a reset, then Bank 0 (File 0 through File h'7F') of the Data store is enabled for access. When RP0 is 1, Bank 1 is enabled; allowing access to File h'80' through File h'FF'.

Most mid-range PIC MCUs have four banks of RAM and use two bits in the Status register, RP1 and RP0 in STATUS[6:5]; see Fig. 5.5 on page 103, to give an effective 9-bit Data store address. This is covered in more detail in the next chapter.

Although banking is an effective solution to overcoming the problem of the limited 7-bit instruction address field, its use can be confusing for the neophyte programmer. As an example, consider a code fragment

---

[3]Think of your PC with hundreds of megabytes of RAM storage!

[4]Presumably because the more obvious Register Bank 0 (RB0) is already used for the pin connected to Port B bit 0.

where it is desired to set the state of File h'86' to the pattern b'00001111'. We are making use here of the movlw (MOVe Literal into W) instruction to load an 8-bit literal (constant) into the Working register.

```
movlw   b'00001111'    ; Put the literal pattern h'0F' into W
movwf   h'86'          ; and copy it into File h'86'
```

This is erroneous as the address h'86', or b'10000110, is too large for the 7-bit instruction address field. Most assemblers will strip off bits above this field to give the address h'06', or b'0000110'. Thus to the assembler the address h'86' and h'06' are the same, although most will print a warning message to the programmer. Here is what the Microchip assembler (anticipating Chapter 8) tells the programmer:

Message[302] Register in operand not in bank 0.
Ensure that bank bits are correct.

In other words, it is up to the programmer to twiddle the RP bits *before* trying to access such a location.

To see how this is done we need to have a sneak preview of the bit twiddling instructions listed in Table 5.2 on page 112. All microcontrollers need to be able to control the state of *single* bits within a File, whether to set an option in a SPR or even to wiggle a port pin. All Microchip PIC MCU families use the following instructions for this purpose[5]:

bcf
Bit Clear File enables the programmer to zero *any* bit in *any* File. For instance, bcf h'20',7 clears bit 7 in File h'20'. All other bits remain unaltered.

bsf
Bit Set File enables the programmer to set *any* bit in *any* File. For instance, bsf h'31',3 sets bit 3 in File h'31'. All other bits remain unchanged.

We now have for our example:

```
bsf    3,5       ; Setting bit 5 (RP0) of STATUS changes to Bank1

movlw b'00001111' ; Put the pattern h'0F' into W
movwf h'86'       ; and copy it into File h'86'

bcf    3,5       ; Clear RP0 to change back to Bank0
```

Actually the PIC16F84 makes little use of Bank 1. All 68 GPRs are mirrored in both banks. This means that File n and File h'n+80' are one and the same. For instance, if the programmers wants to get at the contents of File h'20', it doesn't matter what bank the processor is in as the datum in File h'A0' is the same—not a copy! This is rather unusual, as PIC MCUs

---

[5]The enhanced-range family additionally has a btg Bit ToGgle File instruction.

with larger Data stores do spread their unique GPRs (and SPRs) over available banks. However, the newest devices, such as the PIC16F628, have a small group, usually 16, of GPRs common across all banks. This allows storage and retrieval of critical system data quickly, independent of which bank the processor is in; see page 196.

Most of the more commonly used SPRs are also mirrored across all banks, a typical example being the Status register, which appears both at File 3 and File h'83'. This is because access to the Status register flags and switches is a frequent occurrence, and it would be inefficient to switch back and forth all the time. Indeed, in our code fragment on page 82, when we cleared RP0 to move back from Bank 1 to Bank 0 we were relying on the Status register appearing in both banks—otherwise we could never twiddle RP0 and return from Bank 1!

Most of the SPRs in the PIC16F84 are core for all the mid-range family members—indeed even being located at the same address. Thus the Status register is at File 3/File h'83',[6] and is given the symbolic name or label STATUS.

Apart from the Status register and Program Counter, we will not formally cover most of these SPRs in this chapter. However, for reference it is convenient to list these here in one place, and elaborate on their function at the referenced location.

Indirect Addressing

Normal Direct addressing carries the address of the operand as an integral part of the instruction. In the case of the mid-range PIC MCU family, this is the 7-bit address zone shown in Fig. 3.5 on page 53. In embedded computing, where the instruction codes are stored in some form of read-only memory, this address is *fixed* and therefore cannot be modified.

An alternative approach, used in some form by all computing mechanisms, is to hold this address in a register. In the case of the PIC, this address is generated as the contents of the File Select Register (FSR) at File 04 in the Data store. In order to trigger this mode, the internal logic detects whenever the Data store address is b'0000000'. When an instruction specifies this null address, the 8-bit contents of the FSR is switched onto the Data store's address bus, as shown in Fig. 5.6 on page 104.

With Indirect addressing, the operand location is no longer fixed as code in the Program store, but is a *variable* in the File Select Register. This means that the location of the operand can be altered as the program progresses; that is, at run time. For an example, see Program 5.2 on page 107.

Registers relating to Indirect addressing are:

INDF, File 0

The null address is named the INDirect File. As INDF is simply used to trigger Indirect addressing, this File is not actually implemented as a

---

[6]and in File h'103' and File h'183' in 4-bank devices.

real location. That is, you cannot store data in INDF because it does not physically exist!

FSR, File 4

The File Select Register holds the 8-bit Indirect address used when the instruction actually refers to the null address.

Timer

Most MCUs have facilities to either measure elapsed time and/or to generate digital on/off waveforms with well-defined durations. This is normally based around one or more counters that are incremented either from an external pulse or internal clock. For instance, if an automatic packing machine needs to count cans of beans going along a conveyor belt, then a photoelectric-based transducer could act as the timer input. If a new packing carton needed to be in place every 24 cans, then an internal 8-bit counter would be set to h'E8' (-24). When the counter overflows from h'FF' to h'00' then an interrupt (see Chapter 7) would be generated and the MCU then would take the appropriate action.

All PIC MCU families have at least a basic timer/counter known as Timer 0 (TMR0). The read/write TMR0 counter register at File 1 can be clocked from the outside world via the T0CKI (Timer 0 ClocK In) pin, which is shared with the RA4 Port A pin. Alternatively, the incrementing source can be the internal Q_4 phase clock (Fig. 4.4), which is one-quarter of the crystal frequency. Either clock source can be frequency divided by a buried[7] 8-bit Prescaler counter. This divide ratio is controlled by the lower three prescaling bits of the Option register at File h'81' (see Fig. 13.2 on page 403), labeled PS2:PS1:PS0. The ratio is then 2^{PS+1}. For instance, if PS[2 : 0] = 111 then the counter will increment at $\frac{f}{256}$, where f is the clock source frequency. The Prescaler register can be disconnected by setting bit 3 of OPTION_REG to 1. This will give a direct connection between pulse source and counter. Writing to the Timer 0 register also zeros the Prescaler register (for instance, movlw h'E8', movwf 1) enabling the time period to begin from true base time.

When this PSA (Pre-Scale Assignment) bit is 1 the Prescaler register acts as a postscaler to the **Watchdog timer**; see Fig. 13.1 on page 402. The Watchdog timer is designed to reset the MCU unless periodically preset by the user's program with the instruction clrwdt (CLeaR Watch-Dog Timer). This ensures that the PIC MCU will eventually reset if due to an electrical disturbance or a software bug, the processor malfunctions, perhaps by jumping into an unprogrammed part of the Program store. This will disrupt this periodic preset. If the Prescaler is assigned to the timer then the watchdog circuit will periodically time-out (count down through zero) after approximately 18 ms. With PSA set to 1 then

---

[7]The term "buried" means that the register does not have an address in the Data store and therefore cannot be directly altered by the program.

$2^{PS} \times 18$ ms Watchdog time-outs are required before the processor is reset. Thus, with PS[2 : 0] = 111, $2^7 = 128$ time-outs gives a period to MCU reset of nominally 2.3 s. Thus the software must use the clrwdt instruction before this period elapses to prevent reset. This instruction also clears the Prescale counter. If it does time-out, then the $\overline{TO}$ bit in the Status register will be cleared. If desired the Watchdog timer may be disabled at the same time as code is programmed into the Program store. Various configuration bits (known as fuses) are located by the Flash EEPROM programmer in location File h'2007' (see Fig. 10.6 on page 281), which is not accessible during the normal run mode. Such details are normally hidden from the operator by the EEPROM programmer's operative software.

Registers relating to Timer 0 are:

TMR0, File 1
Sometimes known as the Real-Time Clock/Counter (RTCC), is an 8-bit up-counter register that keeps tally of clock events. It may be preset to any byte value by moving data from W, and read at any time. When it overflows from h'FF' to h'00' it sets the T0IF (Timer 0 Interrupt Flag) in the INTCON (INterrupT CONtrol) register; see Fig. 7.3 on page 191. This may be used to generate an interrupt.

OPTION_REG, File h'81'
Six bits in this register in Bank 1 at File h'81' are used in conjunction with the timer; see Fig. 13.2 at page 403.
- PS2, PS1, PS0 at bits 2,1,0, respectively control the Prescale ratio 2^{PS+1} for the timer or Postscale ratio 2^{PS} for the Watchdog timer.
- T0SE (Timer 0 Set Edge) at bit 4 allows the programmer to select which edge of a pulse at the T0CKI pin will increment the counter; a 0 for $\_\!\!\!\!\diagup\overline{}$ and 1 for $\overline{}\!\diagdown\_$.
- T0CS (Timer 0 Clock Select) at bit 5 allows the programmer to select the clock source as either the internal clock (= 0) or a transition at the T0CKI pin.

The remaining two bits configure external interrupt edge select and electrical properties of Port B inputs.

Program Counter
We have already seen in Fig. 4.2 that the mid-range PIC MCU core has a 13-bit Program Counter (PC) acting as an instruction pointer, for a potential 8 Kbyte Program store. How many bits of the PC are actually used in any particular family member depends on the size of the Program store. Thus the PIC16F84 uses 10 bits (2^{10} = 1 K), the PIC16F628 uses (2^{11} = 2 K), the PIC16F874 uses (2^{12} = 4 K) and the PIC16F877 uses all 13 bits.

Occasionally it may be necessary for a program to modify the state of the PC at run time. To allow for this, the lower byte of the PC is directly accessible as the Program Counter Low (PCL) byte SPR. In order to be able to change all 13 bits an additional SPR is required. The Program

Counter LATch High register (PCLATH)is not actually the upper half of the PC but acts as a buffer. Changing the state of PCLATH will not alter the top byte of the PC, but when the PCL is written to, the new state of the PCLATH *simultaneously* becomes the new top half of the 13-bit PC. Thus, as shown in Fig. 4.8, all 13 bits are updated at the *same* time. This is the principle you will need to use in SAQ 4.2.

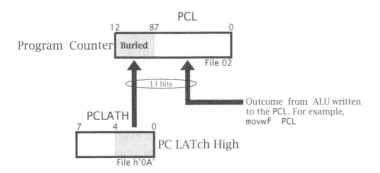

Fig. 4.8 Showing how all 13 bits of the Program Counter are altered at the same time when writing to the PCL.

Registers used in altering the Program Counter are:

PCL, File 2
The Program Counter Low byte is the lower eight bytes of the Program Counter. It may be written to or read from.

PCLATH, File h'0A'
The Program Counter LATch High byte SPR is a holding register carrying data destined to be loaded into the top half of the Program Counter. This occurs at the same time as data is written into the PCL register, allowing a simultaneous updating of the complete 13 bits.

Note that the PCLATH is acting as a surrogate for the upper half of the PC, and thus reading it will not give the actual state of the high byte of the real PC.[8]

Parallel Input/Output Ports
The ability to externally alter or monitor several digital lines at the same time is a virtually universal facility on microcontroller-based systems. Depending on the package size mid-range PIC MCUs range from four up to 52 such external input/output (I/O) lines. For instance, the 40-pin PIC16F877 has potentially 33 I/O pins.

---

[8]In the extended-range PIC18XXXX family, reading the PCLATH does read the state of the appropriate part of the PC.

The PIC16F84 has 13 I/O lines, divided up into two ports. Port A has five I/O lines mapped into the Data store address space at File 5. The remaining eight lines are allocated to Port B at File 6. These ports can be thought of as a "window" into the Data store, in that data written to File 5 or File 6 appear to the outside world on the corresponding pins; pins RA4, · · · , RA0 and RB7, · · · , RB0 respectively—see Fig. 10.2 on page 273. However, the electrical and logical behavior of these ports is more complex than that of a purely internal register File. This will be discussed in Chapter 11, but as an example, a port bit must be configurable as either an output (so that the CPU can control the state of the associated pin) or an input (so that the CPU can read the state of this pin). To do this, each parallel port register has an associated data direction register, which Microchip calls TRISA and TRISB, which map to File h'85' and File h'86', respectively. The term TRIS stands for TRIState; see Fig. 11.3 on page 300. These registers lie in the less convenient Bank 1, as they are usually set up at the beginning of the program and are never subsequently altered.

As an example, consider that we wish to make Port B pins RB[6:0] an input and pin RB7 an output. Then the set-up code would be:

```
        bsf    5,3        ; Change to Bank 1
        movlw  h'7F'      ; Binary pattern 0111 1111
        movwf  h'86'      ; makes RB7 Output, RB6...RB0 Input
        bcf    5,3        ; Back again to Bank 0
```

Although this code fragment is correct and, with the aid of comments its function can be followed, the code is not very human readable. The alternative is more user friendly, but is identical as far as the assembler is concerned; see page 55.

```
STATUS equ   03         ; Status register is at File 03
RP0    equ   05         ; Bank switch bit is 5 in STATUS
TRISB  equ   h'86'      ; Data direction register @ File h'86'
PORTB  equ   06         ; Port B itself is at File 06

       bsf   STATUS,RP0 ; Change to Bank 1
       movlw b'01111111'; Binary pattern 0111 1111
       movwf TRISB      ; makes RB7 Output, RB6...RB0 Input
       bcf   STATUS,RP0 ; Back again to Bank 0
```

Obviously the latter is preferable. Although this might seem to be a cosmetic exercise, clarity reduces the chance of error and makes debugging and subsequent alteration easier. Realistic programs, rather than the code fragment illustrated here, use many variables and register bits, so lucidity is all the more important.

The four header lines of our program illustrate the means whereby the programmer tells the assembler translator program to substitute numbers for names. For instance, the line

```
STATUS   equ   03
```

states that when the programmer uses the name STATUS as an operand, it is to be substituted by the number 3 (that is, File 3). The equ directive means "EQUivalent to." A **directive** is a pseudo instruction in that it does not usually produce actual machine code but rather is a means of passing information from the programmer to the assembler program. From now on we will give our Files and bits names for clarity.

As an example, let us pulse pin RB7 High and then Low as follows:

```
bsf   PORTB,7     ; Pin RB7 High (set bit 7)
bcf   PORTB,7     ; then Low (clear bit 7)
```

with the assumption that the CPU is still in Bank 0.

The registers associated with parallel I/O are:

PORTA, File 05
Only the lower five bits are implemented in this register File, feeding through to pins RA4,...,RA0. Pin RA4 is shared with the Timer peripheral. The phantom upper three bits read as zero. Some members of the family can implement up to eight Port A pins, such as the PIC16F628.

TRISA, File h'85'
This is used to bitwise configure Port A pins as input or output. Setting TRISA[n] to 1 sets pin RA[n] as an Input, and to 0 as an Output.[9] Any type of reset sets the TRIS bits to 1 and thus the associated port pins to input.

PORTB, File 06
A bidirectional 8-bit port connected to pins RB7,...,RB0. Bit RB0 doubles as a Hardware interrupt input.

TRISB, File h'86'
This is used to bitwise configure Port B bits as input or output. Details are the same as TRISA.

Data EEPROM
Most mid/extended-range family members have a block of up to 256 (the PIC16F84 has 64) bytes of data that does not require power to retain its contents. This **non-volatile memory** is not part of the (volatile) Data store and is accessed through SPRs as a peripheral device. Any byte can be addressed and then read from or written to via the EEDATA register, as addressed by the EEADR register and controlled by the EECON1 and EECON2 control File registers. Most Data EEPROM modules have a minimum endurance of 10 million writes and such data is retained for upwards of

---

[9]Aid-mémoire: A 0 configures an Output pin, a 1 configures an Input pin.

40 years. Some typical uses of a non-volatile depository would be to hold the number of pages printed in a laser printer or total miles/kilometers traveled in a car.

Details of the Read-to and Write-from protocols are given in Chapter 15, but are briefly reviewed here for completeness.

Read
1. Put address (h'00 – FF') into EEADR.
2. Set RD (bit 0 of EECON1) to 1 to set to the ReaD mode.
3. Read the addressed contents in EEDATA.

Write
1. Put address into EEADR.
2. Put data into EEDATA.
3. Set WREN (bit 2 of EECON1) to 1 to WRite ENable.
4. Put code h'55' into EECON2.
5. Put code h'AA' into EECON2.
6. Begin the WRite cycle by setting WR (bit 1 of EECON1) to 1.

Writing, which is normally an infrequent act, is deliberately made circuitous to protect against accidental changes to the EEPROM. The register EECON2 does not actually exist, but the interlock writing h'55' followed *directly* by h'AA' is a necessary part of unlocking the target byte. Interrupts can disrupt this sequence and should be inhibited if used. Writing takes around 50 ms to complete, and sets the EEIF (EEPROM Interrupt Flag) bit 4 of EECON1 after this time, and this can be used to interrupt the processor. The WRERR (WRite ERRor) bit 3 of EECON1 is set if a Write cycle is prematurely terminated, say, by an external reset.

Registers associated with the Data EEPROM (with PIC16F84 locations) are:

EEDATA, File h'08'
This contains the addressed data after a Read action or holds data to be written into the addressed byte during a Write action.

EEADR, File h'09'
The address of the target byte is placed here before a Read or Write cycle.

EECON1, File h'88'
This holds the control and status bits that:
- Trigger an EEPROM Read.
- Enable a Write action.
- Trigger an EEPROM Write.
- Signals a premature end to a Write cycle.
- Signals a Write cycle has been completed.

Details are given in Fig. 15.2 on page 486.

EECON2, File h'89'
The EEPROM CONtrol 2 is not a physical register and reads as zero. This address is used as the target for the Write cycle unlocking sequence which is implemented by moving h'55' followed directly by h'AA' into this virtual location.

Interrupts
The Interrupt Control register at File h'0B'[10] holds the mask and status bits controlling the response of the MCU to interrupts. Its operation is described in Chapter 7. Most peripheral devices have local interrupt-related bits in other control registers; for instance, see Fig. 7.5 on page 199.

Examples

Example 4.1
Discuss how the performance of the PIC MCU architecture is improved by incorporating pipelining into the design of the instruction-fetch unit. Do you foresee any problems associated with handling Jump instructions (such as goto) in connection with the Pipeline's structure?

Solution
The Pipeline is a precondition for the parallel operation of the fetch and execution units. That is, in order to allow the execution of instruction n whilst the next instruction $n + 1$ is being fetched from the Program store, internal storage must be provided to present the instruction code to the Instruction decoder. As all instructions are the same size, that is, 14 bits, then the Pipeline's register structure and control is considerably simplified. Most conventional CISC processors have instructions that vary considerably in length. For instance, the 68HC11 MCU core has instructions that cover the range 1 through 4 bytes; that is, the fetch phase can take between 1 and 4 bus transactions. Some more sophisticated processors have multistage pipelines with each stage feeding part of the execution circuitry. Thus several streams of execution activity can occur simultaneously.

The problem with pipelines is that they presuppose that the program instructions will be executed sequentially as they are stored in memory. However, instructions that disrupt this smooth running and move on the Program Counter require that the Pipeline be emptied so that the destination instruction code travels down to the end of the pipe. For instance, if instruction k is goto n, then instruction $k + 1$ will be in the first stage of the Pipeline by the time the processor knows that the next step is actually

---

[10]And images in all the banks.

to be instruction n. Thus a null instruction cycle needs to be executed which simply brings this instruction code into the Pipeline but does not execute instruction $k + 1$ whose code is at the end of the Pipeline. This is sometimes known as **flushing** the Pipeline. Instructions such as goto need two clock cycles to execute. Conditional Skip instructions (see Chapter 5) take two cycles when the skip is implemented and one otherwise. All other instructions always take one cycle.

Example 4.2
Can you determine why, after a subtraction or addition of a negative number, the setting of the **C** flag is the *complement* of the borrow-out. *Hint*: Look at 2's complement arithmetic on page 9.

Solution
Subtract instructions in all PIC MCU families work by 2's complementing the datum byte and then adding; as shown in Fig. 2.9 on page 25. In this situation the resulting carry-out is 0 where a negative outcome is generated and 1 for a positive outcome. For instance:

1. $06 - 0A \rightsquigarrow 00000110 + 11110110 = (0)\ 11111100$ or -4 (no carry).
2. $0A - 06 \rightsquigarrow 00001010 + 11111010 = (1)\ 00000100$ or $+4$ (carry).

In both cases the Carry flag acts as an inverted borrow. This is in keeping with the RISC philosophy of the PIC MCU family, to keep the processor "lean and mean."

Exactly the same borrow inversion occurs if you specify a negative datum with an Add instruction, such as addlw -6. This is translated by the assembler to addlw h'FC', where h'FC is of course the 2's complement of 6.

Example 4.3
A smart alec programmer has decided to copy the contents of the Status register into File h'40' for safekeeping so that it can be returned later without alteration. However, bit 2 of the Status register is invariably forced to 0. Why?

Solution
From page 52 we see that the movf instruction will set **Z** if the contents of the File in question is all zero, else it will clear **Z**. Thus the program fragment

```
movf   STATUS,w  ; Copy contents of File 3 (STATUS) to W
movwf  h'40'     ; and to File h'40'
```

will indeed copy the contents of File 3 into File h'40', but unless the Status register bits are all zero, the **Z** flag will be cleared on the way. The $\overline{\text{PD}}$ and

$\overline{\text{TO}}$ flags are normally 1, so **Z** will end up as 0 no matter what the original state was.

There is a way round this, as will be explained on page 195.

Example 4.4
Show how you would set up the following SPRs in Bank 1 as listed:
- OPTION_REG   b'10101111'
- TRISA   b'00011110'
- TRISB   b'11111111'

Solution
As all three SPRs are in Bank 1, therefore we need to switch banks before the data is sent out and back to Bank 0 after this is done.

```
STATUS      equ   3     ; The Status register @ File 3
RP0         equ   5     ; in which bit 5 is RP0
OPTION_REG  equ   h'81' ; The Option register is at File h'81'
TRISA       equ   h'85' ; The Port A Direction register
TRISB       equ   h'86' ; The Port B Direction register

            bsf   STATUS,RP0 ; Switch into Bank 1

            movlw b'10101111'; First pattern
            movwf OPTION_REG ; into the Option register
            movlw b'00011110'; Second pattern
            movwf TRISA      ; into TRISA
            movlw b'11111111'; Third pattern
            movwf TRISB      ; into TRISB

            bcf   STATUS,RP0 ; Switch back again to Bank 0
```

Example 4.5
Write a program to increment a packed BCD quantity located in data memory at File h'20'.

Solution
Two binary-coded decimal (BCD) digits may be packed into a single byte to represent numbers up to 99. For instance, $\boxed{0100\ 1001}$ $_{\text{File h'20'}}$ represents BCD 49. Incrementing a number stored in this hybrid decimal-binary form using the normal binary addition rules may give an incorrect result. For instance, b'01001001 + 1' (49 + 1) gives b'01001010' (h'4A') after addition, but should give b'01010000' (h'50'). Similarly, b'10011001 + 1' (99+1) gives b'10011010' (h'9A') instead of b'00000000' plus a carry of 1 (h'1 00').

From these examples it can be seen that whenever any of the BCD decades equals ten after incrementation then it should be zeroed and one added to any higher decade. Based on this increment and add algorithm we can formulate the task list.

1. Increment the packed BCD byte using normal binary arithmetic.

2. IF the lower nybble of the outcome is ten then add six to the outcome.

3. IF the upper nybble of the outcome is ten then add six to it.

Program 4.1 Incrementing a packed BCD byte.

```
;****************************************************************
;* FUNCTION: Increments a BCD datum giving a BCD outcome       *
;* ENTRY    : BCD in File h'20'                                *
;* EXIT     : BCD+1 in File h'20'                              *
;* EXAMPLE  : 10011000 (98) + 1 = 10011001 (99)               *
;* ************************************************************
STATUS    equ   3            ; The Status register
C         equ   0            ; Carry flag is bit 0
DC        equ   1            ; Digit Carry flag is bit 1
BCD       equ   h'20'        ; The BCD number is in File h'20'
; -------------------------------------------------------------
BCD_INC   incf  BCD,w        ; Binary inc BCD number and put in W
          addlw 6            ; Add six
          btfss STATUS,DC    ; Needed IF produced a half carry
          addlw -6           ; ELSE not needed
; Now check the upper digit by adding 6 to it and checking carry
          addlw h'60'        ; Add h'60' (i.e. six to upper digit)
          btfss STATUS,C     ; Needed IF caused a carry
          addlw -h'60'       ; ELSE cancel the correction factor
; The incremented and corrected BCD number is now in W
          movwf BCD          ; Put it out in memory
```

Program 4.1 gives an efficient implementation of this task list. After incrementing using normal binary rules, six is added to the previous outcome and the **DC** flag is checked for activity. This flag will only be set when the original nybble is ten (h'0A + 6 = 1 0'). In this case the add six operation is allowed to stand as the necessary correction, otherwise it is canceled by subtraction. The upper nybble (BCD digit) is checked and corrected in the same manner, but this time it is the full Carry flag that is tested. If this is set, then the addition of h'60' is allowed to stand, otherwise it is subtracted. This Carry flag could be used to set a hundreds digit if desired, to show overflow from 99 to 100.

An alternative approach would be to subtract nine *before* incrementation and if the **Z** flag is set then leave the digit at zero and increment the higher digit; otherwise add ten. Repeat for the upper digit.

Self-Assessment Questions

4.1 Where microprocessors are used in a general-purpose computing environment, the program is normally loaded into and run from read/write RAM memory. This means that the system can run a word-processor one minute and a spreadsheet program the next. Of course this means of operation is not applicable to embedded applications, where the program is stored in some variety of non-volatile read-only memory. Discuss why this is so and the virtues of ROM, EPROM and EEPROM implementations of non-volatile storage.

4.2 The PIC16F877 mid-range MCU has a 8-kbyte Program store, which can hold up to 8192 14-bit instructions located in the range h'0000 – 1FFF'. Without using the goto instruction, which has its own limitations (see Fig. 5.1 on page 98), how could you engineer a jump to the instruction located in the Program store at h'1234' from anywhere in the program?

4.3 Given the effect of the movf instruction on the **Z** flag discussed in Example 4.3, how could you use this instruction to determine if the contents of any File is zero?

4.4 From Table 1.1 on page 5 we see that the uppercase letters A through Z differ in coding from their lowercase siblings only in that bit 5 is 0 in the former instance and 1 in the latter. With the instructions we have de facto introduced in this chapter, how could you convert an ASCII character located in File h'20' from lowercase to uppercase?

4.5 Based on the configuration of Example 4.4, write a program to pulse pin RA0 High for $4\,\mu s$ and then Low. You may assume a clock crystal of 4 MHz.

4.6 How could you bring pin RA1 High, then pulse RA0 four times and then RA1 is to go Low again? Your solution should include the setting for TRISA.

4.7 Most digital watches use a 32.768 kHz crystal, commonly known as a watch crystal. Because of high production quantity, such crystals are low cost. Although this slows the processing rate, we shall see in Fig. 10.3 on page 275 that the power dissipation is directly proportional to clocking frequency. Thus a watch crystal is an attractive low-cost proposition for many low-power applications.

Can you determine the instruction cycle time for such a system? What is the significance of the value 32,768 for timing circuits?

CHAPTER 5

The Instruction Set

Writing a program is somewhat akin to building a house. Given a known range of building materials, the builder simply puts these together in the right order. Of course, there are tremendous skills in all this; poor building techniques lead to houses that leak, are drafty and eventually may fall down!

It is possible to design a house at the same time as it is being built. Whilst this may be quite feasible for a log cabin, it is likely that the final result will not remain rainproof very long, nor will it be economical, maintainable, ergonomic or very pretty. It is rather better to employ an architect to design the edifice before building commences. Such a design is at an abstract level, although it is better if the designer is aware of the technical and economic properties of the available building materials.

Unfortunately, much programming is of the "off the cuff" variety, with little thought of any higher-level design. In the software arena, design means devising strategies and designing data structures in memory. Again, it is better if the design algorithms keep in mind the materials of which the program will be built, in our case the machine instructions.

At the level of our examples in this chapter, it will be this coding (building) task we will be mostly concerned with. Later chapters will cover more advanced structures which will help this process, and we will get more practice at devising strategies and data structures.

If you like to think of writing a program as analogous to preparing an elaborate meal, then for any given cooking appliance, such as a microwave oven or electric stove (the hardware) there are a range of processes. These processes—for instance, steaming, frying, boiling—are analogous to the **instruction set** which can be implemented by the CPU. The various ingredients that can be handled by a process are the instruction's data. Such data may lie in internal registers or out in memory. There are several different ways of specifying the **effective address (ea)** of an operand. These are known as **address modes**.

In keeping with the PIC microcontrollers' RISC-like philosophy, the mid-range core have a total of only 33 instructions, plus two legacy instructions from the low-range family; which we will ignore. Each instruction code is contained in a 14-bit word which holds the instruction operation code, address or data and destination bit. We covered a few of these instructions and address modes when discussing our BASIC computer back in Chapter 3; now would be a good time to go over this material. Here we look at the various address modes and the full instruction set in some detail.

After reading this chapter you will:

- Know that an address mode is the way an instruction pin-points its data.
- Understand how Inherent, Literal, Absolute, File Direct, File Indirect and Bit address modes permit an instruction to target an operand for processing.
- Recognize how the binary structure of the instruction word impacts on the usage of instructions.
- Understand how the Status register's IRP switch allows the processor to access a full 4-bank Data store using Indirect addressing.
- Know that movement instructions, copying data in-between the Working register and the Data store, are the most used of the instruction categories.
- Appreciate that the processor can directly implement the common arithmetic operations of addition, subtraction, incrementation, decrementation and bit banging.
- Be able to compare or test data for differences and relative magnitude, and take appropriate action.
- Know that data in the Data store can be rotate-shifted through the C flag.
- Be able to use the four basic logic instructions to invert, set, clear, toggle, bit test and differentiate data.
- Understand how the program flow can be diverted, based on the state of any bit or a zero overall value in a File.

Virtually all instructions act on data, either outside the CPU in its data or program memory space, or in internal registers. Thus the 14-bit instruction code must include bits which inform the CPU's instruction decoder *where* this data is being held. The exceptions to this are the few inherent instructions, such as nop (No OPeration) and return (RETURN from subroutine). Before looking at the instruction set we will discuss the various techniques used to specify the location of any operands.

The general symbolic form of an instruction is:

```
instruction mnemonic <operand A>,<operand B>
```

where operand A is the source datum or its location and operand B the destination. For instance, movf h'20',w (MOVe File) which copies a datum sourced from File h'20' to its destination in the Working register.

There are some variations on this structure. $2\frac{1}{2}$-operand instructions are common. For instance, addwf [FILE],d adds the W register's contents to the specified File's contents and deposits the result either in W or back in the File register itself. Thus addwf h'20',f means "add the contents of W to that of File h'20' and put the outcome in File h'20'." This could be written in shorthand as [f20] <- W + [f20], where the brackets mean "contents of" and <- means "becomes." This notation is called **register transfer language** (**rtl**).

Of course, this is not a true 3-operand instruction, as the destination must be one of the two source locations; that is W or File h'20'. A few instructions have only a destination specified; for example, clrf h'20', and the inherent instructions have no explicit operands.

Instructions can be classified by their address mode.

Inherent

0000000	???????

The instructions clrwdt (CleaR Watchdog timer), retfie (return from interrupt and enable), nop, return and sleep do not explicitly refer to operands in memory. At the binary code level, all these instructions are coded with the upper seven bits zero. For example, clrwdt has a machine code of b'**0000000**0000100'.

Literal

11	????	LLLLLLLL

Literal instructions use the lower eight instruction word bits to specify a source operand which is a *constant* datum, rather than a byte in a File. For instance, addlw 06 is coded as b'**11**1110**00000110**'. The destination of this type of instruction is *always* the Working register, and this is shown in the mnemonic. Thus in our example, the sum W + 6 is copied back into W. In rtl this is expressed as W <- W + #6, where the # (pound or hash) symbol denotes that the following number is a constant or literal rather than a File address.

Some processors call this form of addressing immediate, as the datum is immediately available without having to go out into memory.

Absolute

10	?	AAAAAAAAAAA

Two instructions allow the program to jump to another instruction anywhere in the Program store. These are goto and call (CALL up or go to a subroutine; see Chapter 6). The 14-bit core allocates eleven bits of the instruction word to this absolute instruction address[1] in the Program store. Thus goto h'400' would be coded as b'**10**1 **10000000000**'. Similarly call h'530' is b'**10**0 **10100110000**'.

---

[1]Don't confuse this with the File address in the Data store. In the Harvard structure the two stores are logically distinct with different address spaces.

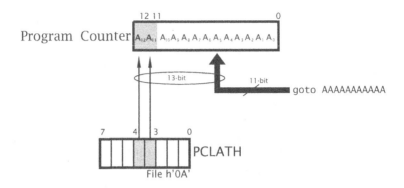

Fig. 5.1 Generating a 13-bit Program-store address from a 11-bit absolute address field for the goto and call instructions.

This 11-bit address can directly locate any instruction in a Program store of up to $2^{11} = 2$ Kbyte capacity. However, the mid-range core has a 13-bit Program Counter which can potentially address a Program store of up to 8 Kbyte instructions; for instance, in the PIC16F877. To cope with this situation, when a goto or call instruction is executed, the absolute 11-bit address is transferred into the PC together with bits 3:4 of the PCLATH (Program Counter LATch High) register to make up an effective 13-bit Program-store address. This process is shown in Fig. 5.1; see also Fig. 4.8 on page 86.

PCLATH is cleared on a Power-on reset, so the Direct goto range is normally h'000 – 7FF'. This covers all the address range for a 2 Kbyte store; for instance, the PIC16F628. For members with larger Program stores, then a far goto and far call (i.e., beyond h'7FF') has to be implemented by twiddling bits PCLATH[4:3]. For instance, in the PIC16F877 a goto h'F00' has to be coded as:

```
bsf   PCLATH,3      ; Make PCLATH(4:3) = 11
bsf   PCLATH,4
goto  h'F00'        ; Go to it!
```

However, once these bits have been altered, they will remain in that state, so care needs to be taken in maintaining these auxiliary bits if the code is subsequently transferred to a processor which has a Program store larger then 2 kByte.[2]

File Direct   `00` `????` `d` `FFFFFFF`
The majority of data which the program will process are located in the Data store. Instructions that specify that their source and/or destination

---

[2]For instance, if working code is transferred from a small program-store device, such as a PIC16F84, to a larger-capacity device, such as a PIC16F877. Thus, it is generally good practice to ensure portability by always clearing the PCLATH register; even if this is strictly unnecessary.

operand lie in a File use this address mode. The address of the File is carried in the lower *seven* bits bits of the instruction code; denoted above as FFFFFFF. For instance, the code for addwf h'26,f' (add the datum in the Working register to that in File h'26' and put the outcome back in the File—or in rtl [f] <- [f] + [W] — is coded as b'*00*0111 *10100110*'.

Most instructions that use Direct addressing can "dump" the outcome either in the Working register or else back in the File. Bit 7 of the instruction code, labeled d (see also Fig. 3.5 on page 53), is used to specify the destination, as in the following example:

```
addwf  h'26',w  ; Coded as 00 0111 0 0100110
addwf  h'26',f  ; Coded as 00 0111 1 0100110
```

In both cases the byte contents of File h'26' are are added to the byte contents of the Working register. In the former instance, illustrated in Fig. 5.2(a), the outcome is put in W, leaving the File contents *unchanged* (D = 0), whilst in the latter, illustrated in Fig. 5.2(b), the original File data is *overwritten* (D = 1) with the sum.

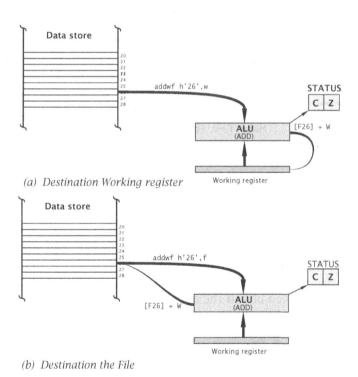

(a) *Destination Working register*

(b) *Destination the File*

Fig. 5.2 Selecting the destination of the outcome from the instruction addwf h'26'.

As we shall see (for instance, Table 5.2) the majority of instructions use Direct addressing. However, there are two limitations to this technique which the programmer needs to be aware of.

The 7-bit Issue

Only seven bits are available in the mid-range family instruction code for the File address; therefore, as it stands, only a bank of Files in the range h'00—7F' can be Directly accessed. In Fig. 4.7 on page 80 and in more detail in Fig. 5.3, we see that the PIC16F84 gets around this by employing the **RP0** bit in the Status register as a surrogate additional most-significant address bit, effectively giving two banks each of up to $2^7 = 128$ Files. This **register page** control, located as bit 5 of the Status register of Fig. 4.6 on page 78, can be altered, like any other read/write File bit, to switch back and forth between Bank 0 (RP0 = 0) and Bank 1 (RP0 = 1).

The PIC16F84 is unusual in that it only has two banks of Data memory. Most mid-range family members have four; for instance, the PIC16F627/8 upgraded PIC16F84, as shown in Fig. 5.4. In order to be able to switch to any bank, *two* register page control bits are required, as shown in the Status register of Fig. 5.5. The two switch bits RP1:RP0, heavily shaded in the diagram, are cleared to zero on any kind of Reset — which gives access to Bank 0. Thereafter, it is the programmer's responsibility to set up these bits if it is necessary to access Files in another bank. For instance, if the contents of File h'120' in Bank 2 are to be copied into the Working register and the system is to be returned to Bank 0, we have:

```
bsf  STATUS,6    ; Make RP1 (bit 6) = 1
bcf  STATUS,5    ; Make RP0 (bit 5) = 0 (move to Bank 2)

movf h'120',w    ; Copy contents of File h'120' into W

bcf  STATUS,6    ; Make RP1 = 0 (back to Bank 0)
```

Program 15.4 on page 492 is an example making use of this extended bank switching.

If the programmer forgot to twiddle the RP1:0 bits before the instruction movf h'120' then the data copied into the Working register would actually be from File h'020' (assuming the processor was in Bank 0), as only the bottom seven bits of the address b'(01)<u>0100000</u>' (h'120') actually made it into the instruction code! The assembler, however, will warn the programmer with a status message, such as shown on page 82.

To reduce the necessity to switch banks too often, the PIC16F84 in Fig. 5.3, has all its **general-purpose registers (GPR)s** shadowed (shown shaded) across both banks. This complete shadowing is rather unusual, but the PIC16F627/8 has a common area of 16 GPRs shown shaded across all four banks in Fig. 5.4. For instance, File h'070', File h'0F0', File h'170'

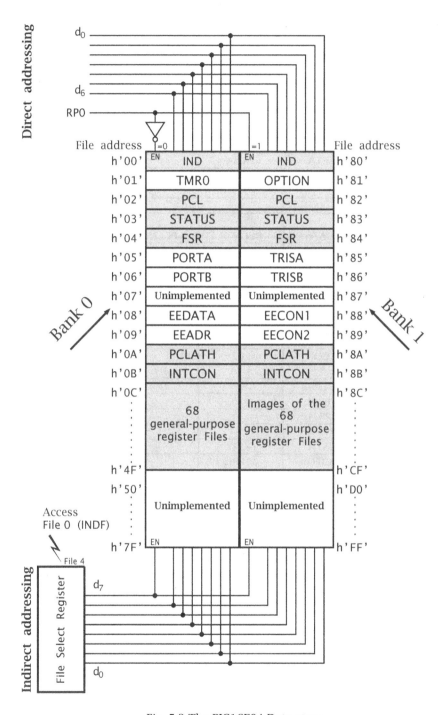

Fig. 5.3 The PIC16F84 Data store.

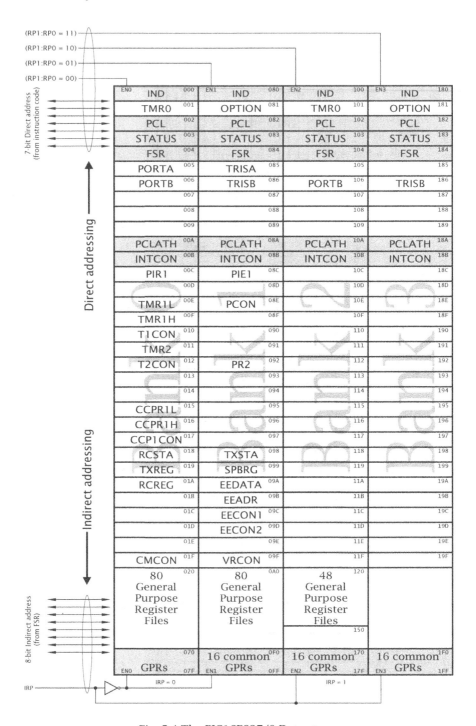

Fig. 5.4 The PIC16F627/8 Data store.

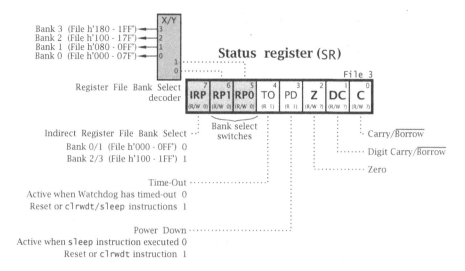

Fig. 5.5 General 14-bit core Status register.

and File h'1F0' are one and the same. Variables which may be required to be accessed from any bank, should if possible be placed within this common pool of Files. In total there are 224 unique GPRs in these devices.

Some of the more commonly used **special-purpose registers (SPRs)** are also shadowed across all banks—for instance, the Status register. It was because of this we were able to tweak the RP1:RP0 bits in the code snippet above and return to Bank 0, even when we were in Bank 2.

Fixed Addresses

The 7-bit address of the operand is *fixed* as an integral part of the instruction code, and thus cannot be changed as execution progresses. Although explicitly specifying its address may seem to be the obvious way to locate an object in the Data store, there are some situations where this restriction is rather onerous.

As an example showing this lack of flexibility, suppose we wished to clear the contents of all File registers in Bank 0 of a PIC16F627/8; that is, File h'20' – File h'7F'. The obvious way to do this is to use the clrf (CLeaR File) instruction 96 times, as shown in Program 5.1.

Although this coding works, it is rather inefficient. Each of the 96 instructions does exactly the same thing, although on a different location. If we were to clear all 224 GPRs then we would need 224 clrf instructions, all to do a rather simple task. As there is only 1024 locations available in the Program store for the PIC16F627 then this represents more than 20% of the entire capacity.

There has got to be a better way!

Program 5.1 Clearing a block of Files the linear way.

```
CLEAR_ARRAY clrf  h'20'  ; Clear File 32
            clrf  h'21'  ; and File 33
            clrf  h'22'  ; Each clrf
            clrf  h'23'  ; uses one instruction
            clrf  h'24'  ; in the Program store
            clrf  h'25'  ; File 37 cleared
            clrf  h'26'  ; and so on
            ....  .....
            clrf  h'7E'  ; Clear File 126; nearly there
            clrf  h'7F'  ; Clear File 127; Phew!
```

File Indirect | 00 | ???? | D | 0000000 |

All processors feature some form of Indirect addressing, where one or more internal registers are used to hold the address of the operand in Data memory. Such address or index registers effectively are used as a **pointer** to the data. The key difference from Direct addressing is that the contents of a pointer register can be altered as execution progresses; that is, the address of the target datum is no longer fixed as bits in read-only (usually) Program memory but is now a *variable*. For instance, the array of data in Program 5.1 may be cleared by using a register to point to the target location and repeating in a loop while incrementing that pointer.

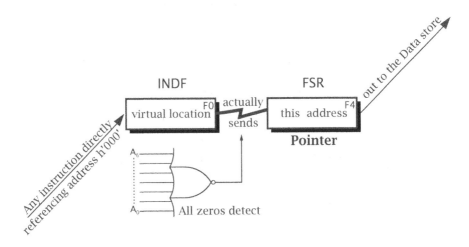

Fig. 5.6 The Indirect addressing mechanism.

PIC MCUs have a rather economical way of triggering this form of addressing—in keeping with their lean-and-mean philosophy. In the low-

and mid-range families[3] a single NOR gate detects the Direct 7-bit address b'0000000' and, as shown in Fig. 5.6, simply switches the contents of File 04 through to the Data store's address bus. In other words, the pointer address is that contained in File 04, known as the **File Select register**, or **FSR**, and the process is triggered when the null address, known as the **INDirect File**, or **INDF**, is used as the target. File 0 is a virtual location, in that it does not physically exist. Its sole use is to trigger the use of the *contents of* the FSR as the operand address. Although this approach to Indirect addressing may seem rather convoluted, it requires very little additional logic in the processor and no extra clock cycles to execute; unlike the alternative techniques used by other MPU/MCUs.

As a simple example, if we assume that the contents of the FSR at any time were h'86', then the instruction clrf 0 (or clrf INDF) would actually clear File h'86' and not File 0! Of course, the contents of the FSR may be altered at any time; for instance, being incremented on each pass through a loop, as in Program 5.2.

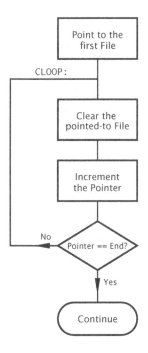

Fig. 5.7 Using a loop to clear an array of data.

---

[3] The enhanced-range family is similar but with several pointer registers and options.

As an example, let us repeat Program 5.1 but folding the linear structure into a **loop,** as shown in Fig. 5.7. A task list description of our program is now:

1. Set the FSR pointer to the initial array address.
2. Clear the pointed-to File by targeting File 0.
3. Increment the FSR pointer.
4. Check. Has the pointer gone over the top; in our case, has it reached h'80'? IF no THEN go to item 2.
5. Continue on to the next part of the program.

This process is perhaps more easily visualised in Fig 5.8.

Fig. 5.8 Walking through an array.

The coding for this scheme is shown in Program 5.2. The linear structure of the previous program has been folded into a loop, shown shaded. The executing path keeps circulating around the clrf instruction, which is "walked" through the array of Files from File h'20' upwards by incrementing the pointer in File 04 on each pass through the loop. Eventually the FSR advances beyond the desired range and the program then exits the loop and continues onto the next section of the code.

Program 5.2 has many new features, as we have yet to review the instruction set.

Task 1
The File Select register is initialized to point to the first File to be cleared, by moving the constant h'20' into the Working register (movlw h'20') and then copying W out to File 4 (movwf FSR). As we shall see, there is no single instruction to directly copy a constant into a File. Nearly all loop routines involve some initialization before entry.

Task 2
The key clearing instruction uses the Indirect address mode by specifying the phantom File 0 (INDF) as the destination address—clrf 0. This line has a **label** associated with it called CLOOP. The assembler knows that this is a label and not an instruction, as it appears in the leftmost column of the source File. Lines without labels should begin with an indent of at least one space.

Task 3
Each pass around the loop involves advancing the pointer by one. This is done by using the incf FSR,f instruction. Notice that the destination here is specified as the File and not the W register.

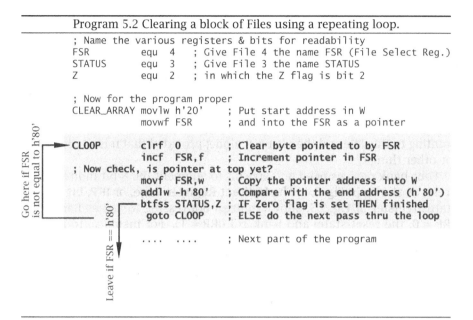

Program 5.2 Clearing a block of Files using a repeating loop.

```
; Name the various registers & bits for readability
FSR          equ  4   ; Give File 4 the name FSR (File Select Reg.)
STATUS       equ  3   ; Give File 3 the name STATUS
Z            equ  2   ; in which the Z flag is bit 2

; Now for the program proper
CLEAR_ARRAY movlw h'20'    ; Put start address in W
            movwf FSR      ; and into the FSR as a pointer

CLOOP       clrf  0        ; Clear byte pointed to by FSR
            incf  FSR,f    ; Increment pointer in FSR
; Now check, is pointer at top yet?
            movf  FSR,w    ; Copy the pointer address into W
            addlw -h'80'   ; Compare with the end address (h'80')
            btfss STATUS,Z ; IF Zero flag is set THEN finished
            goto  CLOOP    ; ELSE do the next pass thru the loop

            ....  ....     ; Next part of the program
```

Annotations around the loop: "Go here if FSR is not equal to h'80'" / "Leave if FSR == h'80'"

Task 4

Unless you wish to go round the loop forever, you need a mechanism to eventually exit. In our case this is done by comparing the contents of the FSR pointer with the constant h'80', that is, with the first File beyond the top target byte; that is, the first address after File h'7F'. The comparison mechanism is to copy the contents of the FSR pointer into W (movf FSR,w) and then subtract W from the literal h'80' using addlw -h'80' (adding a negative number). If they are equal then the Z flag will be set and in this event the next btfss STATUS,Z instruction (see page 133) will skip over the following goto CLOOP instruction. Until this happens, the goto instruction will move the execution back up to the beginning of the loop and the process is repeated with the FSR advanced to the next File to be cleared.

All together our loop version takes eight instructions against the 96 of the linear equivalent, a 12:1 reduction. However, it takes seven times as long to execute, due to the overhead of the various loop control instructions being executed 96 times! Normally the ratio of "housekeeping" to core instructions in a loop is not as extreme as this particular example.

Using the FSR to hold the operand address of course means that we now have an 8-bit variable address to access the Data store instead of the 7-bit Direct address. This means that in a 2-bank Data store, such as in Fig. 5.3, any File can be accessed anywhere. For instance, if we wanted to put the pattern b'01111111' in File h'86', the TRISB SPR in Bank 1, then as an alternative to the code fragment on page 87 we have:

```
movlw  h'86'        ; Set up FSR to point to
movwf  FSR          ; File h'86' (TRISB)

movlw  b'01111111' ; The code pattern
movwf  0            ; Sent out to the pointed-to File
```

without any necessity to twiddle the RP0 switch bit. If a File in Bank 1 needs frequent access, it may be advantageous to leave the FSR set up pointing to this File and then leave it alone; provided that it isn't required for other things.[4]

Four-bank devices need an extra bit to augment the 8-bit Indirect address, as shown in Fig. 5.4. The **Indirect Register Page**, or **IRP**, bit in the Status register of Fig. 5.5, allows Indirect addressing to access Bank 0/1 (IRP = 0, the reset state) and Bank 2/3 (IRP = 1). For instance, to repeat the code fragment on page 100 where we want to copy the contents of File h'120' in Bank 2 of a PIC16F627/8 into W we have:

```
bsf    STATUS,7    ; Make IRP (bit 7) = 1 (Bank 2/3)
movlw  h'120'      ; Initialise the FSR pointer
movwf  FSR

movf   0,w         ; Copy contents of pointed-to File into W

bcf    STATUS,7    ; Make IRP = 0 (Bank 0/1)
```

as W can only hold eight bits, so the top bit will be stripped off when the instruction movlw h'120' is executed; that is W will only hold h'20'. However, the IRP bit set to 1 is the missing ninth bit, so File h'120' will be accessed as required. The assembler will probably warn the programmer that an overlarge number was specified to be moved into W. This can be ignored.

Bit | 01 | ?? | NNN | FFFFFF |

Four instructions (as specified by the two ?? bits above) either alter or test the state of a single bit within a register File. In this situation the instruction word has an embedded 3-bit code NNN defining the bit number from 0 through 7, as well as the File address coded in the normal way. Thus the instruction bcf h'20',7 (Bit Clear bit 7 in File h'20') is coded as b'*01*00 *111*0100000'. The other instructions are bsf (Bit Set in File, coded as 01), btfsc (Bit Test File and Skip if Clear, coded as 10) and btfss (Bit Test File and Skip if Set, coded as 11). We used the latter instruction in Program 5.2 to test bit 2 in File 3 (that is the **Z** flag in the Status register) and skip out of the loop when true.

---

[4]The enhanced-range PIC MCU family has three File Select registers (actually double SPRs to hold 12-bit pointers — see Fig. 16.5 on page 516), which makes this type of thing easier to do.

So far we have classified instructions by the method they pinpoint their operands. The alternative approach is to catalog the instruction set by function. On this basis the 33-instruction 14-bit core PIC MCU's instruction set can conveniently be divided into six groups, of which four will be examined in this chapter. Those relevant to subroutines and interrupts are listed in Chapter 6 and 7, and control instructions pertaining to internal operation of the MCU hardware are left to Chapter 10.

In the instruction tables following; from left to right the instruction's mnemonic is listed, followed by the effect on the three status flags, with a • representing no change and √ normal operation. Finally a shortform description of the operation is given. The complete instruction set is given for reference on page 70. If a more detailed reference is needed, any Microchip data sheet for the appropriate family (see the book's website) gives a detailed description for each instruction. However, because of the RISC nature of the PIC MCU architecture, instructions are rather limited and simple.

Movement Instructions

About one in three instructions in any computer program, regardless of hardware, simply move data around without alteration between memory and internal registers. With this in mind the instructions in Table 5.1 are going to be the most used in the PIC MCU's repertoire.

All three movement instructions *copy* byte data without alteration, either in between the Working register and a specified File or a constant byte into W. In the former instances the source data remains unaltered, it is simply copied into the destination. The swap instruction can also copy a datum from a File to W, but in the process interchanges the higher and lower nybbles.

movlw
Copies the specified 8-bit constant (i.e., **literal**) into W. For instance, movlw h'80' initializes W to b'10000000'.

Table 5.1: Move instructions.

Operation	Mnemonic	Z	DC	C	Description
Move					Copies a datum byte
Literal to W	movlw k	•	•	•	[W] <- #kk
File	movf f,d	√	•	•	[d] <- [f]
W to File	movwf f	•	•	•	[f] <- [W]
Swap					Interchanges File nybbles
File	swapf f,d	•	•	•	[d] <- [F(3:0)][F(7:4)]

•	Flag not affected	√	Flag operates in the normal way
W	Working register	f	File register
[ ]	Contents of	d	Destination, W or a File register
#kk	8-bit constant	<-	Becomes

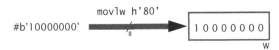

Note that by definition, the target is always the Working register, so another step is necessary to set up a File register to a constant value—see below.

movwf
This instruction is used to copy (or **store**) the contents of W into a File. For instance, movwf h'23' stores the byte in W to File h'23'.

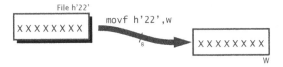

For instance, to initialize File h'23' to, say, b'10000000':

```
movlw  h'80'   ; Set contents of W to b'10000000'
movwf  h'23'   ; and copy to File h'23'
```

movf
This instruction can copy (or **load**) the contents of any File into W. For instance, movf h'22',w loads W with the contents of File h'22'.

Of course the destination of this instruction can also be the File itself, giving rise to a seemingly useless action; in our example movf h'22',f, which copies the contents of File h'22' back on top of itself! However, movf activates the **Z** flag (it is the only instruction in Table 5.1 that has any effect on a flag) and this will be set if the 8-bit File contents is all zero. The instruction movf [File],f does not affect the contents of the specified File, and thus can be used as a TeST File for zero operation; that is the missing tstf [File],f instruction that is commonly available in other MPU/MCUs. Thus the contents of any File can be checked for zero by this means using a single instruction. An alternative technique to test the contents of W for zero is given on page 121.

Given the property of most instructions acting on a File specifying either the same File or the Working register as the destination, then a Move operation can be considered an implicit part of such instructions. As an example, for some situations to increment the contents of a File and then move it to W could be coded either as:

```
incf h'22',f  ; Increment File h'22's contents
movf h'22',w  ; and copy it into W
```

or

```
incf h'22',w  ; Copy the incremented File h'22's contents to W
```

Of course the latter does not actually change the state of the File.

swapf

swapf interchanges the top and bottom 4-bit nybble in a File and places the outcome either back in the File or in W. For instance, swapf h'22',w:

The swapf instruction is useful where the two nybbles are used to hold BCD digits but can also be used as an alternative way of copying the contents of a File into W. Unlike the more obvious movf [File],w the Z flag is not altered. The downside of this transparent equivalent of course is that the two nybbles are swapped around in the process. Program 7.2 on page 202 shows this swap instruction used in this role.

Arithmetic Instructions

The low- and mid-range PIC MCU processors can do little more arithmetic than unsigned adding and subtracting byte operations. Clearing, incremention and decrementation operations are simple extensions of these fundamental processes. As well, Table 5.2 also lists instructions to clear or set any individual bit in the specified File.

Addition and Subtraction

Two addition instructions are available.

addlw

This instruction allows the programmer to add an 8-bit *constant* (literal) to W; for instance, addlw b'10101010':

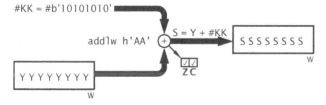

Table 5.2: Arithmetic.

Operation		Mnemonic	Flags Z	DC	C	Description
Add						Binary addition
	Literal to W	addlw k	√	√	√	[W] <- [W] + #kk
	W to File	addwf f,d	√	√	√	[d] <- [W] + [f]
Clear						Zeroes destination byte or bit
	File	clrf f	√	•	•	[f] <- #00
	W	clrw	√	•	•	[W] <- #00
	Bit	bcf f,n	•	•	•	[fₙ] <- #0
Decrement						Subtract one, produce no borrow
	File	decf f,d	√	•	•	[f] <- [f] - #01
Increment						Add one, produce no carry
	File	incf f,d	√	•	•	[f] <- [f] + #01
Set						Sets any bit in a File to one
	Bit	bsf f,n	•	•	•	[fₙ] <- #1
Subtract						Binary subtraction
	W from literal	sublw k	√	√	√	[W] <- #kk - [W]
	W from File	subwf f,d	√	√	√	[d] <- [f] - [W]

#0	Single zero bit	#1	Single one bit
#00	Zero byte	#01	Byte h'01'
#kk	8-bit constant	n	3-bit bit specifier 0–7
fₙ	Bit n of File		

addwf
Adds a *variable* in the Data store to the byte constants of W. Unlike addlw, the destination of the outcome can be specified to be either the Working register or the original File. For instance, addwf h'26',f:

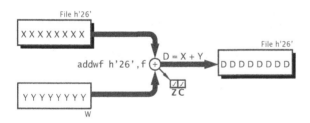

Clearing
Both the Working register and any File may be zeroed directly. In all cases the **Z** flag will be set to show the all zeros outcome.

clrw
This instruction, which zeros the Working register. It is in fact equivalent to movlw 0.

clrf
The contents of any File can be directly zeroed using this instruction; for example, clrf h'26':

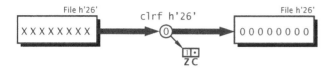

Both addition instructions, and indeed all instructions, act on 8-bit operands. However, operations of any length are possible by breaking down the process into byte-sized chunks. In the case of addition, this involves a sequence of operations from the least to the most significant digits, with any carry from the nth digit byte being added into the $n + 1$th summation. The least significant addition has a presumed carry-in of 0 and carry-out from the most significant becomes the highest bit of the outcome. For instance, $h'FF FF' + h'FF' = h'1 00 FE'$ (65,535 + 255 = 65,790).

To illustrate this process we will write a program to add an 8-bit number (addend) to a 16-bit number (augend) to give a 17-bit sum. The augend from Fig. 5.9 is located in the two memory locations File h'20' and File h'21', in the order high:low byte. The sum is stored as three bytes in the order upper:high:low in File h'30':h'31':h'30'.

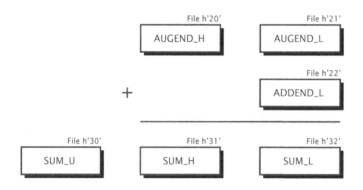

Fig. 5.9 The process.

Given that we need to implement this process as a sequence of steps executable by byte instructions, then we need to produce a task listing.

1. Add the low byte of the augend to the addend, generating the low byte of the sum and carry C1; Fig. 5.10(a).
2. Add the carry C1 to the high byte of the augend to give the high byte of the sum and a new carry-out C2; Fig. 5.10(b).
3. The upper byte of the sum is the last carry-out C2; either 0 or 1; Fig. 5.10(c).

As this is the first program of any substance in this chapter, a detailed visualization, such as shown in Fig. 5.10, has been given. For most instances, detail at this level is not helpful, and a more abstract flow chart can augment a task list.

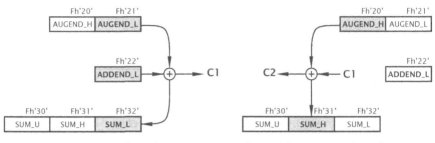

(a) Adding the least-significant bytes

(b) And the most-significant byte

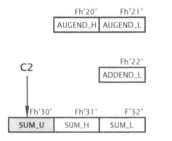

(c) The most-significant sum byte is the last carry-out

Fig. 5.10 Visualisation of the task.

Before discussing the coding of this task list we need to have a sneak preview of two instructions that we will not cover until later in the chapter. incf allows us to directly add one onto the contents of any File and btfsc (Bit Test File and Skip if Clear) will check to see if any bit in any File is clear and if so will skip over the next instruction; see Table 5.4. In our case the File is File 3 (the Status register) and the bit is bit 0 (the Carry flag); i.e., btfsc 5,0 or more symbolically btfsc STATUS,C. We have already used the sister instruction btfss in Program 5.2 to check the Z flag.

In the listing the three tasks are identified by an appropriate comment.

Preamble

All data is named using the equ directive. As discussed in page 87, meaningful names rather than raw File addresses makes for a more readable program, with less chance for error and facilities debugging.

Task 1

The lower byte of the augend is loaded into W, added to the addend and the outcome byte stored in memory as the lower byte of the sum. The C flag is set as appropriate by the addwf instruction and (fortunately) is not altered by the following movement instructions.

Task 2

The higher byte of the augend is fetched into W. If the carry C_1 from task 1 is 0 then the following addition of one (addlw 1) is skipped over, otherwise the byte in W is incremented. The outcome is then copied into the high byte of the sum.

Task 3

If the carry-out C_2 from the addition of Task 2 is 1 then the precleared upper byte of the sum is incremented to h'01'. The clr SUM_U instruction does not affect the Carry flag. If C_2 is 0 then the incf SUM_U,f instruction is skipped over and the upper sum byte is left zeroed.

Program 5.3 The double-precision add program.

```
AUGEND_H equ    h'20'    ; Name the two augend Files
AUGEND_L equ    h'21'
ADDEND_L equ    h'22'    ; Name the addend
SUM_U    equ    h'30'    ; Name the three aum Files
SUM_H    equ    h'31'
SUM_L    equ    h'32'
STATUS   equ    3        ; Naming the Status register as File3
C        equ    0        ; In which bit0 is the Carry flag

; Task 1
DP_ADD   movf   AUGEND_L,w ; Get low byte of Augend
         addwf  ADDEND_L,w ; Add Addend, result in W
         movwf  SUM_L      ; and put away as low byte Sum

; Task 2
         movwf  AUGEND_H,w ; Get high byte of Augend
         btfsc  STATUS,C   ; Was there a carry from last add?
          addlw 1          ; IF yes THEN add one
         movwf  SUM_H      ; Put away as mid byte sum

; Task 3
         clrf   SUM_U,f    ; Zero the upper byte sum byte
         btfsc  STATUS,C   ; Was there a carry from last add?
          incf  SUM_U,f    ; IF yes THEN upper byte sum is 01

         .....  .....      ; Continue with next routine
```

There are two points to note in Program 5.3.

1. Apart from the addition instructions, no instructions alter the **C** flag. This means that **C** can be tested by btfsc even after two intervening instructions have been executed.

2. The instruction following each btfsc instruction has been indented one space. This is simply a matter of style to emphasize that this is optional. It may be skipped over. The assembler will ignore such stylistic flourishes!

Increment and Decrement
The contents of any specified File can be incremented or decremented by one.

incf
One is added to the specified File and the outcome placed either back in the source File or else in W.

decf
One is subtracted from the specified File and the outcome is placed either back in the source File or else in W. For instance, if the contents of File h'26' was h'64', then decf h'26',f would result in a datum byte of h'63' in the File.

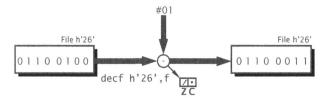

If the destination had been the Working register; i.e., decf h'26',w then the contents of File h'26' would have *remained* h'64' and the contents of W would have been overwritten to h'63'.

One point to note for both of these instructions, is that they do not affect the **C** flag; as opposed to an equivalent add or subtract-one instruction.[5] For instance, this means if you wished to increment a 3-byte datum stored thus | UPPER | MIDDLE | LOWER | (Fh'20' Fh'21' Fh'22'), incrementing the Lower byte and then examining the resulting carry-out to higher bytes and optionally incrementing will not work. The code fragment below uses the btfss (Bit Test File and Skip if Set) counterpoint to the btfsc instruction to skip the next instruction if bit 2 of the Status register (i.e., the **Z** flag) is set to 1.

```
        incf  LOWER,f    ; Add one
        btfss STATUS,Z   ; Is the outcome zero?
        goto  NEXT       ; IF not (Z==0) THEN exit

        incf  MIDDLE,f   ; ELSE increment next higher byte
        btfss STATUS,Z   ; Is the outcome zero?
        goto  NEXT       ; IF not (Z==0) THEN exit

        incf  UPPER,f    ; ELSE increment next higher byte

NEXT ....  .....        ; Next code
```

---

[5]The enhanced family's incf and decf instructions do affect the **C** flag.

The routine increments the lowest byte and if this rolls over to zero (h'FF' →' 00') the next byte is incremented and so on repeating for higher bytes. The chain is broken when incrementing a File gives a non-zero outcome. For instance, h'06 FF FE' → h'06 FF FF' → h'07 00 00'.

Bit Banging

Being able to either Clear or Set an *individual* bit in any File is important, especially to manipulate the settings in the various SPRs controlling the processor and peripheral devices. The File is located using Direct or Indirect addressing.

bcf

This instruction enables the programmer to clear any *one* of the eight bits in the specified File.

bsf

This is similar to bcf but the targeted bit is set to 1. For instance, to set bit 5 of File h'26' we have:

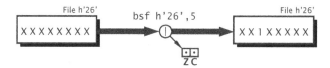

Of especial importance is controlling the various flags and switches in the Status register. We have already used these two instructions for this purpose on page 100 to alter the RP bits and therefore switch banks. Neither instruction affects any Status bits and therefore have a relatively clean action. However, it is important to realize that instructions that appear to modify the contents of data in memory in situ in fact transfer the byte into a temporary register, process the datum (e.g., incf, bcf) using the ALU, transfer the complete byte back to the Data store—a type of **read–modify–write** action — all within a single instruction cycle. Sometimes this can cause unintended side effects; see page 301 for an example.

Subtraction

The two Subtract instructions mirror the Addition instructions.

subwf

This instruction subtracts the contents of the Working register from a byte variable in the Data store. As usual, the destination of the outcome can be specified to be either the Working register or the original File. For instance, subwf h'26',f.

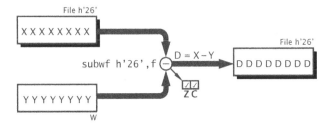

As we discussed on page 78 and Example 4.2 on page 91, the **C** flag acts as the *complement* of the borrow-out after a Subtract instruction. Forgetting the complement is a fruitful source of programming errors!

sublw

The sublw instruction is another prime source of error as the byte in W is subtracted from the constant and not the more obvious subtraction of the literal from W. For instance, if the contents of W were, say, h'64' (d'100'), then the instruction sublw 1 instead of subtracting 1 will give 1 − h'64' = h'9D', which is decimal 157 (actually in 2's complement form −h'63'). Because of the back-to-front way of implementing a subtraction of a literal, I suggest that you use this instruction at your peril! As an alternative, consider addlw h'FF'. This will give in our example h'64' + FF = (1)63' (decimal 99). If we ignore the carry for the moment, the 8-bit outcome in W is one less than the original contents. Of course, knowing that h'FF' is the 2's complement of −1 then our instruction is really addlw -1, which makes more sense in our context. Furthermore, the **C** flag is 1, and treating this as the complement of the borrow-out means that there is no borrow-out.

For the rest of our discussion we will ignore the sublw instruction and use the addlw equivalent. In fact we have already used this approach in Program 5.2, where we needed to subtract the constant h'80' from W. The assembler is happy to convert negative numbers to 2's complement equivalents; for instance, addlw -6 instead of addlw h'FA'.

One of the more important operations is the *comparison* of the magnitude of the two numbers. Mathematically this can be done by *subtracting* the datum (designated below as [f] for either a register File or a literal) from the contents of the Working register [W]. The outcome of [W] − [f] gives the actual magnitude difference between the operands, but in most cases it is sufficient to determine the relative magnitude of the quantities, e.g., is W higher than the datum? This is determined by checking the state of the **C** and **Z** flags in the Status register.

Working register *higher than* datum .... No borrow, non-zero

Working register *equal to* datum .... Zero
Working register *lower than* datum .... Borrow, non-zero

In terms of our processor, the **C** flag represents the *complement* of the borrow after subtraction and the **Z** flag is set on a zero outcome. Thus:

[W] *Higher than or equal* [f] : [W]–[f] gives no borrow; (C = 1).
[W] *Equal to*   [f] : [W]–[f] gives Zero;   (Z = 1).
[W] *Lower than*   [f] : [W]–[f] gives a borrow;   (C = 0).

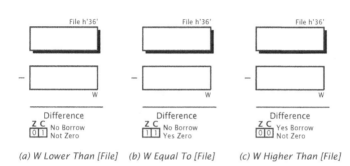

(a) W Lower Than [File]   (b) W Equal To [File]   (c) W Higher Than [File]

Fig. 5.11 Comparing the contents of W with `subwf h'26'`.

Figure 5.11 illustrates this, where the byte in W is to be compared to that in File h'36'. The instruction `subwf h'36',w'` generates the difference and alters the **C** and **Z** flags as shown giving the three magnitude outcomes. The actual difference in W is irrelevant, but overwrites the original contents, which may have to be saved before the comparison.[6]

Consider as an example of a series of comparisons with fixed values, a fuel tank with a capacity of 255 liters, with a sensor at the bottom of the tank indicating the remaining volume of fuel as a linear function of pressure. Assume that the sensor represents the capacity as a byte that can be accessed at Port B (see page 87), which we give the name FUEL. We wish to write a routine that will light an "empty" light (at bit 0 at Port A) if the capacity is below 20 liters and ring an alarm buzzer (bit 1 at Port A) if below 5 liters. Both output peripherals are active on logic 1. This is how it could be coded:

---

[6]The Compare instruction of most MPU/MCUs (such as the PIC18XXXX family) is a subtract which sets the flags in the appropriate way, but which "throws away" the difference outcome, that is, does not overwrite the operand. Thus, it is a type of non-destructive subtract.

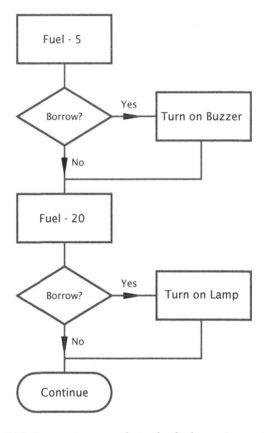

Fig. 5.12 Comparisons made in the fuel warning system.

```
STATUS  equ  3           ; File 3 is the Status register
C       equ  0           ; in which bit0 is the Carry flag
Z       equ  2           ; and bit2 is the Zero flag
FUEL    equ  6           ; Fuel can be read from File 6 (Port B)
DISPLAY equ  5           ; File 5 is Port A
LAMP    equ  0           ; With the warning lamp connected to bit0
BUZZ    equ  1           ; and bit1 drives the buzzer
ALARM   bcf  DISPLAY,BUZZ ; Turn off the Buzzer
        bcf  DISPLAY,LAMP ; Turn off the Lamp
        movf FUEL,w       ; Read fuel gauge into W
        addlw -5          ; FUEL - 5. IF FUEL HIGHER OR SAME
        btfss STATUS,C    ; THEN no borrow (C==1) so skip
         bsf  DISPLAY,BUZZ ; ELSE sound buzzer
        movf FUEL,w       ; Get fuel gauge again into W
        addlw -d'20'      ; FUEL - 20. IF FUEL HIGHER OR SAME
        btfss STATUS,C    ; THEN no borrow (C==1) so skip
         bsf  DISPLAY,LAMP ; ELSE light lamp
NEXT:   .....   .....     ; Continue
```

After each subtraction the Carry/$\overline{\text{borrow}}$ flag will be logic 1 (that is *no* borrow) if the datum in the Working register (the fuel reading) is higher or the same as the literal it is being compared with, that is, by subtraction. Note the use of the bsf (Bit Set in File) instruction to set the appropriate pin in Port A, which we assume to have been initialized as an output. In the same manner the bcf instruction is used to turn off the lamp and buzzer at the beginning of the routine.

Related to comparison is the Test operation, where a datum is checked for zero. We have already seen (see page 52) that the contents of a File can be tested for zero by simply moving its contents back onto itself; for instance, movf h'36',f, which will set the **Z** flag if zero.[7] The equivalent process to this for the Working register would be to add zero, i.e., addlw 0. Again this will set the **Z** flag if the contents of W are zero and leaves the contents of the register unchanged.

Logic and Shifting Instructions

All four basic logic operations of NOT, AND, Inclusive-OR and eXclusive-OR are provided, as shown in Table 5.3.

Table 5.3: Logic instructions.

Operation		Mnemonic	Z	DC	C	Description
AND						Logic bitwise AND
	Literal to W	andlw k	√	•	•	[W] <- [W] · #kk
	W to File	andwf f,d	√	•	•	[d] <- [W] · [f]
Complement						Invert or NOT (1's complement)
	File	comf f,d	√	•	•	[d] <- $\overline{[f]}$
Inclusive-OR						Logic bitwise Inclusive-OR
	Literal to W	iorlw k	√	•	•	[W] <- [W] + #kk
	W to File	iorwf f,d	√	•	•	[d] <- [W] + [f]
eXclusive-OR						Logic bitwise eXclusive-OR
	Literal to W	xorlw k	√	•	•	[W] <- [W] ⊕ #kk
	W to File	xorwf f,d	√	•	•	[d] <- [W] ⊕ [f]
Rotate File shift						Circular shift into Carry
	Left	rlf f,d	•	•	b7	[C]←[7 File 0]←
	Right	rrf f,d	•	•	b0	→[7 File 0]→[C]

·	Boolean bitwise AND	+	Boolean bitwise Inclusive-OR
⊕	Boolean bitwise eXclusive-OR	$\overline{[f]}$	Bitwise inverse of the File contents

---

[7]The Test instruction of most MPU/MCUs (such as the PIC18XXXX) is a more obvious instruction to check for zero.

NOT

The NOT logic function of Fig. 1.1 on page 12 inverts (1's complement) the logic state of the input.

comf

The logic state of any specified File can be inverted. As an example the instruction, comf h'26', f complements the contents of File h'26':

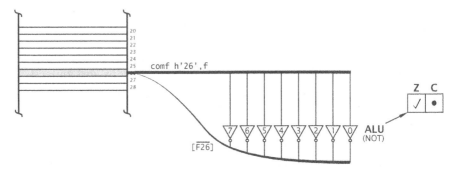

In the normal way the outcome can be placed either in the source File or in W, with the original contents being unchanged; for instance:

10001110	comf h'26',w	
File h'26'	⌇	01110001 W

There is no comw instruction to explicitly invert the contents of the Working register, but this can be achieved in one bus cycle by subtracting W from b'11111111' to give the 1's complement; e.g., sublw h'FF'. For instance (see also page 126):

```
  11111111            Literal h'FF'
- 10001110            Working register
  --------
  01110001            1's complement of W
```

AND

From Fig. 1.2 on page 13 you will recall the following relationship:

- ANDing a bit variable with 0 *always* gives a 0 output.
- ANDing a bit variable with 1 yields an unchanged logic state.

On this basis we can zero a selected group of bits in a datum byte by ANDing with the appropriate pattern.

In the same manner, ANDing a datum with a test pattern to clear all unwanted bits can also be used to check if a selected group of bits in the datum is zero. If this is true, the overall result will be zero and the Z flag will be set accordingly.

andwf

The andwf instruction bitwise ANDs the contents of W together with the contents of any File, with the outcome being placed either in that same File or in W. For example, to AND each bit of W with each corresponding bit in File h'26', with the outcome being put back in File h'26' we have:

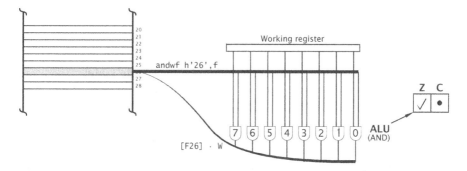

For instance, to clear the upper six bits of File h'26' we have:

```
movlw  b'00000011'  ; Mask pattern
andwf  h'26',f       ; ANDed with contents of File h'26'
```

The alternative would be to use the bcf instruction six times.

To see how the AND function can be used to test for a set of zero bits, consider a controller for a washing machine where the eight switches in the control panel are read via Port B, that is, File 6. It is desired to jump to a routine to implement a Fast Wash if the switches connected to bits 7 and 6 are both zero; that is, the GO and FAST switches are closed. Here is how it could be done:

```
movlw  b'11000000'  ; The test mask
andwf  6,w           ; ANDed with PORTB
btfss  STATUS,Z      ; Skip if Z == 0 (i.e., non-zero outcome)
goto   FAST_WASH     ; ELSE jump to routine called FAST_WASH
.....  .....         ; Next test
```

By ANDing the contents of File 6 with b'11000000', the lower six bits will be cleared. The overall outcome will be all zero only if *both* bits 7 and 6 of Port B are 0 before this action. In this case the **Z** flag will be set and the following Bit Test File Skip on Clear instruction will not be taken and the program will transfer to the instruction located at the label FAST_WASH. If a *single* bit in a File is being tested for zero then btfsc can be used to directly check that bit.

andlw

The contents of W can be bitwise ANDed with a byte literal. For instance:

$$\boxed{10001110}_W \quad \overset{\text{andlw h'0F'}}{\sim\!\sim\!\sim} \quad \boxed{\mathbf{0000}1110}_W$$

where the high nybble of W is zeroed and the low nybble is left untouched.

Inclusive-OR

From Fig. 1.3 on page 13 you will recall the following relationship:

- IORing a bit variable with 0 yields an unchanged logic state.
- IORing a bit variable with 1 *always* gives a 1 output.

On this basis we can set a selected group of bits in a datum byte to 1 by IORing with a suitable pattern.

iorwf

In a similar manner to andwf, the contents of any File can be bitwise Inclusive-ORed with the contents of W. Thus to IOR each bit in W with its corresponding bit in File h'26', with the outcome being put back in the File, we have:

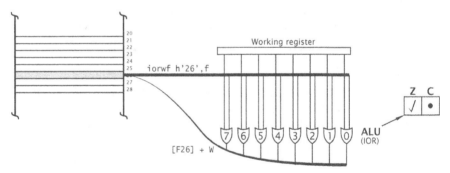

For instance, to set to 1 the top seven bits in File h'36' we have:

```
movlw  b'11111110'; The mask byte
iorwf  h'36',f    ; Set top 7 bits to 1, lowest bit unchanged
```

iorlw

The contents of W can be bitwise IORed with a byte literal. For instance, to set the lower two bits of the Working register to 1:

$$\boxed{10001110}_W \quad \overset{\text{iorlw 03}}{\sim\!\sim\!\sim} \quad \boxed{100011\mathbf{11}}_W$$

eXclusive-OR
From Fig. 1.4 on page 14 you will recall the following relationship.

- XORing a bit variable with 0 yields an unchanged logic level.
- XORing a bit variable with 1 inverts or *toggles* the state of the input logic level.

Another useful property of the XOR logic operator is as a logic differentiator. A close inspection of the truth table shows that the output of an XOR gate is 1 if the two input logic levels are different and 0 if they are the same. Thus bitwise XORing two bytes together will produce a byte output with 0 in locations where the two input bits are the *same* and a 1 where they differ.

xorwf
The contents of any File can be bitwise eXclusively-ORed with the contents of W. Thus to XOR each bit in W with its corresponding bit in File h'26' with the outcome being put back in the File: For instance, to toggle the top bit of File h'36' we have:

```
movlw  b'10000000'   ; The mask byte
xorwf  h'36',f       ; Toggle top bit only of File h'36'
```

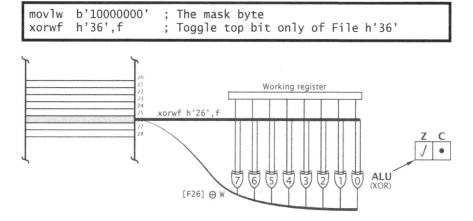

As an example showing the use of XOR to isolate *changes* between two bit patterns, consider a program routine that continually monitors Port B, to which has been connected eight switches as part of the control panel of a washing machine. The routine waits until a switch is moved.

```
START  movf   PORTB,w   ; Get initial state of switches
       movwf  h'20'     ; Put away at File h'20'

S_LOOP movf   PORTB,w   ; Sample switches
       xorwf  h'20',w   ; Check for alterations from original
       btfsc  STATUS,Z  ; Skip out if check gives non-zero
       goto   S_LOOP    ; ELSE check again
```

Two possible scenarios are:

$$\boxed{10011110}_{\text{File h'20'}} \xrightarrow{\text{xorwf h'20',w}} \boxed{10011110}_{\text{W}} = \boxed{00000000}_{\text{W}} \quad Z = 1$$

$$\boxed{10001110}_{\text{File h'20'}} \xrightarrow{\text{xorwf h'20',w}} \boxed{10011110}_{\text{W}} = \boxed{00010000}_{\text{W}} \quad Z = 0$$

The outcome in W reflects any changes. In the first case there are no differences between the latest sample and the original switch settings put away in File h'20'. In the second situation, Switch 4 has just been thrown from 1 to 0. You can determine which switch changed by shifting the outcome (the change byte) right, counting until the residue is zero; see Fig. 5.14. You can also determine the type of change ($0 \rightarrow 1$ or $1 \rightarrow 0$) by ANDing the change byte to the original switch settings in File h'20', i.e., andwf h'20',w. If the outcome at bit 4 is a 0, then the original state must have been 0 and therefore the change must have been $0 \rightarrow 1$, and vice versa.

xorlw

The contents of W can be bitwise XORed with a byte literal. For instance, to invert all bits in W, that is, complement W:

$$\boxed{10001110}_{\text{W}} \xrightarrow{\text{xorlw h'FF'}} \boxed{01110001}_{\text{W}}$$

Shifting

Shifting data left or right is a fundamental operation found in all digital systems. We saw in Fig. 2.22 on page 37 how this could be done in hardware. Without exception, all MPU/MCU devices have ALUs that permit various combinations of Shift Right and Shift Left instructions to be implemented.

All PIC MCU families have two instructions in this category to shift the contents of any File, one for each direction.[8]

rrf

Rotate Right File shifts the byte contents of the specified File once right, with the incoming bit coming from the **C** flag, which is simultaneously replenished with the outgoing bit. This circular action is emphasised in Fig. 5.13. In essence, this is a Shift Right function but with the Carry flag acting as a sort of bit 8.

With this diagram in mind the programmer can do plain shift right with a zero coming in (as in Fig. 2.22) by first clearing the Carry bit before rotating.

```
bcf   STATUS,C    ; Zero the Carry bit in the Status register
rrf   h'30',f     ; Now rotate right ->
```

[8]The enhanced family use mnemonics rlcf and rrcf for Rotate Left/Right thru Carry File, with two extra Rotate instructions that bypass the **C** flag with mnemonics rlnc and rrnc.

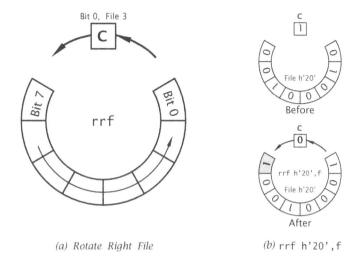

(a) Rotate Right File   (b) rrf h'20',f

Fig. 5.13 Rotating the contents of a File once right.

One use of the shifting operation is to bitwise examine a datum. For example, assume that the state of an array of eight switches from a mobile phone has been copied into File h'26'. You are required to find the leftmost open switch, where you can assume that an *open* switch reads as 1 and a *closed* switch as logic 0. For instance, if the reading was:

0 SW8	0 SW7	1 SW6	0 SW5	1 SW4	1 SW3	1 SW2	1 SW1

then the outcome in W should be 6 (b'00000110').

The coding given in Program 5.4 uses the Working register as a counter. As the Carry flag is cleared each time *before* the shift, logic 0s are brought in from the left.[9] Eventually the residue will become all zeros and the process should then terminate. Thus 00010111 (1) ⤳ 00001011 (2) ⤳ 00000101 (3) ⤳ 00000010 (4) ⤳ 00000001 (5) ⤳ 00000000 (6).

A task list for this problem, also shown diagrammatically in Fig. 5.14, would be:
1. Zero KEY_COUNT
2. WHILE SWITCH_PATTERN is not zero DO
   (a) IF residue is zero THEN break
   (b) Shift left SWITCH_PATTERN once
   (c) Increment KEY_COUNT
3. KEY_COUNT holds the position of the leftmost open switch

Shifting right pops out the rightmost bit into the Carry flag. Replacing btfsc STATUS,Z by btfsc STATUS,C would determine the position of the *rightmost* bit. In many situations repetitively shifting into the Carry

---

[9]MPU/MCUs that have Logic Shift instructions always shift in 0s irrespective of the state of the **C** flag; for instance, Motorola's lsr (Logic Shift Left).

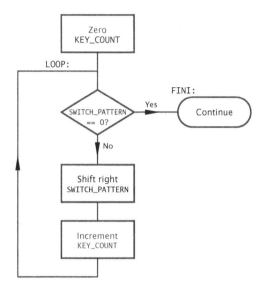

Fig. 5.14 Finding the leftmost 1 by continuous shifting.

Program 5.4 Scanning the File looking for the highest 1.

```
SWITCH_PATTERN equ    h'26'   ; Data is in File h'26'
STATUS         equ    3       ; The Status register is File 3
Z              equ    2       ; in which bit2 is the Z flag

; Task 1
HIGH_BIT       clrw           ; Zero the count

; Task 2: DO Shift right & increment count, WHILE datum isn't zero
; Task 2a
LOOP           movf  h'26',f  ; Is residue zero?
               btfsc STATUS,Z ; IF not zero THEN skip
                goto FINI     ; ELSE break
; Task 2b
               bcf   STATUS,Z ; Carry flag cleared
               rrf   h'26',f  ; Shift datum right
; Task 2c
               addlw 1        ; Continue by adding one to count
               goto  LOOP     ; and do another shift
; Task 3
FINI           ..... ......   ; KEY_COUNT is in W
```

flag can be used to examine the data on a bit-by-bit basis. For instance, we could modify our program to total the number of set bits in the byte, as in Program 5.6.

Notice that Program 5.4 returned a zero outcome if no switch was open. If a test for zero had been done after the Shift operation, then it would not be possible to distinguish between this situation and Switch 1 alone being open. It is important to design your software for robustness, so that limiting conditions, such as this, are dealt with.

rlf

Rotate Left File is similar to rrf but, as shown in Fig. 5.15, the shift direction is from the low to the high bit position.

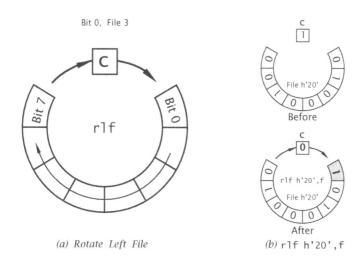

(a) Rotate Left File

(b) rlf h'20',f

Fig. 5.15 Rotating the contents of a File once left.

As an example of the use of rlf we note from page 11 that we can use shifting to the left to multiply by powers of two. For instance:

```
00000110   (6)   <<
00001100   (12)  <<
00011000   (24)  <<
00110000   (48)  <<
     etc.
```

where the C Shift-Left operator << is used to indicate a shift left.

To illustrate the process, assume that if we have the 16-bit number b'00000111 11010000' (which is decimal $1024 + 512 + 256 + 128 + 64 + 16 = 2000$), then this will be stored in two Files; for example:

After shifting once left we have:

00001111		10100000	

File h'30'   File h'31'

which is decimal 4000 (2048 + 1024 + 512 + 256 + 128 + 32 = 4000).

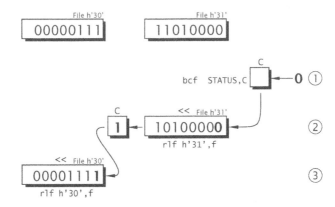

Fig. 5.16 Shifting a double-byte datum once to the left to multiply by two.

The problem is that our rlf instruction can only shift a single byte at a time. Thus we need to break this down to three steps, as illustrated in Fig. 5.16.

1. Clear the Carry flag so that we will rotate in a 0.
2. Rotate the low byte left, with b_7 being popped out into the Carry flag.
3. Rotate in the carry-out of the previous Rotate operation into the high byte.

From the diagram we see that the process is straightforward, with the carry-out from the previous File becoming the carry-in for the second File. The routine is thus:

```
bcf STATUS,C   ; Clear C flag, which will hold the incoming bit
rlf h'31',f    ; Rotate into the low byte, MSB pops out into C
rlf h'30',f    ; Rotate into the high byte
```

Program Counter Instructions

The instructions listed in Table 5.4 modify in some way the setting of the Program Counter.

nop

No OPeration does not alter the state of the system in any way, but the PC will increment as a consequence of the instruction code being fetched from the Instruction store. Thus, its sole outcome is [PC] <- [PC] + 1.

Table 5.4: Program Counter instructions.

Operation	Mnemonic	Flags Z	DC	C	Description
Absolute jump					Goto a fixed instruction
Goto an instruction	goto aaa	•	•	•	[PC] <- aaa
No operation					Do nothing
	nop	•	•	•	[PC] <- [PC] + 1
Bit test and skip					Check bit in File and skip if true
Bit Test in File, Skip if Clear	btfsc f,n	•	•	•	PC++ IF f_n == 0
Bit Test in File, Skip if Set	btfss f,n	•	•	•	PC++ IF f_n == 1
Decrement and skip on zero					Decrement & skip if result is #00
File	decfsz f,d	•	•	•	d <- f--, PC-- IF [f] == #00
Increment and skip on zero					Increment & skip if result is #00
File	incfsz f,d	•	•	•	d <- f++, PC++ IF [f] == #00

++   Increment contents   – –   Decrement contents
aaa   Absolute 11-bit instruction address

This takes one instruction cycle, so its main use is to implement a short delay, 1 μs for a 4 MHz clock rate. For instance, to pulse Port A's pin low for 2 μs and then high we have:

```
bcf  PORTA,0  ; Pin RA0 low
nop           ; for 2 us
nop
bsf  PORTA,0  ; and now high
```

with the assumption that bit 0 of Port A has been set up as an output (see page 87) and that bit 0 (pin RA0) was high before entering the routine.

goto
This instruction allows the programmer to jump to a specified instruction anywhere in the Program store.

In the example shown in Fig. 5.17 the instruction goto h'3F9' is located in the Program store at location h'005'. In the normal course of events the Program Counter has been incremented to h'006' and the Instruction located here has already been fetched into the top of the Pipeline, ready to be executed in the next instruction cycle. However, when the goto h'3F9' instruction at the bottom of the Pipeline is executed, the Program store address h'3F9' is placed in the PC. This means that the next instruction to be executed is the one at this location. To permit this to happen, Instruction 1018 must be fetched down into the Pipeline, overwriting the now unwanted code. This process is known as *flushing*, and takes an extra instruction cycle to implement. Thus goto takes two instruction cycles to execute.

In the diagram, location h'3F9' is labeled FRED (the following colon is optional). Using labels rather than absolute locations is strongly rec-

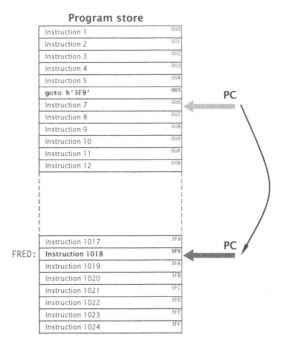

Fig. 5.17 Jumping to instruction 1018.

ommended (see also page 106) as the programmer does not easily know where an instruction is located in the Program store, and in any case, this location may change as the program develops.

btfsc

The crucial role in PIC MCU programming of Bit Test File and Skip if bit is Clear and its mirror image btfss, is shown by their use in virtually every program so far in this chapter. In these programs these instructions have been used as the means to allow decisions to be made on the state of the various flags in the Status register, indicated as an IF-THEN statement in a task list or ◇ symbol in a flow chart. For instance, in Program 5.4, the btfsc STATUS,Z (or the less decipherable btfsc 3,2) instruction allowed the program loop to continue as long as the datum is non-zero, skipping over the exiting goto instruction when Z is 0 (clear).

In fact, btfsc is more flexible than just testing Status flags. *Any* bit in *any* File can be checked, and the *next* instruction skipped over if that bit is 0; see page 108. Figure 5.18 shows the situation where Instruction 6 is btfsc h'20',7. The execution of this instruction checks bit 7 of File h'20' and on the basis of its state implements one of two outcomes:

1. IF bit 7 is 0 THEN skip over Instruction 7 and execute Instruction 8.
2. IF bit 7 is 1 THEN continue on as normal to Instruction 7.

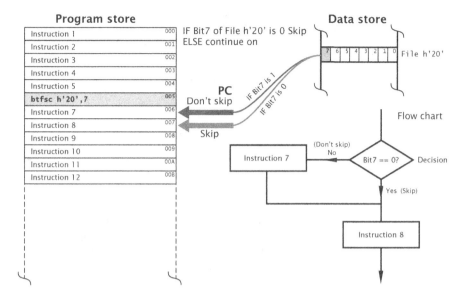

Fig. 5.18 Skipping over the next instruction whenever bit 7 of File h'20' is clear.

Often Instruction 7 is a goto, so the program can react to the state of any bit in the Data store by jumping to an appropriate routine.

When a skip occurs, the Pipeline requires to be flushed, in the same way as a goto, as the linear progression of the program is disrupted. This means that a btfsc instruction takes one instruction cycle to execute if no skip takes place and two instruction cycles if a skip is executed.

btfss

Bit Test File and Skip if bit is Set implements a skip if the specified bit is 1. Other than this reversal, its operation is identical to btfsc.

decfsz

DECrement File and Skip if Zero represents an alternative way of making a decision. As a combination of the instruction pair decf followed by btfss STATUS,Z, this instruction allows the programmer to decrement the contents of any File, and if the outcome is zero, then skip over the next instruction.

A typical use of this instruction is to count the number of passes through a loop. For example, suppose it is necessary to pulse Port A pin RA0 low 20 times. To implement this task, shown in Fig. 5.19, we have the code, assuming a 4 MHz crystal:

```
        movlw   d'20'       ; Put decimal 20 into W and
        movwf   h'3F'       ; copy into File h'3F' as a loop counter
;-------------------------
LOOP bcf        PORTA,0     ; Pin RA0 low
        nop                 ; One extra cycle delay
        bsf        PORTA,0  ; and now high
;-------------------------
        decfsz  h'3F',f     ; Count down
        goto    LOOP        ; Repeat loop if not zero
        .....   ......      ; ELSE break from the loop
```

The original code shown between dashed lines is cocooned by the decrementing test which skips out of the loop whenever the contents of File h'3F' reach zero. Notice the assembler notation d'20' for *decimal* 20; see page 239. This is equivalent to h'14' but more readily understood by the programmer. Incidentally, only one nop is used, as the bsf and bcf between them add an extra cycle delay to the total.

incfsz
INCrement File and Skip if Zero increments rather than decrements the contents of the specified File. If this causes the contents to roll over to zero, e.g., h'FC → FD → FE → FF → 00' then the following instruction will be skipped over.

In the case of our example of Fig. 5.19 if we were to preload File h'3F' with −20 (or h'EC') and replace decfsz h'3F',f by incfsz h'3F',f

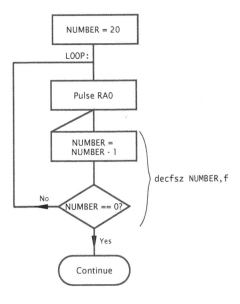

Fig. 5.19 Pulse pin RA0 repeating 20 times.

then this will give the same outcome by counting up rather than counting down.

Examples

Example 5.1

Code a program to decrement a 2-byte number at [MSB] $_{Fh'26'}$ and [LSB] $_{Fh'27'}$. Remember that the instruction decf does not alter the state of the Carry/$\overline{\text{Borrow}}$ flag.

Solution

The task list to implement this job is:

1. IF the least significant byte in File h'27' is zero THEN decrement the most significant byte.

2. Always decrement the least significant byte.

   Program 5.5 gives one possible implementation based directly on this algorithm. Extension to an n-byte word works in the same way, starting from the least byte and moving up, decrementing both the $(n + 1)$th and nth byte if the nth byte is zero, until the chain is broken.

Program 5.5 Double-precision decrement.

```
STATUS   equ   3          ; The Status register
Z        equ   2          ; Bit2 of which is the Zero bit
MSB      equ   h'26'      ; The most significant byte
LSB      equ   h'27'      ; The least significant byte

         movf  LSB,f      ; Test: Is LSbyte zero?
         btfsc STATUS,Z   ; IF not THEN skip MSbyte decrement
         decf  MSB,f      ; ELSE must decrement MSbyte
         decf  LSB,f      ; Always decrement LSbyte
```

Example 5.2

Some early computers used a bi-quinary code to represent BCD digits. This is a 7-bit code with only two bits set to one for any combination:

01	00001	0
01	00010	1
01	00100	2
01	01000	3
01	10000	4
10	00001	5
10	00010	6
10	00100	7
10	01000	8
10	10000	9

Although this is highly inefficient (with only 10 out of a possible 128 code combinations being used) it does have the advantage that it is very easy to decide when an error has occurred. Design an error-detection routine to check the bi-quinary byte in File h'20'. Assume that the most-significant bit is zero. If an error occurs then the Working register is to be set to h'FF', otherwise zero.

Solution
All we need to do here is to determine when there are more or less than two bits set to one. Based on this approach we have the task list:
1. Count the number of ones in the bi-quinary byte.
2. Zero W.
3. IF the count is not two THEN finish with W set to h'FF' to signal an error.

Program 5.6 shows a possible coding implementing this algorithm. Here the loop continually shifts the bi-quinary byte left until the residue

---

Program 5.6 Bi-quinary error detection.

```
STATUS      equ   3          ; Status register is File 3
C           equ   0          ; Carry flag is bit0
Z           equ   2          ; Zero flag is bit2
BI_QUIN     equ   h'20'      ; Bi-quinary byte is in File h'20'
COUNT       equ   h'21'      ; The bit count is put here

BI_QUINARY clrf  COUNT       ; Bit count is cleared
; Task 1
LOOP        bcf   STATUS,C   ; Clear Carry flag
            rlf   BI_QUIN,f  ; Rotate code left
            btfsc STATUS,C   ; IF no carry popped out THEN skip
             incf COUNT,f    ; incrementing the Count
            movf  BI_QUIN,f  ; Test if residue is zero
            btfss STATUS,Z   ; IF zero THEN skip out of loop
             goto LOOP       ; ELSE repeat shift and count
; Tasks 2 & 3
            movf  COUNT,w    ; Get 1's count
            addlw -2         ; Compare with two
            btfss STATUS,Z   ; IF zero break with W = 00
             movlw h'FF'     ; ELSE put h'FF' (-1) in W
            ..... ......     ; and exit
```

---

is zero. After each shift, when the Carry flag is set, the bit count is incremented. On exit from the loop, two is subtracted from the bit tally after moving into W. If it is zero, then the routine is completed and the h'00' setting of W shows a correct outcome. Otherwise h'FF' is placed in W to show an error situation. This is equivalent to decimal −1 and is traditionally used to note an error situation. There are 20 code combinations in all which have two ones, of which only 10 are legitimate. Can you think of a simple extension to the routine to weed out these additional double-one code patterns?

Example 5.3
Low- and mid-range PIC MCUs do not have instructions to directly multiply[10] or divide. However, addition and subtraction can be used to implement these important arithmetic operations.

For instance, to divide a number by ten you can count how many times ten can be subtracted before a borrow-out is generated. The count is then the quotient and the residue is the remainder. Using this technique, write a routine to convert a binary number byte of magnitude no greater than h'63' (decimal 99) in File h'20' to two Binary Coded Digits to be placed in File h'21:22' ordered as Tens:Units; see page 7.

Solution
Dividing the binary number by ten generates a quotient between 0 and 9 (remember the maximum value is 99) and a remainder. The quotient is the number of tens and the remainder is the number of units.

The simplest way of doing this, illustrated in Fig. 5.20, is to keep subtracting ten (addlw -d'10' or addlw -h'0A'). Keeping a count in the TENS File register gives the number of subtractions until a borrow is generated. The required number of tens is one less than this tally; that is, the number of successful subtractions. Adding that one extra ten back again to the residue gives the remainder, which is the units tally.

Example 5.4
Another approach to division is to express the divisor as the sum of fractional powers of two. For instance, the binary approximation to the fraction $\frac{1}{3}$ is:

$$\frac{1}{3} = \frac{1}{2} - \frac{1}{4} + \frac{1}{8} - \frac{1}{16} + \frac{1}{32} - \frac{1}{64} + \frac{1}{128} \cdots$$

---

[10]The enhanced-range instruction set does have an unsigned 8-bit × 8-bit instruction giving a 16-bit product—all in one instruction cycle!

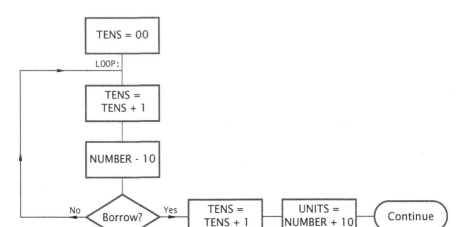

Fig. 5.20 Conversion of a byte up to 99 to BCD.

Program 5.7 Binary to 2-digit BCD conversion.

```
STATUS      equ   3       ; Status register is File 3
C           equ   0       ; Carry/Not Borrow flag is bit0
BINARY      equ   h'20'   ; Binary byte is in File 20h
TENS        equ   h'21'   ; The quotient is put here
UNITS       equ   h'22'   ; The remainder is put here

; First divide by ten
BIN_2_BCD clrf  TENS      ; Zero the loop count
          movf  BINARY,w  ; Get binary byte into W
; DO subtract ten and counting WHILE no borrow is generated
LOOP      incf  TENS,f    ; Record one ten subtracted
          addlw -d'10'    ; Subtract decimal ten
          btfsc STATUS,C  ; IF a borrow (C == 0) THEN exit loop
          goto LOOP       ; ELSE do another subtract/count
; Correct for one subtract too many and hence determine Units
          decf  TENS,f    ; Compensate for one inc too many
          addlw d'10'     ; Add ten to residue
          movwf UNITS     ; which gives the remainder
          ..... ......    ; Next routine
```

Using this series, write a program that will divide a byte N in the Working register by three, with the quotient being in the same register at the end.

You can use File h'20' and File h'21' as temporary storage for the quotient and shifting number respectively.

Solution

Program 5.8 Dividing by three.

```
QUOTIENT equ h'20'        ; Put the final Quotient here
TEMP     equ h'21'        ; Used as a interim location for shifting
DIV_3 clrf  QUOTIENT      ; Zero the outcome
      movwf TEMP          ; Put N into File memory
      bcf   STATUS,C      ; Carry = 0
      rrf   TEMP,f        ; Shift right once to give N/2
      movf  TEMP,w        ; Get it down into W
      movwf QUOTIENT      ; and copy into Quotient = N/2
      bcf   STATUS,C      ; Carry = 0
      rrf   TEMP,f        ; Shift again once right to give N/4
      movf  TEMP,w        ; Copy into W
      subwf QUOTIENT,f;   Subtract to give Q = N*(1/2-1/4)
      bcf   STATUS,C      ; Carry = 0
      rrf   TEMP,f        ; Shift again to give N/8
      movf  TEMP,w        ; Copy into W
      addwf QUOTIENT,f;   Add to give Q = N*(1/2-1/4+1/8)
      bcf   STATUS,C      ; Carry = 0
      rrf   TEMP,f        ; Shift again to give N/16
      movf  TEMP,w        ; Copy into W
      subwf QUOTIENT,f;   Sub to give Q = N*(1/2-1/4+1/8-1/16)
      bcf   STATUS,C      ; Carry = 0
      rrf   TEMP,f        ; Shift again to give N/32
      movf  TEMP,w        ; Copy into W
      addwf QUOTIENT,f;   Add: Q = N*(1/2-1/4+1/8-1/16+1/32)
      bcf   STATUS,C      ; Carry = 0
      rrf   TEMP,f        ; N/64
      movf  TEMP,w        ; Copy into W
      subwf QUOTIENT,f;   Q = N*(1/2-1/4+1/8-1/16+1/32-1/64)
      bcf   STATUS,C      ; Carry = 0
      rrf   TEMP,f        ; N/128
      movf  TEMP,w        ; Copy into W
      addwf QUOTIENT,w;   N*(1/2-1/4+1/8-1/16+1/32-1/64+1/128)
```

The coding shown in Program 5.8 simply zeros the Quotient byte and then copies the datum from W into File h'21'. Once in place, the datum can be shifted right to give the various fractions, which are either added or subtracted from File h'20', gradually building up the final Quotient.

The outcome up to $\frac{1}{128}$ is 0.3359375, which is within 0.78% of the exact value. With an 8-bit datum there is no point in including any further elements in the series.

If greater accuracy is desired, then the original number can be extended to a 16-bit datum, by adding a zero lower byte. The series can then be extended to give a resolution down to one part in 32,768, with double-precision shifting and arithmetic operations.

Example 5.5

As part of a routine to convert from Celsius to Fahrenheit temperature scales, it is necessary to multiply a byte in File h'22' by nine. The resulting 16-bit product is to be located at $\boxed{\text{PRODUCT\_H}}_{\text{F h'21'}}\boxed{\text{PRODUCT\_L}}_{\text{F h'22'}}$.

Solution

The task list implemented in Program 5.9 splits the ×9 task into a ×8+×1 operation. Thus our task list becomes:

1. Multiply Multiplicand by eight (shift left three times).
2. Add the single-byte Multiplicand to the 16-bit partial Product.

The coding copies the 1-byte Multiplicand into the lower byte of the Product to be. Zeroing the upper byte extends the datum to 16 bits. Clearing the Carry flag and then shifting left three times gives the ×8 subproduct. Finally, adding the single-byte Multiplicand to the double-byte partial Product gives the final Product.

The principle of shift-and-add (see page 11) can be used for any multiplication process. For instance, multiplying by ten can be implemented as ×8 + ×2 process. This is a little more complex to code, as a 2-byte workspace is necessary to do the shifting which is added at appropriate points in the shifting process to the 2-byte partial Product. However, it is much faster than repeated addition.

Example 5.6

A certain temperature logging system samples every hour and at the end of a day the 24 samples are to be found in situ in the Data store between

---

Program 5.9 Multiplication by nine.

```
STATUS        equ   3        ; Status register is File 3
MULTIPLICAND  equ   h'22'    ; Multiplicand byte
PRODUCT_H     equ   h'23'    ; High byte of product
PRODUCT_L     equ   h'24'    ; Low byte of product
C             equ   0        ; Carry flag is bit0

; Task1: Multiply multiplicand by eight
MUL_9   movf    MULTIPLICAND,w ; Get Xcand byte which
        movwf   PRODUCT_L      ; becomes the lower product byte
        clrf    PRODUCT_H      ; extended to 16 bits

        bcf     STATUS,C       ; Clear Carry flag
        rlf     PRODUCT_L,f    ; Now shift word left
        rlf     PRODUCT_H,f    ; three times
        rlf     PRODUCT_L,f
        rlf     PRODUCT_H,f
        rlf     PRODUCT_L,f
        rlf     PRODUCT_H,f

; Task2: Add X8 and X1
        addwf   PRODUCT_L,f    ; Add Xcand (still in W!) to LSB
        btfsc   STATUS,C       ; Skip if no carry
        incf    PRODUCT_H,f    ; ELSE add one onto MSB of Product
        .....   ......         ; Next routine
```

File h'30' and File h'47'. Write a program to scan through this array and evaluate the average daily temperature.

Solution

Finding the average involves walking through the array, in the manner of Fig. 5.8, adding each element to a 2-byte grand total. On completion this total is divided by 24 to give the average function:

$$\frac{\sum_{i=0}^{23} \text{Temp}[i]}{24}$$

Based on this approach we have as a task list:
1. Clear Average.
2. Point to Temp[0] (i = 0).
3. DO
   (a) Add Temp[i] to the 2-byte grand total.
   (b) Increment i.
   (c) Repeat WHILE i < 24.
4. Divide by 24.

Program 5.10 directly implements the task list, summing each datum byte by adding to the double-byte location File h'48:49', which has been cleared before entry to the loop. Division is accomplished by repetitively subtracting 24 from the final total. This is similar to the ÷10 routine of Program 5.7 but this time the single-byte constant is taken off the double-byte dividend. The number of successful subtracts is the quotient, which in this case is the truncated Average. Of course it would be more accurate to round up if the remainder is more than half of the divisor.

Program 5.10 Average daily temperature.

```
INDF        equ   0             ; INDirect File register
STATUS      equ   3             ; Status register is File 3
FSR         equ   4             ; File Status Register
TEMP_0      equ   h'30'         ; Array starts @ File h'30'
SUM         equ   h'48'         ; Grand total to be in File h'48:49'
AVERAGE     equ   h'4A'         ; Average byte is to be here
Z           equ   2             ; Zero flag is bit2 of STATUS
C           equ   0             ; Carry flag is bit0

; Task1: Clear grand total and Average
AV_DAILY    clrf  SUM           ; MSbyte sum zeroed
            clrf  SUM+1         ; LSbyte sum zeroed

; Task2: Point to Temp[0]
            movlw TEMP_0        ; Put address of first temp byte
            movwf FSR           ; in the pointer register

; Task3:  DO
; Task3A: Add Temp[i] to the double-byte grand sum
LOOP1       movf  INDF,w        ; Get Temp[i]
            addwf SUM+1,f       ; Add LSB sum to it and put away
            btfsc STATUS,C      ; IF no carry, don't increment MSB
            incf  SUM,f         ; ELSE pass carry on

; Task3B: Increment i
NEXT        incf  FSR,f         ; i++

; Task3C: REPEAT WHILE i < 24
            movf  FSR,w         ; Get pointer address
            sublw TEMP_0+h'18'; Take away end address (Temp[24])
            btfss STATUS,Z      ; IF equal THEN end
            goto  LOOP1         ; ELSE repeat

; Task4: Divide by 24 to give the average
            clrf  AVERAGE       ; Zero the average
; Keep subtracting 24 and keep a count until a borrow-out
LOOP2       movlw d'24'         ; Put the constant 24 in W
            incf  AVERAGE,f     ; Record one subtract 24
            subwf SUM+1,f       ; Take away 24 from the sum LSB
            btfsc STATUS,C      ; IF borrow out, skip to high byte
            goto  LOOP2         ; ELSE do next subtract

            movlw 1             ; Subtract one from high byte
            subwf SUM,f
            btfsc STATUS,C      ; IF a borrow (C==0) THEN exit loop
            goto  LOOP2         ; ELSE do another subtract/count
            decf  AVERAGE,f     ; Compensate for one inc too many
            .....  ......       ; Next routine
```

Self-Assessment Questions

5.1 Can you deduce what function the following code fragment performs on the data byte in the Working register?

```
addwf   FILE,w
subwf   FILE,w
```

5.2 How could you simply, with one instruction, toggle bit 0 of any File register? It is permissible for other bits to be affected.

5.3 Write a program routine that will add two 16-bit numbers giving a 17-bit sum. The augend is located in the two memory locations File h'20':21' in the order high:low byte, thus | AUGEND_H | AUGEND_L | (Fh'20', Fh'21'). The addend is similarly situated | ADDEND_H | ADDEND_L | (Fh'22', Fh'23'). The sum is stored as three bytes in the order high:middle:low, thus | SUM_H | SUM_M | SUM_L | (Fh'24', Fh'25', Fh'26').

5.4 Write a program to subtract the double-byte datum which is located in File h'22:23', called NUM_2, from NUM_1 in File h'20:21'. The double-byte difference is to be in File h'24:25'. Remember, if there is a borrow from the lower byte subtraction then an additional one must be subtracted from NUM_1 in the upper byte subtraction. Assume that NUM_2 is smaller or equal to NUM_1. If this were not so, how could you determine this situation after the routine has been completed?

5.5 How could you extend Example 5.3 to give an outcome packed into a single-byte TENS:UNITS in File h'21'? This is known as packed BCD where each byte holds two decade nybbles rather than one digit per byte. *Hint*: Consider making use of the swapf instruction.

5.6 Develop Example 5.3 to give a 3-digit BCD outcome, removing the restriction that the original binary byte should be limited to decimal 99. The outcome is to be in File h'21:22:23' as HUNDS:TENS:UNITS respectively.

5.7 As part of a Data memory testing procedure, each File in the range File h'20' through File h'4F' is to be set to the pattern b'01010101' (h'55'). Using Program 5.2 as a model, write a suitable coding to implement this task.

5.8 Extend Example 5.1 to decrement a quad-precision 32-bit word located at File h'26:27:28:29', most significant byte first.

5.9 Data from an array of data memory between File h'30' and File h'4F' is to be transmitted byte-by-byte to a distant computer over the Internet. In order to allow the receiver to examine the data and check for transmission errors it is proposed to append a single byte which is the 2's complement (i.e., the negative value; see page 9) of the 8-bit sum of all the data bytes together. If all the received data bytes plus this **checksum** byte are similarly added then the outcome should be zero if no error has occurred. Code a routine to scan through this data, placing this checksum in File h'20'.

5.10 Based on the data logger specified Example 5.6, write a program to evaluate the maximum daily temperature. By the end of the routine this is to be in File h'48'.

5.11 Example 5.6 evaluated the average of an array of hourly temperature samples by summing all bytes and then subtracting 24 until the residue dropped below zero. Write an extension to this program to round the average to the nearest integer; that is, if the remainder is more than 12 then round up.

5.12 Write a routine to multiply a byte in File h'22' by 13. The 2-byte product is to be located at File h'23:24'. The memory map for this is

$$\boxed{\text{OVERFLOW}}_{F\,h'21'}\boxed{\text{MULTIPLICAND}}_{F\,h'22'} \times 13 = \boxed{\text{PROD\_H}}_{F\,h'23'}\boxed{\text{PROD\_L}}_{F\,h'24'}$$

where File h'21' is used to extend the single-byte multiplicand to a 16-bit double byte datum. Note that the solution will require three Shift and Add processes.

5.13 One simple way of encrypting a data byte is to reverse the order of bits. For example b'10111100' $\longrightarrow$ b'00111101'. Write a routine to implement this reversal on a data byte in File h'20'. The encrypted outcome is to be in the Working register. You can use location File h'21' as a temporary workspace and W as a loop counter. *Hint*: Use the Rotate Right and Rotate Left File instruction eight times.

5.14 A simple digital low-pass filter can be implemented using the algorithm:

$$\text{Array[i]} = \frac{S_n}{4} + \frac{S_{n-1}}{2} + \frac{S_{n-2}}{4}$$

where S_n is the nth sample from an eight-bit analog to digital converter located at Port B.

Write a routine assuming that the three byte memory locations to store S_n, S_{n-1} and S_{n-2} are located at File h'20:21:22' respectively. The outcome Array[i] is to be located at File h'48'.

5.15 Consider a 24-bit word stored in the Data store at the three locations

| 24 | File h'30' | 16 | 15 | File h'31' | 8 | 7 | File h'32' | 0 |

. Design a routine to count the number of bits set to 1 in this triple-byte datum.

5.16 A certain television show has eight contestants who are evenly divided into Team A and Team B. Each member has a switch, giving logic 1 when pressed, which may all be read simultaneously by the microcontroller at Port B. Team A switches appear on the lower four bits of the port.

Write a routine that will:

- Decide when a response to the question has been made, any switch closed.
- Determine the team identity that has responded by either clearing File h'20' for Team A, otherwise setting it to any non-zero value to signify Team B.
- Ascertain which team member pressed the switch by putting the member number 0 – 3 in File h'21'.

5.17 Parity is a simple technique to protect digital data from corruption by noise. Odd parity adds a single bit to a word in such a way as to ensure the overall packet has an odd number of 1s. Write a routine that takes an 8-bit byte stored at File h'20' and alters its most significant bit to comply with this specification. You can assume that bit 7 is always 0 before the routine begins. *Hint*: Determine if a binary number is odd or even by counting the number of bits as in Example 5.2 and then examining its least significant bit. All powers of two are even except $2^0 = 1$. Thus if this bit is 1 then the number is odd.

CHAPTER 6

Subroutines and Modules

Good software should be configured as a set of interacting modules rather than one large program working straight through from beginning to end. There are many advantages to modular programming, which is almost mandatory when code lengths exceed a few hundred lines or when a project is being developed by a team.

What form should should such modules take? In order to answer this question we will look at the use of program structures designed to facilitate this modular approach and the instructions associated with it.

After completing this chapter you will:

- Appreciate the need for modular programming.
- Have an understanding of the structure of the stack and its use in the call–return subroutine mechanism.
- Understand the term "nested subroutine".
- See how parameters can be passed to a subroutine and returned to the caller.
- Be able to write a subroutine having a minimal impact on its environment.
- Be able to synthesize a software stack to open and close a frame in the Data store to pass parameters and provide a temporary workspace.

Take a look at the inside of your personal computer. It will probably look something like the photograph in Fig. 6.1, with a motherboard hosting the MPU, assorted memory and other support circuitry, and a variable number of expansion sockets. Into this will be plugged a disk controller card and a video card. There may be others, such as a sound-board, modem or network card. Each of these plug-in cards has a distinct and separate logical task and they interact via the services supplied by the main board, the motherboard.

Advantages of this **modular** construction are:

- Flexibility; that is, it is relatively easy to upgrade or reconfigure by adding or replacing plug-in cards.
- Can reuse from previous systems.
- Can buy standard boards or design specialist boards in-house.
- Easy to maintain.

Of course there are a few disadvantages. A fully integrated motherboard is smaller and potentially cheaper than an equivalent mother/-daughterboard configuration. It is also likely to be more reliable, as input

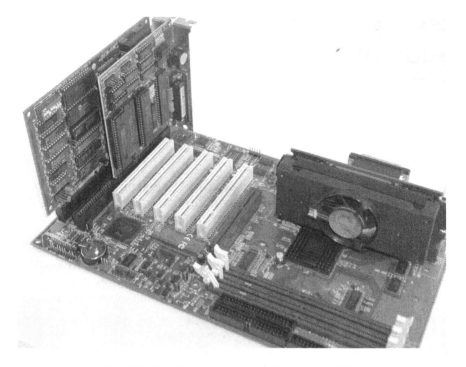

Fig. 6.1 Modular hardware implementing a PC.

and output signals do not have to traverse sockets/plugs. However, when they do occur, faults are often more difficult to track down and rectify.

Modular programming uses the same principle to construct "software circuits," i.e., programs. A formal definition of modular programming is:

> An approach to programming in which separate logical tasks are programmed separately and joined later.[1]

Thus, to write a program in a modular fashion we need to decompose the specification into a number of stand-alone routines, each implementing a well-defined task. Such a module should be relatively short, be well documented and be easy for a human, not necessarily the original programmer, to understand.

The advantages of a modular program are similar to those for modular hardware, but even more compelling:

- Modules can be tested, debugged and maintained on a stand-alone basis; this makes for overall reliability.
- Can be reused from previous projects or bought in from outside.
- Easier to update by changing modules.

---

[1]From *Chambers Science and Technology Dictionary*, Cambridge University Press, 1988.

Deciding how to segment a program into individual stand-alone tasks is where the real expertise lies. The actual coding of such tasks as subprograms is no different than the examples we have given in previous chapters, such as that shown in Program 5.9 on page 140. There are a few additional instructions associated with such subprograms, and these are listed in Table 6.1. We will look at these and some useful techniques in constructing software in the remainder of the chapter.

Table 6.1: Subroutine and interrupt handling instructions.

Operation	Mnemonic	Description
Call		Transfer to subroutine
Call subroutine	call aaa	Push PC on to stack, PC <- <aaa>
Return		Transfer back to caller
from subroutine	return	Pull original PC back from Stack
	retlw	Put literal in W and return as above
from interrupt	retfie	Return with the GIE flag in INTCON[7] set

Program modules may be entered by calling from other software or by a hardware event external to the CPU. This may be a voltage at one of the processor pins or an internal peripheral interface wanting service, such as a Timer module overflowing. In the former case modules at assembly level are universally known as **subroutines,** as they are in some high-level languages such as FORTRAN and BASIC.[2] In the latter they are classified as **interrupt service routines** or **interrupt handlers.** The techniques for writing these interrupt modules and their entry and exit techniques are sufficiently different to warrant a separate treatment in Chapter 7. Here we will look at subroutines.

Subroutines are the analog of hardware plug-in cards. Consider the situation where a 1 ms delay task is to be implemented. This may be needed to generate a 500 Hz tone to alert an aircraft pilot to look at the control panel warning lights for various scenarios, such as low fuel or overheating. In a modular program, this delay would be implemented by coding a 1 ms delay subroutine, which would be *called* by the main program as necessary to, say, continually force a port pin high and low for 1 ms durations. This is represented diagrammatically in Fig. 6.2.

In essence, calling up a subroutine involves nothing more than placing the address of the first subroutine instruction in the Program Counter (PC); that is, doing a goto. Thus, if our initial instruction was located at, say, h'400', then goto h'400' would seem to do the trick. Assum-

---

[2]Other high-level languages use the terms *function* (**C** and Pascal) or *procedure* (Pascal).

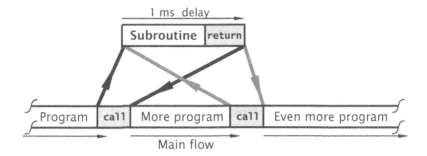

Fig. 6.2 Subroutine calling.

ing the programmer has labeled the subroutine entry point instruction
DELAY_1MS, as in Program 6.1, we have goto DELAY_1MS.

The problem really is how to get back again! Somehow the MCU has to
remember from where in the caller program the subroutine was entered,
so that it can return to the *next* instruction in the caller's sequence. This
can be seen in Fig. 6.2, where the jumping-off point can be from *anywhere*
in the main program, or indeed from another subroutine; see Fig. 6.4.

One possibility is to place this address in a designated Address register or memory location prior to jumping off. As the return mechanism,
this can then be moved back into the Program Counter at the end of the
subroutine. This approach breaks down whenever one subroutine wishes
to call another. Then the secondary subroutine will overwrite the return
address of the first, and the main program can never be regained. To
get around this problem, more than one register or memory location can
be used to hold a stack of return addresses. This **last-in first-out stack**
structure is shown in Fig. 6.3(a).

The 14-bit core PIC MCUs have a stack of eight 13-bit registers, which
are exclusively used to hold subroutine return addresses.[3] This structure,
shown in Fig. 6.3, is known as a **hardware stack**. This stack is outside
the PIC MCU's normal memory map, so its contents cannot be altered by
any normal process.[4]

Associated with this stack is a 3-bit counter which points to the next
available register in the stack. This **Stack Pointer (SP)** register cannot be
explicitly altered by any instruction but it is *automatically* incremented
each time a **call** instruction is executed. The **call** instruction is similar to a goto instruction, but before the specified instruction address is
put into the Program Counter its current value is pushed into the stack.

---

[3]The 12-bit low-range core devices have only two 11-bit stack registers and the
enhanced-range 16-bit core PIC MCUs have a 31-deep 20-bit stack.

[4]Most MPU/MCUs use an area of normal RAM together with a dedicated address register
to implement their stack. This is much more flexible than a dedicated hardware stack but
needs a more complex instruction set to manipulate the Stack pointer and to push and
pull/pop data into and out of the stack.

This is the address of the instruction *after* the call instruction, as the PC has already been incremented and the PIC MCU is fetching this next instruction into the Pipeline at the same time as the call instruction is being executed; see Fig. 4.4 on page 76.

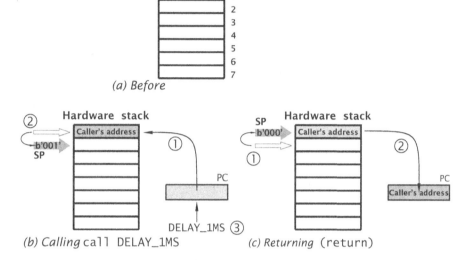

Fig. 6.3 Using the hardware stack to hold return addresses.

In Fig. 6.3(b) the situation is shown after a call to a subroutine labeled DELAY_1MS. The execution sequence of this call DELAY_1MS is:

1. Copy the 13-bit contents of the PC into the stack at the location pointed to by the Stack Pointer. Effectively this stored datum will be the address of the instruction following the call instruction.
2. The Stack Pointer is then incremented.
3. The destination address DELAY_1MS, that is the location of the entry point instruction of the subroutine, overwrites the original state of the PC. Effectively this causes the program execution to transfer to the subroutine.

Apart from the pushing of the return address into the stack in steps 1 and 2, call acts exactly like a plain goto. Thus call requires two instruction cycles for execution, as the Pipeline needs to be flushed to remove the caller's next sequential instruction, which is already lodged in the top of the Pipeline. This similarity also applies to the extension of its 11-bit absolute address in the call instruction word to a 13-bit Program store address using bits 3 and 4 of the PCLATH register, as shown in Fig. 5.17 on page 132. Far calls to subroutines located between h'07FF' and h'1FFF' are only required for 14-bit core PIC MCUs that have Program store capacities greater than 2048 instructions; for instance, the PIC16F877.

The exit point from the subroutine should be a **return** instruction. This reverses the push action of call and pulls the return address back from the stack into the PC, as shown in Fig. 6.3(c). The execution sequence of return is:

1. Decrement the Stack Pointer.
2. Copy the 13-bit address in the stack pointed to by the Stack Pointer into the Program Counter.

Thus no matter where the subroutine was called from, it will return to the instruction just past the original call instruction when the subroutine has been completed.

The **retlw** (RETurn Literal value in W)[5] is similar to the plain return instruction but places the specified constant byte in the Working register. Thus retlw -1 could be used to return with W set to h'FF' (−1 decimal) to, say, indicate an error situation. Both Return instructions flush the Pipeline and therefore take two bus cycles to execute.

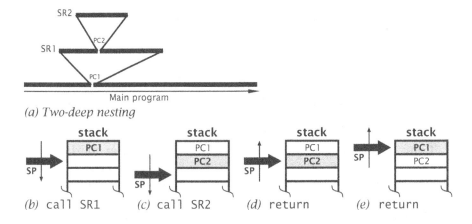

(a) Two-deep nesting

(b) call SR1   (c) call SR2   (d) return   (e) return

Fig. 6.4 Nested subroutines.

The beauty of the stack mechanism is its handling of **nested** subroutines. Consider the situation in Fig. 6.4 where the main program calls the first-level subroutine SR1 which in turn calls the second-level subroutine SR2. In order eventually to get back to the main program, the outward progression sequence must be exactly matched by the inward

---

[5]The 12-bit core PIC MCUs have only this retlw variant.

path. This pattern is matched by the **last-in first-out (LIFO)** structure of the stack mechanism, which can automatically handle any arbitrary nesting sequence up to the depth of the stack — i.e., eight for the mid-range family. It can even handle the (painful) situation where a subroutine calls itself! Such a subroutine is known as **recursive**. As we shall see in Chapter 7, the stack mechanism is also used to handle interrupts. Thus, in a system using both subroutines and interrupts the nesting depth will be somewhat less. The technique is so useful that virtually all MPU/MCUs support subroutines in this manner.

As the stack-Stack Pointer mechanism is part of the PIC MCU's hardware and requires no initialization, from the programmer's perspective only the following points are relevant:

- The subroutine should be invoked using the `call` instruction.
- The entry point to a subroutine should be labeled, and this label is then the name of that subroutine.
- The exit point from the subroutine should be either `return` or `retlw`, with the latter being used when a known constant is to be in the Working register on return; see Program 6.6.

As an example, let us code the 1 ms delay subroutine of Fig. 6.2. Creating a delay in software is simply a matter of doing nothing for the appropriate duration. A common way of doing this is to count down an initial constant to zero, as shown in Fig. 6.5. By choosing an appropriate constant, the delay can be tailored to the desired value. Obviously, this delay will depend on the PIC MCU's oscillator rate. For the examples in this chapter we will assume a clock rate of 4 MHz, giving a bus cycle of 1 μs; see also Program 12.8 on page 360.

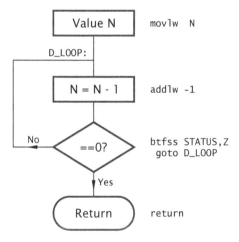

Fig. 6.5 Delaying by counting N times.

Consider the subroutine shown in Fig. 6.1. Here a constant N is placed in the Working register, and this value is decremented down to zero inside a 3-instruction loop. The subroutine then exits using a `return` instruction.

Program 6.1 A 1 ms delay subroutine.

```
; ***********************************************************************
; * FUNCTION: Delays for 1ms with a 4MHz crystal                       *
; * ENTRY    : None                                                    *
; * EXIT     : Flags and W altered                                     *
; ***********************************************************************
N           equ    d'249'    ; Delay parameter computed in the text

DELAY_1MS movlw  N           ; Set up loop count in W,              1˜
; LOOP -----------------------------------------------------------------
D_LOOP      addlw  -1         ; Decrement count                      N˜
            btfss  STATUS,Z  ; Check: Is it zero?                 (N+1)˜
            goto   D_LOOP    ; IF not THEN repeat,             2*(N-1)˜
; ---------------------------------------------------------------------
            return
```

In order to calculate the total number of instruction cycles the program takes, and thus determine a value for N, we need to evaluate an execution cycle budget.

1. The `call DELAY_1MS` instruction used by the caller to jump to the subroutine takes two cycles ($2˜$) to execute.
2. The `movlw` instruction preceding entry into the loop takes one cycle.
3. The `addlw` instruction decrementing the contents of W takes in total N cycles (N times round the loop).
4. The `btfss STATUS,Z` instruction testing the **Z** flag (is W zero after the last decrement?) is also executed N times. However, the very last time the loop exits by skipping out, there is an additional cycle while the Pipeline is being flushed. Thus the total delay is $N + 1$.
5. As the loop exits by skipping over the `goto` instruction, this is only executed $N - 1$ times; each time taking two cycles. This contribution is thus $(N - 1) \times 2$.
6. The final `return` takes two cycles.

The total number of cycles is then:

2 (`call`) $+ 1$ (`movlw`) $+ N$ (`addlw`) $+ (N + 1)$ (`btfss`)
$+ 2 \times (N - 1)$ (`goto`) $+2$ (`return`)

Equating this to 1000 cycles gives:

$$2 + 1 + N + (N+1) + 2 \times (N-1) + 2 = 1000$$
$$4 + (4 \times N) = 1000$$
$$4 \times N = 996$$
$$N = 249$$

Our delay subroutine is pretty limited in that the Working register, like all PIC MCU data registers, is only eight bits wide and thus the maximum value of N is b'11111111' or decimal 255. Actually, in the case of our subroutine in Program 6.1, a value $N = 0$ gives the longest delay! This is because W is decremented *before* being tested for zero. So the sequence would actually go h'00 → FF → FE → ⋯ → 01 → 00'. Thus effectively N acts as if it were d'256', giving a maximum delay of $4 + (4 \times 256) = 1028$ cycles, or 1.028 ms with a 4 MHz crystal.

Slightly longer delays are possible by inserting nop (No OPeration) instructions inside the loop. Each nop adds one instruction cycle but does not affect the Status flags. Thus, putting four nop instructions after the addlw -1 instruction, as shown in Program 6.2, gives a total loop delay of $4 + 8 \times N$ cycles. A value of $N = 249$ will now give $4 + 1992 = 1996$ cycles, or approximately 2 ms with a 1 μs cycle time. How could you use additional nops to increase the number of delay cycles to exactly 2000?

Program 6.2 A 2 ms delay subroutine.

```
; ********************************************************************
; * FUNCTION: Delays for nominally 2ms with a 4MHz crystal          *
; * ENTRY    : None                                                 *
; * EXIT     : Flags and W altered                                  *
; ********************************************************************
N          equ    d'249'    ; Delay parameter computed in the text

DELAY_2MS  movlw  N         ; Set up loop count in W,              1~
; LOOP -------------------------------------------------------------
D_LOOP     addlw  -1        ; Decrement count                      N~
           nop              ; Put in four extra cycles             N~
           nop              ; with No OPeration                    N~
           nop              ; instructions                         N~
           nop              ;                                      N~
           btfss  STATUS,Z  ; Check: Is it zero?               (N+1)~
           goto   D_LOOP    ; IF not THEN repeat,             2*(N-1)~
; -----------------------------------------------------------------
           return
```

Adding nops in this manner can be used to design delay routines that can cope with different clock frequencies. For example, Program 6.2 will give a 1 ms delay with an 8 MHz crystal. Thus adding an appropriate number of nops will allow the programmer to "tweak" our subroutine to cope with crystals between 4 and 20 MHz; see also Program 12.8 on page 360. How many nops would you need to give a 1 ms delay with a 20 MHz crystal?

This technique isn't much use if you need a substantially longer delay. This can be achieved by using a File as a second decrementing counter, effectively encapsulating a kernel comprising our 1 ms delay loop, as shown shaded in Fig. 6.6. If we execute this kernel 100 times, then we will have a 100 ms delay.

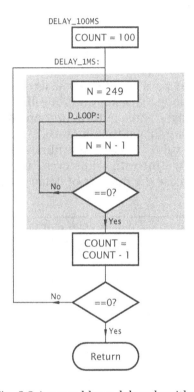

Fig. 6.6 A nested loop delay algorithm.

The coding of our 100 ms delay subroutine is shown in Program 6.3. The File, named COUNT1, is initialized to d'100' on entry and thereafter the inner 1 ms loop is executed. When W reaches zero and the inner loop exits, the File count is decremented in situ using the decfsz COUNT1, f instruction. The outer loop only exits when the count reaches zero, that is, after 100 inner loops. As long as this outer count remains non-zero, the inner 1 ms delay is re-executed.

Our simplified treatment of the timing of Program 6.3 is of course not completely accurate, as we have ignored the time taken by the instructions in the outer loop, such as decfsz. However, to compensate

Program 6.3 A 100 ms delay subroutine.

```
;  *****************************************************************
;  * FUNCTION: Delays for nominally 100ms with a 4MHz crystal    *
;  * ENTRY   : None                                              *
;  * EXIT    : Flags and W altered. File h'30' zero              *
;  *****************************************************************
COUNT1        equ    h'30'     ; Use File h'30' as a loop counter
N             equ    d'249'    ; Delay parameter computed in the text

DELAY_100MS movlw  d'100'      ; Initialize outer loop count to 100
            movwf  COUNT1
; Outer loop ----------------------------------------------------
DELAY_1MS     movlw  N         ; Set up 1ms count in W
; Inner loop ----------------------------------------------------
D_LOOP        addlw  -1        ; Decrement 1ms count
              btfss  STATUS,Z  ; Check: Is it zero?
              goto   D_LOOP    ; IF not THEN repeat
; --------------------------------------------------------------
              decfsz COUNT1,f  ; Decrement outer loop count
              goto   DELAY_1MS ; and repeat until zero
; --------------------------------------------------------------
              return
```

somewhat, the number of cycles in the inner loop has dropped four cycles to $4 \times N$ giving an overall time reduction of 100×4 cycles, as the entry call and exit return instruction now belongs to the outer loop. The actual delay, as calculated in Example 6.2, turns out to be 99,905 cycles, or with a 4 MHz crystal, 99.905 ms; just 95 μs short, or accurate to better than 0.1%. Adding a single nop in the outer loop will give a delay of 100.005 ms, or 5 μs too long in 100,000 μs.

The maximum delay possible with this program is 256,000 cycles, which could give us our 100 ms delay even up to a 10 MHz crystal, or up to 256 ms delay with a 4 MHz crystal. For even longer delays, we could use a triple-loop structure, potentially giving more than a minute delay. For instance, see Example 6.3.

Our 100 ms delay program is an example of a double-void subroutine, in that no parameters (cf. signals in our hardware plug-in card analog) are sent to it and nothing is returned—just the side effect of a delay (and the alteration of a File, W and some Status register flags). Most subroutines process parameters made available at entry time and provide data at return time.

As a simple example, consider the extension of Program 6.3 to give a delay of $K \times 100$ ms, where K is a byte parameter "sent" by the caller. The system view of this function is shown in Fig. 6.7 as a single input signal of range 1 - 256, with no output signal; that is, with a void output. This diagram also documents the location of all **local variables** used internally by the subroutine. This latter attribute is useful in checking for multiple usage of a File register between different subroutines and callers. Notice the double line vertical borders commonly used in flow diagrams to denote modules or subroutines.

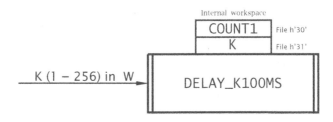

Fig. 6.7 System view of $K \times 100$ ms delay subroutine.

As there is only one input byte-sized parameter, the most convenient place to place K in the calling program is in the Working register. Thus to call up a 5 s delay, the caller could use the sequence:

```
movlw  50           ; 50 x 0.1s gives 5 seconds
call   DELAY_K100MS ; Go to it!
```

The subroutine itself in Program 6.4 implements the task list:

1. DO:
   (a) Delay 100 ms.
   (b) Decrement K.
2. WHILE ($K > 0$).
3. End.

The actual coding simply copies the parameter from W into File h'31' before entering the following delineated coding, which is identical to Program 6.3 and gives a single 100 ms delay. On completion of this fixed delay, K is decremented in situ and the delay block repeated until K reaches zero. Thus the 100 ms block is repeated K times.

As K is tested for zero *after* the 100 ms delay is executed[6] an initial value of $K = 0$ will be treated as $K = 256$, giving a delay range of 0.1 – 25.6 s. Testing *before* the loop[7] would give a range 0 – 25.5 s. Again the actual time calculation is approximate, as we have ignored instructions in outer loops.

As W is needed to set up COUNT1 and time the inner 1 ms loop, it can not be used directly to hold K during the subroutine. In fact, if the caller had known that File h'31' was used by the subroutine to hold K then it could have been passed directly through this File. However, the less the

---

[6]Known to **C** programmers as a DO–WHILE loop.
[7]Known to **C** programmers as a WHILE loop.

caller has to know about the "innards" of its subroutines the better it will be, on the basis that a subroutine should disturb its environment as little as possible. DELAY_K100MS is not very good in this respect, using two Files for its internal use and altering the Working register.

As an example of what could go wrong, Program 6.5 shows an implementation of the task list but calling the 100 ms block as the existing Program 6.3 subroutine; that is, a nested subroutine. Here File h'30' is used as a store for *K* oblivious to the fact that this File is used by subroutine DELAY_100MS as a loop counter. The effect of this interaction is to make *K* zero on return from DELAY_100MS, which when decremented will always give a non-zero outcome. Thus the delay is infinite and the system locks up! Simply changing K equ h'30' to K equ h'31' fixes the problem; but if another member of the team with responsibility for the DELAY_100MS subroutine alters its internal storage map without communicating this to other team members, then catastrophe may occur! Thus even though each subroutine could have been passed when tested on its

---

Program 6.4 A $K \times 100$ ms delay subroutine.

```
;   ****************************************************************
;   * FUNCTION: Delays for around K x 100 ms @ 4MHz               *
;   * EXAMPLE : K = 100, delays 10 seconds                        *
;   * ENTRY   : K in W, range 1 - 256                             *
;   * EXIT    : Flags and W altered; Files 30:1:2h zero           *
;   ****************************************************************
COUNT1 equ    h'30'          ; 100 ms loop counter
K      equ    h'31'          ; Temporary storage for K
N      equ    d'249'         ; Delay parameter

DELAY_K100MS
       movwf  K              ; Put K away in a File

; DO 100ms delay-------------------------------------------------
DELAY_100MS movlw  d'100'    ; Initialize loop count to 100
            movwf  COUNT1

DELAY_1MS     movlw  N        ; Set up 1 ms count in W

D_LOOP        addlw  -1       ; Decrement 1 ms loop count
              btfss  STATUS,Z ; Check: Is it zero?
              goto   D_LOOP   ; IF not THEN repeat

              decfsz COUNT1,f ; Decrement 100 loop count
              goto   DELAY_1MS ; and repeat until zero

; Decrement K---------------------------------------------------

       decfsz K,f

; WHILE K > 0---------------------------------------------------
       goto   DELAY_100MS    ; REPEAT WHILE K > 0

FINI   return
```

Program 6.5 An alternative $K \times 100$ ms delay subroutine.

```
; ***************************************************************
; * FUNCTION: Delays for around K x 100 ms @ 4MHz              *
; * EXAMPLE : K = 100, delays 10 seconds                       *
; * RESOURCE: DELAY_100MS called                               *
; * ENTRY   : K in W, range 1 - 256                            *
; * EXIT    : Flags and W altered; Files h'30:1' zero          *
; ***************************************************************
K          equ   h'30'      ; Temporary storage for K

DELAY_K100MS
           movwf K          ; Put K away in a File

; Task 1: DO 100 ms delay-------------------------------------
DK_LOOP  call   DELAY_100MS

; Task 2: Decrement K-----------------------------------------

           decfsz K,f       ; Decrement K

; Task 3: WHILE K > 0-----------------------------------------
           goto  DK_LOOP    ; REPEAT WHILE K > 0

           return
```

own, certain combinations of calling sequences could cause failure. We will return to this problem later.

Program 6.4 is still void, in that no data was returned to the caller on exit. For our next example we will code a non-void subroutine that will activate a decimal readout. Many numeric electronic displays are based on a selective activation of seven segments in the manner shown in Fig. 6.8. These segments are typically implemented using light-emitting diodes (see Fig. 11.15 on page 325) or electrodes in a liquid-crystal cell.

The system description of our subroutine is shown in Fig. 6.8(a). Here the input signal is a 4-bit binary code representing the ten decimal digits as b'0000 – 1001' in the Working register. The output, also in W, is the corresponding 7-segment code to activate the digit as listed in Table 6.2. This code assumes that a segment is lit/opaque on a binary 1 and is unlit/clear on a binary 0. Depending on the physical connections used, the opposite polarity is possible.

Most MPU/MCUs deal with **look-up tables** by storing the codes as part of the program memory and copying the Nth byte out of the table as the mapping function. In the 12- and 14-bit core PIC families the Harvard structure makes code in the Program store inaccessible as data, but see Program 15.5 on page 493 for an exception. Instead, look-up tables are implemented as a series of **retlw** instructions, each returning a constant

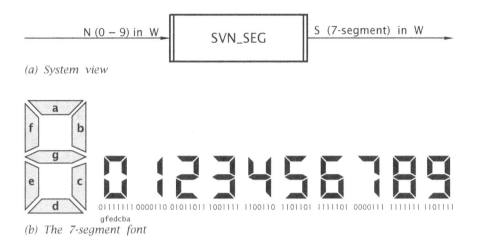

(a) System view

(b) The 7-segment font

Fig. 6.8 The 7-segment display.

byte. This structure is shown in Table 6.2. As each `retlw` places an 8-bit code in W, I have arbitrarily made the unused bit 7 a logic 0.

In developing a coding based on this table structure, the mechanism for element k extraction is to execute the Nth `retlw` instruction. This will place the instruction literal in the Working register and then do a normal return from subroutine back to the caller. In the example shown, if k is six, then the sixth `retlw` is executed, returning with the code b'01111000' for **6** in W.

Table 6.2: The 7-segment look-up table showing byte[k] being extracted.

	PC	Table[n]	Display
	+0	`retlw b'00111111'`	; **0**
	+1	`retlw b'00000110'`	; **1**
	+2	`retlw b'01011011'`	; **2**
	+3	`retlw b'01001111'`	; **3**
	+4	`retlw b'01100110'`	; **4**
	+5	`retlw b'01101101'`	; **5**
$N \Rightarrow$	+6	`retlw b'01111101'`	; **6** = Table[6]
	+7	`retlw b'00000111'`	; **7**
	+8	`retlw b'01111111'`	; **8**
	+9	`retlw b'01101111'`	; **9**

Program 6.6 The software 7-segment decoder.

```
; *****************************************************************
; * FUNCTION: Returns byte[N] in table                          *
; * FUNCTION: where N is the contents of W                      *
; * EXAMPLE : IF W = 06 THEN returns code b'01111101'           *
; * ENTRY   : N range 00 - 09 in W                              *
; * EXIT    : Table entry N in W                                *
; *****************************************************************

PCL      equ   2            ; Low byte of PC is at File 2

SVN_SEG addwf PCL,f         ; Add W to PCL, giving PC + N
;               xgfedcba
        retlw b'00111111'   ; Code for 0; Returned if N = 0
        retlw b'00000110'   ; Code for 1; Returned if N = 1
        retlw b'01011011'   ; Code for 2; Returned if N = 2
        retlw b'01001111'   ; Code for 3; Returned if N = 3
        retlw b'01100110'   ; Code for 4; Returned if N = 4
        retlw b'01101101'   ; Code for 5; Returned if N = 5
        retlw b'01111101'   ; Code for 6; Returned if N = 6
        retlw b'00000111'   ; Code for 7; Returned if N = 7
        retlw b'01111111'   ; Code for 8; Returned if N = 8
        retlw b'01101111'   ; Code for 9; Returned if N = 9
```

The coding shown in Program 6.6 implements this selection mechanism by simply adding N, which is in W, to the lower byte of the Program Counter; that is, PCL in File 2. As the PC is already pointing to the first retlw instruction after the addition it then points to the Nth retlw; as desired.

The code in Program 6.6 takes no account of the possibility that the datum in W is greater than h'09'. Of course it shouldn't be, but robust code should cope with all contingencies even if it is technically erroneous. This is especially true if the code module is to be reusable for general-purpose applications. What would happen if this situation arose and how could you add to the code to gracefully return an error code, say −1, in this eventuality?

This approach of adding a byte number in W to the *low* byte of the Program Counter (PCL) to select one of N Return instructions is deceptively simple. Although it works in most situations where the table is small, for the unwary programmer, it can cause system crashes in seemingly unpredictable situations.

The problem arises, as altering PCL with the instruction addwf PCL,w only alters the lower eight bits of the 13-bit Program Counter. If the addition should cause overflow, then the net effect is to effectively move the Program Counter proper backwards! For instance, if the subroutine of Program 6.6 happened to be located at h'1F8' (that is, the label SVN_SEG was h'1F8') and if the contents of W happened to be h'08', then the outcome of the instruction `addwf PCL,f` would be to leave the Program Counter at $h'(1)F8 + 08 = (1)00'$ rather than $h'200'$. The instruction located at $h'100'$ (if any) is unlikely to be a Return instruction and so we have exited a subroutine illegally and left the state of the Stack unbalanced. The exact position of a subroutine in the Program store is not easy to predict, as the programmer is unlikely to know in advance where the subroutine is located in memory, that is, what value the PC will have at the beginning of the subroutine. Even if he/she checks the assembler listing file (see Table 8.2 on page 222) for the value of SVN_SEG, this can change if subsequent alterations are made to other parts of the program. It is possible to devise code to allow this address boundary to be crossed, but at the expense of complexity; see Example 6.7.

Storing data using a series of `retlw` instructions is rather inefficient in that a 14-bit instruction is being used to store an 8-bit datum. The PIC16F87X group have implemented a technique of being able to read a 14-bit datum directly from the Program store in a rather convoluted way; see Program 15.5 on page 493. The enhanced-range family have instructions, such as `tblrd`, that allow access to individual bytes within each 16-bit Program store word; see Table 16.1 on page 523.

Using W to transfer information to and from a subroutine is limited to a single byte datum each way. Where several pieces of information of byte or greater sizes are to be passed, then Files must be pressed into service for this conduit. An example of this is shown in Program 6.7 where two byte datums, labeled MULTIPLICAND and MULTIPLIER, are to be multiplied giving a 16-bit 2-byte outcome labeled PRODUCT_L:PRODUCT_H.

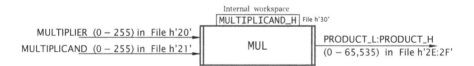

Fig. 6.9 System diagram for the byte multiplication subroutine.

The principle of the multiplication algorithm coded in Program 6.7 is a generalized version of that used by previous multiplication routines, such as Program 5.9 on page 140. There the multiplier nine was decomposed to ×1 + ×8. Similarly multiplying by, say, ten could be implemented by shifting left once (×2) and three times (×8) and adding. In the more general case the multiplicand is continually shifted left and the nth shifted word added to the product if bit n of the multiplier is 1. Doing this eight times, where the << operator denotes shift left, gives:

$$\text{Product} = \sum_{n=0}^{7} (\text{multiplicand} << n) \times \text{bit } n$$

Program 6.7 uses the **shift and add** algorithm outlined in the task list:

1. Zero double-byte product.
2. Extend Multiplicand to 16 bits.
3. DO
    (a) Shift Multiplier right once.
    (b) IF Carry bit is one THEN add 2-byte shifted Multiplicand to 2-byte sub-product.
    (c) Shift Multiplicand right once.
    (d) Repeat WHILE shifted Multiplier is not zero.
4. End with 16-bit Product.

Program 6.7 declares the variables that are passed to and from the subroutine at the beginning of the main program. Keeping all these **global** declarations in one part of the program and using a different File for each overall global variable reduces the possibility of interaction but at the expense of rather extravagant use of scarce Data memory resources. Temporary local storage is declared within each subroutine, as its need will be "thrown away" after the subroutine is terminated. However, interaction can still occur in local storage where nested subroutine structures are used.

The coding follows the task list closely. The decision whether to add the left-shifted 2-byte Multiplicand to the subproduct is dependent on the state of the Carry flag when the Multiplier is shifted right. This implements the conditional addition

$$\text{product} = \text{product} + (\text{multiplicand} << n) \times \text{bit } n$$

Rather than implementing this shift and conditional add process eight times, the summation loop is terminated whenever the Multiplier residue is zero. This means that the execution time of the subroutine is variable, depending on the bit pattern of the multiplier. The worst-case scenario is when the Multiplier is 255 (b'11111111'). This takes 142 cycles, including the 2-cycle call.[8]

In order to use this subroutine, the caller copies the Multiplicand into File h'20' and Multiplier into File h'21'. On return, the 16-bit product can be read at File h'2E:2F'. As an example, consider that the bytes located at File h'42' and File h'46' are to be multiplied.

```
movf  h'42',w   ; Get Number 1
movwf h'20'     ; and copy into MULTIPLIER
movf  h'46',w   ; Get Number 2
movwf h'21'     ; and copy into MULTIPLICAND
call  MULT      ; Go to it!
                ; On return the product is now in File h'2E:2F'
```

---

[8]The enhanced family has instructions mulwf and mullw which do a 8-bit × 8-bit multiplication in a single cycle!

Program 6.7 The byte multiplication subroutine.

```
; Global declarations
STATUS        equ   3          ; Status register is File 3
C             equ   0          ; Carry flag is bit0
Z             equ   2          ; and the Zero flag is bit2
MULTIPLIER    equ   h'20'      ; Multiplier byte
MULTIPLICAND  equ   h'21'      ; Multiplicand byte
PRODUCT_L     equ   h'2E'      ; Low byte of the product
PRODUCT_H     equ   h'2F'      ; High byte of the product

; The MULT subroutine
; ****************************************************************
; * FUNCTION: Multiplies two bytes to give a 2-byte product     *
; * EXAMPLE : MULTIPLICAND = h'10', MULTIPLIER = h'FF'          *
; * EXAMPLE : PRODUCT_H:PRODUCT_L = h'0FF0' (d'16 x 255 = 4080' *
; * ENTRY   : MULTIPLIER = File h'20', MULTIPLICAND = File h'21'*
; * EXIT    : PRODUCT_H = File h'2E', PRODUCT_L = h'2F'         *
; * EXIT    : MULTIPLIER, MULTIPLICAND altered                  *
; * EXIT    : W, Status and MULTIPLICAND_H = File h'30' altered *
; ****************************************************************
; Local declarations
MULTIPLICAND_H equ h'30'       ; Extension byte for Multiplicand

; Task 1: Zero double-byte Product
MUL       clrf   PRODUCT_L
          clrf   PRODUCT_H

; Task 2: Extend Multiplicand to 16 bits
          clrf   MULTIPLICAND_H

; Task 3: DO
    ; Task 3A: Shift Multiplier right once
MUL_LOOP bcf    STATUS,C       ; Clear Carry flag
         rrf    MULTIPLIER,f

     ; Task 3B: IF Carry == 1 THEN add Multiplicand to Product
         btfss  STATUS,C       ; IF C == 1 THEN do addition
         goto   MUL_CONT       ; ELSE skip this task

         movf   MULTIPLICAND,w  ; DO addition
         addwf  PRODUCT_L,f     ; First the low bytes
         btfsc  STATUS,C        ; IF no carry THEN do high bytes
          incf  PRODUCT_H,f     ; ELSE add Carry state
         movf   MULTIPLICAND_H,w; Next the high bytes
         addwf  PRODUCT_H,f

     ; Task 3C: Shift Multiplicand left once (x2)
MUL_CONT bcf    STATUS,C       ; Zero Carry-in
         rlf    MULTIPLICAND,f
         rlf    MULTIPLICAND_H,f

     ; WHILE shifted Multiplier not zero
         movf MULTIPLIER,f     ; Test Multiplier for zero
         btfss  STATUS,Z
          goto  MUL_LOOP       ; IF not THEN go again
         return                ; ELSE finished
```

Most MPU/MCUs have software stacks, which in addition to saving subroutine return addresses allow the programmer to push and pull data to and from memory to pass information between caller and subroutine. As the stack is a dynamic storage entity, growing where necessary to accommodate these passed and temporary variables and shrinking again when the subroutine terminates, this clearly is an efficient method of memory allocation. Furthermore, each call outwards in a nested structure opens a new stack frame for this dynamic storage as an extension to the stack. In this way the possibility of overlap between variable storage when using nested subroutines is virtually eliminated.

High-level languages, such as C (see Chapter 9), are normally based on this stack model. This allows the creation and passing of variables, only restricted by the amount of data memory that can be allocated to this stack.

The downside to this approach is the extra CPU resources necessary to support the creation and maintenance of the stack. One or more dedicated address registers or stack pointers are normally provided and address modes that facilitate access to variables in these stack frames are needed for efficient working. Even then, the outcome is normally slower and coding is longer than models based on fixed memory allocations.

The mid-range PIC MCU core does not explicitly support a software stack.[9] However, it is possible to simulate such a structure using Indirect addressing with the File Select register and the Indirect File; see page 104. As there is no stack pointer register *per se*, in the code fragment below the main routine has allocated File h'40' as a Pseudo Stack Pointer, which we call PSP.

The programmer also has to set aside a block of Data memory to hold the various stack frames. Here we are going to specify that the Top Of Stack (TOS) address is to be at File h'50'. If the range File h'50 – 7F' is kept clear of absolute allocations, then a total of 48 bytes are available for the stack. In the PIC16F62X devices this memory space is common across all banks. As the hardware stack holds the subroutine return address, all locations in this simulated software stack is available for variable passing and local storage inside subroutines. The software stack is initialized by moving the literal h'50', named TOS, into the File holding the Pseudo Stack Pointer.

As an example, consider a stack-oriented version of the multiplication subroutine of Program 6.7. A view of the software stack from the perspective of this new coding is shown in Fig. 6.10. Based on this dia-

---

[9]The enhanced-range family does go some way towards this model, using an enlarged 31-deep hardware stack with accessible Stack pointer.

gram, in order to call up this subroutine the following procedure has to be implemented:

1. Push the Multiplicand and then Multiplier into the stack frame and then call the subroutine.
2. Push zero into the next byte in the frame, which is being used for local storage as an extension byte for the Multiplicand.
3. Push zero out twice more to create an initialized hole for the two bytes making up the Product.

The code fragment in the following page shows how item 1 above is coded.

(a) Transfer the contents of the Pseudo Stack Register to the FSR. This means that the FSR now points to the top of the new stack frame. If the subroutine is a first-level call (that is, not nested from another subroutine) then this will be h'50' in our example.

(b) Copy the Multiplicand from memory (we assume it is at File h'46' as in our last example) into W and then indirectly into the frame using the FSR as the pointer. Decrementing the FSR pointer completes the push action.

(c) In a similar manner, the Multiplicand is pushed into the stack.

(d) Call the subroutine.

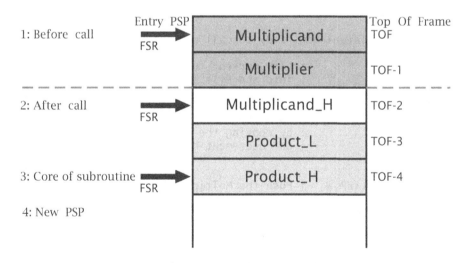

Fig. 6.10 The stack frame viewed from the perspective of subroutine MUL_S.

Coding of the subroutine MUL_S is given in Program 6.8. This implements items 2–4 of Fig. 6.10. Initially MULTIPLICAND_H, the temporary storage of the Multiplicand overflow, is zeroed and then zero is pushed into the next two locations to create the initial value for the Product. In item 4 the Pseudo Stack Pointer is reset to point to the next free byte below the frame. In this way, should the subroutine wish to call another, then there will be a new frame available for that next-level storage, with the new TOF beginning just below the old frame. These two instructions may be omitted in this case as there are no further nested call-outs, although Task 3C will have to be altered. This is done in Example 6.6.

```
; Global declarations

PSP            equ  h'40'  ; Holds the Pseudo Stack Pointer
TOS            equ  h'50'  ; File h'50' is initial Top Of Stack
INDF           equ  00     ; INDirect File
FSR            equ  04     ; File Select Register
STATUS         equ  03     ; Status register is File 3
C              equ  0      ; in which Carry flag is bit0
Z              equ  2      ; and the Zero flag is bit2
MULTIPLIER     equ  h'46'  ; Multiplier byte
MULTIPLICAND   equ  h'42'  ; Multiplicand byte

MAIN
; In the beginning set up Top Of Stack
 movlw   TOS
 movwf   PSP              ; which is stored in File h'40'
;
;
; Some time later when ready to call subroutine
; (a)
movf  PSP,w               ; Copy current top of stack frame address
movwf FSR                 ; into the File Select register

; (b)
movf  MULTIPLICAND,w ; Push the Multiplicand into the stack
movwf INDF                ; by copying the datum out
decf  FSR,f               ; and decrementing the FSR

; (c)
movf  MULTIPLIER,w        ; Push the Multiplier into the stack
movwf INDF                ; by copying the datum out
decf  FSR,f               ; and decrementing the FSR

; (d)
call  MUL_S               ; Call the subroutine
```

The core of the subroutine, that is, Task 3, is similar to Program 6.7 but the FSR has to be moved up and down the frame to access the appropriate level. The only non-obvious use of the FSR is at Task 3C. As there are two ways into this routine, depending on whether the shifted multiplicand is added to the product or not, the state of the FSR is unknown. It can, however, be reset from the PSP which is pointing to just below the frame at this point. By adding five to this PSR value, the FSR will always point to MULTIPLICAND.

Finally the subroutine "cleans up" the stack by updating the Pseudo Stack Pointer to its previous value. Here this is done by adding five, but in general by adding the frame depth n.

Program 6.8 requires 45 instructions as compared to 20 in Program 6.7. Its worst-case execution time of 274 cycles also compares unfavorably with 142 cycles. Thus in all respects, except reusability and robustness, this stack-based model is clearly inferior. It may be more economical in its use of scarce data memory in large software systems. However, programs running on low- and mid-range PIC MCUs are often not very complex. Furthermore, the small Program memory may further restrict the use of this relatively extravagant technique. Where real-time execution time is critical the additional burden of stack handling is unlikely to be worthwhile.

Examples

Example 6.1
Write a subroutine to give a fixed 208 μs delay. Assume a 4 MHz processor clock rate.

Solution
For a short time period like this, the code outlined in Program 6.1 provides adequate delay.

With a 4 MHz clock, one cycle is 1 μs and thus we require 208 cycles. From page 154 we have:

$$
\begin{aligned}
4 + 4 \times N &= 208 \text{ cycles} \\
4 \times N &= 204 \\
N &= 51
\end{aligned}
$$

What value of N would you use if a 20 MHz crystal were used?

Example 6.2
In Program 6.3 we illustrated a nominal 100 ms delay subroutine. Our calculation of this delay was rather simplified in that we took the core 1 ms kernal and multiplied this by the outer loop count of 100. From first principles, calculate the exact delay based on a 4 MHz crystal and deduce the percentage error.

Solution

Repeating the listing below, we can calculate the number of cycles based on each instruction and its position in the loop structure.

Program 6.8 Implementing a byte multiply using a stack model.

```
; ****************************************************************
; * FUNCTION: Multiplies two bytes to give a 2-byte product     *
; * EXAMPLE : MULTIPLICAND = h'10', MULTIPLIER = h'FF'          *
; * EXAMPLE : PRODUCT_H:PRODUCT_L = h'0FF0' (d'16 x 255 = 4080')*
; * ENTRY   : MULTIPLICAND = PSP, MULTIPLIER = PSP-1            *
; * ENTRY   : FSR points to one below MULTIPLIER                *
; * EXIT    : PRODUCT_H = PSP-3, PRODUCT_L = PSP-4              *
; * EXIT    : W, Status altered                                *
; ****************************************************************
;
; Tasks 1 & 2: Extend multiplicand and zero double-byte product
MUL_S     clrf    INDF
          decf    FSR,f       ; FSR ---> PRODUCT_L
          clrf    INDF
          decf    FSR,f       ; FSR ---> PRODUCT_H
          clrf    INDF
          decf    FSR,w       ; Now reset Pseudo Stack Pointer
          movwf   PSP         ; to Bottom Of Frame

; Task 3: DO
   ; Task 3A: Shift multiplier right once
          incf    FSR,f
          incf    FSR,f
          incf    FSR,f       ; FSR ---> MULTIPLIER
MUL_LOOP  bcf     STATUS,C    ; Clear Carry flag
          rrf     INDF,f

   ; Task 3B: IF Carry == 1 THEN add multiplicand to product
          btfss   STATUS,C    ; IF C == 1 THEN do addition
          goto    MUL_CONT    ; ELSE skip this task

          incf    FSR,f       ; FSR ---> MULTIPLICAND
          movf    INDF,w      ; DO addition
          decf    FSR,f
          decf    FSR,f
          decf    FSR,f       ; FSR ---> PRODUCT_L
          addwf   INDF,f      ; First the low bytes
          decf    FSR,f       ; FSR ---> PRODUCT_H
          btfsc   STATUS,C    ; IF no carry-out THEN do high bytes
          incf    INDF,f      ; ELSE add Carry state
          incf    FSR,f
          incf    FSR,f       ; FSR ---> MULTIPLICAND_H
          movf    INDF,w      ; Next the high bytes
          decf    FSR,f
          decf    FSR,f       ; FSR ---> PRODUCT_H
          addwf   INDF,f
```

(continued on the next page)

Program 6.8 (*continued*).

```
; Task 3C: Shift multiplicand right once
MUL_CONT movf   PSP,w        ; Reset FSR to the bottom of frame
         addlw  5
         movwf  FSR          ; FSR ---> MULTIPLICAND
         bcf    STATUS,C     ; Zero Carry-in
         rlf    INDF,f
         decf   FSR,f
         decf   FSR,f        ; FSR ---> MULTIPLICAND_H
         rlf    INDF,f

; Task 3D: WHILE multiplier not zero
         incf   FSR,f        ; FSR ---> MULTIPLIER
         movf   INDF,f       ; Test multiplier for zero
         btfss  STATUS,Z
         goto   MUL_LOOP     ; IF not THEN go again

; Task 4: End and clean up stack
         movlw  5            ; FSR ---> Top Of Frame by adding 5
         addwf  PSP,f        ; to the Pseudo Stack Pointer
         return              ; Finished
```

Program 6.9 Coding a $208\,\mu$s delay.

```
N          equ    d'51'     ; The delay parameter is decimal 51

DELAY_208 movlw  N         ; The delay parameter,              1˜

D_LOOP    addlw  -1        ; Decrement count,                  N˜
          btfss  STATUS,Z  ; Skip IF reached zero,           N+1˜
          goto   D_LOOP    ; ELSE repeat,                 2*(N-1)˜

          return           ; Finish,                           2˜
```

```
; Two cycles getting from the caller                           2˜
DELAY_100MS movlw  d'100'  ;                                   1˜
            movwf  COUNT1  ;                                   1˜
; Outer loop ---------------------------------------------------------
DELAY_1MS   movlw  d'249'  ; A hundred  times thru this, 100*1˜
; Inner loop ---------------------------------------------------------
D_LOOP      addlw  -1      ; 249 times 100,              249*100˜
            btfss  STATUS,Z ; plus one when skips out,   250*100˜
            goto   D_LOOP  ; 248 times 2˜ times 100, 248*2*100˜
; ---------------------------------------------------------------------
            decfsz COUNT1,f ; 100 plus one skip out,       100+1˜
            goto   DELAY_1MS; Skip over means 99 passes,     2*99˜
; ---------------------------------------------------------------------
            return         ;                                   2˜
```

which gives a grand total of 99,905 cycles. This misses the target by 95 cycles, or $-\frac{95}{100,000} \times 100 = -0.95\%$ error.

A single nop instruction just before the return instruction will add 100 cycles, giving a total of $100,005\,\mu s$, which is five parts in 100,000 or +0.005% error.

Example 6.3
At the other end of the spectrum write a subroutine to give a delay of one minute.

Solution
Sixty seconds can be implemented as $240 \times 250\,ms$. Our solution Program 6.10 closely follows the coding in the triple-loop Program 6.4 which carries out a $K \times 100\,ms$ delay. The maximum value of K is 255, which would only give 25.5 s, but we can increase the middle loop to 250 ms and thus give increments of $\frac{1}{4}$ s. If we now use an outer loop with a 240 count, we have our 60 s delay.

Program 6.10 A 1-min delay program.

```
COUNT1      equ     h'30'     ; Counter at File h'30'
COUNT2      equ     h'31'     ; and File h'31'

; **************************************************************
; * FUNCTION: Delays for approx a minute for a 4 MHz XTAL     *
; * ENTRY    : None                                          *
; * EXIT     : Status & W altered, Files h'34:35:36' zero     *
; **************************************************************
DELAY_1_MIN
            movlw   d'240'    ; Put 240 as the MS count,            1˜
            movwf   COUNT2    ;                                     1˜

DELAY_250MS
            movlw   d'250'    ; Put 250 as the mid count,           1˜
            movwf   COUNT1    ; for a 250ms delay,                  1˜
; 1ms inner core
DELAY_1MS
            movlw   d'249'    ;                              250*240˜

D_LOOP addlw    -1            ;                          249*250*240˜
            btfss   STATUS,Z  ;                      (249+1)*250*240˜
            goto    D_LOOP    ;                  (2*(249-1)*250*240)˜

            decfsz  COUNT1,f  ;                          (250+1)*240˜
            goto    DELAY_1MS ;                        2*(250-1)*240˜

            decfsz  COUNT2,f  ;                               240+1˜
            goto    DELAY_250MS ;                          2*(240-1)˜
            return            ;                                   2˜
```

Comments in our listing give the full delay calculation, which totals 59.821088 s, accurate to approximately 0.3%. Once again the routine can be padded out with nop instructions. Each nop after the first decfsz adds $250 \times 240 = 60,000$ cycles, so three will change the shortfall to an excess of 1088 cycles in 60,000,000 cycles, or better than +0.002%.

Example 6.4
Design a subroutine to convert a binary byte passed in W to a 3-digit BCD equivalent in HUNDREDS (File h'30'), TENS (File h'31') and UNITS (File h'32').

Solution
We have already coded a routine to implement this binary $\mapsto$ BCD mapping in Example 5.3 on page 137. However this was restricted to a range 0 – 99, that is, two digits. Nevertheless we can extend the technique used there by first subtracting and counting hundreds from the original binary byte. After this has been computed, then the residue will be less than 100 and the rest of the coding will be the same, as shown in Program 6.11. Thus a suitable task list would be:
1. Divide by 100; the remainder is the hundreds digit.
2. Divide the quotient by ten; the remainder is the tens digit.
3. The quotient is the units digit.

Example 6.5
Write a subroutine to evaluate the square root of a 16-bit integer located in File h'26:27' and return the 8-bit outcome in the Working register.

Solution
The crudest way of doing this is to try every possible integer k from 1 upwards, generating k^2 by multiplication and checking that the outcome is no more than Number. An equivalent but slightly more sophisticated approach is based on subtracting the series 1, 3, 5, 7, 9, 11,..., from Number until underflow occurs. Counting the number of subtractions gives the nearest square root. This series comes from the relationship:

$$k^2 = \sum_{I=0}^{k} (2 \times I) + 1$$

On this basis a possible structure for this function is:
1. Zero the loop count.
2. Set variable I (the magic number) to 1.
3. DO forever:
    (a) Take I from Number.
    (b) IF the outcome drops below zero THEN BREAK out.
    (c) ELSE increment the loop count.
    (d) Add 2 to I.
4. Return loop count as $\sqrt{\text{Number}}$.

Program 6.11 Binary to 3-digit BCD conversion.

```
; ****************************************************************
; * FUNCTION: Converts a binary byte in W to three BCD digits*
; * EXAMPLE : Binary = h'FF' (d'255'), HUNDREDS = h'02'       *
; * EXAMPLE : TENS = h'02', UNITS = h'05'                     *
; * ENTRY   : Binary in W                                     *
; * EXIT    : HUNDREDS = Hundreds digit, TENS = Tens digit    *
; * EXIT    : UNITS = Units digit. W holds units              *
; ****************************************************************
;
; First divide by a hundred
BIN_2_BCD  clrf  HUNDREDS  ; Zero the Hundreds loop count

LOOP100    incf  HUNDREDS,f ; Record one hundred subtracted
           addlw -d'100'    ; Subtract decimal hundred
           btfsc STATUS,C   ; IF a borrow (C==0) THEN exit loop
           goto LOOP100      ; ELSE do another subtract/count

           decf  HUNDREDS,f ; Compensate for one inc too many
           addlw d'100'     ; by adding a hundred to residue

; Next divide by ten
           clrf  TENS        ; Zero the Tens loop count

LOOP10     incf  TENS,f      ; Record one ten subtracted
           addlw -d'10'      ; Subtract decimal ten
           btfsc STATUS,C    ; IF a borrow (C==0) THEN exit loop
           goto LOOP10       ; ELSE do another subtract/count

; Retrieve last remainder for units
           decf  TENS,f      ; Compensate for one inc too many
           addlw d'10'       ; by adding ten to residue
           movwf UNITS       ; which gives the remainder
           return            ; and return to caller
```

An example giving $\sqrt{65} = 8$ is given in Fig. 6.11(a) using this series approach. A flowchart visualizing the task list is also given in Fig. 6.11(b). The coding in Program 6.12 follows the task list closely. The maximum value of the loop count is h'FF', as $\sqrt{65535} \approx 255$. Thus a single byte at File h'35' is reserved for this local variable. Similarly the maximum possible value of the magic number is 511 (h'1FF') and so the two registers File h'36:37' are reserved for this local variable. This of course means that Task 3(a) entails a double-byte subtraction. If there is a borrow-out from the least significant byte subtraction, one is added onto a copy of the high byte of I, that is I_H before the high-byte subtract to return the borrow. As I_H will in practice never be more than h'01' this procedure will never give an overflow. If a borrow is generated from this high-byte subtraction the outcome is under zero and the loop is exited. Otherwise COUNT is incremented and I augmented by two. Actually the loop Count is always half of I less one, so COUNT is not needed. Instead, on return the 16-bit value I can be shifted once right. This divides by 2 and by throwing away the one that pops out into the Carry flag, effectively subtracts one; I is always odd and so its least significant bit is always 1. Try coding this alternative arrangement.

Program 6.12 Coding the square root subroutine.

```
; Global declarations
STATUS        equ    3       ; Status register is File 3
C             equ    0       ; Carry flag is bit0
NUM_H         equ    h'26'   ; Number high byte
NUM_L         equ    h'27'   ; Number low byte

; ****************************************************************
; * FUNCTION: Calculates the square root of a 16-bit integer    *
; * EXAMPLE : Number = h'FFFF' (65,535), Root = h'FF' (d'255')  *
; * ENTRY   : Number in File h'26:27'                           *
; * EXIT    : Root in W. Files h'26:27' and h'35:36:37' altered*
; ****************************************************************

; Local declarations
COUNT      equ    h'35'    ; The loop count
I_H        equ    h'36'    ; Magic number high
I_L        equ    h'37'    ; Magic number low

; Task 1: Zero loop count
SQR_ROOT clrf    COUNT

; Task 2: Set magic number I to one
         clrf    I_L
         clrf    I_H
         incf    I_L,f

; Task 3: DO
; Task 3(a): Number - I
SQR_LOOP movf    I_L,w      ; Get low byte magic number
         subwf   NUM_L,f    ; Subtract from low byte Number
         movf    I_H,w      ; Get high byte magic number
         btfss   STATUS,C   ; Skip if No Borrow out (C==1)
          addlw  1          ; Return borrow
         subwf   NUM_H,f    ; Subtract high bytes

; Task 3(b): IF underflow THEN exit
         btfss   STATUS,C   ; IF No Borrow (C==1) THEN continue
         goto    SQR_END    ; ELSE the process is complete

; Task 3(c): ELSE increment loop count
         incf    COUNT,f

; Task 3(d): Add two to the magic number
         movf    I_L,w
         addlw   2
         btfsc   STATUS,C   ; IF no carry-out THEN done
          incf   I_H,f      ; ELSE add Carry bit to upper byte I
         movwf   I_L
         goto    SQR_LOOP

; Task 4: Return loop count as the square root
SQR_END  movf    COUNT,w    ; Copy into W
         return
```

Example 6.6
Design software to multiply the byte contents of File h'46' by ten (×2 + ×8). Use a software stack for data storage and parameter passing.

Solution

The global declarations for the subroutine of Program 6.13 and calling procedure is:

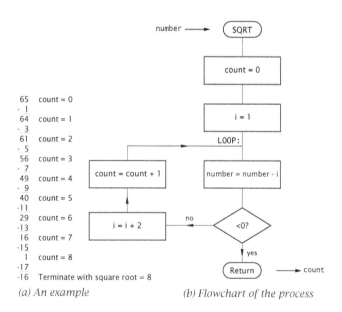

65	count = 0
· 1	
64	count = 1
· 3	
61	count = 2
· 5	
56	count = 3
· 7	
49	count = 4
· 9	
40	count = 5
-11	
29	count = 6
-13	
16	count = 7
-15	
1	count = 8
-17	
-16	Terminate with square root = 8

(a) An example   *(b) Flowchart of the process*

Fig. 6.11 Finding the square root of an integer.

```
PSP     equ   h'40' ; Holds the Pseudo Stack Pointer
TOS     equ   h'50' ; File h'50' is the initial Top Of Stack
INDF    equ   0     ; INDirect File
FSR     equ   04    ; File Select Register
XCAND   equ   h'46' ; Multiplicand byte
STATUS  equ   3     ; Status register is File 3
C       equ   0     ; Carry flag is bit0
; The main routine sets up the Pseudo Stack Pointer (PSP)
MAIN    movlw TOS   ; Set up the PSP
        movwf PSP   ; to the initial Top Of Stack
; ....................and so on
; Get ready to call up the X10 subroutine
        movf  PSP,w ; 1st point to current stack position
        movwf FSR
; Now copy multiplicand onto the stack
        movf  XCAND,w ; Copy Multiplicand into W
        movwf INDF    ; and onto the stack
        decf  FSR,w   ; Point down one
        call  X10     ; Now call subroutine
; On return PSP is returned to original position
; and product is at PSP+3:PSP+2
NEXT_MAIN  ..... ... ; Continuation of main routine
```

Program 6.13 first shifts the multiplicand left once to multiply by two and then two further shifts multiplies by eight. The two resulting 16-bit data are then added to give the Product. In the same manner as Program 6.8, the File Select Register is moved up and down to point to the appropriate datum as the program progresses. The double-byte Product can be accessed relative to the Pseudo Stack Pointer by the caller. Unlike Program 6.8, this PSP is not altered when pushing out the Multiplicand nor in the subroutine. This is because the subroutine is a dead end in that it can never call another subroutine. Thus a new stack frame need not be formed.

Example 6.7
In order to ensure that the 7-segment decoder subroutine of Program 6.6 does not cause the PCL register to overflow when the offset is added, a programmer has used the directive org (ORiGin ; see page 218) to tell the assembler to locate the subroutine beginning at the absolute instruction address h'700'; as shown in Program 6.14. When the subroutine is tested by calling from another part of the program in the store somewhere lower than h'700' the system fails and performs unpredictably. What has gone wrong?

Solution
The system goes berserk because on reset the PCLATH register is zeroed. When summoning the subroutine using call h'700' the value of the PC becomes h'0700' but the value of the PCLATH register remains unaltered. Later when the addwf PCL,f instruction is executed the full 13-bit Pro-

gram Counter is updated with the bottom eight bits from the PCL register *and* the top five bits from the PCLATH register, as described in Fig. 4.8 on page 86. Thus, instead of the execution branching to one of the retlw instruction it will jump somewhere in Program memory in the area h'0000

Program 6.13 Using a software stack to pass parameters and to provide a workspace.

```
; ************************************************************
; * FUNCTION: Xs byte XCAND by 10 giving double-byte product *
; * EXAMPLE : h'64 x 0A = 3E8' (d'100 x 10 = 1000')          *
; * ENTRY   : Multiplicand pushed into software stack at PSP  *
; * EXIT    : Product_H:_L in PSP-3:PSP-2                      *
; ************************************************************
X10       movf    PSP,w     ; Point FSR at current stack position
          movwf   FSR
          clrf    INDF      ; Zero XCAND overflow

; Now multiply by two by shifting 16-bit XCAND left once
          bcf     STATUS,C  ; Clear Carry-in
          incf    FSR,f     ; Point to the XCAND LSB
          rlf     INDF,f    ; Shift left LSB
          decf    FSR,f     ; Point to MSB
          rlf     INDF,f    ; Shift left MSB

; Add to 16-bit subproduct
          incf    FSR,f     ; Point to XCANDx2_L
          movf    INDF,w    ; Get it
          decf    FSR,f     ; Point at PROD_L
          decf    FSR,f
          movwf   INDF      ; Update it with XCANDx2_L
          incf    FSR,f     ; Point to XCANDx2_H
          movf    INDF,w    ; Get it
          decf    FSR,f     ; Point at PROD_H
          decf    FSR,f
          movwf   INDF      ; Update it with XCANDx2_H

; Now shift left twice more to give x8
          incf    FSR,f     ; Point to XCANDx2_L
          incf    FSR,f
          incf    FSR,f
          bcf     STATUS,C  ; Clear Carry-in
          rlf     INDF,f    ; Shift left LSB
          decf    FSR,f     ; Point to MSB
          rlf     INDF,f    ; Shift left MSB
          incf    FSR,f
          rlf     INDF,f    ; Shift left LSB
          decf    FSR,f     ; Point to MSB
          rlf     INDF,f    ; Shift left MSB
```

(continued on the next page)

Program 6.13 (*continued*).

```
; Add to 16-bit subproduct
        incf   FSR,f      ; Point to XCANDx8_L
        movf   INDF,w     ; Get it
        decf   FSR,f      ; Point at PROD_L
        decf   FSR,f
        addwf  INDF,f     ; Update it with XCANDx8_L
        incf   FSR,f      ; Point to XCANDx8_H
        btfsc  STATUS,C   ; IF Carry set THEN inc XCANDx8_H
        incf   INDF,f
        movf   INDF,w     ; ELSE get it
        decf   FSR,f      ; Point at PROD_H
        decf   FSR,f
        addwf  INDF,f     ; Update it with XCANDx8_H

        return
```

Program 6.14 The software 7-segment decoder revisited.

```
        org    h'700'      ; Start the subroutine at h'700'
SVN_SEG addwf  PCL,f       ; Add N to PCL giving PC + N
        retlw  b'00111111' ; Code for 0
        retlw  b'00000110' ; Code for 1
        retlw  b'01011011' ; Code for 2
        retlw  b'01001111' ; Code for 3
        retlw  b'01100110' ; Code for 4
        retlw  b'01101101' ; Code for 5
        retlw  b'01111101' ; Code for 6
        retlw  b'00000111' ; Code for 7
        retlw  b'01111111' ; Code for 8
        retlw  b'01101111' ; Code for 9
```

– 00FF'! This will happen even though adding the offset in W will not cause overflow of the PCL register, as was the original intention.

One way of avoiding this error would be to set the PCLATH to h'07' (page 7) prior to the call. This causes the Program Counter to be advanced to location h'07NN' as desired, instead of h'00NN'.

```
movlw  h'07'    ; Prepare to point PCL
movwf  PCLATH   ; to page7 of Program store
movf   NN,w     ; Get the decimal number NN into W
call   SVN_SEG  ; Call up the subroutine
```

Even with this kludge, the table is limited to 255 entries before the addition overflows the PCL register, causing a malfunction. In any case it is considered bad practice for the programmer to specify the absolute location of sections of program code, as it is possible to overwrite code that the assembler places itself automatically. With large programs it is error prone to try and keep track of the location of a myriad of modules. One way around the problems of both large tables and ensuring that the PCLATH register is correctly set, is for the caller to calculate the proper addition of the 13-bit value of the beginning of the subroutine SVN_SEG to the offset and place the top byte of the outcome in the PCLATH register. Of course the PIC MCU is only capable of doing 8-bit arithmetic at a time and so we need to be able to find the values of both the bottom and top bytes of the label SVN_SEG. Fortunately the Microchip assembler has the directives high and low which can be used to dismember a 13-bit address to facilitate address arithmetic.

```
movlw high SVN_SEG+1 ; Get high byte of table start address
movwf PCLATH         ; Which is the correct Program store page
movlw low SVN_SEG+1  ; Get the low byte of the table address
addwf NN,w           ; Add the offset in File NN to it
btfsc STATUS,C       ; Did this cause a carry-out?
 incf PCLATH,f       ; IF so then meant boundary overflowed
movf  NN,w           ; Get the offset
call  SVN_SEG
```

In the code segment above the address of the start of the table (that is, SVN_SEG+1) is used, as this will be the value of the PC after the opening addwf PC,f instruction. Of course the org h'700' instruction used in Program 6.14 can be dispensed with.

This code segment can easily be extended to deal with offsets which are greater than one byte by doing a double-byte addition to update PCL. As before, only the lower byte of the offset is sent to the subroutine. In this way look-up tables of any size and located anywhere in Program memory can be implemented, subject to the limited size of the Program store. More details are given in the Microchip application note AN556, *Implementing a Table Read*.

Self-Assessment Questions

6.1 A certain student has coded his 1 ms delay subroutine of Program 6.1 thus:

```
DELAY_1MS movlw  d'249'    ; Set up loop count in W
D_LOOP    addlw  -1        ; Decrement count
          btfss  STATUS,Z ; Check: Is it zero?
           goto  DELAY_1MS; IF not THEN repeat
          return
```

What will be the outcome?

6.2 Create a subroutine that will read Port B every hour. You can base it on a 60-minute version of Program 6.10. Say why this may not be a good use of the PIC MCU's resources.

6.3 Code a subroutine with the following specification:
- To divide a double-byte Dividend by a byte Divisor.
- Input DIVIDEND_H:DIVIDEND_L to be passed in File h'2E:2F'.
- Input Divisor to be passed in the Working register.
- Output QUOTIENT_H:QUOTIENT_L to be returned in File h'29:2A'.
- Output Remainder to be returned in W.

Implement the division by subtracting until underflow occurs. You can refer to Program 5.10 on page 142 for a similar task. Comment on the problem of doing this division in this way.

6.4 Extend Program 6.6 to display A through F. Your solution can use a combination of lower- and uppercase glyphs and should be robust.

6.5 Program 6.15 is claimed to be a 30 s delay subroutine. Calculate the timing and thus its actual delay.

Program 6.15 A 30 s delay subroutine.

```
;  *********************************************************
;  * FUNCTION: Delays for approx 30 s for a 4 MHz XTAL     *
;  * ENTRY    : None                                       *
;  * EXIT     : Status & W altered: Files h'34:35:36' zero *
;  *********************************************************
; Local storage
COUNT0    equ     h'34'           ; 3-byte counter at File h'34'
COUNT1    equ     h'35'           ; and File h'35'
COUNT2    equ     h'36'           ; and File h'36'
H         equ     d'153'          ; The delay constant

DELAY_30S
          movlw   H               ; Put decimal 153 as the MS count
          movwf   COUNT2
          clrf    COUNT1
          clrf    COUNT0
D_LOOP
          decfsz  COUNT0,f        ; Dec LSB count
          goto    D_LOOP          ; to zero
          decfsz  COUNT1,f        ; Then dec NSB count
          goto    D_LOOP          ; to zero and then repeat
          decfsz  COUNT2,f        ; Then dec MSB count
          goto    D_LOOP          ; to zero and then repeat
          return
```

6.6 Readings of the state of a mechanical switch can be erratic as the contacts will bounce for several milliseconds when closed, thus giving a series of 1s and 0s. Similar considerations apply to electronic devices such as phototransistors when passing through a shadow. Although this problem can be fixed with hardware, it is usually more cost effective to use a software solution.

Devise a subroutine that will return with the stable state of a switch connected to Port B pin RB7 as bit 7 of the Working register. Stability is defined as 5000 (h'1388') reads all giving the same value. The other bits of W on return are undefined.

6.7 An analog-to-digital converter is connected to Port B. Repeat SAQ 6.6, but this time defining stability as 1000 identical reads, and returning with the stable digitized analog voltage in W.

6.8 The subroutine in SAQ 6.7 returns the stable value of a noisy digitized signal, assuming 1000 identical values. Using this subroutine, code a main routine that will generate how this stable reading differs from a previous value previously stored in location File h'40'. Each bit that differs is to be logic 1. Generate the position of the rightmost change bit in File h'41'.

6.9 The subroutine of SAQ 6.7 will not return a value when relatively high-frequency noise is present on the analog signal, as the resulting digital jitter will ensure that 1000 identical readings rarely occur. As an alternative, noise reduction can be obtained by taking the average of multiple readings. If the noise is random then n readings will give a noise improvement of $\sqrt{n}$. Devise a subroutine that will read Port B 256 times and return the 8-bit average in W, which will give an increase in signal to noise ratio of 16.

6.10 The circuit diagram of Fig. 6.12 shows a 7-bit pseudo-random number generator (PRNG) based on a shift register with an Exclusive-OR gate feedback. Devise a routine to continually send these 127 binary random numbers to Port B. The routine must initialize the PRN to any non-zero value. For instance, if the initial value were 01, then the first 32 hexadecimal values output are:

```
02 04 08 10 20 41 83 06 0C 18 30 61 C2 85 0A 14
28 51 A3 47 8F 1E 3C 79 F2 E4 C8 91 22 45 8B 16 …
```

The sequence will repeat after 127 output values.

What would happen if the initial value of the random number was zero?

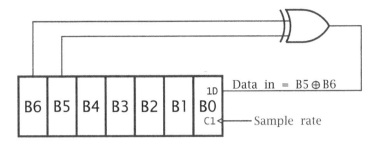

Fig. 6.12 A 7-bit pseudo-random number generator.

6.11 Mathematically to convert from Celsius to Fahrenheit we have:

$$F = C \times \frac{9}{5} + 32$$

Write a subroutine where the temperature is passed via the Working register, ranging from 0°C through 100°C, with the Fahrenheit equivalent being returned in the same location.

CHAPTER 7

Interrupt Handling

The subroutines discussed in Chapter 6 are predictable events in that they are called up whenever the program dictates. Real-time situations, defined as where the processor interacts in concert with external physical events, are not as simple as this. Very often something happens beyond the core CPU which necessitates urgent action from the processor. The vast majority of controllers have the capability to deal with a range of such events that disrupt their smooth running. In the case of a micro-controller, requests for service may come from an internal peripheral device, such as a timer overflowing, or from the outside world from a source entirely external to the device. At the very least, on an external reset (a type of outside hardware event) the MCU must be able to get (vector) to the first instruction of the main program. In the same manner an external service request, or interrupt, when answered must lead to the start of the special subroutine, known as an interrupt service routine.

Although members of the mid-range family have different mixes of internal peripherals, such as analog, serial and timer ports, all respond to interrupts in the same manner. In this chapter we will concentrate on the basic sources of such real-time requests common to all devices, in particular, external interrupts. We will also lay the foundations of how these internal peripheral devices interact with the interrupt system; with individual instances being discussed as appropriate in Part 3 of the text.

After reading this chapter you will:

- Be aware of the need for interrupt handling.
- Appreciate the concept of a vector table as a jumping-off point for reset and interrupt events.
- Follow the sequence of events when the PIC microcontroller recognizes an interrupt request.
- Understand the principle of latency.
- Recognize the function of the Global Interrupt Enable mask is to enable the interrupt system.
- Understand the operation of the local interrupt mask and flag pairs, corresponding to the various sources of interrupts.
- Be able to write a simple interrupt handler according to the principles:
  - Context switching.
  - Determination of interrupt source.
  - Return via the `retfie` instruction.

A simple example of a time-sensitive situation is shown in Fig. 7.1. Here we wish to measure the elapsed time between the R points of an electrocardiogram (EKG)[1] signal, which by definition is an external real-time event. The time resolution is to be 0.1 ms and the maximum peak-to-peak duration is likely to be no more than 1.5 seconds. In order to measure this time a free-running 16-bit counter clocked at 10 kHz can be used as the time base. As we shall see in Chapter 13, all mid-range PIC MCUs have an internal 8-bit counter at File 01 and Fig. 7.1 shows File h'3F' used as an extension byte to give a total 16-bit count. The details of this configuration are discussed in Program 13.2 on page 411. Here we will assume that the state of the count can be read at any time from these two specified Files. If the count at the last R point is stored in two spare Files, then subtraction of the count at the current R point will give the required beat-to-beat duration.

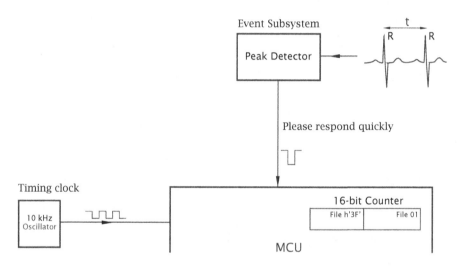

Fig. 7.1 Detecting and measuring an external event.

The next problem is how to detect the signal peak, as by definition the patient's heart is not synchronized to the MCU! One technique is to continually read this signal and perform a peak-detection algorithm to determine the R point. Now this **polling** technique will have to be carried out 10,000 times each second in order to keep to the specified resolution. Taking a nominal human heart rate of 60 beats per minute, 99.99% of the time no peak will be detected. Essentially, this means that the processor will spend the vast majority of its processing power just looking out for one event in 10,000.

---

[1]From the German *Elecktrocardiogramm*; ECG in UK.

The alternative approach is to use external hardware whose task is to find the peak signal. That peak-picking hardware could be an analog circuit or even a MCU with an analog-to-digital converter dedicated to this one task; see Example 14.2 on page 472. Whatever the implementation, the peak-picker sends a signal to the main processor when an R point has been detected. This signal **interrupts** the MCU, which must drop whatever it is doing and read the counter within $100\,\mu s$, if a counter tick is not to be missed.

In the situation where external processes happen in their own good time and are in no way synchronized to the processor, there has to be some way for certain events to interrupt the process and direct it to attend to their immediate need. **Polling** a series of external events is adequate where nothing much happens quickly outside and/or there are few parameters to monitor and little processing to do. The possibility of missing anything important can be reduced by increasing the polling rate, but there comes a time when the processor does little else but read peripheral data. This resource burnout is especially a problem when there are many signals to poll in a short period of time.

The downside of interrupt-driven real-time monitoring is additional hardware complexity and the greater intricacy of the hardware – software interface. If you are confused, consider the telephone system. It would be possible to have a telephone network where the subscriber would pick up the phone every, say, 5 minutes and ask "Is there anyone there?" Apart from the bother (processing overhead) of doing this,[2] the caller might get bored and hang up. You could reduce the chance of this happening by increasing the polling rate to, say, once per minute. But you would then end up spending all your time on the phone and, depending on how popular you are, getting only a few hits a day. That is, 99% of your effort would be wasted.

This is obviously ridiculous, and in practice an interrupt-driven technique is used so that you only respond when the bell/buzzer sounds. This is highly efficient, but at the cost of a lot more complexity for the phone company, as the signaling side of the system can be more demanding than the speech side. There is another problem too, in that you (cf. the processor) have no idea when the phone will ring. And it surely will be at the most inconvenient time. Thus you have to (unless you have an iron will) break off what you are doing at the drop of a hat. For instance, if you happen to be in the middle of solving a problem in your head you should save your partial results before responding, so, when finished, you can return to where you left off.

Microcontrollers can respond to interrupt requests from a wide range of sources, either physically outside the chip or from the various ports and peripheral devices supported by that particular family member. For

---

[2]It would of course make it easier just to ignore the phone!

instance, the PIC16F874/7 has up to 13 separate interrupts originating from these modules, plus one from outside via the pin labeled INT (pin 6 in Fig. 4.1); which is shared with bit 0 of Port B; that is pin RB0. The programmer can choose to enable/disable each of these sources individually or indeed disable the complete interrupt system. Because the response to an interrupt request is essentially the same, no matter whence it arises, we will in this chapter refer mainly to this external, or **Hardware**, interrupt.

Keeping in mind the randomness of an external event, the response of a processor to an **interrupt request** will normally be something like:

1. Finishing the current instruction.
2. Automatically saving, at the very least, the state of the Program Counter (PC)—which is needed to get back. Some processors (such as the enhanced-range PIC MCUs) can also automatically save the Status register and other internal registers at this point.
3. Entering the appropriate interrupt service routine.
4. Executing the defined task.
5. Restoring the processor state and returning to the point in the background program whence control was first transferred.

Essentially, signaling an interrupt causes the PIC MCU to drop whatever it is doing, save its position in the interrupted **background program** on the stack and go to a special subroutine known as an **interrupt service routine (ISR)**. This **foreground program** is just a subroutine entered at the behest of an external happening.

The minutiae of the response to an interrupt request varies somewhat from processor to processor. As detailed in Fig. 7.2, that of the mid-range family is[3]:

1. The processor checks once in each instruction if an interrupt request from an enabled source has been received. Even if this request is active, the instruction continues to completion; that is, execution does not break part way through the instruction, even in a 2-cycle instruction.
2. If there is no valid request, the PIC MCU simply continues on into the next instruction and the process is repeated.
3. If there is a valid enabled request, then the next three instruction cycles are involved in moving execution to the interrupt service routine. This comprises a dummy cycle,[4] plus two more cycles to flush the Pipeline. This 3- to 4-cycle delay from the instant of the external

---

[3]The low-range family has no interrupt handling capability, whilst the enhanced-range family has low- and high-priority responses, but otherwise is virtually identical to the mid-range family.

[4]Alternatively this could be the final cycle of a 2-cycle instruction, e.g., a Skip instruction.

signal to the INT pinand beginning the execution of the first instruction of the ISR is known as **latency**. It is impossible to be more precise due to the time-random nature of the external request signal, which can occur anywhere in the instruction cycle.

4. During this latency period the PIC MCU does three things:

   (a) The complete interrupt system is disabled, to ensure that once an interrupt response is in train, any further interrupt requests are locked out, This is done by clearing bit 7 of the the INTerrupt CONtrol register INTCON, which is labeled in Fig. 7.3 as the General Interrupt Enable (GIE). GIE is an example of an **interrupt mask**, as it is able to mask out interrupt activity. After reset, GIE is clear, so by default interrupt activity is disabled.

   (b) The state of the 13-bit Program Counter is pushed into the stack in exactly the same manner as for a `call` instruction; see Fig. 6.3 on page 151. As for subroutines, this is to allow the processor to return to the interrupted background program after the interrupt service routine. As the mid-range PIC MCUs have an 8-deep hardware stack, subroutines nested to depth of seven can be called from an ISR.

   (c) The first instruction of the ISR is *always* in location h'004' in the Program store. Thus the final step of the sequence is to overwrite the PC with this instruction address, known as the **Interrupt vector**. If the interrupt handling software is elsewhere in Program memory then this entry instruction can of course be a `goto` instruction; see Program 7.1.

5. Like a subroutine, an ISR must be terminated by a Return instruction. However, in this case not only has the PC to be pulled back out of the hardware stack to move execution back to the interrupted program but the GIE bit in the INTCON register must be set to re-enable the interrupt capability. This counteracts the resetting of this bit in 4(a) above on entry to the ISR. The Return instruction relevant to this situation is `retfie` (RETurn From Interrupt and Enable); see Table 6.1 on page 149. Thus on re-entry to the background program any pending or future interrupts can be serviced.

An ISR differs from a subroutine in more subtle ways than the use of the `retfie` instruction of item 5 above. Some of these differences relate to the logic of the interrupt system and some are due to the pseudo random nature of the interrupts. Discussing the former first, let us examine the logic circuitry relating to the interrupt process.

Although most members of the mid-range PIC MCU families support interrupts from a multitude of sources; all devices handle a kernel of three sources using the INTCON register, as shown in Fig. 7.3. These core sources are:

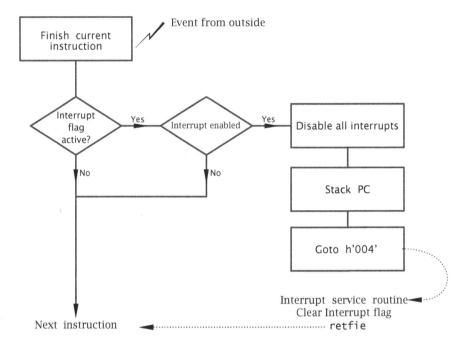

Fig. 7.2 Responding to an interrupt request.

- An external signal at the pin labeled INT. This Hardware interrupt can be activated by either a rising edge _/‾ or a falling edge ‾_ at this input, as selected using the INTEDG bit in the Option register. This is accomplished with an eXclusive-OR gate acting as a programmable inverter, as described on page 14.
- An input change at any of the top four Port B (File 6) pins since the last read of this port.
- The Timer counter TMR0 (File 1) overflowing h'FF → 00'.

The PIC16F84 MCU's INTCON register is shown in Fig. 7.3. Each of its four interrupt sources can set an associated **interrupt flag** bit. For instance, after a Reset a falling edge ‾_ on pin 6 will set INTF (bit 1) to 1. This will happen, irrespective of whether the interrupt system is enabled or not. In cases where more than one source of interrupt requests are enabled, these flags can be monitored by software by polling to check the origin of the request; see the listing on page 196. You can also poll these bits even if the interrupt system is disabled. Although a flag is set by an external (to the CPU) event, it has to be cleared in software, as part of the program. If being used to create an interrupt response, it is *vital* that it is cleared in the interrupt service routine before returning, e.g., bcf INTCON, INTF.

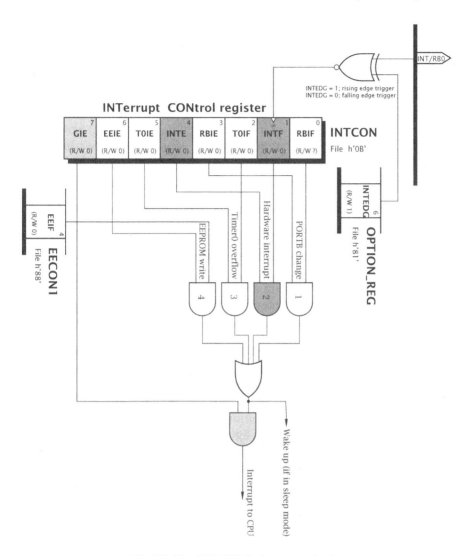

Fig. 7.3 The PIC16F84's interrupt logic.

Each interrupt flag is paired with its own personal local Enable bit; thus INTF is associated with INTE. This allows the programmer to selectively mask out any combination of interrupt sources. Effectively, each interrupt flag is AND'ed with an **interrupt mask**. In Fig. 7.3, AND gate 2 shows the mechanism for the Hardware interrupt. As an example, if we wished to enable interrupts from either the INT pin or from Timer 0, we would set bits 7, 5 and 4; e.g., movlw b'10110000', movwf INTCON. The local mask bit can be written to in the normal way by software. On reset it is zeroed and thus the interrupt from the affiliated source is disabled.

In the specific case of the PIC16F84, an interrupt can be generated when an internal Data EEPROM write-to action has been completed. There is not enough room in INTCON to hold the interrupt flag EEIF and so it has been banished to bit 4 of the EEPROM CONtrol register 1 EECON1. We will look at how the more generously endowed members of the family handle additional interrupts later in the chapter in Fig. 7.5.

As there are four sources of interrupt, each flag:mask AND gate must be ORed to give a composite request signal, which when active initiates the CPU's interrupt response. In Fig. 7.3 this ORing process is further gated with the Global mask bit GIE, which is located in bit 7 of INT-CON. However, the raw, i.e., preglobally masked, request signal is used to awaken the processor if it is in a power-down or **Sleep** state. As we will see in Chapter 10, the current consumption of the device can be considerably reduced to typically less than $1\,\mu A$ if processing is stopped and the PIC MCU is put in a state of suspended animation. For instance, monitoring the temperature profile at the bottom of a lake over a period of a year at one-hour intervals using a battery-powered data logger requires processing for a tiny proportion of the time. Placing the PIC MCU in this power-down mode after each sample has been taken and stored will reduce the necessary battery capacity. The sleep instruction initiates this mode. An interrupt from an outside source, in this case a low-power hourly oscillator, is used to wake the PIC MCU up. As shown, this awakening is independent of the setting of the General mask GIE.

To illustrate the software aspects of interrupt handling, consider the problem of counting customers coming into a small shop. One approach would be to use a low-power laser light beam/photocell across the entry door. When the shopper breaks the beam, the resulting pulse _/‾_ requests service from the monitoring microcontroller, as shown in Fig. 7.4, which is away in its main background routine doing something else; maybe handling the communication link between the point of sale (POS) terminal and the main inventory computer.

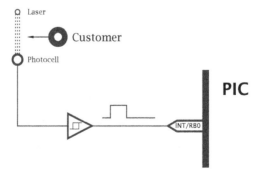

Fig. 7.4 Monitoring customers entering the shop.

From the software point of view we specify that each time the customer enters the shop a File called EVENT is incremented. Of course customers will also be leaving, but if the entrance is relatively narrow we can divide the number of breaks by two to get the actual number of bodies. This will limit the customer count, but we can easily increase the total by using extra Files. We assume that the system is reset at the beginning of business each day, so we are not expecting a customer base greater than 126.

Program 7.1 People counting.

```
STATUS   equ   3      ; The Status register
INTCON   equ   h'0B'  ; The INTerrupt CONtrol register in which
INTF     equ   1      ; bit1 is the Hardware interrupt flag
INTE     equ   4      ; and bit4 is the associated mask bit
GIE      equ   7      ; and bit7 is the Global mask bit

_work    equ   h'4E'  ; Place to save the background Working reg.
_status  equ   h'4F'  ; Place to save the background Status reg.

EVENT    equ   h'20'  ; Keeps tally of passing customers
; Reset vector ---------------------------------------------------
         org   000    ; PIC resets to location 000 in Program store
         goto  MAIN   ; Go to start of background routine
; Interrupt vector -----------------------------------------------
         org   004         ; PIC goes to 004 if interrupt accepted
         goto  PERS_COUNT  ; go to start of foreground ISR

; Background program starts by intialization ------------------
MAIN     bsf   INTCON,INTE; Enable Hardware interrupts
         bsf   INTCON,GIE ; Enable interrupt system overall
         clrf  EVENT       ; Zero the customer count
; Main endless loop ----------------------------------------------
M_LOOP                     ; Do this
                           ; Do that
                           ; Do the other
         goto  M_LOOP

;
; ****************************************************************
; * FUNCTION: ISR increments EVENT count on entry              *
; ****************************************************************
;
PERS_COUNT
         movwf _work       ; Save the entry state of Working reg
         swapf STATUS,w    ; Get Status reg. don't change flags
         movwf _status     ; Save it in the Data memory
; ==============================================================
         bcf   INTCON,INTF; Clear the Hardware interrupt flag

         incf  EVENT,f     ; Record one more event
; ==============================================================
         swapf _status,w   ; Untwist & get original flags
         movwf STATUS      ; back into the Status register
         swapf _work,f     ; Now restore the original state of W
         swapf _work,w     ; without changing the Z flag

         retfie            ; and return to interrupted background
```

Program 7.1 lists the two vectors at the top of the listing. At location h'000', as specified by the directive org 000 (see page 218), which is where the PIC MCU resets, is the instruction goto MAIN. Similarly the instruction goto PERS_COUNT actions a switch to the ISR should the Program Counter alight at location h'004'. Thus if the PIC MCU responds to an interrupt, we have the sequence interrupt ⤳ h'004' ⤳ PERS_COUNT.

The main program itself simply sets both the INTF and GIE mask bits to enable external interrupts and zeros the customer Event count. The following endless loop represents the processor's background tasks.

Interrupts happen randomly as viewed by the software and thus, unless masked out, may happen at any part of the background software, including in the middle of a subroutine. An ISR foreground routine uses the internal processor registers in the same way as any other software, so conflict over such resources will exist. For instance, the background program could just be testing the contents of a File when an interrupt occurs. The Skip instruction which follows the test could be dependent on, say, the state of the Zero flag in the Status register. However, the ISR will in all probability alter Z and thus on return the background program will execute the skip, oblivious of the fact that execution has been transferred in the interregnum. Any change to Z would cause an erroneous branch in the background program. Trying to debug this sort of problem is virtually impossible because the effect of such an interrupt is sporadic, as the particular bug depends on the interrupt occurring at just this wrong time and wrong place—something it may do perhaps once a week—and thus is difficult to reproduce.

Corrupting the Status register in this way may have even more serious consequences; for example, in the polling listing on page 196. Here on entry to the ISR, bit RP0 of the STATUS register (see Fig. 4.7 on page 80) was set to allow access to Bank 1 SPRs. This was necessary as the EEP-ROM control registers only appear in this bank, whilst both the STATUS and INTCON SPRs are shadowed in both banks. At the exit point, RP0 is cleared to move back to Bank 0. However, this assumes that the background program was in Bank 0 when interrupted. Clearly this is erroneous if an interrupt occurs during an access to Bank 1.

Thus in all but the most elementary ISR, we will need to, at the very least, save both the Status and Working registers. Generally the programmer sets aside Files as temporary storage and for no other use. Traditionally such locations are named with a leading underscore to show that they are used for system purposes and are not to be tampered with by the user's application program. In Program 7.1 File h'4E' is labeled _work and File h'4F' as _status to conform to this convention.

With this need in mind, ISRs can usually be divided into three distinct phases.

Context Switching

A copy is made of the contents of the Working register in _work. The movwf instruction does not alter any of the Status bits. The Status register is then saved in h'4F' in the Data store labeled _status. The obvious approach is to copy it into W and then out to _status. However, the movf instruction alters the Z flag. Instead, we use swapf to copy the datum into W. The swapf instruction does not affect the flags but does of course interchange the top and bottom halves of the byte. However, we can untwist them on restoration.

This process of saving and restoring the state of internal registers (this internal state is called the context) on entry and exit is known as **context switching**. Of course, care must be taken that the ISR does not use these safehouse locations in Data memory for any other purpose.

Core Function

The interrupt flag, INTF in INTCON, is cleared to ensure that on return to the background program another interrupt request is not immediately auctioned—indefinitely. In situations where there is more than one source of interrupts, the various interrupt flags would be polled at this stage of the program; see page 196. There would then be several core routines, each of which would commence by clearing the appropriate interrupt flag.

The functional sector of the core function simply increments the datum EVENT. This, of course, is the chief function of the ISR.

Restoring the Context

The exit process first restores the original state of the Status register by swapping out of memory into W. This cancels the original swapf, which was used to save the Status register on entry to the ISR. It is then copied from W into STATUS.

The original value of W is then pulled out of temporary storage in _work using two swapf instructions in series. This cancels out the twist but does not affect any Status bits.

Finally, the exit instruction retfie does not alter the flag state.

Although our example showed a context change involving only the Status and Working registers,[5] other SPRs can be saved in the same way. For instance, Example 7.3 shows the File Select Register, being saved, as it is used by both background and foreground routines. In such instances these additional registers should also be saved and retrieved at the beginning and end of the ISR. In general, if the ISR alters any SPR then its original state should be restored on exit. In all cases W needs to be saved first, as it is used as an intermediary for the other transfers, and similarly restored last.

---

[5]In the enhanced-range family, these two registers can be saved automatically; see page 520.

Where possible the locations chosen to save the context should be independent of what bank the processor is in when interrupted. In the PIC16F84 all GPRs are shadowed across both banks and thus the actual location is irrelevant. Nevertheless, this is not usually the situation; with additional banks being used to increase the number of unique GPRs. Newer devices will often have a small number of common addresses; for instance, the top 16 bytes for the PIC16F627/8, as shown in Fig. 5.4 on page 102. Many older processors, such as the PIC16C74, have no common GPRs. In such cases, the programmer either must ensure that no interrupts can occur when the processor is in a bank other than where the copies are stored, or else check the RP bits on entry to the ISR and change anyway to the system bank (usually Bank 0) before saving the Status register. The RP bits in the copy _status are then altered to reflect their original state.

Our deliberately simple example assumed that only Hardware interrupts were enabled to be serviced. In the majority of instances, interrupts from several sources may be enabled, and as there is only one common interrupt vector, i.e., at h'004', then one of the first tasks the ISR has to do is check which peripheral is calling for help. All interrupt flags can be read, so these can be polled in turn until the one that is set is found. For the extreme case where all four PIC16F84 interrupt sources are active and remembering that the EEPROM CONtrol Register is in Bank 1, we have a typical polling sequence:

```
          bsf    STATUS,RP0  ; Change to Bank 1 registers
          btfsc  INTCON,1    ; Check for external interrupt
          goto   EXTERNAL    ; IF set THEN go to INT handler
          btfsc  INTCON,2    ; Check for Timer0 interrupt
          goto   TIMER0      ; IF set, go to TMR0 handler
          btfsc  INTCON,0    ; Check for change at PortB int
          goto   CHANGE_B    ; IF set, go to correct handler
          btfsc  EECON1,4    ; Check EEPROM write-to interrupt
          goto   EEPROM_WR   ; IF set, go to EEPROM handler

IRQ_EXIT  bcf    STATUS,RP0  ; Return to Bank 0 registers
          retfie             ; and return
```

The order of polling gives a priority level if more than one interrupt request should coincide. Thus if both the external Hardware and Timer 0 interrupts are active, the former will be processed first. In this case, on return the pending Timer 0 interrupt request will then force the processor back to the ISR, where it will be processed—unless another higher-priority interrupt request has occurred. In all instances the appropriate interrupt flag should be cleared, otherwise the interrupt will be generated indefinitely! This clearing should be in an appropriate handling routine.

Where masks are set, this same polling technique can be used to check on the status of events without using the PIC MCU's interrupt processes. For instance, when a byte is written to the Data EEPROM (see Program 15.2 on page 488) the program typically checks the state of EEIF (bit 4 of EECON1) until it is set, then clears it and continues on.

```
W_LOOP  btfss  EECON1,EEIF  ; Poll the state of the EEIF flag
        goto   W_LOOP       ; IF still zero THEN try again

; ELSE continue after clearing the write-to EEPROM flag
        bcf    EECON1,EEIF
```

In common with all interrupt-driven systems, special care has to be taken where multiple sources of interrupt are being handled. As an example, consider a certain system receiving interrupt requests from the Timer at, say, 1000 times per second and externally through the INT pin at an irregular rate. If the INT handler takes, say, 4 ms to execute, then on return three Timer requests will have been lost! Some processors[6] have interrupt priority logic so that a higher-priority request (the Timer here) will interrupt a lower-priority process (the INT handler). Here the only solution is to ensure that the latter never takes more than 1 ms to execute.[7] In situations where interrupts from several sources occur, a worst-case scenario budget of execution times (including latency) and interrupt rates must be made. As some of these parameters are related to external events beyond the control of the processor, this can be a non-trivial exercise.

Another problem which frequently arises is dealing with events where multiple-precision data are monitored and changed by both background and foreground routines. Consider as an example a real-time clock (RTC) which updates four Files holding time in the 4-byte multiple-precision format HOURS:MINUTES:SECONDS:JIFFY, where the JIFFY byte holds tenths of seconds; see Example 7.3. We assume an external 10 Hz oscillator interrupts the PIC MCU ten times per second, and the ISR updates the time-array.

Consider now that this RTC is part of a central heating controller. At 09:00:00:00 hours the water pump is to be toggled from on to off by the background program. One day this has been done and the time is now 09:59:59:09. The background program, which spends most of its time just looking at the time, reads the hours as 09. Getting interested, it is just going to read the Minutes variable when the Jiffy oscillator "ticks." The MCU is interrupted and the RTC now is updated to 10:00:00:00. On return the background program now reads in succession 00, 00, 00. Thinking that it is now 09:00:00:00 it toggles the pump off and thereafter the on and off periods are interchanged indefinitely!

---

[6]Such as the enhanced-range family.

[7]Rather inelegantly the latter could poll the Timer's interrupt flag T0IF at regular intervals.

Of course it is bad design practice to use a toggle action; instead the pump should be switched on at 9 am rather than toggled. At least in this latter case the harm done would be time limited. In general, the interrupt handler should be disabled by clearing the appropriate mask where such multiple-precision data manipulation routines are being executed in the background. Any interrupts occurring during this time will be acknowledged when the mask is subsequently set, although events could be missed if the masked-out period is too long.

In conclusion, ISRs are similar to subroutines, but keep in mind the following points:

- The ISR should be terminated by retfie instead of return.
- W and any SPRs that are to be altered should be saved on entry and retrieved on exit if they are also used in the background program.
- Parameters cannot be passed to and from the ISR via the Working register. Instead, global variables (data in known memory locations) should be used as required.
- ISRs should be as short as possible, with minimal functionality. This helps in debugging, and helps ensure that other events are not missed.
- Where multiple-byte data objects are being processed by an ISR, consideration should be given to disabling the interrupt system (by clearing GIE) during any background access.

The INTCON register of Fig. 7.3 is only capable of handling the General Interrupt Enable bit plus the three core interrupt sources; namely Hardware, Timer 0 and Port B change. The bit still left in the PIC16F84 is used to hold the EEPROM module's Enable mask bit EEIE, located in INTCON[6]. Due to lack of space, the corresponding interrupt flag EEIF is located separately as bit 4 of the EECON1 register. Eighteen-pin mid-range devices introduced in the same era frequently used the same technique. For instance, the PIC16C71 Analog-to-Digital Convertor module's mask flag ADIE is located in INTCON[6] and the matching interrupt flag ADIF resides in the Analog Digital CONtrol register 0 ADCON0[1].

Apart from these few original family members, the interrupt system has to cope with a multitude of peripheral modules, each of which can request one or more interrupt service. One way of implementing this is to devolve the interrupt and enable mask bits to local peripheral control and status registers, in the same manner as we have described for the PIC16F84's EEPROM module. A more consistent approach is to consider the interrupt system as a single entity and introduce additional registers to hold all these interrupt flags and mask bits together in one place.

Figure 7.5 shows the logic of the PIC16F627/8, which has five peripheral modules beyond the basic core, generating in total seven distinct interrupt requests. The right-hand side of the diagram is virtually identical to Fig. 7.3, with the only difference being INTCON[6], which is labeled

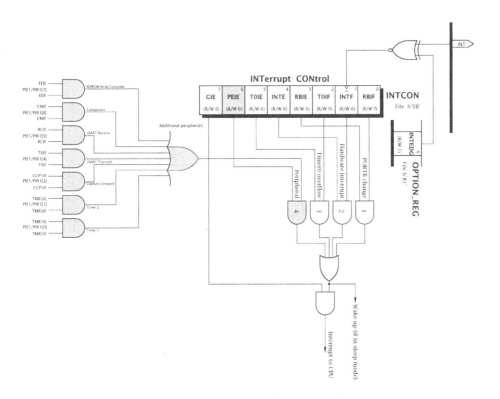

Fig. 7.5 The PIC16F627/8 MCU's interrupt logic.

PEIE for PEripheral Interrupt Enable.[8] Shown shaded to the left of the main diagram are the seven additional interrupt sources and their enable gates. These are ORed together to give a single line, which is in turn gated with the PEIE mask bit using AND 4. Thus INTCON[6] now performs the role of group Peripheral module enable.

Mid-range devices have differing mixes of peripheral devices, but all use PEIE as a second-level mask bit to enable/disable these modules as a set. If enabled, any activity within this group can awaken the PIC MCU if in the Sleep mode irrespective of the state of the GIE overall mask bit.

As seen from a register perspective, Fig. 7.6 shows two additional registers. Peripheral Interrupt Register 1 (PIR1) in Bank 0 which holds the seven interrupt flags and Peripheral Interrupt Enable 1 (PIE1) which holds the corresponding Enable mask bits. Shown as an example is the logic of an interrupt generated when a serial character (see Fig. 12.20 on page 375) arrives at the serial port, which sets the ReCeive Interrupt Flag (RCIF), shown shaded in the diagram in PIR1[5]. To enable this event (i.e., RCIF

---

[8]Some data sheets more sensibly label this GIEL for Global Interrupt Enable Low, with the plain GIE being labeled GIEH for Global Interrupt Enable High.

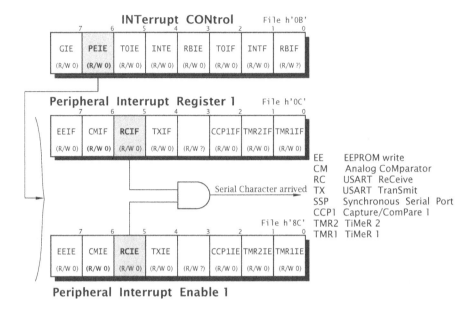

Fig. 7.6 A register view of the PIC16F627/8 MCU's interrupt system, showing the logic of the USART's Serial Character arrived interrupt request enable.

going to 1) to generate an interrupt and force the processor to go to the ISR, we need to set three Enable mask bits.

```
bsf   STATUS,RP0    ; Change to Bank 1 (RP[1:0] = 01)
bcf   STATUS,RP1    ; to get at the PIE1 register
bsf   PIE1,RCIE     ; Enable the ReCeive serial Interrupt
bsf   INTCON,PEIE   ; Enable the Peripheral group
bsf   INTCON,GIE    ; Enable the complete interrupt system
bcf   STATUS,RP0    ; Back to Bank 0
```

Several points should be made about this listing. Firstly, like most mid-range PIC MCUs, the PIC16F627/8 has four banks of register Files, as selected by switch bits RP1 and RP0 in the Status register; see Fig. 5.4 on page 102. On Reset both switches are zeroed; i.e., Bank 0, and so we didn't need to explicitly clear bit RP1 above. PIE1 is always in Bank 1, as once it is set up it usually needs no further change, whereas for convenience PIR1 is in Bank 0 as it will frequently be polled and altered. INTCON is shadowed across all four banks.

Some processors, such as the PIC16F87X, have too many sources of interrupts even for this scheme. Such devices have a second pair of registers PIR2 and PIE2 in the same banks as the first equivalent registers, to hold the additional enable mask bits and interrupt flags.

Examples

Example 7.1

Consider a conveyor belt in a pea-canning factory. As part of the automatic packing system, a photocell generates a single short pulse for each passing can, in the manner of Fig. 7.4. After each batch of 24 cans, a nominal 1 ms pulse _/‾_ is to be generated using Port A's pin 0 (RA0) and this triggers the packing mechanism's electronics. Assume that the PIC16F84 is being clocked using a 4 MHz crystal.

Solution

The software is shown in Program 7.2. The Reset vector at h'000' actions a jump to the Main background routine and the Interrupt vector at h'004' causes execution to transfer to the foreground ISR labeled CAN_COUNT.

As interrupts are automatically disabled on a Power-on reset, the various Files and ports are normally set to their initial value at the beginning of the background program before interrupts are enabled. This eliminates the possibility of servicing an interrupt before the initialization code has been completed. The initialization schedule is:

1. Clearing bit 0 of Port A will ensure that pin RA0 starts low after Reset.
2. All parallel port lines are configured as inputs on Reset. To change Port A bit 0 to an output, the associated bit in the TRISA SPR must be cleared. As TRISA is located in Bank 1, RP0 in STATUS is used to switch banks; see page 82. More details are given in Chapter 11.
3. File EVENT recording the photocell pulse count and BATCH; which is set to non-zero in the ISR whenever a batch of 24 cans has passed, are both zeroed.
4. Clearing all bits in the INTCON register clears all interrupt flags that may have been set since Reset. This is important as such flags may be set irrespective of the state of the associated mask bits. Setting the Global Interrupt Enable mask bit now enables the interrupt system and specifically setting INTE enables interrupts from the INT pin.

The core of the main background routine simply repetitively checks the content of BATCH. This is normally zero, but the foreground ISR sets this whenever each batch of 24 cans have passed. When this is the case BATCH is zeroed and RA0 is brought high and a 1 ms delay subroutine[9] called DELAY; see Program 6.1 on page 154. The loop is then repeated. In general, the background routine in embedded systems is an **endless loop** of this form, although executing rather more tasks than this simple example. For instance, a 7-segment multidigit display may be continually

---

[9]Of course the delay subroutine can be interrupted, which will randomly slightly lengthen the delay. In time-critical situations GIE should be zeroed before calling the delay subroutine and set on return.

Program 7.2 Program for the pea-canning packer.

```
_work    equ   h'7E'; Place for the background Working register
_status  equ   h'7F'; and the background Status register
EVENT    equ   h'20'; Keeps count of cans of peas
BATCH    equ   h'21'; Signals when a lot of 24 cans have passed
; ------------------------------------------------------------
         org   000       ; Resets here
         goto  MAIN      ; Go to start of background routine
; ------------------------------------------------------------
         org   004       ; The Interrupt vector
         goto  CAN_COUNT ; Go to start of foreground ISR
; ------------------------------------------------------------
; Background program starts by setting up and initialization
MAIN     bcf   PORTA,0   ; Ensure that starts with pin RA0 low
         bsf   STATUS,RP0 ; Change to Bank1
         bcf   TRISA,0   ; Make bit RA0 an o/p by clearing TRIS0
         bcf   STATUS,RP0 ; Go back to Bank0
         clrf  BATCH     ; Zero the Batch signal
         clrf  EVENT     ; and the can count
         clrf  INTCON    ; Zero any set interrupt flags
         bsf   INTCON,GIE ; Enable all interrupts
         bsf   INTCON,INTE; Enable external INT-pin interrupts
; ------------------------------------------------------------
; WHILE Batch signal is zero DO nothing
M_LOOP   movf  BATCH,f   ; Check BATCH == 0?
         btfsc STATUS,Z  ; Skip out if No
         goto  M_LOOP    ; IF not THEN try again
; Pulse on the 24th can ---------------------------------------
         clrf  BATCH     ; Zero the Batch signal
         bsf   PORTA,0   ; Bring line RA0 high
         call  DELAY     ; Wait for one millisecond
         bcf   PORTA,0   ; and go low again
         goto  M_LOOP    ; DO forever
; ------------------------------------------------------------

; ************************************************************
; This is the interrupt handler foreground program        ***
CAN_COUNT
         movwf _work     ; Save current W reg. in Data memory
         swapf STATUS,w  ; Get current Status, don't change flags
         movwf _status   ; and put away in Data memory
; ============================================================
         bcf   INTCON,INTF ; Clear the Hardware interrupt flag
         incf  EVENT,f   ; Record one more event
         movf  EVENT,w   ; Get count
         addlw -d'24'    ; Compare with 24 (EVENT - 24)
         btfss STATUS,C  ; IF EVENT is higher or same THEN skip
         goto  CAN_EXIT  ; ELSE jump to exit point
         clrf  EVENT     ; Zero can count and
         incf  BATCH,f   ; tell the world that 24 cans have passed
; ============================================================
CAN_EXIT
         swapf _status,w ; Untwist & get original Status from mem
         movwf STATUS
         swapf _work,f   ; Now get original Working register from
         swapf _work,w   ; Data memory without altering flags
         retfie          ; and return to interrupted background
```

scanned in the manner described in Fig. 11.16 on page 326, showing, say, the total number of cans since the start of business.

When an interrupt occurs, as triggered by a can breaking a beam then execution will be transferred to the ISR, that is, Interrupt ⤳ h'004' → CAN_COUNT. This foreground routine is divided into three phases in the usual manner.

Context Switching
Both the Working and Status registers are saved in the Data store, as described on page 194.

Core Function
The Hardware interrupt flag, INTF in INTCON[1], is cleared to ensure that on return to the background program another interrupt request is not immediately auctioned indefinitely. In situations where there is more than one source of interrupt, the various interrupt flags would be polled at this stage of the program; see page 196.

The functional sector of the core function simply increments the datum EVENT. By subtracting 24 from a copy of this count and checking for no borrow (i.e., C is 1), the program determines when the Event tally rises to 24, or even, due to a software glitch, above. When this occurs BATCH is incremented to signal the background program that 24 cans have passed and EVENT is zeroed to give a module-24 count.

Restoring the Context
The exit process swaps out of memory the state of the Status and Working registers, as described on page 194. The final exit instruction `retfie` does not alter the flag state.

Example 7.2
In a food processing factory, cans of baked beans on a conveyer belt continually pass through a tunnel oven, as shown at the top of Fig. 7.7, where the contents are sterilized. Photocell detectors are used to sense cans, both entering and leaving the oven. The output of the sensors are logic 1 when the beam is broken.

You are asked to design an interrupt-driven interface for this system, combining the two signals to activate the PIC MCU's *one* INT input. A buzzer connected to Port B's bit RB0 is to be sounded if the number of tins in the oven exceeds four, indicating that a jam has occurred.

Solution
The hardware aspect of this example presents two problems. The first of these involves distinguishing which cell, IN or OUT, generates a request. In Fig. 7.7 each cell clocks an associated D flip flop when the beam is broken. As the D input is permanently tied to logic 1, the clocked flip flop output goes to logic 1. ORing both of these interrupt flags together generates a falling edge at the INT pin if *any* beam is broken.

Both the IN and OUT external flags can be read at Port A pins RA0 and RA1, and this allows the ISR software to distinguish between the two events (can-in and can-out). The appropriate flag can then be reset by toggling the appropriate flip flop reset using two further port lines RA2 and RA3 for Cancel_in and Cancel_out, respectively.

To show how this operates, consider a can has just broken the Out cell beam, as shown in the diagram. The following sequence occurs.

1. The resulting pulse clocks the OUT flag.
2. The flip flop goes high which in turn brings pin RA1 high and via the OR gate pin INT/RB0. This requests a Hardware interrupt.
3. When the PIC MCU transfers to the interrupt handler it checks the state of both flip flops by testing pins RA1 and RA2. In this case it finds RA1 high and in software pulses pin RA3 low.
4. This resets the OUT flip flop and hence cancels the interrupt request from this source.

One problem remains: If one event follows another before the ISR software has time to reset the appropriate external flip flop, that second event will be missed, as the OR gate will hold INT low. In this situation

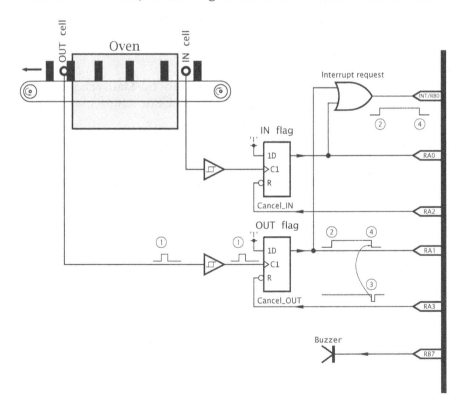

Fig. 7.7 Oven safety hardware.

no further edge can occur and the interrupt system will be permanently disabled! This can be circumvented in software by polling both external flags before exiting the ISR and taking the appropriate action if both bits are not clear.

The interrupt service routine for this hardware configuration is given in Program 7.3. As described on page 194 the context is saved on entry and restored on exit.

The meat of the code simply resets the internal INTF flag and checks each of the external flip flops in turn. Depending on the state of these flip flops, one of three paths through the code is followed:

Program 7.3 Foreground ISR for oven safety.

```
OVEN    movwf   _work          ; Save current W reg. in data memory
        swapf   STATUS,w       ; Get status, don't change flags
        movf    _status        ; and put away in data memory
; -----------------------------------------------------------------
CHECK   bcf     INTCON,INTF    ; Clear the Hardware interrupt flag
        btfsc   PORTA,0        ; Check, IN signal?
        goto    IN             ; IF non-zero, a can has just gone in
        btfsc   PORTA,1        ; Check for OUT signal
        goto    OUT            ; IF non-zero, a can has just gone out
; -----------------------------------------------------------------
; The exit point
        swapf   _status,w      ; Untwist & get old status from memory
        movwf   STATUS
        swapf   _work,f        ; Now get original W register from
        swapf   _work,w        ; Data memory without altering flags
        retfie                 ; and return to interrupted background
; -----------------------------------------------------------------
; The ISR core
IN      incf    EVENT,f        ; Record a can gone in (count up)
        bcf     PORTA,2        ; Clear external IN flag
        bsf     PORTA,2        ; by pulsing its reset
        goto    ALARM          ; and check for alarm situation

OUT     decf    EVENT,f        ; Record a can gone out (count down)
        bcf     PORTA,3        ; Clear external OUT flag
        bsf     PORTA,3        ; by pulsing its Reset

ALARM   movf    EVENT,w        ; Get Can count
        addlw   -5             ; Can count - 5
        btfss   STATUS,C       ; IF no borrow THEN sound the alarm
        goto    BUZ_OFF        ; ELSE OK, turn the buzzer off
        bcf     PORTB,0        ; Turn buzzer alarm on
        goto    CHECK          ; and repeat poll of cell flags
BUZ_OFF
        bsf     PORTB,0        ; Turn buzzer off
        goto    CHECK          ; and repeat poll of cell flags
```

1. If pin RA0 is high then a can has broken the IN beam and one is added to the Event counter kept in a File labeled EVENT. The external IN flip flop is reset. If the total is greater than four, the buzzer is turned on by bringing RB0 low, otherwise it is turned off. Repeat check.
2. If pin RA1 is high then a can has broken the OUT beam and one is taken away from the Event counter. This time the external OUT flip flop is reset. Again the total is checked against the boundary of four and the buzzer set to its appropriate state. Repeat check.
3. If neither flip flop is set then the ISR exits after restoring the context. This sequence is repeated whenever actions 1 or 2 have been completed. This ensures that the situation where both beams are broken simultaneously or within a short time window, will be properly serviced.

The main background program is not shown here. It will be similar to that of Program 7.2 in that the various ports will be set up, the Event counter File cleared and interrupts enabled. It is likely that this background program will be in charge of sounding the alarm and other consequental tasks rather than implementing this as part of the ISR, in keeping with the philosophy of reducing the size of the foreground code. In a practical system the background program would probably drive a numeric display showing the aggregrate of cans (four was a ridiculous value, chosen for illustrative purposes only) in the oven. Also some means of resetting to a non-zero value after a jam and some sign in the (erroneous) event of a subzero count being computed must be facilitated.

Example 7.3
On page 197 a central heating real-time clock was discussed. Write an ISR to add one onto the array of Files holding the four time bytes in a 24-hour time representation, on each 0.1 s interrupt. Each byte location is to hold two binary-coded decimal (BCD) digits; for instance BCD 40 in the File labeled MINUTES is represented as b'0100 0000'. This format is known as **packed binary-coded decimal** format.

Solution
Each time the PIC MCU enters the ISR one Jiffy must be added to the array of bytes HOURS:MINUTES:SECONDS:JIFFY. The base of each byte count differs in that JIFFY rolls over at a count of ten (i.e., modulo-10), SECONDS and MINUTES have a modulo-60 count and HOURS is modulo-24. Based on this scenario we have as a task list:
1. Add one onto the JIFFY count.
2. IF this gives 10 THEN zero JIFFY and add one onto the SECONDS count; ELSE goto EXIT.
3. IF this gives 60 THEN zero SECONDS and add one onto the MINUTES count; ELSE goto EXIT.
4. IF this gives 60 THEN zero MINUTES and add one onto the HOURS count; ELSE goto EXIT.
5. IF this gives 24 THEN zero HOURS.
6. EXIT

Program 7.4 Coding the real-time clock ISR.

```
_work     equ    h'4D'      ; Space for a copy of W
_status   equ    h'4E'      ; Space for a copy of STATUS
_fsr      equ    h'4F'      ; Space for a copy of the FSR
HOURS     equ    h'20'      ; Space for the 2-digit Hour count
MINUTES   equ    h'21'      ; Space for the 2-digit Minute count
SECONDS   equ    h'22'      ; Space for the 2-digit Seconds count
JIFFY     equ    h'23'      ; Space for the 0.1s predivision

; First save the context ----------------------------------------
RTC       movwf  _work      ; Put away W
          swapf  STATUS,w   ; and the Status register
          movwf  _status
          movf   FSR,w      ; and the File Select Register
          movwf  _fsr
          bcf    INTCON,INTF ; Clear the Hardware Int flag
; The core code ================================================
; Task1
          incf   JIFFY,f    ; Add one onto Jiffy count
; Task2
          movlw  d'10'      ; Compare to ten
          subwf  JIFFY,w
          btfss  STATUS,Z   ; IF equal THEN continue
          goto   EXIT       ; ELSE finished
          clrf   JIFFY      ; ELSE clear Jiffy count
; Task3
          movlw  SECONDS    ; Point FSR to Seconds count
          movwf  FSR
          call   BCD_INC    ; and increment in BCD
          movlw  h'60'      ; Compare with 0110 0000 (60 BCD)
          subwf  SECONDS,w
          btfss  STATUS,Z   ; IF equal THEN continue
          goto   EXIT       ; ELSE finished
          clrf   SECONDS    ; ELSE clear Seconds count
; Task4
          decf   FSR,f      ; Point FSR to Minutes count
          call   BCD_INC    ; and increment in BCD
          movlw  h'60'      ; Compare with 0110 0000 (60 BCD)
          subwf  MINUTES,w
          btfss  STATUS,Z   ; IF equal THEN continue
          goto   EXIT       ; ELSE finished
          clrf   MINUTES    ; ELSE clear Minutes count
; Task5
          decf   FSR,f      ; Point FSR to Hours count
          call   BCD_INC    ; and increment in BCD
          movlw  h'24'      ; Compare with 0010 0100 (24 BCD)
          subwf  HOURS,w
          btfsc  STATUS,Z   ; IF not equal THEN continue
          clrf   HOURS      ; ELSE zero Hours count
; Retrieve the context ========================================
EXIT      movf   _fsr,w     ; Get the original FSR back
          movwf  FSR
          swapf  _status,w  ; Untwist the original Status reg
          movwf  STATUS
          swapf  _work,f    ; Get the original W reg back
          swapf  _work,w    ; leaving STATUS unchanged
          retfie            ; and return from interrupt
```

Coding for this task list is given in Program 7.4. Saving and restoring the context is implemented in the normal way. However, as the File Select Register is used in the core of the ISR, it too is saved in the File labeled _fsr and retrieved on exit.

Program 7.5 Incrementing a packed-BCD byte with maximum value of 99.

```
;  ****************************************************************
;  * FUNCTION: Adds one onto packed BCD byte, maximum value 99   *
;  * ENTRY    : FSR points to byte                               *
;  * EXIT     : BCD byte incremented; W and STATUS altered       *
;  ****************************************************************
BCD_INC   incf   INDF,f      ; Add one onto pointed-to BCD byte
          movf   INDF,w      ; Get it down
          addlw  6           ; Add six
          btfss  STATUS,DC   ; Check Decimal half Carry
           goto  BCD_EXIT    ; IF none THEN OK to exit
          movwf  FSR         ; ELSE corrected value put away
BCD_EXIT  return
```

The core of the ISR is sectioned as shown to follow the task list. After each incrementation, the base literal is subtracted from the datum. If they are equal, then the datum is zeroed and the next datum incremented. The alternative of checking the Carry/Not Borrow flag would implement this task if the datum was equal or higher than the base literal, btfss STATUS,C.[10]

The example specified that the datum format should be packed BCD. Thus, 59 minutes should be stored as b'0101 1001' or h'59'. This means that the incrementation process has to preserve this BCD format. This can be done after a normal increment by checking that the least significant nybble has not gone above nine. If it has, then six is added to correct the situation. As no datum should be above 59 this process is not needed for the upper nybble. See Example 4.5 on page 92 for a task list for a complete packed BCD increment.

As this process needs to be carried out three times (for all except the Jiffy byte which is never greater than nine) then it is best implemented as a subroutine. This is shown in Program 7.5. Here the FSR is pointing to the packed-BCD datum that has to be incremented. This datum is simply binary incremented in situ using Indirect addressing. It is then corrected as described. The subroutine assumes that the pointed-to datum is already in a packed-BCD format on entry; i.e., it does not convert a natural binary byte to BCD.

---

[10]This is more robust than equality as it is conceivable that due to a software bug a time datum could be set to a value outside the legitimate range.

Example 7.4

A certain vending machine channels coins of various denominations past one of six microswitches connected to Port B. Any coin will close one switch and pull the appropriate pin low, as shown in Fig. 7.8.

Write the foreground ISR so that the appropriate quantity is added to a File called MONEY. You can assume that the background routine has set up the INTCON register to enable Hardware interrupts via the RB0/INT pin.

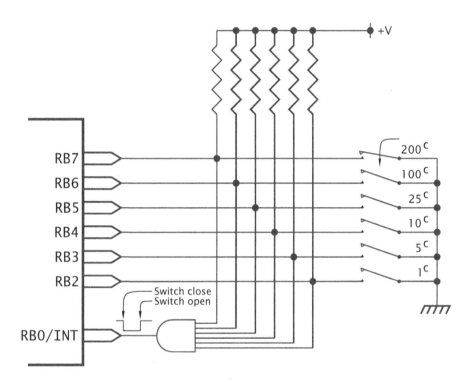

Fig. 7.8 Coin entry for a vending machine.

Solution

As shown in Program 7.6, after saving the context and clearing INTF, each switch is tested in turn. Any pin which is low reflects a logic 0 in the corresponding Port B bit. With the coin mechanism outlined, only one switch will be closed at any time, so the scanning need not exit after a successful find.

Program 7.6 Interrupt handler for the vending machine.

```
VEND   movwf   _work        ; Save current W reg. in Data memory
       swapf   STATUS,w     ; Get Status, don't change flags
       movwf   _status      ; and put away in data memory
; ================================================================
CHECK bcf     INTCON,INTF  ; Clear the Hardware interrupt flag
       movf    MONEY,w      ; Get current money tally

       btfss   PORTB,7      ; Check for $2
       addlw   d'200'       ; IF 0 THEN add 200
       btfss   PORTB,6      ; Check for $1
       addlw   d'100'       ; IF 0 THEN add 100
       btfss   PORTB,5      ; Check for 25c
       addlw   d'25'        ; IF 0 THEN add 25
       btfss   PORTB,4      ; Check for 10c
       addlw   d'10'        ; IF 0 THEN add 10
       btfss   PORTB,3      ; Check for 5c
       addlw   5            ; IF 0 THEN add 5
       btfss   PORTB,2      ; Check for 1c
       addlw   1            ; IF 0 THEN add 1

       movwf   MONEY        ; Copy sum to File MONEY
; ================================================================
; The exit point
       swapf   _status,w    ; Untwist & get old Status from memory
       movwf   STATUS
       swapf   _work,f      ; Now get original W register from
       swapf   _work,w      ; Data memory without altering flags
       retfie               ; and return to interrupted background
```

Self-Assessment Questions

7.1 Rewrite Program 7.2 to deal with a packing quantity of one gross (144). The count is to be kept in packed BCD (Hundreds and Tens:Units) which can be used by the background software to display the can tally.

7.2 What changes to Example 7.2 would you have to make to allow for a maximum value in the oven of 1000?

7.3 Based on Fig. 7.1, design an ISR to perform the following tasks:
   - Copy the 16-bit count into two GPR Files labeled TEMP_H and TEMP_L.
   - Deduct from the previous count reading located in LAST_COUNT_H and LAST_COUNT_L and place the difference in DIFFERENCE_H and DIFFERENCE_L.
   - Update the previous count with the new count.
   - Set a GPR labeled NEW to a non-zero value to signal the background software that a new reading is available. The background routine will clear NEW when it has processed the data.

7.4 The speed of a rotating shaft can be measured by using a coded disk to generate a pulse on each angular advance of 10°, which can be used to interrupt a PIC MCU. If the top speed is 20,000 revolutions per minute, what is the absolute maximum duration of the ISR in this worst-case situation to avoid missing pulses? You may assume a crystal frequency of 4 MHz.

7.5 An electronic tape measure determines distance by pulsing an ultrasonic transmitter and detecting the time it takes for the echo return. The hardware for this echo sounder is shown in Fig. 7.9 and is based on that of Fig. 7.7.

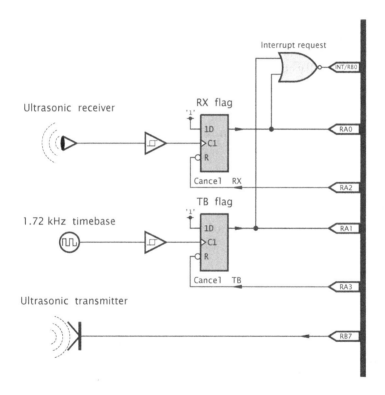

Fig. 7.9 Echo sounding hardware.

The maximum range is specified as 2.5 m with a resolution of 1 cm. The speed of sound in air is 344 m/s at 20°C, which gives a go-return time for one meter of 5.813 ms. Using a 1.72 kHz oscillator as a time base gives one interrupt per 5.813 ms; that is, a Jiffy per centimeter.

Based on this hardware, the software must implement the following task list:

- Background routine
  1. Zero Jiffy count and New flag.
  2. Pulse the sounder.
  3. Wait until New flag is non-zero.
  4. Display reading.
  5. Repeat forever.
- Foreground routine.
  1. IF oscillator THEN increment Jiffy count.
  2. IF receiver THEN set New flag to non-zero to tell background program that the Jiffy count is the final value.
  3. Repeat until neither is active.
  4. Return

Code the foreground ISR tasked above using a GPR as a flag labeled NEW to tell the background program that the echo has returned and to read the Jiffy count as the required value. Use Program 7.3 as your model.

7.6 It is proposed to increase the range of the digital echo sounder to 10 m and resolution to 1 mm. What change in the hardware and software would be required?

7.7 The system in SAQ 7.6 has been built and tested. However, readings seem to shift slowly with time. Drift is suspected but the oscillator has been proven to be stable. Thinking laterally, one student wonders if the speed of sound varies with atmospheric conditions. After some research the student arrives at the formula for temperature dependence as

$$V_t = V_0 \sqrt{1 + \frac{\Delta t}{273}}$$

where V_0 is the propagation velocity at 20°C and V_t is the velocity at a temperature of t. How much change in temperature Δt will there be to cause an error of 1 mm with the sounder measuring at its maximum range?

CHAPTER 8

Assembly Language Code Building Tools

We have now been writing programs with abandon since Chapter 3. For clarity these listings have been written in a human-readable form. Thus instructions have been represented as a short mnemonic, such as `return` instead of b'00000000001000'; the Files similarly have names, such as INTCON; lines have been labeled and comments attached. Such symbolic representations are only for human consumption. The MCU knows nothing beyond the binary codes making up operation codes and data, such as shown on page 49.

With the help of the device's instruction set (see page 70), it is possible to translate by hand from the human-readable symbolic form to machine-readable binary. This is not particularly difficult for a device such as a PIC MCU that has a reduced set of instructions (RISC) and few address modes. However, it is slow and tedious, especially where programs of a significant length are being coded. Furthermore, it is error prone and difficult to maintain whenever there are changes to be made.

Computers are good at doing boring things quickly and accurately; and translating from symbolic to machine code definitely falls into this category. Here we will briefly look at the various software packages that aid in this machine-level translation process. In the following chapter we will look at a high-level language alternative.

After reading this chapter you will:

- Know what assembly-level language is and how it relates to machine code.
- Appreciate the advantages of a symbolic representation over machine-readable code.
- Be aware of the function of the assembler.
- Understand the difference between absolute and relocatable assembly.
- Understand the role of a linker.
- Appreciate the process involved in translating and locating an assembly-level language program to absolute machine code.
- Understand the structure of a machine-code file and the role of the loader program.

- Be aware of the role of a simulator.
- Appreciate the use of the integrated development environment to automate the interaction of the various software tools needed to convert source code into a programmed MCU device.

The essence of the assembly-level conversion process is shown in Fig. 8.1. Here the program is prepared by the tame human in symbolic form, digested by the computer and output in machine-readable form. Of course, this simple statement belies a rather more complex process, and we want to examine this in just enough detail to help you in writing your programs.

```
incf   COUNT,f
movf   COUNT,w
addlw  6
btfsc  STATUS,DC
movwf  COUNT
return
```

Translate

```
00101010100000
00100000100000
11111000000110
01100100000101
00000010100000
00000000001000
```

Fig. 8.1 Conversion from assembly-level source code to machine code.

In general, the various translator and utility computer packages are written and sold by many software companies, and thus the actual details and procedures differ somewhat between the various commercial products. In the specific case of PIC MCU devices, Microchip Technology, Inc. as a matter of policy, has always provided their assembly-level software tools free of charge, a large factor in their popularity. For this reason commercial low-level software for the PIC MCU is relatively rare and what there is usually conforms to the Microchip syntax. For this reason we will illustrate this chapter with the Microchip suite of computer-aided code building tools.

Using the computer to aid in translating code from more user-friendly forms (known as **source code**) to machine-friendly binary code (known as **object code** or **machine code**) and loading this into memory began in the late 1940s for mainframe computers. At the very least it permitted the use of higher-order number bases, such as hexadecimal.[1] In this base the code fragment of Fig. 8.1 becomes:

```
0AA0
0820
3E06
1905
00A0
0008
```

---

[1] Actually base-8 (octal) was the popular choice for several decades.

A **hexadecimal loader** will translate this into binary and put the code in designated memory locations. This loader might be part of the software in your PIC-EPROM programmer. Hexadecimal coding has little to commend it, except that the number of keystrokes is reduced—but there are more keys—and it is slightly easier to spot certain types of errors.

As a minimum, a symbolic translator, or **assembler**,[2] is required for serious programming. This allows the programmer to use mnemonics for the instructions and internal registers, with names for constants, variables and addresses. The symbolic language used in the source code is known as **assembly language**. Unlike high-level languages, such as **C** or PASCAL, assembly language has a *one-to-one relationship* with the generated machine code, i.e., *one line* of source code produces *one instruction*. As an example, Program 8.1 shows a slightly modified version of Program 6.12 on page 175. This subroutine computes the square root of a 16-bit variable called NUM, which has been allocated two bytes in the Data store, and returns the 8-bit integer square root in the Working register.

Giving names to addresses and constants is especially valuable for longer programs, which typically comprise several thousand lines of code. Together with the use of comments, this makes code easier to debug, develop and maintain. For instance, in most of our programs up to now we have had statements such as:

```
STATUS   equ  3    ; Status register is File 3
C        equ  0    ; in which the Carry flag is bit 0
```

The pseudo instruction **equ** is a simple example of an **assembler directive**. A directive does not generate code like a processor instruction; rather, it is a command giving information from the programmer to the assembler concerning its operation. In this case, stating that whenever the name STATUS is encountered in an instruction operand field, it is replaced by the number 3 and that the name C is likewise is to be replaced by the number 0.

The **equ** directive is best suited to listing names of the SPRs and bits within. As these are fixed for a given member of the PIC MCU family, and therefore are not unique to any particular program, Microchip provide .inc files for each device. These can be *included* in user programs as a Header file.[3] For instance, Table 8.1 shows the first part of the file p16f84a.inc.[4]

In Program 8.1 the directive **include**[5] has been used to make known the SPR register set to the program. In addition to saving the programmer

---

[2]The name is very old; it refers to the task of translating and *assembling* together the various modules making up a program.

[3]Of course you can make your own version with additional information.

[4]The PIC16F84A introduced in 1999 is a slightly updated version of the standard PIC16F84 introduced in 1994.

[5]Microchip recommend the usage #include.

Program 8.1 Absolute assembly-level code for our square-root module.

```
; Global declarations
          include "p16f84a.inc"  ; Header file

          cblock h'26'    ; Begin block of variables @ File h'26'
          NUM:2           ; High byte (NUM), low byte (NUM+1)
          endc            ; End of block
; Dummy main loop ------------------------------------------------
MAIN      call   SQR_ROOT ; Dummy main loop
          sleep           ; Stop computing
; --------------------------------------------------------------
;
; ************************************************************
; * FUNCTION: Calculates the square root of a 16-bit integer *
; * EXAMPLE : Number = h'FFFF'; (65,535), Root = h'FF' (255) *
; * ENTRY   : Number in File NUM:NUM+1                        *
; * EXIT    : Root in W. NUM:NUM+1; I:I+1 and COUNT altered   *
; ************************************************************
; Local declarations
          cblock
          I:2, COUNT:1    ; Magic number hi:lo byte & loop count
          endc

          org    h'200'   ; Code to begin @ h'200' in Program store
SQR_ROOT  clrf   COUNT    ; Task 1: Zero loop count

          clrf   I        ; Task 2: Set magic number I to one
          clrf   I+1
          incf   I+1,f

; Task 3: DO
SQR_LOOP  movf   I+1,w    ; Task 3(a): Number - I
          subwf  NUM+1,f  ; Subtract lo byte I from lo byte Num
          movf   I,w      ; Get high byte magic number
          btfss  STATUS,C ; Skip if No Borrow out
          addlw  1        ; Return borrow
          subwf  NUM,f    ; Subtract high bytes

; Task 3(b): IF underflow THEN exit
          btfss  STATUS,C ; IF No Borrow THEN continue
          goto   SQR_END  ; ELSE the process is complete

          incf   COUNT,f  ; Task 3(c): ELSE inc loop count

          movf   I+1,w    ; Task 3(d): Add 2 to the magic number
          addlw  2
          btfsc  STATUS,C ; IF no carry THEN done
          incf   I,f      ; ELSE add carry to upper byte I
          movwf  I+1
          goto   SQR_LOOP

SQR_END   movf   COUNT,w  ; Task 4: Return loop count as the root
          return
          end
```

having to type in a set of **equ** directives for each program, any subsequent change in the processor, say from a PIC16F84A to a PIC16F627, can be accomplished by replacing the Header file to **p16f627.inc**. We will use this technique from now on. Although we have used **include** to insert

Table 8.1: Part of Microchip's file p16f84a.inc.

```
; This Header file defines configurations, registers, and other
; useful bits of information for the PIC16F84A microcontroller.
; These names match the data sheets as closely as possible.
;----- Register Files-----------------------------------------
INDF                       EQU        H'0000'
TMR0                       EQU        H'0001'
PCL                        EQU        H'0002'
STATUS                     EQU        H'0003'
FSR                        EQU        H'0004'
PORTA                      EQU        H'0005'
PORTB                      EQU        H'0006'
EEDATA                     EQU        H'0008'
EEADR                      EQU        H'0009'
PCLATH                     EQU        H'000A'
INTCON                     EQU        H'000B'
OPTION_REG                 EQU        H'0081'
TRISA                      EQU        H'0085'
TRISB                      EQU        H'0086'
EECON1                     EQU        H'0088'
EECON2                     EQU        H'0089'
;----- STATUS Bits ----------------------------------------------
IRP                        EQU        H'0007'
RP1                        EQU        H'0006'
RP0                        EQU        H'0005'
NOT_TO                     EQU        H'0004'
NOT_PD                     EQU        H'0003'
Z                          EQU        H'0002'
DC                         EQU        H'0001'
C                          EQU        H'0000'
;----- INTCON Bits ----------------------------------------------
GIE                        EQU        H'0007'
EEIE                       EQU        H'0006'
TOIE                       EQU        H'0005'
INTE                       EQU        H'0004'
RBIE                       EQU        H'0003'
TOIF                       EQU        H'0002'
INTE                       EQU        H'0001'
RBIF                       EQU        H'0000'
```

a Header file, it may be used to insert any relevant type of file, such as a
subroutine—for example, see Program 12.8 on page 360.

The directive **equ** has also been used to name variables stored in GPRs
in the data memory. Thus in Program 6.12 on page 175 we have:

```
NUM_H          equ    h'26'   ; Number high byte
NUM_L          equ    h'27'   ; Number low byte
```

Such names and locations are of course unique to the program rather
then any specific device. Program 8.1 uses the alternative directive pair
cblock-endc which lets the assembler take over the job of allocating
variables to specific Files, within given constraints. Sandwiched inside

these directives are listed the names of the variables and how many bytes each occupies. In our example we have:

```
cblock  h'26'    ; Begin block of variables at File h'26'
  NUM:2          ; Reserve two bytes for NUM
endc             ; End of block
```

where the colon-delimited number following specifies the number of bytes to be reserved for that name. Individual bytes within the variable can subsequently be accessed by using the arithmetic + operator; for instance, with a 3-byte variable SUM:3 byte 1 is SUM, byte 2 is SUM+1 and byte 3 is SUM+2.

The first code block in Program 8.1 is directed to begin at File h'26' by the programmer. In any subsequent cblock this specification can be omitted, in which case the new variables simply follow on. Thus I:2 is located at File h'27:28' and COUNT:1 at File h'29'. This approach is much more flexible than the programmer allocating locations by hand, as whenever modules are altered or new elements added, the allocations are automatically altered. In addition, changing any specific code block location, say from File h'26' to File h'20', will automatically alter the complete program variable set to the new set of locations.

A third way of naming entities is to use the #define directive. For example:

```
#define  6,7  BUZZER
```

enables us to use the string bsf BUZZER instead of bsf 6,7 to turn on a Buzzer connected to pin 7 of Port B (File 6).

For illustrative purposes the programmer has asked the assembler to place the subroutine beginning at location h'200' in the Program store. This is done using the **org** directive—see also Program 7.1 on page 193. Effectively the program label SQR_ROOT has been given the value h'200'.

The last line of Program 8.1 is the **end** directive. This command tells the assembler to ignore any following text, that is, to cease translation.

Of course symbolic translators demand more computing power than simple hexadecimal loaders, especially in the area of memory and backup store. Prior to the introduction of personal computers in the late 1970s, either mainframe, minicomputers or special-purpose MPU/MCU development systems were required to implement the assembly process. Such implementations were inevitably expensive and inhibited the use of such computer aids, and hand-assembled coding was relatively common.

Translation software essentially implements two tasks:

- conversion of the various instruction mnemonics and labels to their machine-code equivalents;
- allocation of the instructions and data to the appropriate memory location.

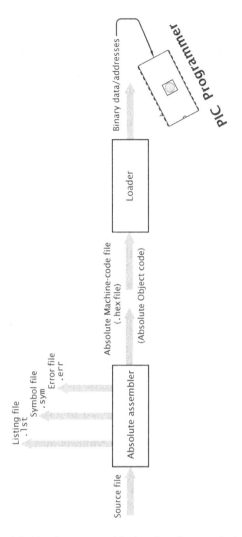

Fig. 8.2 Absolute assembly-level code translation.

Most programs running on the low- and mid-range PIC MCUs are adequately handled by an absolute assembler. To clarify the process outlined in Fig. 8.2, we will take our program through from the creation of the source file to the final absolute machine-code file. We will examine relocatable assemblers later on.

Editing

Initially the source file must be created using a **text editor**. A text editor differs from a word processor in that no embedded control codes, giving formatting and other information, are inserted. For instance, there is no line wrapping; if you want a new line then you hit the [ENT] key. Most operating systems come with a simple text editor; for example, notepad for Microsoft's Windows. Third-party products are also available and most word processors have a text mode which can double as a program editor. Microchip-compatible assembly-level source file names have an extension .src.

The format of a typical line of source code looks like:

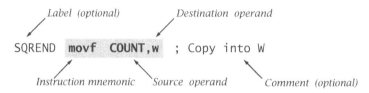

With the exception of comment-only lines, all lines must contain an instruction (either executable by the MCU or a directive) and any relevant operand or operands. Any label must begin in column 1, otherwise the first character must be a space or a tab to indicate no label. A label can be up to 32 alphanumeric, underline or question mark characters, with the proviso that the first character be a letter or underscore. Labels are usually case sensitive. A line label names the Program store address of the first following executable instruction. A space, colon or even new line should separate a label from the following instruction or directive.

An optional comment is delineated by a semicolon, and whole-line comments are permitted; see lines 9 – 16 of Program 8.1. Comments are ignored by the assembler and are there solely for human-readable documentation. Such notes should be copious and should explain what the program is doing, and not simply repeat the instruction. For instance:

```
movf I,w    ; Move I into W
```

is a waste of energy:

```
movf I,w    ; Get high byte of magic number
```

is rather more worthwhile. Not, or doing so only minimally, commenting source code is a frequent failing. A poorly documented program is difficult to debug and subsequently to alter or extend. The latter is sometimes known as program maintenance.

Space should separate the instruction from any operand. Where there are two operands, the source and destination fields are delineated by a comma. In instructions where the destination can be the Working register or the addressed File, the predefined names w or f should appear in the destination fields or numbers 0 or 1 respectively. The assembler will default to destination File if this is omitted, but warn the programmer.

Assembling

The assembler program will scan the source file, checking for syntax errors. If there are no such errors, the process goes on to translate to absolute object code; which is basically machine code with information concerning the locations in which it is to be placed in program memory. Syntax errors include such things as referring to labels that don't exist or instructions that are not recognized. The output will include an error file giving any such *faux pas*. If there are no syntax errors, a listing file and machine-code file are generated.

In the case of our example, the translation was invoked by entering:

```
mpasmwin /aINHX8M /e+ /l+ /c+ /rhex /p16f84a root.asm
```

where mpasmwin.exe is the the assembler program and root.asm is the specified source file. The flags are of the form /<option> and may be followed by + or – to enable or disable the option. Thus /e+ orders the production of an error file, /l+ likewise for a listing file, /c+ makes labels case sensitive, /rhex specifies the default base radix to be hexadecimal. The flag /p16f84a tells the assembler to treat the source file as pertaining to the PIC16F84A device. mpasmwin can translate code for all PIC microcontrollers; whether for 12-, 14- or 16-bit cores.

Listing

The **listing file** shown in Table 8.2 reproduces the original source code, with the addition of the hexadecimal location of each instruction and its code. The file also provides a symbol table enumerating all symbols/labels defined in the program; for instance, NUM is listed as File h'26'. The memory usage map gives a graphical representation of program memory usage. Any warning messages are embedded in the file where they are applicable. For instance, if the destination operand w or f is omitted, the assembler will default to the latter and embed a warning message at that point in the listing file.

This file has only documentation value and is not executable by the processor.

Executable Code

The concluding outcome of any translation process is the **object file**, sometimes known as the **machine-code file**. Once the specified code is in situ in the Program store, it may be run as the executable program.

As can be seen in Table 8.3, such files consist essentially of lines of hexadecimal digits representing the binary machine code, each preceded by the address of the first byte location of the line. This file can be used by the PIC MCU programmer to put the code into Program ROM memory at the correct place. Because the location of each code byte is explicitly specified, this type of file is known as **absolute object code**. The software component of the PIC MCU programming hardware (see Fig. 17.4 on page 547), reading, deciphering and placing this code is sometimes called an **absolute loader**.

In the MPU/MCU world there are many different formats in common use. Although most of these de facto standards are manufacturer-specific,

Table 8.2: The listing file root.lst.

```
MPASM 03.00 Released          ROOT_ABS.ASM    11-22-2003  10:58:58      PAGE   1
LOT OBJECT CODE LINE SOURCE TEXT
 VALUE
            01 ; Global declarations
            02         include "p16f84a.inc"  ; Header file
            01         LIST
            02 ; P16F84A.INC Standard Header File, V2.00 Microchip Tech, Inc.
            03
            04         cblock  h'26'     ; Begin block of variables @ File h'26
00000026 05            NUM:2             ; High byte (NUM), low byte (NUM+1)
            06         endc              ; End of block
            07 ; Dummy main loop -------------------------------------------
0000 2200  08 MAIN     call    SQR_ROOT ; Dummy main loop
0001 0063  09          sleep            ; Stop computing
            10 ; -----------------------------------------------------------
            11
            12 ; ***************************************************************
            13 ; * FUNCTION: Calculates the square root of a 16-bit integer *
            14 ; * EXAMPLE : Number = h'FFFF'; (65,535), Root = h'FF' (255) *
            15 ; * ENTRY   : Number in File NUM:NUM+1                       *
            16 ; * EXIT    : Root in W. NUM:NUM+1; I:I+1 and COUNT altered  *
            17 ; ***************************************************************
            18
            19 ; Local declarations
            20         cblock
00000028 21            I:2, COUNT:1      ; Magic number hi:lo byte & loop count
            22         endc
            23
0200       24         org     h'200'    ; Code to begin @ h'200' in Program sto
0200 01AA  25 SQR_ROOT clrf   COUNT     ; Task 1: Zero loop count
            26
0201 01A8  27          clrf    I        ; Task 2: Set magic number I to one
0202 01A9  28          clrf    I+1
0203 0AA9  29          incf    I+1,f
            30
            31 ; Task 3: DO
0204 0829  32 SQR_LOOP movf   I+1,w     ; Task 3(a): Number - I
0205 02A7  33          subwf   NUM+1,f  ; Subtract lo byte I from lo byte Num
0206 0828  34          movf    I,w      ; Get high byte magic number
0207 1C03  35          btfss   STATUS,C ; Skip if No Borrow out
0208 3E01  36          addlw   1        ; Return borrow
0209 02A6  37          subwf   NUM,f    ; Subtract high bytes
            38
            39 ; Task 3(b): IF underflow THEN exit
020A 1C03  40          btfss   STATUS,C ; IF No Borrow THEN continue
020B 2A13  41          goto    SQR_END  ; ELSE the process is complete
            42
020C 0AAA  43          incf    COUNT,f  ; Task 3(c): ELSE inc loop count
            44
020D 0829  45          movf    I+1,w    ; Task 3(d): Add 2 to the magic number
020E 3E02  46          addlw   2
020F 1803  47          btfsc   STATUS,C ; IF no carry THEN done
0210 0AA8  48          incf    I,f      ; ELSE add carry to upper byte I
0211 00A9  49          movwf   I+1
0212 2A04  50          goto    SQR_LOOP
            51
0213 082A  52 SQR_END  movf   COUNT,w   ; Task 4: Return loop count as the roo
0214 0008  53          return
            54         end
```

(continued on next page)

Table 8.2: (*continued*).

```
SYMBOL TABLE
   LABEL                                VALUE

C                                    00000000
COUNT                                0000002A
I                                    00000028
MAIN                                 00000000
NUM                                  00000026
SQR_END                              00000213
SQR_LOOP                             00000204
SQR_ROOT                             00000200
STATUS                               00000003
__16F84A                             00000001

MEMORY USAGE MAP ('X' = Used,   '-' = Unused)

0000 : X--------------- ----------------- ----------------- --------------
0200 : XXXXXXXXXXXXXXXX XXXXXX---------- ----------------- --------------

All other memory blocks unused.

Program Memory Words Used:     23
Program Memory Words Free:   1001

Errors   :     0
Warnings :     0 reported,    0 suppressed
Messages :     0 reported,    0 suppressed
```

in the main they can be used for any brand of microcontroller. The format of the machine-code file shown here is known as an 8-bit Intel hex and was specified with the flag /a INHEX8M.

Let us look at one of the lines in root.hex in more detail.

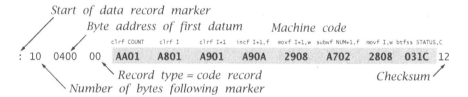

The loader recognizes that a record follows when the character : is received. The colon is followed by a 2-digit hexadecimal number representing the number of machine-code bytes in the record; h'10' = d'16'

Table 8.3: The absolute 8-bit Intel format object-code file root.hex.

```
:020000040000FA
:040000000022630077
:10040000AA01A801A901A90A2908A7022808031C12
:10041000013EA602031C132AAA0A2908023E031859
:0A042000A80AA900042A2A0808000F
:00000001FF
```

in this case. The next four hexadecimal digits represent the starting data address. As the PIC MCU's Program store addresses 2-byte words, instruction address h'200' translates to byte address h'400'. The following 2-digit number is h'00' for a normal record and h'01' for the end-of-file record; see the last line of Table 8.3.

The core of the record is the machine code, with each instruction taking two 2-digit hexadecimal bytes ordered low:high byte. The loader reads this lower byte first (e.g., h'AA') and then "tacks on" the upper byte (e.g., h'01') giving a 12-, 14- or 16-bit program word as appropriate to the target PIC MCU core, e.g., h'01AA' for a 14-bit core clrf h'2A'.[6]

The final byte is known as a **checksum**. The checksum is calculated as the 2's complement of the sum of all preceding bytes in the record; that is, $-$sum; ignoring any overflow. As a check-up on transmission accuracy, the loader adds up all received bytes including this checksum for each record. This received count should give zero if no download error has occurred.

Assemblers are very particular that the syntax is correct. If there are **syntax errors**[7] then an **error file** will be generated. For instance, if line 50 was mistakenly entered as:

```
got   SQRLOOP
```

then the error file of Table 8.4 below would be generated.

Table 8.4: The error file.

```
Warning[207]   ROOT.ASM  50 : Found label after column 1. (got)
Error[122]     ROOT.ASM  50 : Illegal opcode (SQRLOOP)
```

The assembler does not recognize got as an instruction or directive mnemonic and erroneously assumes that it is a label mistakenly not beginning in column 1. On this basis it assumes that SQRLOOP is an instruction/directive mnemonic and again does not recognize it.

Most assemblers allow the programmer to define a sequence of processor instructions as a **macro instruction**. Such macro instructions can

---

[6]Locating the multibyte code in memory in the Intel way, formatted low:high byte, is known as little-endian (working up from low to high address, the low byte end comes first) whereas the high-endian arrangement is favored by, amongst others, Motorola.

[7]If the assembler announces that there are no errors then there is a tendency to think that the program will work. Unfortunately a lack of syntax errors in no way guarantees that the program will do anything of the sort!

subsequently be used in a similar manner to native instructions. For example, the following code defines a macro instruction called `Delay_1ms`[8] that implements a 1 ms delay when executed on a PIC MCU running with a 4 MHz crystal. The directive pair **macro** - **endm** is used to enclose the sequence of native instructions which will be substituted when the mnemonic `Delay_1ms` is used anywhere in the subsequent program. The mnemonic will be replaced by the assembler with the defined code. Note that this will be in-line code, unlike calling up a subroutine.

```
Delay_1ms       macro
                local   LOOP

                movlw   d'250'    ; Count from 250
LOOP            addlw   -1        ; Decrement
                btfss   STATUS,Z  ; to zero
                goto    LOOP

                endm
```

Where labels are used within the body of the macro, they should be declared using the **local** directive. This means that any conflict with labels where a macro instruction is evoked more than once is avoided.

This example is unusual in that the "instruction" did not have any operands. Like native instructions, macros can have one or more operands. To see how this is done, consider a macro instruction called Bne for Branch if Not Equal (to zero).[9] Thus the instruction Bne NEXT causes execution to transfer to the specified label if the Z flag is zero, otherwise continue on as normal. The definition of Bne is:

```
Bne     macro   destination

        btfss   STATUS,Z
        goto    destination

        endm
```

Macro names should not be the same as for a real instruction; even from a different family.

Macros can be of any arbitrary complexity and can have any number of comma-separated operands. For instance, Microchip has available a large number of macros implementing arithmetic operations such as 16-bit × 16-bit and 32-bit × 32-bit multiplication. However, extensive use of macros can make programs difficult to debug, especially when an apparently simple macro instruction hides a number of side effects which

---

[8]I have capitalized the first letter of all macro instructions to distinguish them from native instructions.

[9]This is a native instruction for the PIC18XXXX family with the mnemonic bnz; see Table 16.1 on page 523.

alter register contents and flags. A frequent source of error is to precede a macro instruction with a Skip instruction, intending to go around it on some condition. As the macro instruction is in fact a structure of several native instructions, this skip will actually be into the middle of the macro—with dire consequences.

Macro definitions, whether commercial or/and in-house may be collected together as a single file and *included* in the user program using the include directive. Thus if your file is called mymacros.mac then the line at the beginning of your program

```
include   "mymacros.mac"
```

will allow access by the programmer to all macro definitions in the file. Any macros defined in the included file but not used will have no effect on the final machine code.

The process outlined up to here is known as absolute assembly. Here the source code is in a single file (plus maybe some included files) and the assembler places the resulting machine code in known (i.e., absolute) locations in the Program store. Where many modules are involved, often written by different people or/and coming from outside sources and commercial libraries, some means must be found to *link* the appropriate modules together to give the final single absolute executable machine-code file. For example, you may have to call up one of the modules that Fred is busy writing at the moment. You do not know exactly where in memory this module will reside or where its variables are stored, until the project is nearing its conclusion. What can you do? You should be able to call module FRED and refer to its component objects without knowing exactly what address they will be allocated.

The process used to facilitate this is shown in Fig. 8.3. Central to this modular tie-up is the **linker** program, which satisfies such external cross-references between the modules. Each module's source-code file needs to have been translated into **relocatable object code** prior to the linkage. "Relocatable" means that its final location and various addresses of external labels have yet to be determined. This translation is done by a **relocatable assembler**. Unlike absolute assembly, it is the linker that determines where the machine code is to be located in memory, rather than the human programmer, although absolute locations, say of the Ports, can still be specified.

Treating the linker as a type of task builder, its main functions are:

- To concatenate code and data from the various input module streams.
- To allocate values to symbolic labels which have not been given explicit fixed values by the programmer using equ and similar directives.
- To generate the absolute machine-code executable file together with any symbol, listing and link-time error files.

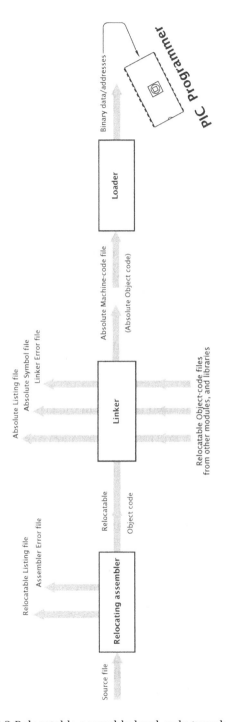

Fig. 8.3 Relocatable assembly-level code translation.

In order to allow the linker to do its job, it must have knowledge of the memory architecture of the target processor; basically where the array of general-purpose register Files start and end, where the vectors reside in program memory and where the code begins and ends. In the case of Microchip's mplink.exe linker this information is supplied in the form of a **linker command file**.

Table 8.5: The rms.lkr linker command file.

```
// File: rms.lkr
// Simple linker command file for PIC16F627 Created 23/11/2003
CODEPAGE   NAME=vectors  START=0x0    END=0x4
CODEPAGE   NAME=program  START=0x5    END=0x3FF

DATABANK   NAME=gprs     START=0x20   END=0x4F
DATABANK   NAME=auto     START=0x50   END=0x6F

SECTION    NAME=STARTUP  ROM=vectors  // Reset and int vectors
SECTION    NAME=TEXT     ROM=program  // ROM code space
SECTION    NAME=BANK0    RAM=gprs     // Bank0 static storage
SECTION    NAME=TEMP     RAM=auto     // Temporary auto storage
```

A simple example of such a command file for a PIC16F627 is given in Table 8.5. Three directives are used in the file.[10]

codepage

The codepage directive is used for code in the Program store. Here these directives are used to define two regions, one for the Reset and Interrupt vectors in between h'000' and h'004' called vectors, and the other called program to be used for executable code from h'005' through h'3FF'. Notice the use of the prefix 0x to denote the hexadecimal base. This is the notation used in the **C** language.

databank

This is similar to codepage but is used for variable data in RAM. Here the File array between File h'20' and File h'4F' is called gpr0 and between File h'50' and File h'6F' is called auto. The former is for general storage in Bank 0 and the latter is an area that the programmer can use for general storage in subroutines and reuse on return.

section

This linker directive names two code streams for the Program store. The first called STARTUP will be used by the programmer to store the two vector goto instructions while TEXT is used for the core program code. The source code assembler directive **code** with the appropriate label tells the linker into which stream any following code is to be placed; for example,

---

[10]Microchip's *MPASM™ Assembler Users Guide with the MPLINK™ Object Linker and MPLIB™* gives a full list of linker directives.

see Program 8.2. As many code sections from any codepage can be created as desired. For instance, all subroutines may be placed together in program memory by modifying the linker file thus:

```
SECTION  NAME=TEXT        ROM=program  // ROM code space
SECTION  NAME=SUBROUTINES ROM=program  // ROM subroutine stream
```

Sections can be made from DATABANK memory with RAM replacing the ROM attribute. In our case we have defined two streams. One called BANK0 is for storage that will persist for the entire program, and one called TEMP is for storage that can be thrown away when a subroutine is complete. The assembler directive **udata** (Uninitialized DATA) allows space to be reserved for labels in the general-purpose register array. The directive **udata_ovr** (Unitialized DATA OVeRlay) is used to tell the assembler that the programmer considers the following data storage is reusable between subroutines; see Program 8.4.

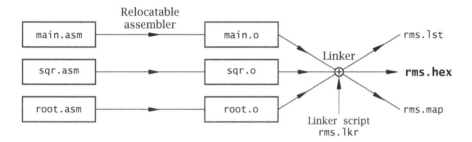

Fig. 8.4 Linking three source files to implement a root mean square program.

To illustrate the principle of linking we will implement the mathematical function $\sqrt{NUM\_1^2 + NUM\_2^2}$, known as root mean square. There are three teams working on this problem.[11] Tasks have been allocated by the project manager (a fourth person?) as follows:

1. The main function which sequences the steps:
   (a) Square NUM_1.
   (b) Square NUM_2.
   (c) Add $NUM\_1^2 + NUM\_2^2$
   (d) Square root item (c).
2. Design of a subroutine to square a byte number in the Working register to give a double-byte outcome in two GPRs.
3. Design of a subroutine to evaluate the square root of a double-byte sum and return it in W.

---

[11] Obviously this is a ridiculously simple problem for teamwork, but it illustrates the principle in a manageable space.

The process based on this decomposition of the task is shown diagrammatically in Fig. 8.4.

The main function is shown in Program 8.2. The program commences with the Reset goto instruction and is located in the STARTUP code stream. From the MAIN label onwards, code is located in the TEXT code stream using the directive TEXT code. We see from the map file output by the linker in Table 8.6 that MAIN is located at h'005'.

The main routine uses four variables located in data stream BANK0. These are placed in uninitialized RAM with the directives udata and **res** (REServe). A single File is reserved for each of the two input variables NUM_1 and NUM_2, respectively. Two bytes are reserved for SUM which is used to hold the sum NUM_1 + NUM_2. As this is to be the input for the subroutine SQR_ROOT, it is declared **global** at the end of the file. This means that the location is public; that is, additional files that are linked together can use the label SUM by declaring it **extern** (i.e., external to the

---

Program 8.2 The main relocatable source file main.asm.

```
            include    "p16f627.inc"
            extern     SQR_ROOT, SQR, SQUARE
; --------------------------------------------------------------
BANK0       udata                      ; Static data
NUM_1       res        1               ; The first number
NUM_2       res        1               ; The second number
SUM         res        2               ; Two bytes HI:LO for the sum
RMS         res        1               ; One byte for the outcome
; --------------------------------------------------------------
STARTUP     code
            goto       MAIN            ; The Reset vector
; --------------------------------------------------------------
TEXT        code
MAIN        movf       NUM_1,w         ; Get Number 1
            call       SQR             ; Square it
            movf       SQUARE+1,w      ; Get lower byte
            movwf      SUM+1           ; Is the low byte of sum
            movf       SQUARE,w        ; Get upper byte
            movwf      SUM             ; Is the high byte of sum

            movf       NUM_2,w         ; Now get Number 2
            call       SQR             ; Square it
            movf       SQUARE+1,w      ; Get lower byte
            addwf      SUM+1,f         ; Add to the low byte of sum
            btfsc      STATUS,C        ; Check if produces carry
            incf       SUM,f           ; Add the carry
            movf       SQUARE,w        ; Get upper byte
            addwf      SUM,f           ; Add to the high byte of sum

            call       SQR_ROOT        ; Work out the square root
            movwf      RMS             ; which is the root mean square

            sleep                      ; Stop executing

            global     SUM
            end
```

---

file). Variables not declared thus are "hidden" from the outside world, i.e., are private (or local) variables. In this manner the directive extern at the head of Program 8.2 allows the main routine to call the subroutines SQR_ROOT and SQR without knowing in advance where they are. In the same way the variable SQUARE is used by subroutine SQR to return the square of the byte sent to it in W. Space for this is reserved in a GPR in subroutine SQR and its exact location in the Data store is not known by main.asm but will be allocated later by the linker. From the map file of Table 8.6 it is finally located in File h'25' (high:low byte).

The main body of the code follows the task list enumerated above. The value NUM_1^2 is placed in Files SUM:SUM+1 to which the computed NUM_2^2 is added. The outcome is then used as input to subroutine SQR_ROOT to return the root-mean square byte in W. Finally this is copied to the File named RMS, for which a single byte has been reserved in the BANK0 Data stream.

The subroutine sqr.asm of Program 8.3 is based on the subroutine of Program 6.7 on page 165, which multiplies two byte numbers. In this case, on entry the contents of the Working register are copied to a File labeled X and a 16-bit version constructed in X_COPY_H:X_COPY_L. The shift and add algorithm then evaluates $X \times X = X^2$. These three Files are put in the data stream TEMP, with the directive udata_ovr to indicate to the linker that these Files can be reused by other modules. In the map file of Table 8.6 we see that X has been allocated File h'50' as has I, a variable in subroutine SQR_ROOT; see Program 8.3. This will make more efficient use of available data memory. Variables that are only alive within the subroutine that they are declared in are known in the C language as **automatic**, as their space is automatically reallocated as needed. The situation where variable space is preserved is known as **static**. Global variables, such as SQUARE are always static. In this case the variable SQUARE is created by reserving two bytes using the udata directive in the BANK0 stream. It is also published using the global directive, as is the name of the subroutine.

The final source file of the trio is the subroutine coded in Program 8.4. This is virtually identical to the absolute equivalent described in Program 8.1. Comparing the two, the org directive has been replaced by TEXT code and cblock by TEMP udata_ovr for the automatic local data. The data is passed to the subroutine SQR_ROOT via the external 2-byte global variable SUM, space for which has been allocated in main.asm. The subroutine name SQR_ROOT is published as global to make it visible to main.asm.

Like all source files, root.asm makes use of SPRs such as STATUS. For this reason the file p16f627.inc has been included at the head of each of the source file. Because this file comprises a set of equ directives, the names thus published are absolute and are not allocated or changed in

Program 8.3 The relocatable source file sqr.asm.

```
          include  "p16f627.inc"
; The SQR subroutine
; ************************************************************
; * FUNCTION: Squares one byte to give a 2-byte result       *
; * EXAMPLE : X = h'10' (16), SQUARE = h'0100' (256)         *
; * ENTRY   : X in W                                          *
; * EXIT    : SQUARE:2 (Low:High) in shared unitialized data *
; ************************************************************

BANK0     udata                 ; Static data
SQUARE    res     2             ; High:Low byte of square
; -----------------------------------------------------------
TEMP      udata_ovr             ; Auto data
X         res     1             ; Place for X
X_COPY_L  res     1             ; Holds a copy of X
X_COPY_H  res     1             ; Copy X overflow high
 byte
; -----------------------------------------------------------
TEXT      code

; Task 1: Zero double-byte square
SQR       clrf    SQUARE        ; High byte
          clrf    SQUARE+1      ; Low byte

; Task 2: Copy and extend X to 16-bits
          movwf   X             ; Put X away into data memory
          movwf   X_COPY_L      ; Copy of X
          clrf    X_COPY_H      ; and extend to double byte

; Task 3: DO
  ; Task 3A: Shift X right once
SQR_LOOP  bcf     STATUS,C      ; Clear carry
          rrf     X,f           ; Shift

    ; Task 3B: IF Carry == 1 THEN add 16-bit shifted X to square
          btfss   STATUS,C      ; IF C == 1 THEN do addition
          goto    SQR_CONT      ; ELSE skip this task

          movf    X_COPY_L,w    ; DO addition
          addwf   SQUARE+1,f    ; First the low bytes
          btfsc   STATUS,C      ; IF no carry THEN do high bytes
          incf    SQUARE,f      ; ELSE add carry
          movf    X_COPY_H,w    ; Next the high bytes
          addwf   SQUARE,f

    ; Task 3C: Shift 16-bit copy of X right once
SQR_CONT  bcf     STATUS,C      ; Zero Carry-in
          rlf     X_COPY_L,f
          rlf     X_COPY_H,f

    ; WHILE X not zero
          movf    X,f           ; Test multiplier for zero
          btfss   STATUS,Z
          goto    SQR_LOOP      ; IF not THEN go again
FINI      return                ; ELSE finished

          global SQUARE, SQR
          end
```

Program 8.4 The relocatable source file `root.asm`.

```
          include  "p16f627.inc"
          extern   SUM    ; The 2-byte number high:Low

TEMP      udata_ovr       ; Auto variables
I         res 2           ; Magic number high:low
COUNT     res 1           ; Loop count
; -----------------------------------------------------------
TEXT      code

SQR_ROOT clrf    COUNT    ; Task 1: Zero loop count

          clrf    I       ; Task 2: Set magic number I to one
          clrf    I+1     ; Low byte
          incf    I+1,f   ; = one

SQR_LOOP movf    I+1,w    ; Task 3(a): Number - I
          subwf   SUM+1,f  ; Subtract lo byte I from low byte Num
          movf    I,w     ; Get high byte magic number
          btfss   STATUS,C ; Skip if No Borrow out
          addlw   1       ; Return borrow
          subwf   SUM,f    ; Subtract high bytes

          btfss   STATUS,C ; IF No Borrow THEN continue
          goto    SQR_END  ; ELSE the process is complete

          incf    COUNT,f  ; Task 3(c): ELSE inc loop count

          movf    I+1,w    ; Task 3(d): Add 2 to the magic number
          addlw   2
          btfsc   STATUS,C ; IF no carry THEN done
          incf    I,f     ; ELSE add carry to upper byte I
          movwf   I+1
          goto    SQR_LOOP

SQR_END  movf    COUNT,w  ; Task 4: Return loop count as the root
          return

          global SQR_ROOT
          end
```

any way by the linker. Thus the linker map of Table 8.6 does not list such fixed symbols. They are, however, enumerated in the listing file produced by the linker.

In order to link the three source files together, the linker program must be given a command line listing the names of the input object files output by the relocatable assembler, the linker command file and the names of the output map and machine code file. In the case of our example this was:

```
mplink rms.lkr main.o sqr.o root.o /m rms.map /o rms.hex
```

which names the output map file rms.map and the absolute machine-code file rms.hex.

For documentation purposes the linker generates a composite listing file, similar (but more comprehensive) to that of Table 8.2 and an optional map file. The map file of Table 8.6 shows two lists. The first displays information for each section. This includes its name, type, start address, whether the section resides in program or data memory and its size in bytes. The Program Memory Usage table shows that 63 bytes of

Table 8.6: The output linker map file rms.map.

```
MPLINK 2.40.00, Linker
Linker Map File - Created Sun Nov 23 15:11:16 2003

                       Section Info
      Section       Type     Address    Location Size(Bytes)
    ---------   ---------   ---------   ---------   ---------
      STARTUP        code    0x000000     program    0x000002
       .cinit     romdata    0x000001     program    0x000004
         TEXT        code    0x000005     program    0x000078
        BANK0       udata    0x000020        data    0x000007
         TEMP       udata    0x000050        data    0x000003

              Program Memory Usage
                Start          End
              ---------    ---------
              0x000000     0x000002
              0x000005     0x000040
63 out of 1024 program addresses used, memory utilization is 6%

                Symbols - Sorted by Name
         Name    Address    Location    Storage  File
    ---------   ---------   ---------   ---------   ---------
         FINI    0x000040     program      static  SQR.ASM
         MAIN    0x000005     program      static  MAIN.ASM
          SQR    0x00002b     program      extern  SQR.ASM
     SQR_CONT    0x00003a     program      static  SQR.ASM
      SQR_END    0x000029     program      static  ROOT.ASM
     SQR_LOOP    0x00001a     program      static  ROOT.ASM
     SQR_LOOP    0x000030     program      static  SQR.ASM
     SQR_ROOT    0x000016     program      extern  ROOT.ASM
        COUNT    0x000052        data      static  ROOT.ASM
            I    0x000050        data      static  ROOT.ASM
        NUM_1    0x000020        data      static  MAIN.ASM
        NUM_2    0x000021        data      static  MAIN.ASM
          RMS    0x000024        data      static  MAIN.ASM
       SQUARE    0x000025        data      extern  SQR.ASM
          SUM    0x000022        data      extern  MAIN.ASM
            X    0x000050        data      static  SQR.ASM
     X_COPY_H    0x000052        data      static  SQR.ASM
     X_COPY_L    0x000051        data      static  SQR.ASM
```

program memory are used, including the two bytes of the Reset vector goto instruction — or around 6% of the possible total.

The second table shows information about the symbols in the composite program. Each symbol's location in either the Program or Data store is given together with the source file where it is defined. Global symbols are noted as extern. Local variables are all labeled static, including automatic reusable variables such as COUNT and X_COPY_H both at File h'52'.

Table 8.7: The resulting absolute object file rms.hex in INHEX8 format.

```
:02000000528D1
:040002000034003492
:06000A0020082B2026084F
:10001000A3002508A20021082B202608A307031807
:10002000A20A2508A2071620A4006300D201D0016D
:10003000D101D10A5108A3025008031C013EA202BB
:10004000031C2928D20A5108023E0318D00AD10005
:100050001A2852080800A501A601D000D100D2013B
:10006000C0310D00C031C3A285108A6070318A50A50
:100070005208A5070310D10DD20DD008031D30285A
:02008000080076
:00000001FF
```

The final outcome, shown in Table 8.7, is a normal executable machine code file. The format of this file is described for Table 8.3 and can be loaded into absolute program memory and run in the normal way.

Developing, testing and debugging software requires a large number of software tools, many of which we have discussed earlier, such as an editor, assembler and linker. In practice there are many other packages such as high-level language compilers (see Chapter 9), simulators and EPROM programmers; shown diagrammatically in Fig. 8.5. Setting up these tools and interacting on an individual basis can be quite complex, especially where products from differing manufacturers are involved. In this latter case, ensuring compatibility between the various intermediate file formats can be a nightmare.

Many software houses designing code development tools provide a graphical environment which integrates and sequences the process in a logical and easy to use manner. Of relevance to the PIC MCU family, Microchip Technology provides a Microsoft Windows-based **Integrated Development Environment** (IDE) called MPLAB®, which brings all compatible code development tools under one roof. Like all Microchip software tools (except **C** compilers) the MPLAB IDE is supplied free of charge.

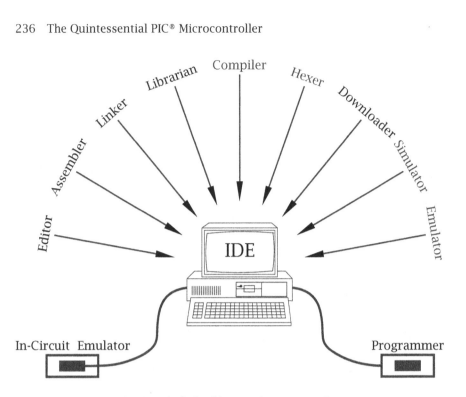

Fig. 8.5 Code building and testing tools.

MPLAB integrates Microchip-compatible tools to form a complete software development environment. Among its features are:

- A project manager which groups the specific files related to a project; for instance, source, object, simulator, listing and hex files.
- An editor to create source files and linker script files.
- An assembler, linker and librarian to translate source code and create libraries of code, which can be used with the linker.
- A simulator to model the instruction execution and I/O on the computer; see Fig 8.7.
- A downloader to work in conjunction with device programmers via the PC's serial or USB ports; see Fig. 17.4 on page 547.
- Software to emulate PIC MCUs in real time in the target hardware. This is accomplished by driving an In-Circuit Emulator (ICE)[12] or debugger via the PC's serial or USB port, replacing the target processor.

The Microchip manual *MPLAB® IDE User's Guide* gives a MPLAB IDE tutorial and reference details, which are beyond the scope of this book. However, for illustrative purposes two screen shots taken during the

---

[12]This is a hardware "pod" that replaces the PIC MCU in the target circuit and allows the PC to take over the running of the system.

development of our previous example linking main.asm, sqr.asm and root.asm are reproduced in Figs. 8.6 and 8.7.

Figure 8.6 shows the window displaying the Project file rms.mcw after the set-up wizard. The three source files, which have already been created using the editor, are specified. Also the name of the linker script file rms.lkr, which has also been previously created and saved. The resulting machine code file will be called rms.hex.

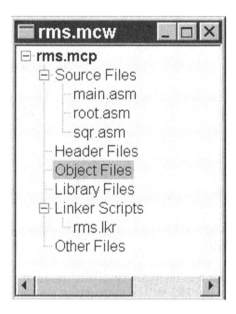

Fig. 8.6 Versions 6 and 7 MPLAB Project window, showing files selected to assemble, link and simulate Program 8.4.

Once the project specified in this manner, the sequence of operations, namely:

1. Assemble main.asm to give main.o.
2. Assemble sqr.asm to give sqr.o.
3. Assemble root.asm to give root.o.
4. Use rms.lkr to link together object files 1, 2 and 3.
5. If no syntax errors, create the absolute executable file of Table 8.7.

can be initiated by choosing from the Project menu (top fourth left in Fig. 8.7) Make Project. If there are syntax errors an Error window will appear listing errors. Double clicking on any specific error will bring up the relevant Source-code window with the cursor set to the line in question.

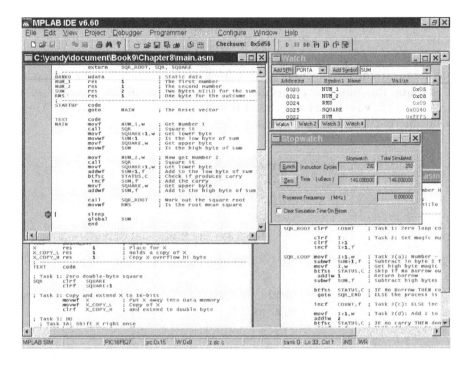

Fig. 8.7 MPLAB Version 6 & 7 screen shot showing the programs selected in Fig. 8.6 being simulated.

Once the program has been successfully made, it may be simulated. Here the PC models the PIC MCU's instruction set and I/O ports and allows the user to reset the (simulated) PIC, set break points, single step or run continuously. During this process user-selected File registers or the whole of data memory can be monitored, as can execution time. Of course, simulated execution time by the PC will be several orders of magnitude slower than a real PIC MCU.

A simulation can be actioned via the Degugger menu. This brings up the Debugger toolbar, top right of Fig 8.7. This allows the operator to:

- Reset the virtual processor by clicking on the ⬚ icon.
- Run ▷ and pause ❚❚ the simulation at top speed.
- Continuously execute, i.e., animate ▷▷ , at a rate of several steps per second.
- Single step in three modes; one line being performed on a single click.
  - Step-in ⟰ steps through all code, including subroutines.
  - Stop-over ⟱ runs through subroutines at top speed.
  - Step-out ⟰ only single steps inside a subroutine.

Figure 8.7 shows the end result of a simulation of our root-mean square example. As well as windows showing the three source files, a Watch window has been opened using the <u>V</u>iew menu. This allows the operator to add any named GPRs, such as NUM_1, the value of which can be displayed in binary, decimal or hexadecimal in bit, single, double or triple byte format. These values are updated as the simulation proceeds in a Single-step or Animate mode. If top speed is used, the Watch window is updated when the simulation is paused or stops at a breakpoint.

Also shown in the diagram is the Stop-watch window. This shows that the program took 292 cycles to execute with initial values for NUM_1 and NUM_2 of 0x05 and 0x08, respectively. With a simulated 8 MHz crystal, execution time is shown as 146 μs.

As the simulation proceeds, the currently executed instruction is marked by a $\Rightarrow$ in the left pane of the relevant Source-code window. In Fig. 8.7, this is pointing to the final sleep instruction and overlays the breakpoint 🅱 symbol. Breakpoints can be set or cleared by right-clicking on the relevant instructions. Each click of the ▷ icon will run the simulation at top speed to the next breakpoint.

Simulation will not catch all problems, especially those involving complex hardware/software interaction. However, over 95% of problems are caused by purely software design faults and simulation is a good technique for testing and debugging such code.

For instance, our code will fail if the total $NUM\_1^2 + NUM\_2^2 > 65,535$, as SUM is only double-byte; see SAQ 8.5. Debugging should always, at a first iteration, try largest and smallest values of variables. However, correct operation is by no means guaranteed by this test for all possible combinations and sequences of input.

Finally, we review some general information specific to Microchip-compatible assemblers as an aid to reading programs in the rest of the book:

- Number representation.
  - Hexadecimal: Denoted by a leading h with the number delineated by quotes, e.g., h'41' or a following h, e.g., 41h, or a 0x prefix, e.g., 0x41. The assembler normally defaults to this base so some programs show no hexadecimal indicators. However, it is better not to rely on the default behavior.
  - Binary: Denoted by a leading b with a quote delimited number; e.g., b'01000001'.
  - Decimal: Denoted by a leading d with a quote delineated number; e.g., d'65' or a leading period prefix; .65 in our example.
  - ASCII: Denoted by a quote delimited character; e.g., 'A'.

- Label arithmetic.
  - Current position: $; e.g., goto $+2.
  - Addition: +; e.g., goto LOOP+6.
  - Subtraction: -; e.g., goto LOOP-8.
  - Multiplication: *; e.g., subwf LAST*2.
  - Division: /; e.g., subwf LAST/2.
- Directives.
  - org: Places the following code in program memory starting from the specified address; e.g., org h'0100'. If no org is used, the default reset point is h'000'. Can only be used for absolute assembly.
  - code: Counterpart to org for relocatable assembly. The actual address of the code stream is defined in the linker's command file. More than one code stream may be defined in the linker script file and in this case its name appears in the label field; for instance SUBROUTINES code.
  - equ: Associates a value with a symbol; e.g., PORTB equ 06. The #define directive may be used instead; #define PORTB 06.
  - cblock - endc: Used in absolute assembly to allocate program variables in data memory; e.g.,

```
        cblock  h'20'
        FRED            ; One byte at h'020' for FRED
        JIM:2           ; Two bytes at h'021:22' for JIM
        ARRAY:10        ; Ten bytes for ARRAY at h'023 - 02C'
        endc
```

    The address is optional after the first cblock use.
  - udata: Counterpart to cblock for relocatable assembler. The start address for data memory streams are in the linker's script file. There may be more than one Data stream defined in this script file in which case its name is published in the label field; e.g.,

```
SCRATCHPAD   udata           ; Uninitialized data stream
        FRED    res 1   ; Reserve one byte for FRED
        JIM     res 2   ; Reserve two bytes for JIM
        ARRAY   res 10  ; Reserve ten bytes for ARRAY
```

  - udata _ovr: OVeRlay Uninitialized DATA is similar to udata but the linker tries to reuse Files for the specified named variables.
  - res: Used with udata to REServe one or more bytes for a variable in a Data stream.
  - extern: Publishes the named variables as defined outside the current file, to be subsequently resolved by the linker.
  - global: Publishes the named variables that have been defined (i.e., space reserved) in the file and that are to be made visible to the linker.
  - macro - endm: Used to allow the specified enclosed sequence of instructions to be replaced by a new macro instruction; e.g.,

```
Addf    macro  N,datum
        movf   datum,w
        addlw  N
        movwf  datum
        endm
```

adds the literal N to the specified File datum. For instance, to add five
to File h'20' the programmer can use the invocation Addf 5,h'20'.

- include: Used to include the specified file at this point; for instance,
  include "myfile.asm". The #include directive is identical.
- end: Normally the last line of an assembly-level source file. Tells the
  assembler to ignore anything following.

Examples

Example 8.1
The following routine effectively exchanges the byte contents of W and a
File F without needing an additional intermediate File.

```
xorwf   F,f    ; [File] <- W^F
xorwf   F,w    ; W <- W^(W^F) = 0^F = F
xorwf   F,f    ; [File] <- F^W^F = 0^W = W
```

where ^ denotes eXclusive-OR.

Wrap the given code within a macro to generate a new instruction
Exgwf F where F is the designated File; e.g., Exgwf h'20'.

Solution
Wrapping the code inside a macro gives:

```
Exgwf   macro   FILE
        xorwf   FILE,f
        xorwf   FILE,w
        xorwf   FILE,f
        endm
```

Note that this macro instruction will not affect the C flag and activate the
Z flag according to the datum that was in the Working register at entry.

Example 8.2
The PIC18XXXX family has an instruction bnc (Branch if No Carry) which
transfers the program execution to the specified destination if the Carry
flag is zero. Devise a macro instruction to simulate this for the 12- and
14-bit core families.

Solution
The code fragment below follows the macro on page 225 but with the
C flag replacing the **Z** flag. The macro name has been changed to Bcc
(Branch if Carry Clear) as Bcn is recognized by the Version 3+ assemblers
as a reserved name; that is, a PIC18XXXX instruction mnemonic.

```
Bcc    macro   destination
       btfss   STATUS,C
       goto    destination
       endm
```

Example 8.3
Write a macro that will create a delay of n instruction cycles, where n is
an integer up to 1024. Thus, for instance; Delay_cycles d'400' will be
replaced by a delay of 400 instruction cycles.

Solution
The macro code below takes four instruction cycles to execute per loop,
and so the macro operand is divided by four to give the loop count. In
the example given, the operand of 400 becomes d'100' in the Working
register.

```
Delay_cycles
        macro cycles
        local LOOP

        movlw  cycles/4  ; There are 4 cycles in this macro
LOOP    addlw  -1        ; Decrement
        btfss  STATUS,Z  ; Zero?
        goto   LOOP

        endm
```

The macro label is qualified with the local directive to ensure that each
time a macro is used it does not inject LOOP into the assembler's symbol
table. If not so qualified, then an Address label duplicated assembler error will occur.

Example 8.4
Macros may be nested, that is, a macro may use other macros in its definition. For example, consider a macro to create a countdown process
that initializes a given GPR and then decrements to zero. Assuming that
the macro Movlf has already been defined:

```
Movlf macro    literal,destination

        movlw  literal              ; Put the literal in W
        movwf  destination          ; & out to the dest file

        endm
```

write a suitable macro definition.

Solution

One possible solution is:

```
Countdown macro    literal,counter
          local   C_LOOP            ; A macro label

          Movlf  literal,counter    ; Initialize counter File
C_LOOP    decfsz counter,f          ; Decrement
          goto   C_LOOP             ; REPEAT UNTIL zero

          endm
```

The specified File, called count, is first initialized to literal using the macro Movlf. The actual countdown uses the decfsz instruction to both decrement the contents of count and break out of the loop on zero. Thus the invocation Countdown d'100',h'40' will initialize File h'40' to decimal 100 and decrement to zero. The process give $(3 \times \text{count}) + 1$ cycles delay, that is 301 cycles in our example.

Note that both the Working register and STATUS are altered by this macro as well as the target GPR. Side effects are a hazard in using macro instructions, especially if the macro has been designed by someone else and hidden in an Include file. At the very least assume that W and STATUS are altered unless known otherwise. Altering banks in a macro is also potentially hazardous.

Example 8.5
The PIC16F84 is unusual in that its GPRs are mirrored across both Data banks; see Fig. 4.7 on page 80. More commonly, different banks hold unique GPRs. For instance, the PIC16F627/8 devices have 80 unique GPRs in Bank 0, another 80 unique GPRs in Bank 1, 48 unique GPRs in Bank 2 as well as 16 common GPRs across all four banks; see memory map Fig. 5.4 on page 102.

In order to select a File in Bank 1, the RP0:RP1 bits must be set as appropriate. For instance; to copy W into File h'E0' we have:

```
    bsf    STATUS,RP0   ; Change to Bank 1
    bcf    STATUS,RP1

    movwf h'E0'         ; Copy W to File h'E0'

    bcf    STATUS,RP0   ; and move back to Bank 0
```

When using a relocatable assembler, the programmer will not necessarily know into which bank the linker has placed a variable. Furthermore, as the suite and mix of component source files changes, the bank may switch back and forth as different phases of the project evolve!

To get around this problem, Microchip-compatible assemblers provide the **banksel** (BANK SELect) directive. This automatically keeps track of the location of the named variable and issues code to make the appropriate change-over. Show how this directive should be used when storing the decimal literals 1, 10, 100 in three GPRs called var_0, var_1 and var_2, respectively.

Solution
A possible sequence of instructions is shown below. The directive issues the appropriate combination of bsf STATUS,RPX and bcf STATUS,RPX instructions as appropriate before the following instruction.

```
    movlw    1      ; The first literal
    banksel  var_0  ; Change to the appropriate bank
    movwf    var_0  ; Do it

    movlw    d'10'  ; Literal ten
    banksel  var_1  ; Change to the appropriate bank
    movwf    var_1  ; Do it

    movlw    d'100' ; Literal hundred
    banksel  var_2  ; Change to the appropriate bank
    movwf    var_2  ; Do it
```

Where Indirect addressing is used for 4-bank PIC MCUs the IRP bit in STATUS must be 0 for Banks 0:1 and 1 for Banks 2:3; see page 108. This can be implemented using the **bankisel** (BANK Indirect SELect) directive in a similar manner to banksel.

Self-Assessment Questions

8.1 Design macros to simulate the PIC18XXXX family relative conditional Branch instructions bc (Branch on Carry) and bz (Branch if Zero).

8.2 Design a macro of the form Mul XPLIER,XCAND,PRODUCT that will implement the function PRODUCT:2 = VAR1 × VAR2. *Hint*: Check Program 6.7 on page 165. What do you think are the advantages and disadvantages of using a macro instead of a subroutine in a long implementation like this?

8.3 The goto and call instruction op-codes use an 11-bit address suitable for transfer anywhere within a 2 Kbyte Program store; as illustrated in Fig. 5.17 on page 132. As shown in this diagram, the 13-bit Program Counter is overwritten by the instruction's 11-bit address together with PCLATH[4:3] (PCLATch High byte) File h'0A', to give a 13-bit destination address.

Some mid-range PIC MCUs have a 4- or 8-Kbyte Program store, such as the PIC16F74 and PIC16F876, respectively. These require 12- or 13-bit destination addresses for goto and call instructions making use of the PCLATH[4:3] bits (see page 98) and effectively partitioning up the Program store into corresponding two or four pages. The programmer needs to manipulate these bits before the goto or call instructions to specify the page. For instance, for the PIC16F876 to call a subroutine beginning at FRED which is at address h'0B00' (i.e., in Page 1) we have:

```
bcf    PCLATH,3     ; Change to Page1 of Program store
bsf    PCLATH,4
call   FRED         ; Go to it
```

In a relocatable program the location of a label, such as FRED is uncertain and in a multipage PIC MCU may be placed by the linker in any page. To allow the assembler to alter the PCLATH[4:3] bits as appropriate, Microchip-compatible assemblers have a directive **pagesel** which must precede any goto or call instruction; rather in the manner of the banksel directive of Example 8.5. Show how you could use this to support a series of calls to subroutines named SUB_0, SUB_1 and SUB_2.

8.4 The banksel approach to selecting a bank is inefficient in that an extra instruction is issued even if the PIC MCU is already in the correct bank. Consider how in a time- or space-critical subroutine this inefficiency can be avoided.

8.5 To be safe, determine a maximum value that NUM_1 and NUM_2 should not exceed to guarantee correct working for our program to calculate the root mean square of the two variables.

8.6 Rewrite the routine `main.asm` of Program 8.2 and the subroutine `root.asm` of Program 8.4 to allow for *all* values of NUM_1 and NUM_2. This will require a 3-byte sum and square root subroutine.

8.7 The following routine based on the macro instruction Mov1f of Example 8.4 does not work as intended. COUNT is altered seemingly at random and not consistently with the desired literal 32. Why is this?

```
movf    COUNT,f     ; Test COUNT for zero
btfsc   STATUS,Z    ; IF not Zero THEN skip
   Mov1f d'32',COUNT ; ELSE re-initialize it to 32
```

8.8 A programmer with expertise in the Motorola 68HC05 MCU has been converted to the PIC MCU family and wishes to design macros to simulate, amongst others, the following 68HC05 instructions. Note that the Accumulator register in the 68HC05 family is the equivalent to the Working register of the PIC.

lda memory
LoaD Accumulator with data from memory.

lda #data
LoaD Accumulator with literal data.

sta memory
STore Accumulator data into memory.

tst memory
TeST memory for zero

tsta
TeST Accumulator for zero

Code suitable macros. Why do you think this approach might not be such a good idea?

CHAPTER 9

High-Level Language

All the programs we have written in the last six chapters have been in symbolic assembly language. Whilst assembly-level software is a quantum step up from pure machine-level code (see page 214) nevertheless there is still a one-to-one relationship between machine and assembly-level instructions. This means that the programmer is forced to think in terms of the MCU's internal structure—that is, of registers and memory—rather than in terms of the problem algorithm. Although most assemblers have a macro facility, whereby several machine-level instructions can be grouped to form pseudo high-level instructions, this is only tinkering with the difficulty. What is this difficulty with machine-oriented language? In order to improve the effectiveness, quality and reusability of a program, the coding language should be mostly independent of the underlying processor's architecture and should have a syntax more oriented to problem solving.

We are not going to attempt to teach a high-level language in a single short chapter. However, after completing this chapter you will:

- Understand the need for a high-level language.
- Appreciate the advantages of using a high-level language.
- Understand the problems of using a high-level language for embedded microcontroller applications.
- Be able to write a short program in **C**.

The difficulty in coding large programs in a computer's native language was clearly appreciated within a few years of the introduction of commercial systems. Apart from anything else, computers became obsolete with monotonous regularity, and programs needed to be rewritten for each model introduction. Large applications programs, even at that time, required many thousands of lines of code. Programmers were as rare as hen's teeth and worth their weight in gold. It was quickly deduced that for computers to be a commercial success, a means had to be found to preserve the investment in scarce programmers' time. In developing a universal language, independent of the host hardware, the opportunity would be taken to allow the programmer to express the code in a more natural syntax related to problem-solving rather than in terms of memory, registers and flags.

Of course, there are many different classes of problem tasks which have to be coded, so a large number of languages have been developed since.[1] Amongst the first were Fortran (FORmula TRANslation) and COBOL (COmmon Business Oriented Language) in the early 1950s. The former has a syntax that is oriented to scientific problems and the latter to business applications. Despite being around for over 40 years, the inertia of the many millions of lines of code written has made sure that many applications are still written in these antique languages. Other popular languages include Algol (ALGOrithmic Language), BASIC, Pascal, Modula, Ada, **C**, **C++** and Java—the latter three forming a related family.

Although writing programs in a high-level language may be easier and more productive for the programmer; the process of translation from the high-level source code to the target machine code is rather more complex than the assembly process described in Chapter 8. The translation package for this purpose is called a **compiler** and the process, **compilation**.

The complexity and cost of a compiler was acceptable on the relatively powerful and extremely expensive mainframe computers of that time. However, until the mid-1980s the use of high-level languages as source code was virtually unknown for microprocessor-controlled circuitry. In the last decade the easy availability of relatively powerful and cheap personal computers and workstations, capable of running compilers, together with the growing power of MPU/MCU hardware and financial importance of this market, is such that the majority of software written for such targets is now in a high-level language.

If you are going to code a task in a high-level language to run in a system with an embedded MCU; for instance, a washing-machine controller, then the process is roughly as follows.

1. Take the problem specification and break it up into a series of modules, each with a well-defined task and set of input and output data.
2. Devise a coding to implement the task for each module.
3. Create a source file using an editor in the appropriate high-level syntax.
4. Compile the source file to its assembly-level equivalent.
5. Assemble and link to the machine-code file.
6. Download the machine code to the target's program memory.
7. Execute, test and debug.

This is virtually identical to the process outlined in Fig. 8.3 on page 227, but with the extra step of compilation. Some compilers go directly from the source file to the machine-code file; however, the extra flexibility of going through the assembly-level phase, as shown in Fig. 9.1, is nearly universal when embedded MPU/MCU circuitry is targeted.

The choice of a high-level language for embedded targets is crucial. Of major importance is the size of the machine code generated by a

---

[1]A popular definition of a computer scientist is one who, when presented with a problem to solve, invents a new language instead!

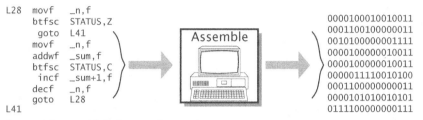

```
                                                      L28  movf   _n,f
                                                           btfsc  STATUS,Z
  while(n>0)              Compile                           goto   L41
  {                                                         movf   _n,f
  sum = sum + n;                                           addwf  _sum,f
  --n;                                                     btfsc  STATUS,C
  }                                                         incf   _sum+1,f
                                                           decf   _n,f
                                                           goto   L28
                                                      L41
```

(a) First, compile to assembly-level code.

```
L28  movf   _n,f                                       0000100010010011
     btfsc  STATUS,Z                                   0001100100000011
     goto   L41          Assemble                      0010100000001111
     movf   _n,f                                       0000100000010011
    addwf  _sum,f                                      0000100000010011
     btfsc  STATUS,C                                   0000011110010100
      incf   _sum+1,f                                  0001100000000011
     decf   _n,f                                       0000101010010101
     goto   L28                                        0111100000000111
L41
```

(b) Second, assemble-link to machine code.

Fig. 9.1 Conversion from high-level source code to machine code.

high-level language implementation, as compared with the equivalent assembly-level solution. Most embedded MCU circuitry is lean and mean, such as the remote controller for your television. Lean translates to physically small and mean maps to low processing power and memory capacity—and cost! Most low-cost MCUs have a low-capability processor with a few hundred bytes of RAM and a few kilobytes of ROM Program store at best. Thus to be of any use, the high-level language and the compiler must generate code that, if not as efficient as assembly-level (low-level), at least is in the same ball park.[2]

By far the most common high-level language used to source code for embedded MPU/MCU circuitry is **C**. Historically **C** was developed as a language for writing operating systems. At its simplest level, an operating system (OS) is a program which makes the detailed hardware operation of the computer's terminals, such as keyboard and disk organization, invisible to the operator. As such, the writer of an OS must be able to poke about the various registers and memory of the computer's peripherals and easily integrate with assembly-level driver routines. As conventional high-level languages and their compilers were profligate with resources, depending on a rich and fast environment, assembly language was mandatory up to the early 1970s, giving intimate machine contact and tight fast code. However, the sheer size of such a project means that it is likely to be a team effort, with all the difficulties in integrating the code and foibles of several people. A great deal of self-discipline and skill is demanded of such personnel, as is attention to documentation. Even

---

[2]In the author's experience a code size increase factor of $\times 1.25 \cdots \times 2.5$ is typical.

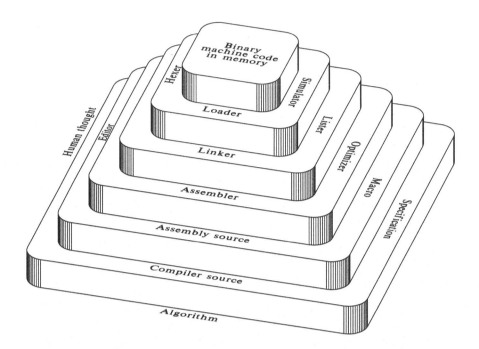

Fig. 9.2 Pyramid view of the steps leading to an executable program.

with all this, the final result cannot be easily transplanted to machines with other processors, needing a nearly complete rewrite.

In the early 1970s, Ken Thompson—an employee at Bell Laboratories—developed the first version of the UNIX operating system. This was written in assembler language for a DEC PDP7 minicomputer. In an attempt to promote the use of this operating system (OS) within the company, some work was done in rewriting UNIX in a high-level language. The language CPL (Combined Programming Language) had been developed jointly by Cambridge and London universities in the mid-1960s, and had some useful attributes for this area of work. BCPL (Basic CPL) was a somewhat less complex but more efficient variant designed as a compiler-writing tool in the late 1960s. The language B (after the first letter in BCPL) was developed for the task of rewriting UNIX for the DEC PDP11 and was essentially BCPL with a different syntax.

Both BCPL and B only used one type of object, the natural size machine word—16 bits for the PDP-11. This typeless structure led to difficulties in dealing with individual bytes and floating-point computation. **C** (the second letter of BCPL) was developed in 1972 to address this problem, by creating a range of objects of both integer and floating-point types. This enhanced its portability and flexibility. UNIX was reworked in **C** during the summer of 1973, comprising around 10,000 lines of high-level code and 1000 lines at assembly level. It occupied some 30% more storage than the original version.

Although **C** has been closely associated with UNIX, over the intervening years it has escaped to appear in compilers running under virtually every known OS, from mainframe CPUs down to single-chip MCUs. Furthermore, although originally a systems programming language, it is now used to write applications programs ranging from Computer Aided Design (CAD) packages down to the intelligence behind smart egg-timers!

For over 10 years the official definition was the first edition of *The C Programming Language*, written by the language's originators Brian W. Kernighan and Dennis M. Ritchie. It is a tribute to the power and simplicity of the language that over the years it has survived virtually intact, resisting the tendency to split into dialects and new versions. In 1983 the American National Standards Institute (ANSI) established the X3J11 committee to provide a modern and comprehensive definition of **C** to reflect the enhanced role of this language. The resulting definition, known as Standard or ANSI **C**, was finally approved during 1990 by the International Standards Organisation (ISO).

Apart from its use as the language of choice for embedded MPU/MCU circuits, **C** (together with its **C++** and Java object-oriented offspring) is without doubt the most popular general-purpose programming language at the time of writing. It has been called by its detractors a high-level assembler. However, this closeness of **C** to assembly-level code, together with the ability to mix code based on both levels in the one program, is of particular benefit for embedded targets.

The main advantages of the use of high-level language as source code for embedded targets are:

- It is more productive, in the sense that it takes around the same time to write, test and debug a line of code irrespective of language. By definition, a line of high-level code is equivalent to several lines of assembly code.
- Syntax is more oriented to human problem-solving. This improves productivity and accuracy, and makes the code easier to document, debug, maintain and adapt to changing circumstances.
- Programs are easier to port to different hardware platforms, although they are rarely 100% portable. Thus they are likely to have a longer productive life, being relatively immune to hardware developments.
- As such code is more or less hardware-independent, the customer base is considerably larger. This gives an economic impetus to produce extensive support libraries of standard functions, such as mathematical and communication modules, which can be reused in many projects.

Of course there are disadvantages as well, specifically when code is being produced to run in poorly resourced MPU/MCU-based circuitry.

- The code produced is less space-efficient and often runs more slowly than native assembly code.
- The compiler is much more expensive than an assembler. A professional product can cost several thousand pounds/dollars.
- Debugging can be difficult, as the actual code executed by the target processor is the generated assembler code. The processor does not execute high-level code directly. Products that facilitate high-level debugging can be expensive.

Program 9.1 is an example of a **C** function (a function is **C**'s counterpart to a subroutine) that evaluates the relationship:

$$\text{sum} = \sum_{k=1}^{n} k$$

for example, if $n = 5$ then we have:

$$\text{sum} = 5 + 4 + 3 + 2 + 1$$

In the implementation n is the integer passed to the function, which computes and returns the integer sum as defined. The program implements this task by continually adding n to the pre-cleared sum, as n is decremented to zero.

Let us dissect it line by line. Each line is labeled with its number. This is for clarity in our discussion and is not part of the program.

Line 1: This line names the function (subroutine) summation and declares that it returns an unsigned long integer (a 16-bit unsigned object in the compiler used to illustrate this chapter) and expects an unsigned integer (an 8-bit unsigned object) to be passed to it called n.

Line 2: A left brace { means begin. All begins must be matched by an end, which is designated by a right brace }. It is good practice to indent each begin from the immediately preceding line(s). This makes it easier to ensure each begin is paired with an end. However, the compiler is oblivious of the style the programmer

---

Program 9.1 A simple function coded in **C**.

```
 1:  unsigned long summation(unsigned int n)
 2:      {
 3:      unsigned long sum = 0;
 4:      while(n>0)
 5:          {
 6:          sum = sum + n;
 7:          --n;
 8:          }
 9:      return sum;
10:      }
```

uses. In this case line 10 is the corresponding end brace. Between lines 2 and 10 is the body of the function summation().

Line 3: There is only one variable that is local to our function. Its name and type are defined here. Thus sum is of type unsigned long. In C all objects have to be defined before they are used. This tells the compiler what properties the named variable has; for example, its size (16 bits), to allocate storage and its arithmetic properties (unsigned). At the same time sum is given an *initial* value of zero. The complete statement is terminated by a semicolon, as are all statements in C.

Line 4: In evaluating sum we need to repeat the same process as long as n is greater than zero. This is the purpose of the while construction introduced in this line. The general form of this loop construct is:

```
while(true)
    {
    do this;
    do that;
    do the other;
    }
```

The body of the loop, i.e., the set of statements that appears between the following left and right braces of lines 5 and 8, is continually executed as long as the expression in the brackets evaluates as non-zero. Anything non-zero is considered true by C. This test is done before each pass through the body. In our case the expression n>0 is evaluated. If true, then n is added to sum. n is then decremented and the loop test repeated. Eventually n>0 computes to false (zero) when n reaches zero and the statement following the closing brace is entered (line 9).

Line 5: The opening brace defining the while body. Notice that for style it is indented.

Line 6: The expression to the right of the assignment = is evaluated to sum + n and the resulting value given to the left variable sum. In adding an 8-bit to a 16-bit variable, C will automatically extend to 16-bits—see Table 9.1, instructions located at locations h'000E - 0011'.

Line 7: The value of n is decremented, as commanded by the -- Decrement operator.[3] This is equivalent to the statement n = n - 1; As an alternative, most C programmers would incorporate this into the while test expression thus: while(--n > 0).

Line 8: The end brace for the while body. Again note how the opening (line 5) and closing braces line up. The compiler does not give

---

[3] The analogous Increment operator ++ has given the name C++ to the next development of the C language.

a hoot about style; this is solely for human readability and to reduce the possibility of errors.

Line 9: The `return` instruction passes one parameter back to the caller, in this case the completed value of `sum`. The compiler will check that the size of this parameter matches the prefix of the function header in line 1, that is `unsigned long`. This returned parameter is the value of the function, i.e., the function can be used as a variable in the same way as any other. Thus, if we had a function called `sqr_root()` that returned the square root of an integer passed to it (see Program 9.2), then the statement in the calling program:

```
x = sqr_root(y);
```

would assign the returned value of `sqr_root(y)` to `x`.

Line 10: The closing brace for function `summation()`.

We see from Fig. 9.1 that the output from the compiler is assembly-level code, which can then be assembled and linked with other modules[4] in the normal way. To illustrate this process, Table 9.1(a) shows the assembly-level code generated when the **C** code of Program 9.1 is passed through the Custom Computer Services (CCS), Inc cross-C compiler.[5] This is a low-cost **C** compiler ($\approx$ $125) that can be integrated with MPLAB; see Fig. 9.3. This listing file shows each line of **C** source code as a comment together with the resulting assembly-level code. Two minor changes were made to the source code to generate this illustrative listing:

- The function was renamed `main()` from `summation()`, because each **C** program *must* at the very least have a `main()` function. This root function is similar to any other **C** function but causes the compiler to set up the software environment; see below.
- The directive `#include` is used to insert a header file to give the compiler information concerning the memory structure and other pertinent information for the PIC16F627.

It is instructive to look at how the compiler has translated this program.

`long main(int n)`

Entry to the `main()` function is always at the Reset vector h'000'. First the PCLATH SPR (File h'0A') is zeroed to ensure that any following code is located at the bottom of the Program store. Execution is then moved to the Interrupt vector at h'004'. As there is no interrupt service function in our example, the compiler locates code for the `main()` function here.

---

[4]Some of which can be functions hand-coded in native assembly-level language for efficiency, and from libraries supplied with the compiler or bought in.

[5]See `http://www.ccsinfo.com/picc.shtm`

Table 9.1: Resulting assembly-level CCS compiler output after linking.

```
CCS PCM C Compiler, Version 3.180, 20941

                  Filename: SUM.LST

                  ROM used: 25 (2%)
                            Largest free fragment is 999
                  RAM used: 8 (5%) at main() level
                            8 (5%) worst case
                  Stack:    0 locations

*
0000:  MOVLW    00
0001:  MOVWF    0A
0002:  GOTO     004
0003:  NOP
...................  #include <16f627.h>
...................  // Standard Header file for the PIC16F627 device
...................  #device PIC16F627
...................  #list
...................
...................  long main(int n)
...................  {
0004:  CLRF     04
0005:  MOVLW    1F
0006:  ANDWF    03,F
0007:  MOVLW    07
0008:  MOVWF    1F
...................  long sum = 0;
0009:  CLRF     22
000A:  CLRF     23
...................  while(n>0)
...................      {
000B:  MOVF     21,F
000C:  BTFSC    03.2
000D:  GOTO     014
...................      sum = sum + n;
000E:  MOVF     21,W
000F:  ADDWF    22,F
0010:  BTFSC    03.0
0011:  INCF     23,F
...................      --n;
0012:  DECF     21,F
...................      }
0013:  GOTO     00B
...................  return sum;
0014:  MOVF     22,W
0015:  MOVWF    78
0016:  MOVF     23,W
0017:  MOVWF    79
...................  }
...................
...................
0018:  SLEEP
```

(a): Assembly-level code listing file generated by the CCS compiler.

(continued on the next page)

Table 8.2: (*continued*)

```
:1000000000308A000428000084011F308305073077
:100010009F00A201A301A108031914282108A20727
:100020000318A30AA1030B282208F8002308F900EB
:0200300063006B
:00000001FF
;PIC16F627
```

(b): Executable Intel machine code file.

The main() function begins by clearing the File Select register (h'004'). Then IRP, RP1 and RP0 bank switching bits in the Status register are zeroed, to ensure that Bank 0 is the active bank at entry. Finally, specifically for the PIC16F627, the analog Comparator module is turned off by setting the bottom three bits of the ComParator CONtrol register (CMCON at File h'1F') to 1; see Fig. 14.6 on page 443.

This initialization phase is a feature of the main() function, so that the "useful" code can run from Reset in a known software state. A **C** program typically comprises many functions, but only main() will set up the starting environment.

long sum = 0;
The CCS compiler reserves two bytes for a long object. In this case File h'22:23' stores main.sum low:high bytes. To zero these two GPRs the compiler has generated two clrf instructions:

```
clrf   h'22'    ; Clear sum_low
clrf   h'23'    ; Clear sum_high
```

while(n>0){
The compiler has allocated File h'21' for the single-byte int object main.n. Normally this will have been given a value by the calling function. The while statement is implemented by testing main.n for zero and if true jumping to the exit return statement.

```
movf   h'21',f    ; Test for zero
btfsc  STATUS,Z   ; IF not Zero THEN skip
goto   h'014'     ; ELSE go to instruction in h'014' (return)
```

sum = sum + n;
This is implemented as an "add a single byte to a double byte" operation thus:

```
movf   h'21',w    ; Get main.n
addwf  h'22',f    ; Add and update low byte sum
btfsc  STATUS,C   ; Skip over if no Carry
incf   h'23',f    ; ELSE increment high byte sum
```

Many **C** programmers use the alternative statement sum+=n; which states sum *augmented by* n.

```
--n;
```

Now decrement the single byte in File h'21'.

```
decf    h'21',f  ; Decrement main.n
```

In more complicated expressions the placement of the -- decrement operator (and the analogous ++ operator) before or after the object can affect the outcome. Where it appears before, such as in:

```
number = -n + 4;
```

then the value of n is *first* decremented before being added to 4. In the following case:

```
number = n- + 4;
```

4 is added to n and only then is n decremented.

In our example the logic of the program is unaffected if the operator is pre-decrement or post-decrement. However, the compiler in the latter case adds an extra instruction to bring main.n down into the Working register before it is decremented in situ as it thinks that some computation involving the original value of main.n is to be performed.

```
}
```

The while loop is repeated by going back to the loop test, which is located starting at h'00B'.

```
goto    h'00B'
```

return sum;

At the end of a function returning a long object the CCS compiler places the two bytes in the fixed GPRs File h'78:79' ordered low:high. Thus this code fragment simply copies the two bytes in File h'22:23'; i.e., main.sum, into the return locations.

```
movf  h'22',w ; Copy low byte sum
movwf h'78'   ; and put in return slot low
movf  h'23',w ; Copy high byte sum
movwf h'79'   ; and put in return slot high
```

Specifically the main() function is terminated by the sleep instruction; see page 277. Normally a function is terminated by a return to the caller function.

The final machine code file is shown in Table 9.1(b) and gives a total length of only 24 instructions, including the one-off environment settings.

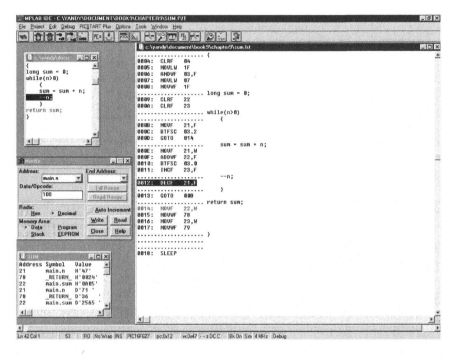

Fig. 9.3 Simulating our example program in MPLAB Version 6.60

C-level programs can be compiled and simulated in the IDE environment of Microchip's MPLAB; see page 237. The screen shot of Fig. 9.3 shows windows into both the C-level source code and the resulting assembly-level code. Although simulation is at the latter level, the C code is arrowed in the appropriate place corresponding to the arrowed simulated assembly-level instruction. The Watch window shows the state of the two C objects int n (corresponding to the assembly symbol main.n in the symbol list) and long sum (i.e., main.sum). The compiler generates the system symbol _RETURN_ to label the two GPRs File h'78:79'. The Watch window can be utilized in the usual manner to monitor the state of the C level objects, and in the diagram n and sum are shown both in hexadecimal and in decimal; the latter often being more relevant to high-level objects. Any base can be chosed by right clicking the variable's value and chosing [Properties]. The value can also be set by clicking on the value entry and in our example n was set to decimal 100 before the run. The screen shot shows the situation where n reached 71 in the process of decrementation. At the end of the simulation n reached zero and sum showed decimal 5050.

Using C to implement source code gives the programmer access to structures, operators and library functions appropriate to a modern high-level language. Nevertheless, to be of use in a microcontroller environment it is necessary to permit the programmer to easily access specified locations in the Data store and individual bits within. In this manner Special-Purpose Registers, such as the parallel ports, may be initialized, monitored and controlled in order to allow the processor to interact with its peripheral modules and the outside world. It is possible to do this using standard C operators. However, many compilers targeted to microprocessors and microcontrollers have non standard extensions to facilitate this "bit twiddling." As we are using the CCS product as the exemplar for this text, we will use the syntax appropriate to this compiler.

As an example, consider a routine that is to continually pulse Port A pin 0 (that is pin RA0) as long as pin 7 of Port B is high; see page 131. This is how it might be coded in standard C, where the 0x prefix is the language's way of indicating hexadecimal.[6]

```
#define  PORTA *(unsigned int *)0x05
#define  PORTB *(unsigned int *)0x06

while(PORTB & 0x80)            /* Isolate bit7; is it non-zero?*/
    {
    PORTA = PORTA | 0x01;   /* IOR with 00000001; RA0 -> hi */
    PORTA = PORTA & 0xF7;   /* AND with 11111110; RA0 -> lo */
    }
```

Note the use of the pair /*....*/ to denote comments in C.

Of particular note, even for veteran C programmers, is the use of pointers to name an absolute Data store address. For instance:

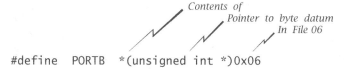

Contents of
Pointer to byte datum
In File 06

```
#define  PORTB  *(unsigned int *)0x06
```

defines the name PORTB as synonymous with the contents of File 6. For the CCS compiler an unsigned int is a byte, but some compilers either use a unsigned short int or unsigned char to hold an 8-bit datum. The named object can then be used as a normal global C int.

The routine above ANDs (&) Port B with b'10000000' to determine if bit 7 is set, which if so will give a non-zero (true) outcome; see page 123. If this is the case, the body of the while loop will be executed. This body uses Inclusive-OR (|) to set a bit in Port A (see page 124) and AND

---

[6]Decimal is the default base in C but beware, because a leading zero is interpreted as octal; e.g., 026 is octal 26 (which is decimal $2 \times 8 + 6 = 22$).

to clear a bit in Port A. As we see from the following assembly-level code generated by the CCS Version 3 compiler, this has been interpreted as a *single-bit* set or cleared respectively, and correctly uses the btfss, bcf and bsf instructions.

If several bits had been tested set or cleared then the appropriate ior and and instructions would have been used.[7]

```
        btfss   6,7     ; Test bit7 of Port B
        goto    NEXT    ; IF 0 THEN break out of loop
        bsf     5,0     ; Pin RA0 high
        bcf     5,0     ; Pin RA0 low
NEXT    .....   .....
```

This executable code above is exactly the same as a hand-coded assembly version.

In the specific case of the CCS compiler the non-standard directive **#byte** can be used to name the contents of a fixed Data store address; e.g., #byte INTCON = 0x0B assigns to File h'0B' the name name INTCON. In a similar manner, an individual bit can be named in the CCS compiler using the **#bit** directive. For instance, #bit INTF = 0x0B.1 names bit 1 of File h'0B'. Alternatively, if INTCON has already been named as above #bit INTF = INTCON.1 does the same thing. Such defined objects can only have the value 0 and 1.[8] Thus the statement INTF = 0; will clear bit 1 of the INTCON File.

Using this CCS syntax gives us the equivalent code:

```
#byte   PORTA = 5       /* Port A is File 5                       */
#byte   PORTB = 6       /* Port B is File 6                       */
#bit    RA0 = PORTA.0   /* Bit 0 of File 5 is now named RA0       */
#bit    RB7 = PORTB.7   /* Bit 7 of File 6 is now named RB7       */

while(RB7)
    {
    RA0 = 1;            /* Pin RA0 high                           */
    RA0 = 0;            /* pIN RA0 low                            */
    }
```

which generates exactly the same executable code as our rather more long-winded standard **C** version. However, where a compiler has a special notation like this it gives a stronger message that efficient "bit twiddling" instructions should be generated; at the expense of portability. We will use this notation from now on.

A table of all standard **C** operators is given in Appendix C for reference.

---

[7]Many compilers are unable to distinguish between single bits which use efficient "bit twiddling" instructions and use the less efficient logic instructions; for instance, Version 2 of this compiler.

[8]A 2-valued object of this kind is sometimes called a Boolean. In the CCS compiler a short int is a Boolean, but this is unusual.

Examples

Example 9.1
Write a **C** function to return the square root of a positive 16-bit integer based on the algorithm of Fig. 6.11 on page 176.

Solution
Modifying the task list of Example 6.5 to suit the structure of the **C** while loop gives:
1. Zero the loop count
2. Set variable i (the magic number) to 1
3. WHILE i is less than or equal to the number
   (a) Take i from Number
   (b) Add 2 to i
   (c) Increment the loop count
4. Return loop count as $\sqrt{\text{Number}}$

   The function heading gives it its name sqr_root and defines the parameters to be passed to the function and the outcome. The script unsigned int sqr_root(unsigned long number) declares that it will return an unsigned int value and one unsigned long int object will be passed to it, which will be known as number within the function. On this basis the coding of Program 9.2 directly implements the task list. Since the square root of a 16-bit object will fit into an 8-bit byte, the loop count is declared unsigned int. The magic number i will, however, be twice (plus one) that of count and is therefore defined as a unsigned long object. At the same time as these internal function variables are defined they are given their initial values.

   The while loop is repeated until the value of the reducing number drops below the increasing value of i, at which point any further subtraction will drop the outcome below zero. The value of the loop count is the square root and is returned to the caller at the end of the function.

Program 9.2 Coding the square root function.

```
unsigned int sqr_root(unsigned long number)
      {
      unsigned int count = 0;
      unsigned long i = 1;
      while(number>=i)
             {
             number = number - i;
             i = i + 2;
             count++;
             }
      return count;
      }
```

Using the CCS **C** Version 3.18 compiler gives 29 executable instructions compared to 21 in the original assembly-level implementation of Program 6.12 on page 175. This gives an efficiency ratio of 72%.

Example 9.2
A K-type thermocouple is characterized by the equation:

$$t = 7.550162 + 0.0738326 \times v + 2.8121386 \times 10^{-7} v^2$$

where t is the temperature difference across the thermocouple in degrees Celsius and v is the generated emf spanning the range $0 - 52,398\,\mu V$, represented by a 14-bit unsigned binary number, for a temperature range of $0 - 1300°C$. Write a **C** function which will take as its input parameter a 14-bit output from an analog to digital converter and return the integer temperature in Celsius measured by the thermocouple.

Solution
Our function, named thermocouple() in line 1 of Program 9.3, takes one unsigned long integer (16-bit) parameter, named emf, and returns a similar 16-bit value. The internal variable temperature is defined in line 3 to be a floating-point object[9] to cope with the complex fractional mathematics of line 6. Because we are told that only the 14 lower bits of emf have any meaning, line 5 ANDs the 16-bit object with h'3FFF' (0x3FFF) to clear the upper two bits. Finally, an unsigned long version of the float object temperature is made and returned in line 8.

Program 9.3 Linearizing a K-type thermocouple.

```
unsigned long thermocouple(unsigned long emf)
    {
    float temperature;
    unsigned long outcome;
    emf = emf & 0x3FFF;              /* Clear upper two bits */
    temperature = 7.550162+0.073832605*emf+2.8121386e-7*emf*emf;
    outcome = (unsigned long)temperature;
    return outcome;
    }
```

The resulting executable code running on a mid-range PIC MCU core takes 653 program words; that is, around $\frac{2}{3}$ of the Program store of a PIC16F627 device! Because of the size penalty of using floating-point objects, fixed-point arithmetic is used wherever possible in embedded microcontroller implementations.

---

[9]Having a mantissa and exponent of the form $m \times 10^e$.

Example 9.3

On page 229 we implemented a root mean square program to evaluate the mathematical relationship $\sqrt{NUM\_1^2 + NUM\_2^2}$. Write a C function to implement this relationship, where the two 8-bit objects num_1, num_2 are passed to the function which returns the 8-bit value rms.

Solution

The solution shown in Program 9.4 uses the internal unsigned long 16-bit variable sum to hold the addition of the two squared 8-bit variables. The squaring operation is simply implemented using the C multiplication operator * rather than coding a squaring function of the manner of Program 8.3 on page 232. However, the programmer needs to force the compiler to do its arithmetic in 16-bit precision to match the 16-bit sum. This is done by **casting** one of each of the multiplication operands thus (unsigned long). The function developed in Program 9.2 is used to generate the square root of the 16-bit sum object and is called from line 6 of the function variance() with the return value being assigned to the variable sum as part of the call. In compiling the source code using the CCS C compiler, 94 machine-level instructions are needed to implement this problem. This compares to 62 instructions for the assembly-code version of Chapter 8. This gives an efficiency ratio of 66%.

Program 9.4 Generating the root-mean-square value of two variables.

```
unsigned int variance(unsigned int num_1, unsigned int num_2)
    {
    unsigned long sum;
    unsigned int rms;
    sum = (unsigned long)num_1*num_1 + (unsigned long)num_2*num_2;
    rms = sqr(sum);
    return rms;
    }
```

Example 9.4

Write a function that will shift each bit in File h'20' out of pin RA0 in turn, working right to left. As each bit is presented in turn to RA0, pin RA1 is pulsed _/‾_ to tell the outside world that a new bit is ready.

Solution

The code given in Program 9.5 uses a for loop to shift the contents of File h'20' (named DATUM) right one place eight times using the >> Shift Right C operator. Before each shift, pin RA0 (named SER_OUT) is either set or cleared, depending on the state of bit 0 of DATUM, (named LSB) using a if...else decision structure. In either case pin RA1 (named CLOCK) is pulsed. The resulting code shown in Program 9.5 implements a simple synchronous serial communications link; see Chapter 12.

Program 9.5 A simple serial data transmitter.

```
#byte   DATUM   = 0x20      /* Name File h'20'                      */
#bit    LSB     = DATUM.0   /* in which bit0 is named               */
#byte   PORTA   = 5         /* Port A is File 5                     */
#bit    SER_OUT = PORTA.0   /* in which pin0 is named               */
#bit    CLOCK   = PORTA.1   /* and pin1 is named                    */

void put_char(void)         /* Returns & accepts no data (void) */
{
int i;                      /* Loop count                          */
for(i=0; i<8; i++)          /* DO eight times                      */
    {
    if(LSB) {SER_OUT = 1;}/* IF Datum bit0 is 1, make RA0 = hi*/
    else    {SER_OUT = 0;}/* ELSE make RA0 low                   */
    CLOCK = 1;              /* Pulse pin RA1 high                  */
    CLOCK = 0;              /* and then low                        */
    DATUM = DATUM >> 1;     /* Shift Datum right one place          */
    }
}
```

Example 9.5
Repeat the interrupt-driven batch can counter of Example 7.1 on page 201 but using a CCS **C** implementation.

Solution
As in the assembly-level solution there are two functions in Program 9.6. The main() function uses the CCS built-in function enable_interrupts() to set both the INTE and GIE Enable mask bits (see Fig. 7.3 on page 191) and clearing Port A ensures that the initial state of pin RA0 is low.

Inside a DO forever loop the state of the variable BATCH is continually monitored for non-zero (truth) and if this is the case it is reset and RA0 is pulsed high once for 1 ms. The CCS built-in function delay_ms() provides an easy means of of generating precise delays of up to 65,535 ms in this implementation of the language. To facilitate this, the programmer must be able to tell the compiler, as shown, what the clock frequency is. The delay_us() and delay_cycles() functions can be used for shorter delays. For compilers without similar non-standard functions, assembly-level delay subroutines can be used with care!

The function can_count() is signaled to the compiler as an interrupt service routine using the prefix directive #int_ext, and there are simi-

Program 9.6 Program in **C** for the pea-canning packer.

```c
#include <16f84.h>
#use delay (clock=8000000)      /* Tell the compiler 8MHz clock */
#bit  RA0   = 5.0               /* Bit0 of PortA                */
/* Declare function can_count() accepts and returns nothing     */
void can_count(void);
int EVENT, BATCH;               /* Two global variables         */

main()
{
enable_interrupts(INT_EXT);     /* Set the INTE bit in INTCON   */
enable_interrupts(GLOBAL);      /* and the GIE bit likewise     */
RA0 = 0;                        /* Ensure that RA0 starts low   */
while(1)                        /* DO forever                   */
    {
    if(BATCH)                   /* IF the BATCH variable is non-zero*/
        {
        BATCH = 0;              /* THEN zero it                 */
        RA0 = 1;                /* and pulse the RA0 pin high   */
        delay_ms(1);            /* for one millisecond          */
        RA0 = 0;
        }
    }
}
/***************************************************************/
/* This is the Interrupt service function                   */
#int_ext                        /* For external (hardware) interrupts*/
void can_count(void)
    {
    if(++EVENT == 24)           /* Increment count and IF 24    */
        {
        EVENT=0;                /* THEN zero count              */
        BATCH++;                /* AND make the BATCH variable non 0 */
        }
    }
```

lar directives for each of the various sources of interrupt request. The compiler handles setting up an appropriate Interrupt response to handle interrupts from multiple sources and context saving and retrieval.

As can_count() is an ISR, values are not passed in the normal way, as indicated by the void keyword. Instead, any variable monitored or changed is global. In our program, both BATCH and EVENT are defined *outside* a function and are therefore known to all functions, both foreground and background.

The kernel of can_count() *first* increments EVENT—the ++ appears before the variable. If this produces a value of 24 then it is zeroed and BATCH is incremented to tell the background function that a batch of 24 cans has been recorded.

Compared to the 40 instructions of the assembly-level coding, this high-level implementation generated 94 executable instructions. However, the latter's interrupt handling would be much more flexible if multiple sources of interrupt requests were to be handled and the delay function handled longer delays, which would probably be the case in a real situation.

Example 9.6
Arrays of identically sized objects can be defined in **C** using the notation
`fred[n]`, where `fred` is the name of the array (actually the address of
the first element) and `n` is the *n*th element. The compiler will allocate 16
sequential Files in the Data store.

It is possible to give each element of an array an initial value; e.g., for
`svn_seg[10]`:

```
unsigned int svn_seg[10] = {0x3f, 0x06, 0x5b, 0x4f, 0x66,
                            0x6d, 0x7d, 0x07, 0x7f, 0x6f};
```

defines an array of ten bytes initialized with the 7-segment patterns described in Fig. 6.8 on page 161.

These ten values for `svn_seg[0]` through `svn_seg[9]` will be placed
in ten consecutive Files. Most PIC MCUs have a severely limited Data
store capacity and, as in this example, where the values will not be subsequently changed, it makes more sense to place these ten constants in
Program ROM as a table of `retlw <constant>` instructions; in the manner of Table 6.6 on page 162. This can be done by qualifying the array
with the keyword `const`; giving the definition:

```
unsigned int const svn_seg[10] = {0x3f, 0x06, 0x5b, 0x4f, 0x66,
                                  0x6d, 0x7d, 0x07, 0x7f, 0x6f};
```

Using the techniques outlined here, write a program to implement an
electronic die, with seven LEDs connected to the top seven pins of Port B,
as shown in Fig. 9.4(a). The main routine is simply going to increment
a global integer as rapidly as possible. The throw switch is connected
to the `INT/RB0` pin and when it pulses the program is to transfer to an
interrupt service function; see Example 9.5. This ISR is to display one of
six die patterns and after 10 s blank out the display to save battery life.
Running the PIC MCU using a Watch crystal frequency of 32,768 Hz also
reduces the energy requirements, as illustrated in Fig. 10.3 on page 275.

Solution
The LED patterns are listed as a global array `const array[6]` of six patterns in Program 9.7, following the tabulation of Fig. 9.4(b). The patterns
in the array are shifted left one place compared to the truth table, to align
with the top seven bits in Port B.

The main routine simply increments the byte variable `throw` and continually resets the count to zero when it goes beyond five to give a modulo-
6 count; that is 0, 1, 2, 3, 4, 5, 0....

The CCS version 3.18 compiler generates eight instructions for `main()`,
including some Gotos and Skips, giving an incrementation rate of about

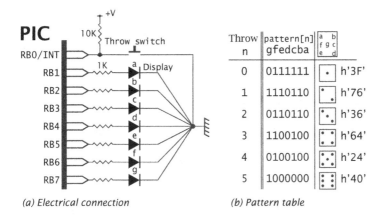

(a) Electrical connection

(b) Pattern table

Fig. 9.4 The active-low die patterns.

1000 per second at the stated clock frequency. This effectively gives a random selection when the outside human throws the switch connected to the INT pin and thus causes an interrupt.

Program 9.7 The electronic die.

```
#include <16f84.h>
#use delay (clock=32768)
#byte PORTB = 6

void die(void);
unsigned int const array[6] = {0x7e, 0xec, 0x6c, 0xc8, 0x48, 0x80};
unsigned int throw;

main()
{
enable_interrupts(INT_EXT);      /* INTE set to 1                 */
enable_interrupts(GLOBAL);       /* GIE  set to 1                 */

while(1)                         /* Forever DO                    */
    {
    PORTB = 0;                   /* LEDs off                      */
    if(++throw > 5) {throw=0;}   /* Increment modulo-6            */
    }
}

#int_ext                         /* Hardware ISR                  */
void die(void)
    {
    PORTB = array[throw];        /* Display nth element pattern   */
    delay_ms(10000);             /* for 10,000 ms                 */
    }
```

The ISR function die() copies the nth element of our array of constants to Port B and then delays 10 s before returning to the background function. As Port B is cleared in main(), the display will then be blanked.

Self-Assessment Questions

9.1 Driving the die of Example 9.6 requires seven parallel port lines and a particular electronic game needs to drive two die displays. By inspection of the patterns of Fig. 9.4, how could you reduce the requirement to four bits per die?

9.2 As part of an electronic game, a function is to be written to return the next pseudo random number in the 127 sequence defined by the generator configuration of Fig. 6.12 on page 183. The current number is to be passed to the function and the next number in the sequence returned. Assume that this passed datum is non-zero.

How could you modify the function to send the entire sequence of random numbers out of Port B beginning with the passed number?

9.3 A PIC MCU-based digital thermometer is to display temperatures between 0°C and 100°C. To be able to market the device to USA the thermometer is to have the option to display the temperature in degrees Fahrenheit. Write a function for a PIC-MCU based thermometer that is to convert Celsius integers to the equivalent Fahrenheit integer. The input is to be an unsigned int byte representing Celsius and the return Fahrenheit is also to be an unsigned int datum. The relationship is:

fahrenheit = (celsius × 9)/5 + 32

and 16-bit arithmetic should be forced to avoid overrange errors.

9.4 A cold-weather indicator in an automobile dashboard display comprises three LEDs, which are connected to the lower three bits of Port A. Bit 2 of this location is connected to the red LED, which is to light if the Fahrenheit temperature is less than 34. Bit 1 is the yellow LED for temperatures below 40°F, and bit 0 is the green LED. You may assume that the appropriate port pins have already been set as outputs and that a LED is illuminated when the driving pin is low. Write a function, whose input is °F, that activates the appropriate LED.

PART III

The Outside World

Apart from our brief discussion of the Harvard structure in Chapter 3, we have confined our deliberation to the internal structure of the PIC mid-range microcontroller family and its software, with some limited allusion to parallel ports. This final part looks at how the MCU core reacts with the environment physically beyond the confines of its pins. This process involves consideration of the interaction of the software and hardware of its integrated ports and devices, ending up with a case study which builds a complete stand-alone embedded controller.

Figure 4.1 on page 73 illustrates the internal architecture and pinning of the PIC16F84. Externally this 18-pin mid-range device is characterised by two parallel ports, a count input to an 8-bit timer shared with bit 4 of Port A and a Hardware interrupt shared with bit 0 of Port B. In addition to these external peripheral devices, an internal 64-byte EEPROM module and Watchdog timer completed the inventory of communicatable modules.

The PIC16F84 was one of the first of the mid-range members, and as such inherited the parallel ports and timers of the earlier low-range family, which was augmented with the introduction of interrupts and the EEPROM data module. As new family members were introduced, the range of peripheral modules became much more extensive. In this part we will examine a selection of peripheral modules; mainly using the 8-pin PIC12F629/75, 18-pin PIC16F627/28/48 and 28/40-pin PIC16F873/74-/76/77 devices as exemplars. However, the function of a module is largely independent of the device in which it is embedded, although in some cases the functionality has been augmented in newer members. On the way you will:

- Look at support issues such as the power supply, clock, power management and device configuration.
- Consider parallel and serial digital data input and output.
- Examine the Timer and Watch-dog subsystems.
- See what is involved in dealing with analog signals.
- Design an embedded MCU-based viva timer.
- Consider how a system may be tested and debugged.

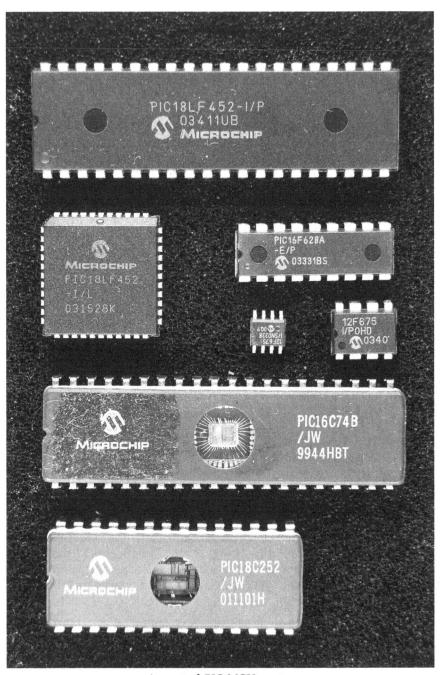

Assorted PIC MCU parts.

CHAPTER 10

The Real World

Up to this point we have mainly concentrated on how the software has interacted with the processor's internal registers and data memory. Now, as a prelude to how the MCU relates to its internal peripheral devices and hence monitors and controls its external environment, i.e., the *real world* outside its pins, we need to look at external support issues, such as power supply requirements, clocking and resetting.

After reading this chapter you will:

- Be familiar with the permitted range of power supply and input/output voltages.
- Distinguish between quiescent and dynamic power dissipation and recognize that the latter is directly proportional to both frequency and to the square of the supply voltage.
- Be aware of how the Sleep mode is invoked and exited, and its effect on the processor.
- Understand the basics of the integral clock oscillator.
- Know how the PIC MCU's configuration can be set-up during programming.
- Understand the nuances of the various reset processes.

As a prelude to our discussion on real-world issues, Fig. 10.1 shows the architecture of the PIC16F874 and 16F877 MCUs, which we are going to use as one of our exemplars for most of the rest of the book. Apart from the latter's larger Program store, Data store and EEPROM module, the two devices are identical and so we will concentrate on the PIC16F877. The PIC16F873/6 MCUs are corresponding 28-pin variants and therefore support a somewhat truncated inventory of peripherals. We will refer to these four devices as the PIC16F87X group.

Except for issues relating to memory capacity, the core of these processors are very similar in all mid-range devices, as is the 33-instruction set described in Chapter 5. In comparing the PIC16F84 of Fig. 4.1 on page 73 to Fig. 10.1 we see that the main difference is in the latter's rich set of peripheral modules which we will be describing in the following chapters. Of course, even 40 pins is not enough to go round and give each peripheral its own separate I/O connection to the outside world. Thus the majority of pins are a shared resource. For instance, pin RA3 is bit 3 of Port A but can also be used as ANalog channel 3 AN3 or even as an

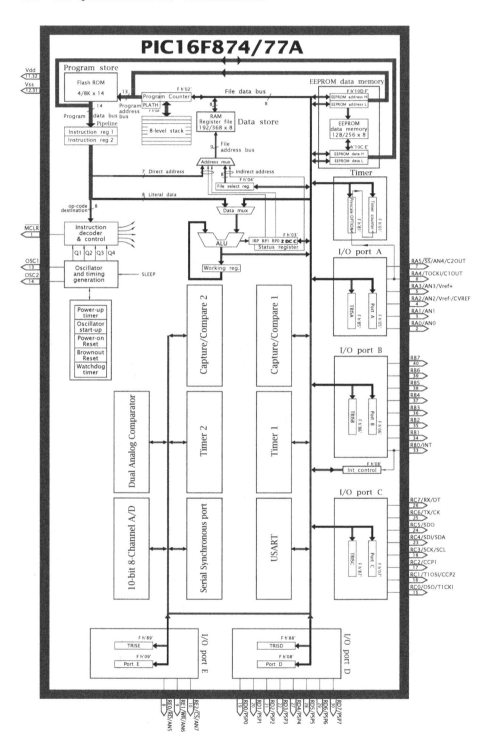

Fig. 10.1 Architecture of the PIC16F874/77A devices.

external positive reference voltage input V_{ref+} for the analog-to-digital converter module. PIC MCUs with smaller form factors, such as the 18-pin PIC16F627[1] and 8-pin PIC12F675 shown in Fig. 10.2 can still support multiple peripheral modules, but in such cases pin sharing is more extensive with a consequently more severe restriction on what can be used in any given application. In small packaged family members, the designer can usually opt to use a completely internal clock oscillator and drop the External reset input to save precious pin resources; see Table 10.2.

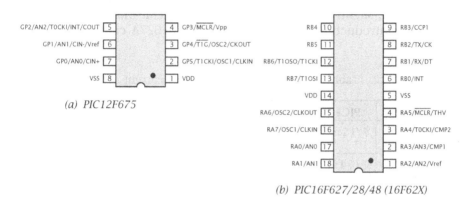

(a) PIC12F675

(b) PIC16F627/28/48 (16F62X)

Fig. 10.2 Pinout for a variety of PIC MCU family members.

All members of the PIC MCU family will operate typically with a supply voltage V_{DD} of nominally 5 V. The standard PIC16F87X can run up to 20 MHz over the range 5 ± 0.5 V. The voltage can drop to 4 V if the clock frequency is limited to 16 MHz. Many family members have a low-voltage variant. For example the PIC16LF87X can be clocked up to 10 MHz in the range 3 – 5.5 V and down to 2 V for a top frequency of 4 MHz. However, the PIC12F629/675[2] can run over a voltage range 2 V – 5.5 V in the same device version.

To the outside world, the electrical characteristics of a PIC MCU are similar to any other electronic digital circuit. In terms of voltages, a pin configured to be an output which has been set to the Low state by the PIC MCU will normally be no more than $V_{OL} = 0.6$ V if sinking (accepting) a current up to 8.5 mA, over the temperature range $-40°C$ to $+85°C$. A pin set to the High state by the PIC MCU can source (supply) up to 3 mA and not drop more than 0.7 V below the supply; e.g., a V_{OH} of 4.3 V with a 5 V supply.

A port pin configured to be an input will generally recognise a voltage less than 15% (20% for Schmitt trigger buffered inputs) of the supply

---

[1]The PIC16F628 is identical but with twice the program memory at 2K and the PIC16F648 has twice as much again. Where appropriate we will refer to this threesome as the PIC16F62X group.

[2]The PIC12F629 is identical to the PIC12F675, but without the analog-to-digital converter module.

voltage as being a Low-state input; for instance, $V_{IL} = 0.75$ V for a 5 V supply. With exceptions[3] an input pin will normally recognise a voltage more than 25% plus 0.8 V of the supply (80% for Schmitt trigger inputs) as being in the High state; for instance, $V_{IH} = 2$ V for a 5 V supply.

Table 10.1 shows the supply current for our three exemplar devices over a range of conditions. The first figure is the typical value which is quoted for 25°C and the last is the maximum value for the industrial temperature range device (-I) of −40°C to +85°C. Extended devices (-E) can operate over a range of −40°C to +125°C. The A suffix denotes a later release, slightly enhanced or debugged specification, compared to the initial introduction; for example, the PIC16F627A compared to the PIC16F627.

Table 10.1: Power supply operating current.

Oscillator	PIC16F87XA-I	PIC16F62XA-I	PIC12F629/675-I
20 MHz HS mode	7/15 mA @ 5.5 V	3/3.3 mA @ 5 V	2.4/3 mA @ 5V
4 MHz XT mode	1.6/4 mA @ 5.5 V 0.6/2 mA @ 3 V*	670/780 μA @ 5 V 240/300 μA @ 2 V*	0.6/1.4 mA @ 5 V 70/110 μA @ 2 V
32 kHz LP mode	20/35 μA @ 3 V*	38/48 μA @ 5 V 12/15 μA @ 2V*	35/54 μA @ 5 V 9/16 μA @ 2 V
Sleep base	1.5/16 μA @ 4 V 0.9/5 μA @ 3 V*	200/950 nA @ 5 V 100/800 nA @ 2 V*	2.9/995 nA @ 5 V 0.99/700 nA @ 2 V

* PIC16LF variant.

Many microcontroller applications are battery powered and in such situations power consumption is critical. Specifically, the PIC12F629/675, described by Microchip as part of the nanoWatt family, with low current requirements and a large supply range, is particularly suitable for such uses. These bare figures from the data sheets show a variation range of more than 10 million, so it is important that the factors influencing current consumption be understood.

The relationship between the PIC MCU's clocking frequency and current is graphed in Fig. 10.3. Clearly power dissipation $V_{DD} \times I_{DD}$ is directly proportional to operating frequency. For instance, 100 times more current is required at 10 MHz as compared to 100 kHz.

To see why this is so, consider a switch charging and discharging a capacitive load C, as in Fig. 10.4. The switch is implemented by a transistor and the load is due to the stray capacitance of the connection to the next field-effect transistor and its input gate. R_S represents the resistance of the switching transistor.

---

[3]The main exceptions are the input to $\overline{MCLR}$ ($\overline{Master\ CLeaR}$) which requires a voltage V_{IH} of above $0.85V_{DD}$ before coming out of reset, and $0.7V_{DD}$ for any oscillator connected to the OSC1 input as an external clock.

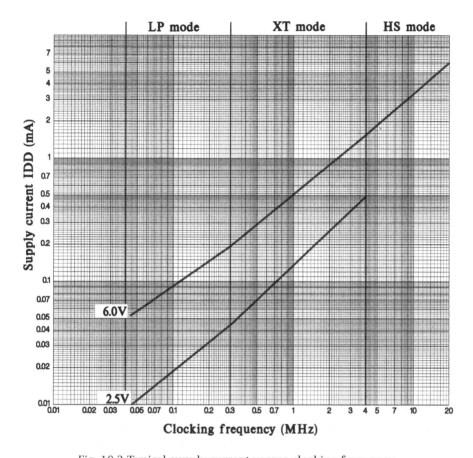

Fig. 10.3 Typical supply current versus clocking frequency.

When the switch opens, the capacitance charges up exponentially to V volts with a time constant $\tau = CR_L$. In steady state $\frac{1}{2}CV^2$ Joules of energy is stored. Energy is dissipated in the load by this charging current as follows:

Initial charging current ($V_C = 0$) : $i_0 = V/R_L$

Instantaneous current $\qquad$: $i_c = i_0 e^{-\frac{t}{\tau}}$

Instantaneous power in $R_L \qquad$: $i_c{}^2 R_L = i_0{}^2 R_L e^{-2\frac{t}{\tau}} = (V^2/R_L)e^{-2\frac{t}{\tau}}$

Total energy dissipated in $R_L \quad$: $E = V^2/R_L \int_0^\infty e^{-2\frac{t}{\tau}}\, dt$

$$= V^2/R_L \left| -\frac{\tau}{2}e^{-2\frac{t}{\tau}} \right|_0^\infty$$

$$= V^2/R_L(\tfrac{\tau}{2}) = \tfrac{1}{2}CV^2$$

Thus in going high, $\frac{1}{2}CV^2$ Joules are dissipated in the load resistance (irrespective of its value R_L!) and $\frac{1}{2}CV^2$ Joules are stored in the capacitor's electric field. On discharge, this stored energy is dissipated in $R_S//R_L$

(once again irrespective of value). The energy dissipated in one switching cycle thus CV^2 Joules. The total power is this figure multiplied by the number of cycles per second ($CV^2 f$), plus any quiescent dissipation due to leakage through the switches.

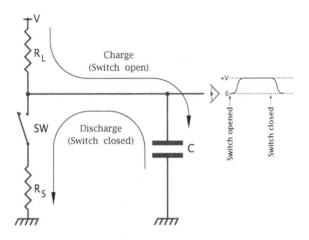

Fig. 10.4 Equivalent output circuit, where C represents both intrinsic and external load capacitance.

The preceding relationship $CV^2 f$ shows that dissipated power is proportional to frequency for any given supply voltage. Furthermore, it is proportional to the square of the supply voltage, so halving V_{DD} from 5 V to 2.5 V should quarter the power dissipation $V_{DD} \times I_{DD}$.[4]

The dynamic power dissipation derived above should be added to that due to the quiescent current the device consumes when the clocking rate is dropped to zero. From the lower row of Table 10.1 this base or Power-Down current, listed in data sheets as I_{PD}, is typically less than $1\,\mu A$. These figures assume that peripheral modules that (sometimes optionally) have their own private clock oscillator, such as the Watchdog timer, and any Brown-out reset circuitry are disabled.

Of course, not clocking the processor is rather unproductive, in that nothing happens! However, many embedded systems only need a processing capability on a sporadic basis, and it would be advantageous to be able to put the processor in a standby mode when no action is required. For instance, a MCU-based radio telemetry transducer at the bottom of a lake may need to measure the temperature only once an hour and have a battery life of a year.

---

[4]This is why most current microprocessors used as the PC's CPU, such as the Intel Pentium IV, are powered at under 3 V rather than the standard 5 V of older devices.

To facilitate situations like this, all PIC MCU families feature a **Sleep mode** which effectively turns off the internal clock oscillator. This switch is actioned in software using the sleep instruction. Once asleep, the contents of the Data store are retained provided that the supply voltage remains above 1.5 V (V_{DR} in the Data sheet). The PIC MCU can be awakened either by resetting the device (see page 284), by an enabled interrupt request or if the enabled Watchdog timer overflows.

When the processor executes a sleep instruction it will clear the $\overline{PD}$ (**Power Down**) bit in the Status register (see Fig. 4.6 on page 78) and the internal clock oscillator is turned off. If the Watchdog timer (see page 402) is enabled at that time then it will be cleared, including its Prescaler, but will continue to run, as it has its own private internal oscillator. At this time the $\overline{TO}$ (**Time Out**) flag will be set (i.e., no Time Out). All File contents, including the various port settings, remain unchanged.

In the case of an interrupt-actioned awaking, the relevant interrupt flag needs to be cleared and the corresponding interrupt mask bit set to enable requests from that source. If the Global Interrupt Enable mask (GIE; see Fig. 7.3 on page 191) is set to enable the entire interrupt system, then *after* the instruction following sleep is executed, the processor will go to the interrupt service routine as a normal interrupt response. However, if GIE is clear, hence disabling the interrupt response, then the processor will not vector to the ISR, but will simply execute the instruction following sleep and continue on as normal. However, the programmer should clear the relevant interrupt flag following the sleep instruction.

In this latter situation if an enabled interrupt occurs *before* the sleep instruction is executed—as signaled by the associated interrupt flag being set—then sleep is executed as a nop (No Operation). In this situation the $\overline{PD}$ bit will not be cleared, so the program can determine, if necessary, after a sleep instruction if the PIC MCU really did go through a dormant period. The software can also determine if the processor was awakened by the Watchdog timing out, by checking to see if the $\overline{TO}$ bit in the Status register has been cleared. Normally in Watchdog-enabled applications, the sleep instruction is followed by a clrwdt (CLeaR WatchDog Timer) instruction. Checking the appropriate interrupt flag in the INTCON register will determine if the source of the awakening was an interrupt.

Whatever the source of the awakening, there will be a delay of 1024 clock cycles f_{osc} before processing of the instruction following the sleep breakpoint. This is to ensure that the crystal clock oscillator has started up and stabilized. This oscillator startup delay, illustrated in Fig. 10.10, is not implemented if the PIC MCU is using a resistor-capacitor clock mode; see Fig. 10.5(b) and Table 10.3.

The powered-down current I_{PD} is lower when the Watchdog timer is not enabled; for instance, for the PIC16F87XA-I devices I_{PD} is quoted as typically 1.5 μA (16 μA maximum) and 10.5 μA (42 μA maximum) with the Watchdog timer disabled/enabled, respectively. These values are for a V_{DD} of 4 V and I/O ports set to input with pins tied to either V_{DD} or V_{SS}—usually ground.

All members of the PIC MCU family have an integral clock generator, which when augmented with timing elements provide the internal clocking waveforms shown in Fig. 4.4 on page 76. All family members can be operated in one of four standard modes, as listed in Table 10.2. Three of these modes are based on the use of a quartz crystal or ceramic resonator as the timing element, connected across the OSC1 and OSC2 pins. A low-cost option is available for cost-sensitive non-critical applications, using an external capacitor and resistor as the clocking elements.

Table 10.2: Oscillator operation modes.

Standard modes for all devices	
LP	Low Power for frequencies up to 200 kHz
XT	Crystal (XTAL) for frequencies between 200 kHz and 4 MHz
HS	High Speed for frequencies between 4 MHz and 20 MHz
RC/EXTRC	External Resistor-Capacitor. Pin CLKOUT gives clock output
Additional modes for the PIC12F629/675	
INTOSC1	Internal 4 MHz oscillator. Pin CLKOUT used for I/O function
INTOSC2	Internal 4 MHz oscillator. Pin CLKOUT gives clock output signal
EC	External Oscillator. Pin CLKOUT used for I/O function
RC2	External Resistor-Capacitor. Pin CLKOUT used for I/O function

We see from Fig. 10.5(a) that the three crystal-mode oscillator configurations comprise an inverting amplifier, which is disabled by the sleep instruction, together with the user-supplied timing elements. The only difference between modes is the value of the inverting amplifier's gain. In the LP mode the gain is lowest and power consumption is minimized. The HS mode is used for high frequencies and has the largest current requirement. In general the oscillator option with the lowest possible gain should be used. The target device's data sheet will give details of range and component values. The maximum clocking rate for the mid-range family is 20 MHz,[5] although some devices come in variants with a lower maximum rate; typically 4 MHz.

A typical 10 MHz system uses a 10 MHz AT-cut crystal with a C_1 of 22 pF and a C_2 of 33 pF in the HS mode. A 32 kHz crystal needs a C_1 of 68 pF and a C_2 of 100 pF in the LP mode. Although both capacitors may have the same value, making C_2 larger improves the oscillator start-up characteristics after reset and awakening from the Sleep state. Some

---

[5]The extended-range PIC18XXXX family can be clocked up to 40 MHz with a 10 MHz crystal using a phase-locked loop mode to frequency multiply.

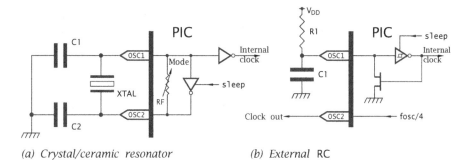

(a) Crystal/ceramic resonator   (b) External RC

Fig. 10.5 Typical oscillator configurations.

crystals in the HS mode may require a series resistor at the OSC2 pin. Details are given in Microchip's application note AN588 *PIC16/17 Oscillator Design*. Ceramic resonators are less expensive than crystals but have an inferior frequency accuracy of the order of 0.5%, and temperature stability is poorer. Ceramic resonators may come with integral capacitors to reduce the part count. Microchip's application note AN588 gives a comparison between ceramic resonators and crystals used in this application.

The fourth standard mode makes use of an external resistor and capacitor as an alternative to precision crystal timing. In this situation, shown in Fig. 10.5(b), the OSC2 pin provides a buffered output clock signal, which can be used to synchronize external digital circuits, or even other PIC MCUs. The RC mode is useful for low-cost applications where the actual clocking rate and stability is not of importance. The rate is dependent on the external resistor R_1 and C_1 and supply voltage V_{DD} in a complex manner. Generally, the chosen device's data sheet will give tables and graphs showing typical frequencies against these variables. For example, the PIC16F87X devices will have an average clocking rate of 1.7 MHz $\pm$ 10% for a V_{DD} of 5 V, R_1 of 3.3 kΩ, and C_1 of 100 pF at 25°C. Of course the tolerance and temperature variation of the timing components and V_{DD} must be considered.

It is possible to clock a PIC MCU from an external oscillator. This can be useful if several devices are to be synchronized to the one clock. In such cases, the external oscillator should drive the OCS1 pin and OSC2 is either left open or grounded via a resistor to reduce noise. The oscillator should have a Low level V_{IL} below $0.3V_{DD}$ and a High level above $0.7V_{DD}$.[6] The PIC MCU should be set to the crystal mode (as opposed to RC) appropriate to the frequency.

Using up two pins for the timing elements is rather extravagant in an 8-pin device, especially when two have already been used up for the power supplies! Because of this, small footprint family members nor-

---

[6]If using a TTL-compatible oscillator then a pull-up resistor may be needed to ensure a high enough V_{IH}.

mally have additional clock modes designed to release one or even both pins for parallel port access. Table 10.2 shows the particular example of the PIC12F629/675. Here the RC timing elements can be entirely internal to the device, giving a nominal 4 MHz clocking frequency. Like the external RC mode, the actual frequency is only approximate, but the programmer can fine-tune the frequency for any particular production device by changing the bit pattern in the lower four bits of the OSCAL (OScillator CALibrate) Special-Purpose File Register (SPR) to give 16 slightly different values. The best value is programmed into the top Program store word in the factory at production time, as a retlw n instruction. This can be called as part of the user program's startup initialisation, with the returned value in W being copied into OSCAL.

The INTOSC2 mode uses the OSC2 pin to furnish a clock signal to drive external circuitry, as shown in Fig. 10.5(b). The other three modes release OSC2 for use as a parallel port pin. In particular, the RC2 mode is identical to the standard RC (sometimes called the EXTRC) mode, but with no output clock supplied. The External Clock mode allows the use of an outside oscillator, but compared to using a crystal mode, does not prevent pin OSC2 being used as a Port pin.

The various oscillator modes are just one of the various options which can be selected at the same time as the software is blasted into the Program store. The actual electrical process involved in this programming process is not an issue unless you are designing your own Device programmer. Normally you will use a commercial Device programmer, such as the Microchip PICSTART® of Fig. 17.4 on page 547.

For background information, the High-Voltage Programming process is shown in Fig. 10.6(a). This special Program/Verify state is initiated by raising the $\overline{\text{Master CLear}}$ ($\overline{\text{MCLR}}$) pin to +13 V whilst holding both RB7 and RB6 pins Low. Subsequently the Programming data may be read in from the former as synchronized by the incoming clock signal on the latter pin. This data may be command instructions or machine code. Conversely, the contents of unprotected (not code protected) Program store may be read out and compared with the original code for correctness.

Most newer family members have an alternative Low-Voltage Programming (LVP) mode does not require the use of a 13 V supply. This is especially useful for In-Circuit Serial Programming (ICSP™) where the Program store can be reprogrammed in situ on the circuit board. In this instance pin RB3 actions entry to this state. Initially held Low during reset, when RB3 is brought High, programming via RB7 and RB6 can begin. The problem with this mode is that pin RB3 cannot subsequently be used as a port pin.

With the PIC MCU in its special Programming state, the Device programming has access to the Program store and can burn in the application code. The Device programmer also has access to certain private

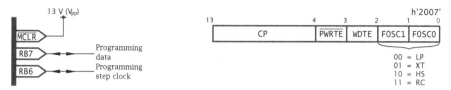

(a) High-voltage Programming mode   (b) Configuration word for the PIC16F84

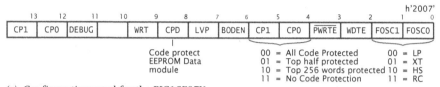

(c) Configuration word for the PIC16F87X

Fig. 10.6 Configuring some representative mid-range PIC microcontrollers.

Program store locations which are not visible to the software when the PIC MCU is running normally. Specifically, the mid-range PIC MCU family reserve "secret" location h'2007' as their **Configuration word**.[7]

Setting each bit, sometimes known as a **fuse**, in the Configuration word to the appropriate value ensures that when the MCU is in its normal Running state the clock oscillator and other facilities will be set-up appropriately. As a simple example, Fig. 10.6(b) shows the Configuration word for the PIC16F84, which illustrates the basic set of options available across the range. Four options are available.

- Four oscillator modes as listed in Table 10.2, as set by bits FOSC1:0.
- The Watchdog timer of Fig. 13.1 on page 402 can be disabled or activated by bit WDTE.
- The Power-up timer, discussed on page 286, is controlled with the setting of bit $\overline{PWRTE}$.
- Program memory code can normally be read out serially when the device is in its Program/Verify state. This is intended to allow the device programmer to verify the correct state of the code that has just been burnt into the Program store; see Fig. 17.4 on page 547. If all the CP fuse bits are cleared then this facility is blocked. This gives a measure of security protection against any attempt to copy software. Once programmed, the CP bits cannot be subsequently erased *even* in windowed or EEPROM Program store devices. For this reason Microchip do *not recommend using this feature* for such devices when being used for prototyping.

---

[7]The area of program memory beyond the user Program store space belongs to the special Test/Configuration memory space h'2000 – 3FFF' which can be accessed only in the Programming/Verify state.

Newer members of the family have additional selectable options. For example, the PIC16F87X whose Configuration word is shown in Fig. 10.6(c), can code protect either all the Program store, the top half or only the top 256 instructions. New options offered in these devices are:

- Brown-out reset to restart the PIC MCU if the power-supply voltage dips, as described in Fig 10.11, is enabled when the BODEN fuse is 1.
- When the LVP fuse is 1 the Low Voltage Programming mode is selected.
- The EEPROM Data module's contents (See Chapter 15) are code protected when the CPD fuse is 0.
- It is possible to internally write to unprotected Program store areas, as described in Fig. 15.4 on page 491. Clearing the WRT fuse disables this facility.
- These devices have a special Debugger mode which is disabled if the DEBUG fuse is 1.

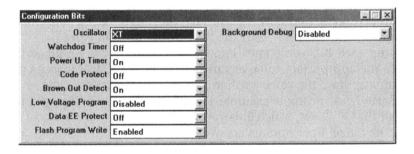

Fig. 10.7 Manually setting up the configuration menu in MPLAB Version 5 for the PICSTART programmer.

Most device programmers' software will allow the operator to set the required fuses "manually"; for instance, as shown in Fig. 10.7, before beginning the actual Program store burn process. However, it is better to embed this desired fuse state in the program code to automatically action this every time the device is programmed. As an example, consider a PIC16F87X device which is to have the following configuration:

Oscillator in XT mode
Bits 1:0 = 01

Watchdog timer off
Bit 2 = 0

Power-up timer on
Bit 3 = 0

No Program code protection
Bits 5:4, 13:12 = 11

Brownout Reset on
Bit 6 = 1

Low-Voltage Programming off
Bit 7 = 0

Then the directive

```
__config b'11111101110001'    ; or h'3F71'
```

in the assembly-level source file will create the line of machine code:

```
:02 400E 00 713F 00
```

to the format described on page 223.[8] At programming time this will set
the fuses in h'2007' accordingly. The default state of the Configuration
word is all ones, so an unconfigured PIC16F87X device will be in the RC
oscillator mode with no code protection of any type nor Power-up timer,
and the Watchdog timer, Brown-out reset, Low-Voltage Programming and
writing to the Program store will be enabled.

The Header file supplied by Microchip for each of their devices, and
described in Table 8.1 on page 217, has mnemonics for the bit patterns
for each configuration mode supported by that PIC device. These are designed to be ANDed together to give the composite 14-bit Configuration
word. Using this technique gives for our example:

```
__config   _XT_OSC & _WDT_OFF  & _PWRTE_ON & _CP_OFF
           & _BODEN_ON & _LVP_OFF
```

which gives exactly the same machine code but is more obvious to the
programmer and therefore less error prone. It is also more portable in
that altering the Header file is all that needs to be done when changing
to an alternative processor, which may have a different arrangement of
bits in its Configuration word.[9] If the incorrect Header file is included,
then the wrong fuse bits may be programmed.

C compilers will have a similar mechanism for programming the configuration fuses. For instance, the CCS compiler uses the directive #fuses
at the top of the file. For our example this is:

```
#fuses   XT, NOWDT, PUT, NOPROTECT, BROWNOUT, NOLVP
```

---

[8]Remember the byte address h'400E' is twice the word address equivalent h'2007' and
words are presented least-significant byte first.

[9]Even such close relatives as the PIC16F87X and PIC16F87XA have differing fuse dis-
positions (see Fig 10.6 and Fig. 15.6), so it is essential to use the exact correct Header
file!

All sequential digital systems must come out of its non-powered or other malfunctioning state in an orderly manner; as it were, up and running. The PIC MCU can be reset in several different ways.

- Manually by using an external switch connected to the $\overline{\text{MCLR}}$ pin, as shown in Fig. 10.8(a).
- On application of power, as shown in Fig. 10.10.
- Where the power supply of a normally running PIC MCU dips below a threshold; as shown in Fig. 10.11.
- Where the Watchdog timer of Fig. 13.1 on page 13.1 times out due to a software bug or perhaps a glitch in the power supply.

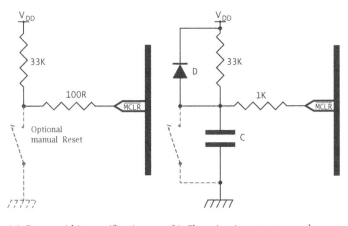

(a) Power within specification   (b) Slow-rise time power supply

Fig. 10.8 Externally resetting the PIC MCU.

Looking at these four mechanisms in turn, we first have the External reset. All PIC MCUs have, at least optionally, a $\overline{\text{MCLR}}$ pin which can be used by an external switch or other digital circuit in the manner of Fig. 10.8(a) to restart the device at the instruction residing in the Reset vector. Provided that $\overline{\text{MCLR}}$ remains below $0.2V_{DD}$, the device will remain halted (in Phase Q_1 of the internal clock cycle; see Fig. 4.4 on page 76). In order to be recognized as a legitimate reset action, $\overline{\text{MCLR}}$ must be Low for at least $2\,\mu s$; see Example 10.2. The value $33\,k\Omega$ is the maximum recommended pull-up resistor to ensure that leakage current flow from V_{DD}, when the switch is open, will not drop the pin voltage below $0.8V_{DD}$. The maximum leakage I_{IL} into $\overline{\text{MCLR}}$ is given as $\pm 5\,\mu A$ for an input voltage range $V_{SS} \leq \overline{\text{MCLR}} \leq V_{DD}$. The $100\,\Omega$ resistor gives a measure of protection, by limiting current if a negative-going noise spike breaks down the input protection diodes.

When $\overline{\text{MCLR}}$ is logic 1 (i.e., $\geq 0.8V_{DD}$) the processor will begin running normally, with the Program Counter and PCLATH zeroed to point to the first instruction at h'000'; the **Reset vector**. In addition, the three Status register bank page bits (IRP, RP1 & RP0) are zeroed, forcing the processor to see data in Bank 0. If $\overline{\text{MCLR}}$ is used to awaken the processor from its Sleep state; $\overline{\text{TO}}$ will be 1 (no Watchdog time out) and $\overline{\text{PD}}$ will be 0 (processor was powered down), otherwise these bits will be unchanged. In all cases the Status register's code condition flags remain unchanged. The effect on the SPRs of resetting is summarized in Appendix B.

Power CONtrol register

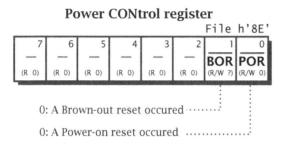

Fig. 10.9 The PCON register for the mid-range PIC16XXXX family.

In addition to the External $\overline{\text{MCLR}}$ initiated reset, all PIC MCUs have a **Power-on reset**. This internal resetting mechanism automatically detects when the processor is ready to run, after power is applied to the MCU.

Whenever a Power-on reset occurs, the **POR** (**Power-On Reset**) bit in the **Power CONtrol** SPR,[10] shown in Fig. 10.9, is cleared to enable the software to determine that a Power-on reset has occurred; see Table 10.4. Reading $\overline{\text{POR}}$ does not set it back to 1; so it must be set in software.

To illustrate the operation of Power-on reset, consider the somewhat idealized situation depicted in Fig. 10.10, where power is turned on at t_0 and V_{DD} rises exponentially towards +5 V. If this initial rate of change is ≥ 0.05 V/ms, then when V_{DD} rises to somewhere in the range 1.5 V–2.1 V an Internal reset signal is generated. This initiates the following sequence of operations.

1. A fixed delay T_{PWRT} Power-up timer period of nominally 72 ms is generated by clocking an integral 10-bit counter with an internal oscillator. This delay can be by-passed if the $\overline{\text{PWRTE}}$ fuse in the Configuration word of Fig. 10.6 is set to 1.
2. If one of the crystal modes is used, on completion of T_{PWRT} a further delay of 1024 main clock pulses is launched. This Oscillator Start-up timer comprises a 10-bit counter clocked from the internal crystal

---

[10]Early members of the family, such as the PIC16F84, do not have a Brown-out reset and thus have no need for a PCON register.

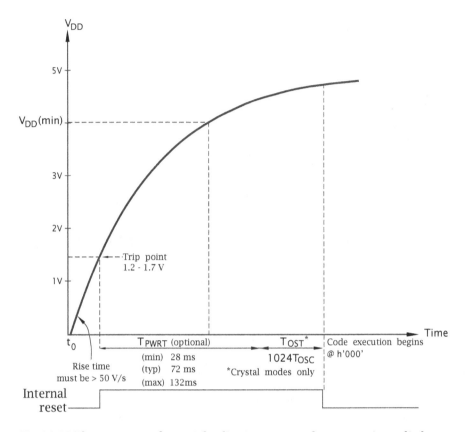

Fig. 10.10 The sequence of events leading to startup when power is applied to a 5 V device.

oscillator circuit. It ensures that the main oscillator has started up and is functioning correctly before processing begins. T_{OST} is dependent on the crystal frequency; for example, a 32 kHz crystal will give a minimum 32 ms delay whilst a 10 MHz configuration gives a 102 μs delay. If the oscillator has not yet started up [11] there will be a further indeterminate delay. This delay is not implemented whenever the PIC MCU is in an RC clock mode. The T_{OST} delay is also invoked when the MCU awakens from a Sleep state, again to ensure that the crystal oscillator restarts and is running normally before processing commences.

3. Just as in the case of an External reset, code execution commences from the Reset vector h'000'. However, unlike an External reset, which does not alter the $\overline{TO}$ and $\overline{PD}$ bits, a Power-on reset sets both Status bits to their inactive state.

---

[11] 32 kHz crystal oscillators have a typical start-up time of 1–2 seconds. Crystal oscillators $\geq$ 100 kHz have a typical start-up time of less than 10–20 ms and ceramic resonators are typically less than 1 ms. Times are voltage dependent.

The Power-on sequence for various situations is summarized below in Table 10.3.

Table 10.3: Power-on reset and sleep instruction timeouts.

Oscillator	Power-on		Wake up
mode	PWRT Enabled	PWRT Disabled	from sleep
Crystal	72 ms + $1024T_{osc}$	$1024T_{osc}$	$1024T_{osc}$
RC and internal	72 ms	—	—

Where a system does not need a Manual reset, $\overline{MCLR}$ may be tied directly to V_{DD} via a current-limiting resistor. In a small footprint device, such as the 8-pin PIC12F629/675, this pin can be configured as a general-purpose port line by setting its MCLRE fuse in the Configuration word to 0.

It is possible that the onset of the PIC MCU's power supply is so slow that either the internal Power-on reset pulse is not generated; or even if it is, V_{DD} does not reach its specified operating level after the T_{PWRT} and T_{OST} delays. This is generally 4 V for normal 5 V devices not operating in the HS crystal mode and 4.5 V for this high-speed operation. In this case the PIC MCU may start execution in an erratic manner or not at all. Where the reliability of the internal Power-on circuitry is in doubt, additional circuitry may be added to hold $\overline{MCLR}$ Low when the power is first applied for long enough to ensure that the part does not come out of reset until V_{DD} has reached its operating range. The circuit in Fig. 10.8(b) is designed to hold $\overline{MCLR}$ in the Low state long enough to allow the supply to settle. The value of capacitor should be chosen so that the time constant CR is several times greater than that taken by the power supply to stabilize. With the resistance given, a 2.2 μF capacitor will give a time constant of approximately 100 ms. More details are given in Microchip's application notes AN522: *Power-up Considerations* and AN607: *Power-up Trouble Shooting*.

A normally running MCU can malfunction if its power supply falls below its rated value. This could be due to a momentary blip on V_{DD} when switching in a large current load or due to battery exhaustion. In either case, the PIC MCU may operate erratically due to this **brownout**.[12] This may have serious consequences; for instance, a dishwasher's heating element may be turned on with no water in the reservoir!

---

[12]The term is from the same phenomena in the mains supply that causes the lights to dim and give a brownish hue to the surroundings!

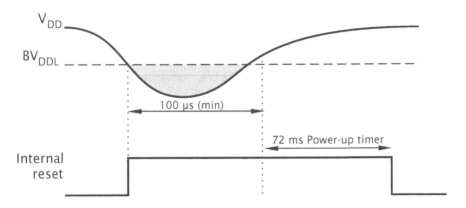

Fig. 10.11 A Brown-out reset due to a blip on the power supply.

From Fig. 10.11 we see that if the Brown-out timer is enabled (i.e., if the BODEN fuse in the Configuration word of Fig. 10.6(c) is 1) the Internal reset will be generated if V_{DD} falls below the threshold voltage BV_{DD}.[13] In a nominal 5 V rated device, this is typically 4 ± 0.3 V. For a wider-range device, such as the PIC12F629/675, this can be as small as 2 V.

The diagram shows the supply subsequently rising back a little above the threshold trip voltage. Provided that the shaded time is more than $100\,\mu s$, the Power-up timer, if enabled, kicks in for the nominal 72 ms shown in Fig. 10.10, before the processor comes out of reset. The processor then executes the instruction at the Reset vector h'000'. Some devices, such as the PIC16F627, automatically override the setting of the PWRTE fuse to ensure that this extension is always present on Brown-out, and some, such as the PIC12F675, leave it to the programmer to do this; so read the data sheet carefully! Enabling the Power-up timer reduces the possibility that a slowly rising V_{DD} may give rise to multiple triggers due to noise on the supply line.

A Brown-out reset initialises the SPRs in the same way as a Power-on reset, except that the **$\overline{BOR}$ (Brown-Out Reset)** flag in the PCON register is cleared, so that the software can determine which type of reset occurred. The state of $\overline{BOR}$ is not known after Power-on, and so should be set at the beginning of the user program.

One problem in enabling the Brown-out reset facility, is the high value of operating current. For instance, ΔI_{BOR} is typically $85\,\mu s$ for the PIC16F87X group, with a maximum of $200\,\mu A$. This has to be added to the base current of Table 10.1 and is considerable compared to the quiescent Sleep value. However, it is only $0.3\,\mu A$ typical/$1.5\,\mu A$ maximum at a V_{DD} of 2 V for the PIC12F629/675 nanoWatt device.

---

[13]If the supply falls too far and then recovers, then a Power-on reset will occur.

Table 10.4: Reset conditions.

Reset	Sleep	Execution commences at	$\overline{\text{TO}}$	$\overline{\text{PD}}$	Status register	$\overline{\text{POR}}$	$\overline{\text{BOR}}$
External	No	h'000'	U	U	000U UUUU	U	U
External	Yes	h'000'	1	0	0001 0UUU	U	U
Power-on	—	h'000'	1	1	0001 1???	0	?
Brown-out	—	h'000'	1	1	0001 1000	1	0
Watchdog	No	h'000'	0	1	0000 1UUU	1	1
Watchdog	Yes	PC+1	0	0	UUU0 0UUU	1	1
Interrupt	Yes	PC+1	1	0	UUU1 0UUU	1	1

? Not known: U Unchanged

It is also possible to reset the PIC MCU with the Watchdog timer timing out. In this situation the processor will immediately begin code execution from the Reset vector and also clear the $\overline{\text{TO}}$ Status flag (active) and set the $\overline{\text{PD}}$ flag (not active). A Watchdog time-out when the processor is asleep will cause code execution to commence at the instruction following the sleep instruction; after a delay of T_{OST} if in a crystal mode. This time both $\overline{\text{TO}}$ and $\overline{\text{PD}}$ Status bits will be zeroed (active).

A summary of the various reset conditions is given in Table 10.4, which also includes for completeness the response to an awakening from the Sleep state by an interrupt. A Power-on reset will set both $\overline{\text{TO}}$ and $\overline{\text{PD}}$ flags (inactive), whereas an External reset will leave these bits unchanged. $\overline{\text{TO}}$ will be activated (0) when a Watchdog time-out occurs and deactivated (1) when a clrwdt or sleep instruction is executed. clrwdt also deactivates $\overline{\text{PD}}$, which is active after a sleep instruction. Both these status flags are read-only; that is they cannot explicitly be altered by instructions such as bsf.

Resetting zeros the Program counter (the Reset vector) and the various banking bits, such as RP0. The three status bits **Z**, **C** and **DC** are unknown on Power-on, otherwise are unchanged.

Examples

Example 10.1
Figure 10.12 is based on Fig 10.5(b), but using a chain of two resistors. Given that the clock frequency for a given value of capacitor is proportional to the resistance seen by the OSC1 pin, justify the designer's statement that the circuit enables the software to control its execution rate without any external intervention.

Can you think of a use for such a facility?

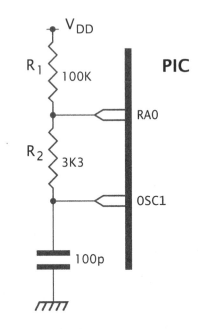

Fig. 10.12 Variable oscillator frequency.

Solution

The center tap of the series resistors is connected to a Port pin. Port pins may be configured to be either input or output, as will be described in the next chapter. On a reset of any type, all port pins are set as input. In this situation, apart from a small leakage current, pin RA0 will not affect the resistors connected to it. Thus, the total resistance will be a little over $10\,\text{K}\Omega$. With a 100 pf capacitor, the data sheet gives an oscillator frequency of nominally 100 kHz, with a V_{DD} of 5 V. This gives an instruction rate of 25,000 per second—actually a little less, as instructions that flush the Pipeline take two instruction cycles.

If the software reconfigures RA0 to be an output with bit 0 of Port A a 1, then the junction of the two resistors is pulled up to V_{DD}. In this situation OSC1 only sees the $3.3\,\text{k}\Omega$ resistor. With this value the data sheet gives an oscillation frequency of 1.7 MHz, a 17-fold increase in instruction rate to 450,000 per second. Thus where speed is important, the software can increase it own rate, otherwise slow down to conserve energy.

Example 10.2

The data sheet for the PIC16F84 indicates that the minimum duration of the Low state on the $\overline{\text{MCLR}}$ pin that will be recognized as a valid External reset is 100 ns. Can you think of problems that might arise as a consequence of this time sensitivity?

Solution
In a noisy environment erratic operation may occur with narrow pulses occasionally resetting the device seemingly at random. In such situations, low-pass filtering should be placed on the $\overline{\text{MCLR}}$ pin. Typically, a 1 nF high-frequency capacitor physically adjacent to $\overline{\text{MCLR}}$ together with a 10 kΩ pull-up resistor will suffice. The power supply should be well decoupled at the PIC MCU's power supply pins. Because of this problem, newer family members, even the PIC16F84A, have increased this minimum duration to 2 μs. Even so, the same principles apply, especially in a noisy environment.

Self-Assessment Questions

10.1 What would happen if due to a software bug in the situation shown in Fig. 10.12, a 0 was output at pin RA0? This may inadvertently happen if the pin is configured an an output before Port A's bit 0 is made 1.

10.2 In what manner will the circuit shown in Fig. 10.13 affect the oscillator frequency?

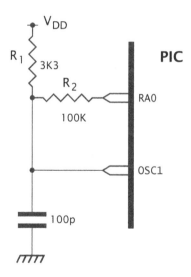

Fig. 10.13 An alternative oscillator frequency control circuit.

10.3 In an attempt to reduce the current consumption of the circuit when reset, a student has used a 1 MΩ resistor as a pull-up resistor in the

Manual reset circuit of Fig. 10.8. Why does the PIC MCU not come out of reset?

10.4 The current consumption of a PIC MCU operating at 4 MHz and a V_{DD} of 5 V is measured as 550 μA with no loading at the port pins. What will be the current consumption if the device were to be clocked at 100 kHz and powered by a 4 V supply?

CHAPTER 11

One Byte at a Time

The ability of the software to activate or monitor the state of pins connected to circuitry in the outside world is the most fundamental of the
various input and output capabilities provided by a microprocessor or microcontroller. These input/output pins are generally gathered in groups
of up to the size of the internal Data bus. In the PIC MCU family these
parallel ports allow up to eight bits of external data to be directly read
into or sent out of the processor core *one byte at a time*. The total number
of such parallel lines available on any specific family member depends on
the package footprint and on how much shared resources are used. This
parallel line budget varies from up to four for the 6/8-pin PIC10FXXX
family to a maximum of 52 for the 64-pin PIC16C924.

When you have completed this chapter you will:

- Appreciate the function of a parallel input/output (I/O) port.
- Be able to configure an I/O port line.
- Understand the structure of a parallel I/O port and differentiate be-
  tween an active and passive pullup.
- Comprehend how read–modify–write instructions interact with parallel
  I/O ports.
- Appreciate the electrical and power characteristics of an I/O port.
- Know how to enable internal port pull-up resistors.
- Understand how the function of the Change in Port B interrupt oper-
  ates.
- Be able to expand the number of I/O lines using external hardware.

Conceptually a parallel I/O port can be considered as a File with its
contents visible to the outside world. This somewhat simplified view is
represented in Fig. 11.1, which is based on a magnified section of the
PIC16F84 Data store shown in Fig. 4.7 on page 80. All PIC16XXXX mid-
range family members have at the very least the depicted 13-I/O lines.
The PIC16F87X group has a permanent additional RA5 Port A pin, whilst
the PIC16F62X group has up to the three extra lines shown dotted in
the diagram, if the external OSC1, OSC2 and $\overline{\text{MCLR}}$ pins are sacrificed.[1]
The diminutive PIC10FXXX/12XXX families have a single General-Purpose

---

[1] A port pin substituting a sacrificed $\overline{\text{MCLR}}$ pin can only ever be an input.

parallel I/O (GPIO) port which combines many of the properties of Port A and Port B, with a maximum of six I/O pins.

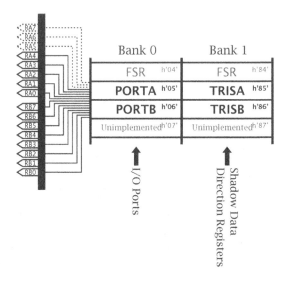

Fig. 11.1 A boiled-down view of the mid-range PIC16XXXX series parallel Ports A and B.

Mid-range 28-pin+ devices have extra ports, as listed in Table 11.1. Such parts will have a larger repertoire of on-chip peripheral devices which share the parallel I/O lines, so the increased parallel I/O capacity may be largely illusionary. For instance, the PIC16F87X shares five of the Port A lines (RA5, RA[3:0]) and the 3-bit Port E as analog inputs to its 8-channel A/D converter.

Table 11.1: Summary of mid-range PIC MCU parallel I/O provision.

Port	Size	Characteristics
A	3 – 8 I/O	RA4 is open-drain output and common with Timer 0's input. Shared with analog modules.
GP	3 – 6 I/O	6/8 pin PIC10FXXX/12XXXX General-Purpose I/O port.
B	8 I/O	RB0 is shared with Hardware interrupt. Weak pull-up resisters. RB7:4 can generate a Changed interrupt.
C	8 I/O	28 pin+ PICs shared with Serial ports.
D	8 I/O	40 pin+ PICs shared with Parallel Slave port or LCD segments.
E	3 – 8 I/O	40 pin+ PICs shared with A/D converter. 64 pin+ PIC16C92X 8-bit Input shared with LCD segments.
F	8 Input	68 pin+ PIC16C92X shared with LCD segments.
G	8 Input	68 pin+ PIC16C92X shared with LCD segments.

Despite the depiction of Fig. 11.1, an I/O port does not behave quite like any other internal File. For instance, it has to be configured either to read the voltages on its associated pins (input) or to be able to write to these pins (output). Furthermore, we need to determine how this configuration interferes with the action of software that tries to alter or read the state of the port.

In Fig. 11.1 each parallel port in Bank 0 is shown paired with a TRIS register in Bank 1. In Appendix B this pairing is seen to be a characteristic of all such ports. Each bit n in a parallel port has a shadow bit n in its TRIS register whose function is to configure the associated pin either as an input (TRIS[n] is 1) or an output (TRIS[n] is 0).[2] Most microcontrollers label these as Data Direction registers, but Microchip enigmatically use the term TRIS as short for TRI-State, for reasons we will see later in the chapter.

As an example, consider a situation where pin RA0 and pins RB[7:0] are to be outputs and the rest of Port A are to be inputs. The following code fragment normally appears at the beginning of the main routine; see Program 11.1(a).

```
bsf     STATUS,RP0      ; Change to Bank 1

movlw b'1111110'        ; Pin RA0 = Output
movwf TRISA             ; Rest of pins are Inputs
clrf  TRISB             ; All Port B are Outputs

bcf     STATUS,RP0      ; Back to Bank 0
```

In **C** we could ape the assembly-level code above; for instance, in our exemplar CCS **C** language:

```
#bit BANK_SWITCH = 3.5 /* The RP0 bit in the Status register*/
#byte TRISA       = 0x85/* The TRISA Data Direction register */
#byte TRISB       = 0x86/* The TRISB Data Direction register */
main()
{
BANK_SWITCH = 1;        /* Change to Bank 1                   */
 TRISA = 0xFE;          /* Pin RA0 = Output, rest are Inputs*/
 TRISB = 0;             /* All PortB are Outputs            */
BANK_SWITCH = 0;        /* Back to Bank 0                   */
```

However, some compilers may come with built-in functions to support port set-up and usage. In the case of the CCS compiler we have the function set_tris_x() for Port X, giving:

---

[2] Aide-mémoire: 0 for Output and 1 for Input.

```
main()
{
/* Define any variables first                              */
set_tris_a(0xFE);      /* Pin RA0 = Output, rest are Inputs  */
set_tris_b(0);         /* All PortB are Outputs              */
```

Any reset will set all TRIS bits to 1; that is, after a reset all pins will come on stream as inputs. The choice of a starting configuration as input is deliberate, as if a pin were set to output before the software has had a chance to set the port pin to its initial state, then such a pin will come out of reset with an unpredictable voltage state. This could activate the driver circuitry in an undesirable manner. For instance, a latch actuating a switch turning on the heating element of a washing machine may be triggered before any water is in the tank. Where this kind of catastrophe could occur, the state of the appropriate port bits should be set to their appropriate initial value *before* configuring the TRIS registers.

Once the directional properties of a port's pins have been setup, then the software can either read from or write to a port in a comparable manner to a normal File and hence interact with the outside world. Specifically:

- To *monitor* the state of any pin set as an input, use the btfss or btfsc instruction. For instance, btfss PORTA,1 skips if pin RA1 is High (that is, if PORTA[1] is set to logic 1). Several pins at a time can be read by copying the complete File into W; e.g., movf PORTA,w. If required, the byte can then be copied into a GPR File for further processing.
- To *change* the state of any pin set as an output, use the bcf or bsf instruction. For instance, bcf PORTA,0 will force pin RA0 to its Low state (that is, PORTA[0] is cleared to logic 0). Several pins at a time can be changed by copying the contents of W to the port. For instance, if all Port B's pins are set as outputs, then to bring pins RB[7:6] to its High state and RB[5:0] Low we have movlw b'11000000' : movwf PORTB.

We will look at the electrical characteristics of ports later in the chapter; for instance, what happens if you read a pin that is set to be an output? First we will illustrate the usage of parallel ports with an example.

Consider the situation shown in Fig. 11.2 where any external peripheral device (maybe a printer) wishes to read the byte contents of File h'20' via Port B on request, every time it brings pin RA1 to its Low state. This signal from the peripheral is labeled $\overline{\text{RFD}}$ (Ready For Data). When the PIC MCU responds some time later, it copies the datum to Port B and then it pulses ‾\/‾ pin RA0, which is labeled $\overline{\text{DAV}}$ (Data AVailable), to inform the peripheral that the datum is now available. On a Power-on reset, Port B pins are to be in their Low state and RA0 is to be High.

Signaling in this manner using semaphores, such as $\overline{\text{RFD}}$ and $\overline{\text{DAV}}$, is known as **handshaking**. The term comes from the protocol when

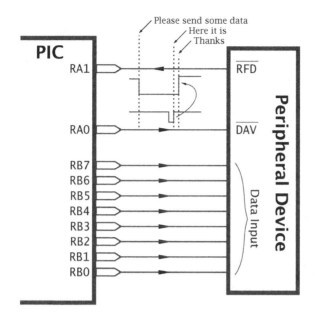

Fig. 11.2 Outputting data from PortB using a handshake transfer.

beginning a conversion and ending it. Handshaking allows separate non-synchronized devices to converse with each other without missing data.

Program 11.1(a) shows how the handshake is implemented in assembly code, based on the port setup outlined on page 295. Preceding this configuration code, the initial state of Port A and B are setup by writing the appropriate pattern to each port. The Include file has defined the names PORTA and PORTB for File 05 and File 06, respectively. Once the Data store is switched back to Bank 0 the port pins set to output will take on the initial state of RA0 High (corresponding to bit 0 of PORTA being 1) and pins RB[7:0] Low (corresponding to all PORTB bits being 0).

After this initialization code, the state of pin RA1 is checked for a Low-state voltage, which reflects as a 0 in bit 1 of PORTA. When this occurs, the datum in File h'20' is copied out to PORTB via the Working register and pin RA0 dropped Low. A single nop intervenes before it is brought High again, to give an extra cycle's delay. The specification did not give a duration of the $\overline{DAV}$ pulse, so in practice nop would be replaced by a call to a delay subroutine.

Finally, the state of pin RA1 is again monitored (as imaged in bit PORTA[1]) until the peripheral brings its $\overline{RFD}$ line High to indicate that the transaction is over. Of course, this is a potential hazard, as if, if the peripheral fails to respond, the PIC MCU will hang. It would be safer to have a time-out; perhaps if there is no response after 65,536 tries then go to some error handling subroutine.

Program 11.1 Implementing a parallel port handshake data transfer.

```
            include  "p16f84a.inc"
DATUM       equ    h'20'
; Initialize ports and setup the pins -----------------------
MAIN        clrf   PORTB      ; Initial state of Port B is zero
            bsf    PORTA,0    ; Initial state of DAV is 1

            bsf    STATUS,RP0 ; First Change to Bank 1
            movlw  b'1111110' ; Pin RA0 = Output
            movwf  TRISA      ; Rest of pins are Inputs
            clrf   TRISB      ; All Port B pins are Outputs
            bcf    STATUS,RP0 ; Back to Bank 0
; Monitor pin RA1 looking for a Low-state voltage -----------
RFD_YES     btfsc  PORTA,1    ; Bit 1 of Port A: Is it 0?
            goto   RFD_YES    ; IF not THEN try again
; Copy the requested Datum to Port B ------------------------
            movf   DATUM,w    ; Copy into W
            movwf  PORTB      ; and to the outside world
; Now pulse the DAV pin RA0 Low to signal "Here it is" ------
            bcf    PORTA,0    ; DAV (pin RA0) Low
            nop               ; for a short time
            bsf    PORTA,0    ; and then High
; Now hang around until the RFD signal goes High ------------
RFD_NO      btfss  PORTA,1    ; Skip if RA1 is High
            goto   RFD_NO     ; IF not THEN keep trying

            goto   RFD_YES    ; Repeat forever
```

(a) Assembly code (17 instructions).

```
#include <16f84a.h>
#byte   PORTB = 6           /* Port B is File 6             */
#byte   DATUM = 0x20        /* File 0x20 holds the datum byte */
#bit    DAV = 5.0           /* Pin RA0 is the DAV line      */
#bit    RFD = 5.1           /* Pin RA1 is the RFD line      */

main()
{
DAV   = 1;                  /* and the DAV line not active @ 1 */
PORTB = 0;                  /* Start with Port B all zero   */

set_tris_a(0xFE);           /* Pin RA0 (DAV) is Output      */
set_tris_b(0);              /* All Port B pins are Outputs  */

while(TRUE)                 /* DO forever loop              */
    {
    while(RFD) {;}          /* Wait until RFD goes FALSE (Low) */

    PORTB = DATUM;          /* Copy Datum out to Port B     */
    DAV = 0;                /* DAV (pin RA0) Low            */
    delay_cycles(1);        /* For a short time            */
    DAV = 1;                /* and then High                */

    while(!RFD) {;}         /* Wait until RFD goes TRUE (High) */
    }
}
```

(b) CCS C code (26 instructions).

Program 11.1(b) gives an equivalent routine using CCS C. This follows a similar structure, but notice in particular how an input pin is tested using constructions like while(RFD) {;} which does nothing ({;} is a null statement) as long as the pin named RFD is True; that is, non-zero. When RFD does go to 0, that is, pin RA1 goes Low, then the loop exits to the next statement. Later on the while(!RFD) {;} test uses the C NOT operator ! to wait until RFD goes to 1 (!RFD goes to 0) to exit.

The built-in function delay_cycles() gives an additional 1-cycle delay, and it can be replaced by an appropriate delay function if this is not satisfactory.

These programs may seem rather useless, as the datum in File h'20' is never setup or changed. However, in a real situation the value could be changed if an interrupt occurs, maybe on a regular basis as dictated by an internal or external timer. This could trigger an analog-to-digital converter module, which dumps its outcome in this holding File. We will be looking at Timer and ADC modules in subsequent chapters.

Our program used the PIC16F84A as the exemplar. Other devices may be programmed in an equivalent manner, but care must be taken with family members which have analog modules. In these situations, ports with pins shared with analog functions will come out of reset with such shared pins set to read signals into the analog module. This is because an analog voltage generated by an external peripheral device could damage input transistors which were configured to expect one or other digital voltage levels. In such situations, where the programmer wishes to use such pins for digital purposes, the analog module's control registers should be setup to switch such pins from the analog module to the digital port. This will normally be done as part of the setup routine, along with the TRIS settings. For instance, the PIC16F87X group use Port A as input channels for the ADC module. To set pins RA[3:0] and RA5 to be digital port lines, the pattern b'00000110' needs to be loaded into the ADC CONtrol register 1 ; see Fig. 14.12 on page 456. As this is in Bank 1, the instructions:

```
movlw  b'000000110'   ; All pins are digital
movwf  ADCON1
```

need to be inserted along with the rest of the Bank 1 sandwich instructions. In CCS C the function setup_adc_ports(NO_ANALOGS); will do the same sort of thing.

In order to fully understand the characteristics of parallel I/O ports we need to look at its hardware implementation. A somewhat simplified version of a single I/O port bit n together with its associated Data Direction bit is shown in Fig. 11.3. The two key elements in this circuit are the Data D flip flop and Data tristate (3-state) buffer.

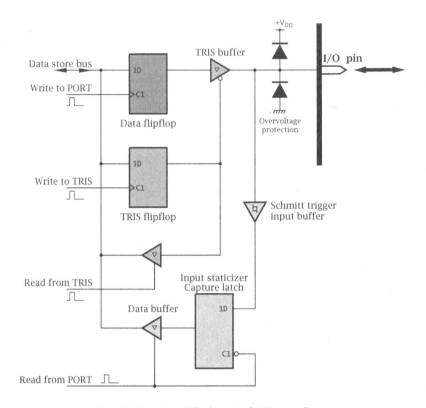

Fig. 11.3 A simplified typical I/O port line.

- Writing to this port will trigger the Data D flip flop, causing the data on the internal Data store line to be clocked in and held as long as the MCU is powered; see Fig. 2.16(c) and (d) on page 31. For instance:

```
movlw  b'11111111'    ; Working register all ones
movwf  h'06'          ; Send to Port B (File 06)
```

will set all eight Data flip flops in Port B to logic 1.

Setting the port bits will occur irrespective of whether its associated I/O pins are configured as input or output. However, to pass the flip flop's state through to the I/O pin, the TRIS (tristate) buffer must be enabled. In this situation, as shown in Fig. 11.4(b), the Data flip flop is directly connected to the outside world.

- Reading from this port enables the Data buffer, causing the state of the staticizer latch[3] to be gated through to the internal Data store line. When the port is idling, i.e., not being read, the D latch is transparent and its output follows the state of the pin; see Fig. 2.16(a) and (b) on

---

[3]There is no staticizer latch in the low-range 12-bit family.

page 31. When the port is being read, the D latch clock enable goes High and the data into the 3-state Data buffer is frozen, effectively holding its state constant while being read; that is, staticizing it. The Data latch's input is isolated from the I/O pin using a buffer with hysteresis (a Schmitt trigger) for noise immunity. For instance, to read the state of Port B we have:

```
movf    h'06',w    ; Read all eight input PortB lines into W
```

This reading action, shown in Fig. 11.4(a), will occur independently of whether the associated I/O pin is configured as an input or output.

The first-generation low-range PIC16C5XX series 12-bit core PIC micro-controllers have no explicit TRIS registers; that is, the TRIS flip flops are not mapped into the Data store. Instead they use the tris instruction, which copies the contents of the Working register to an internal control register that is not mapped into the Data store. Thus, for our example:

```
movlw   b'00001111'  ; Top bits to be output, bottom to be input
tris    h'06'        ; Do it on PortB
```

When the 14-bit mid-range core devices were introduced with explicit TRIS registers, Microchip kept the tris instruction but did not guarantee that it would be implemented for future devices. However, many pro-grammers still use tris and some C compilers, such as the CCS compiler, retain its use.

From Fig. 11.3 we see that a TRIS bit can be read from as well as written to. Although this may be rather useless, consider a programmer wishing to alter pin RB7 to an output (see Example 11.4).

```
bcf     h'86',7    ; Clear bit 7 of TRISB
```

bcf (Bit Clear File) is an example of a **read–modify–write** instruction (see page 117) whereby the state of TRISB is *read* into the processor, modified and then *written* out to TRISB. To do this the processor needs to be able to both read from and write to the File.

Because a parallel port may be configured as an input, or output, or a mixture of both, it is important to know what restrictions are introduced when reading or altering the state of such special Files. For instance, what would happen if the software read from a port bit which has been configured as an output? The four possibilities enumerated in Fig. 11.4 are:

(a) **Reading from a port pin set as input (TRIS = 1)**

Here the TRIS buffer is disabled and the state of the Data flip flop remains unchanged. For instance, movf h'06',w reads the state of Port B input pins into the Working register.

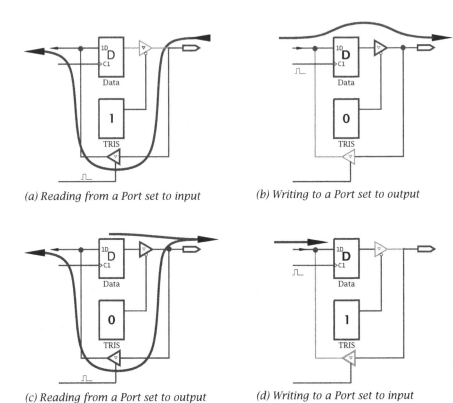

(a) Reading from a Port set to input

(b) Writing to a Port set to output

(c) Reading from a Port set to output

(d) Writing to a Port set to input

Fig. 11.4 Reading from and writing to a port bit with linked I/O pin set to input or output.

(b) **Writing to a port pin configured as an output (TRIS = 0).**
Here the TRIS buffer is enabled and the Data flip flop altered by the processor writing to the port. The state of this flip flop appears on output pins. For instance, if all Port B pins RB[7;0] are set as output, movlw b'10101010' followed by movwf h'06' sets the Port B pins to HLHLHLHL (H = High, L = Low).

(c) **Reading from a port pin configured as an output (TRIS = 0).**
In this situation the TRIS buffer is enabled and so the applicable I/O pins are connected to their associated Data flip flop. In most situations reading port pins set to output will effectively copy their flip flop and associated pin states into the CPU; however, this is not always the case. If the current taken by the device connected to an I/O pin is large the logic voltage at the pin may deviate significantly from the normal logic levels. For instance, connecting a bipolar transistor directly to a port pin, as in Fig. 11.5(a), will take sufficient current

from the TRIS buffer to drag the pin voltage to $\approx 0.7\,V$; the forward conducting voltage of a typical transistor base-emitter.[4]

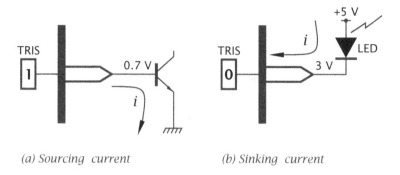

(a) Sourcing current   (b) Sinking current

Fig. 11.5 Sinking and sourcing current.

The situation in Fig. 11.5(b) is similar, with current flowing through the light-emitting diode (LED) into the port pin,[5] and the TRIS buffer will be pulled up to $\approx 3\,V$, assuming a conducting LED offset of $2\,V$. In these situations the outcome of reading a port pin set as output is often not the state of that port bit's Data flip flop, due to the improper voltage levels. For instance, `btfsc PORTB,7` in purporting to skip if bit 7 of Port B is zero may fail to function as expected if the linked pin RB7 is sinking or sourcing too much current.

(d) **Writing to a port pin configured as input (TRIS = 1).**
In this situation the state of the Data flip flop will be altered in the proper manner. However, as the TRIS buffer is disabled, any change will not be reflected at a linked I/O pin until the direction of the port pin is subsequently changed to output. This ability to setup the state of a port in a manner invisible to the outside world was used in Program 11.2 to initialize the parallel ports after reset and before any pins are set to output. Remembering that on reset, all ports are set to input; in other words, all TRIS registers are set to b'11111111'.

Most port pins configured as inputs have Schmitt trigger buffers which trigger at 80% of the supply when the input voltage is going from a Low to High state, and 20% of the supply voltage when the direction is opposite. This considerably improves the reliability in sensing the voltage state in noisy environments. The single General-Purpose GPIO I/O Port in 8-bit devices (except GP2), Port B and in some older devices, such as the PIC16F84, Port A (except RA4) have normal non-hysterisis buffers, with a lower threshold of $V_{IL} \leq 0.5\,V$ and upper point of $V_{IH} \geq 2\,V$.

---

[4] Typically somewhere between 25 mA and 35 mA (see Example 11.1).
[5] Typically around 60 mA; see Fig. 11.17.

Any pin setup to be an output has to be able to carry an appropriate current to activate the driven load. In most situations a port pin configured as an output will only be required to source or sink a few milliamps of load current. Nevertheless, it is important to be aware of the limitations of the drive capabilities of port output pins.

Generally two situations are tabulated in a device's data sheet.

1. Sink current I_{OL} into a pin when an output is in a Low state should not exceed +8.5 mA if the Low voltage V_{OL} is not to rise above 0.6 V.
2. Source current I_{OH} out of a pin which is High should not exceed −3 mA if the High-state voltage is not to drop more than 0.7 V below V_{DD}. The negative current denotes source; i.e., out of the device.

Larger currents may be sourced or sunk, as in Fig. 11.5, if degradation of logic levels are acceptable, subject to an absolute limitation that it must be within the range ±25 mA for any single I/O pin to avoid damage. Where more than one I/O pin is involved in driving current, an overall global limit must be observed. Eight-pin devices limit their single Port GP (which is a combination of Port A and Port B in its bigger cousins) to a grand total of 125 mA source and 125 mA sink current. Larger footprint family members limit the group Port A, Port B, and, if available, Port C to a global maximum of ±200 mA. Similarly the group Port D and Port E have an additional global limit of ±200 mA.

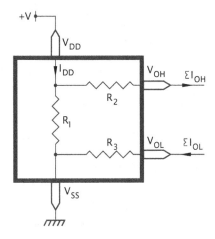

Fig. 11.6 Power dissipation model.

Each output pin in sourcing or sinking current will dissipate energy, which appears in the package as heat. From the simplified model of Fig. 11.6 we see that there are three components to this dissipation that are modeled as resistors.

1. From the V_{DD} power pin we have a current I_{DD}. However, the current through the body resistor R_1 is less by the source currents from the port pins, giving a $V \times I$ dissipated power of $V_{DD} \times (I_{DD} - \sum I_{OH})$.
2. The voltage drop across the equivalent resistance R_2 between output pins and the power pin is $\Delta V = V_{DD} - V_{OH}$. Thus the power dissipated is $\Delta V \times \sum I_{OH}$.
3. Current sunk into the output pins through R_3 to ground via the V_{SS} pin dissipates $V_{OL} \times \sum I_{OL}$.

Adding these components gives the formula quoted in the data sheet:

$$P_{DIS} = V_{DD} \times \left(I_{DD} - \sum I_{OH}\right) + \sum \left((V_{DD} - V_{OH}) \times I_{OH}\right) + \sum (V_{OL} \times I_{OL})$$

taking into account that the output voltages at each pin will differ with different currents. The figure given in the data sheets for small package devices for P_{DIS} is 800 mW and 1 W for 40-pin family members. In all cases the maximum current into the V_{DD} power pin should not exceed 250 mA and 300 mA out of the V_{SS} pin.

In practice, the equivalent resistance R_2 is not linear and varies in a rather complex way; as illustrated in Fig. 11.13. That is, V_{OH} does not drop with current in a straight line. Data sheets show graphs of this current–voltage relationship; for instance, see Figs. 11.13 and 11.17. However, a worst-case scenario with large currents would be to assume that V_{OH} had dropped to zero and V_{OL} had been pulled up to the supply V_{DD}. In this situation the excess of I_{DD} over $\sum I_{OH}$ supplying the CPU and other peripheral modules would be minimal and could be ignored. The total power dissipated would then be:

$$P_{DIS} = V_{DD} \times \left(\sum I_{OH} + \sum I_{OL}\right)$$

The block diagram of Fig. 11.1 is a typical representation of a parallel port I/O bit. Specific ports may vary in ways that can affect the electrical performance in a significant manner; especially Ports A and B. The majority of ports have TRIS buffers implemented as shown in Fig. 11.7 which uses a series N-channel/P-channel field effect transistor totem-pole structure.

- When the TRIS flip flop is logic 1 the lower AND gate has a logic 0 output and the upper OR gate has a logic 1 output. In this situation, both transistors TRN and TRP are non-conducting and the state of the Data flip flop is isolated from the I/O pin. In this situation the port pin is configured as an input.
- When the TRIS flip flop is logic 0 then the complement state of the Data flip flop is gated through to both totem-pole transistor. With D Low, TRN conducts and TRP is off, giving a Low pin voltage. With D High, TRP conducts and TRN is off giving a High pin voltage. In this situation the pin follows the state of the Data flip flop, with current being

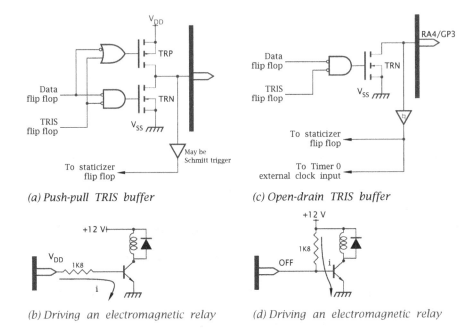

(a) Push-pull TRIS buffer   (c) Open-drain TRIS buffer

(b) Driving an electromagnetic relay   (d) Driving an electromagnetic relay

Fig. 11.7 Output driver structures.

sourced or sunk through the relatively low resistance active conducting transistors.

As an example, consider the situation where an electromagnetic relay is to be activated, requiring a 200 mA activation current at 12 V. For currents and voltages of this magnitude we need external buffering. In Fig. 11.7(b) a bipolar transistor acts as an external switch. If the minimum gain of this transistor is 100 then a 1.8 kΩ resistor will give a base current of 2 mA, assuming a base-emitter conduction voltage of 0.7 V and a PIC MCU V_{OH} of at least 4.3 V.

The output of RA4/GP3 shown in Fig. 11.7(c) is somewhat different in that only the bottom totem-pole transistor is implemented. As opposed to the 3-state structure of Fig. 11.7(a), this structure has only two states; that is, active logic 0 and open-circuit. This type of output is known as **open drain** (or open-collector); see Fig. 2.3 on page 20.

- When the TRIS flip flop is logic 1 (its reset state) then the AND output is Low and TRN is off with the output pin high resistance. RA4/GP3 is then set to input.
- When TRIS is logic 0 the output transistor conducts when the Data flip flop is logic 0, giving an active-Low output. When the Data is logic 1, TRN is off and the output pin floats.

An open-drain output cannot source current; either the load itself must be connected from the output pin to a positive voltage or an external pull-up resistor used as a load for the on-chip transistor. This is the case in Fig. 11.7(d), where the base current for the external transistor is derived from the 1.8 kΩ pull-up resistor when RA4/GP3 is off.

If RA4 is to be used as the Timer 0 input, it is usually configured as an input. If configured as an output, then in this situation RA4 must be set to logic 1, which will disable the open-drain transistor and prevent interaction between it and the external clock input to the Timer.

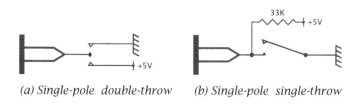

(a) Single-pole double-throw   (b) Single-pole single-throw

Fig. 11.8 Interfacing switches to a port pin.

Many applications involve reading the state of arrays of switches. Rather than use the relatively more expensive single-pole double-throw (SPDT) switch arrangement of Fig. 11.8(a) to give the two logic states, most switches—for instance, those in the keypad of Fig. 11.10—are single-throw (SPST) types. In these situations an external pull-up resistor is needed to convert the open-circuit state to a High-state voltage, as shown in Fig. 11.8(b). A similar situation arises when open-drain/collector electronic devices, such as phototransistors, are to be read by a port. The value of such pull-up resistors should not be too low, as a large current will flow through the switch when closed, nor too high or else noise will be induced by electromagnetic means from external sources. A good compromise is in the range 10 – 100 kΩ.

In order to simplify the interface of such devices, Port B *inputs* have optional internal pull-up resistors. These internal resistors are called **weak pull-ups**, as their typical equivalent values of around 20 kΩ is high enough not to interfere with devices being read which have "normal" logic Low and High outputs.

We see from Fig. 11.9 that the internal pull-up resistors (actually a P-channel FET) are switched in only if $\overline{\text{RBPU}}$ (Register B Pull Up, bit 7) of the Option register is 0. Although all eight pull-ups are qualified by $\overline{\text{RBPU}}$, those pins configured as outputs (TRIS[n] = 0) will have the resistor switched off. $\overline{\text{RBPU}}$ resets to 1, and so the pull-up resistors default to off.

Eight-pin devices with their single Port GP have a similar arrangement. In the specific case of the PIC16F629/75, each internal pull-up resistor (GP3 does not have an associated pull-up resistor) can be individually switched in or out using the WPU (Weak Pull-Up) SPR, controlled in turn by bit 7 of OPTION_REG, which here is called $\overline{\text{GPPU}}$.

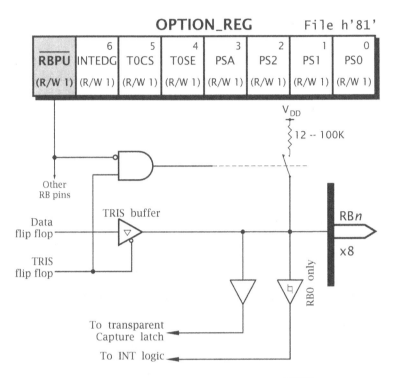

	6	5	4	3	2	1	0
RBPU	INTEDG	T0CS	T0SE	PSA	PS2	PS1	PS0
(R/W 1)	(R/W 1)	(R/W 1)	(R/W 1)	(R/W 1)	(R/W 1)	(R/W 1)	(R/W 1)

Fig. 11.9 Port B's weak pull-up resistors controlled by $\overline{\text{RBPU}}$ in OPTION_REG.

As a typical application of weak pull-ups, consider the problem of reading a keypad, such as that illustrated in Fig. 11.10(a). In this particular instance there are 12 switches, and rather than use up all these scarce I/O pins it is hardware efficient to connect these switches in the form of a 4 × 3 matrix, as illustrated in Fig. 11.10(b). This 2-dimensional array reduces the I/O pin count to 7. Larger keypads show an even greater efficiency gain, with a 64-contact 8×8 keyboard needing only 16 I/O pins.

Although there are variations on this theme, the topology shown here is typical. The three columns are read in via RB[7:5], with internal pull-up resistors enabled. The four rows connected to RB[3:0] can be individually selected in turn by driving the appropriate pin Low, thus scanning through the matrix. The sequence is shown in Fig. 11.10(c). The switch contacts are normally open and, because of the pull-up resistors, read as logic 1. Should a switch connected to a Low row line be closed then the appropriate column line will be driven Low. This means that once the closed key column has been detected the column:row intersection is known. The 330 Ω resistors limit the current through the switch should one of the RB[7:5] pins accidentally give a High-state output due to erroneous software.

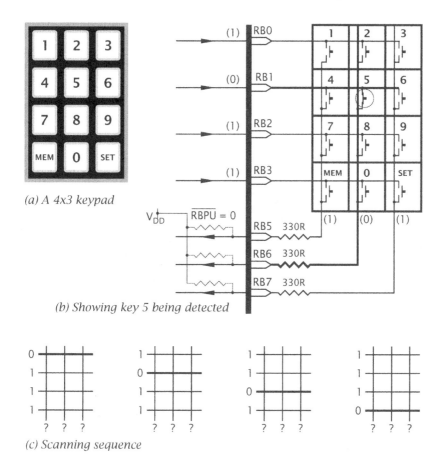

(a) A 4x3 keypad

(b) Showing key 5 being detected

(c) Scanning sequence

Fig. 11.10 Interfacing to a keypad.

In order to tie these concepts together, consider a subroutine to interrogate the keypad and return either with the key pressed (or at least the first key found if more than one) or if no key then -1 (i.e., h'FF'). Before looking at the coding we can assume that somewhere in the main software Port B has been configured appropriately with the correct input and outputs assigned and that bit $\overline{\text{RBPU}}$ in the Option register has been cleared. Something like:

```
      include "p16f627.inc"
MAIN  bsf   STATUS,RP0             ;Change to Bank1 where
      movlw b'11110000'            ;TRISB & OPTION_REG lie
      movwf TRISB                  ;RB[7:4] inputs, RB[3:0] outputs
      bcf   OPTION_REG,NOT_RBPU    ;Activate internal pull-ups
      bcf   STATUS,RP0             ;Go back to Bank0
```

The listing of Program 11.2 is based on the task list:

1. Set KEY_COUNT to one.
2. For i = 0 to 3.
   - Activate row i.
   - For j = 0 to 2.
     - Check column j.
     - IF zero THEN BREAK to step 4.
     - ELSE increment KEY_COUNT.
3. Set KEY_COUNT to −1 if no key found.
4. Return KEY_COUNT.

Basically the sequence of operations is to begin with a count of one; i.e., key[1], and bring row[0] Low. As each column is checked for a zero, the count kept in the Working register is incremented. If no closure (that

---

Program 11.2 Scanning the keypad.

```
;***************************************************************
; * FUNCTION: Scans 4x3 keypad & returns with a key identifier*
; * ENTRY    : None                                           *
; * EXIT     : Key in W [MEM]=10, [0]=11, [SET]=12            *
; * EXIT     : Return -1 (h'FF') if no key detected           *
; * ENVIRON  : KEY, PATTERN byte vars                         *
;***************************************************************

        cblock                  ; Two global variables
        KEY_COUNT:1, PATTERN:1
        endc

SCAN_IT clrf  KEY_COUNT         ; Key 1 is the first key
        incf  KEY_COUNT,f
        movlw b'11111110'       ; The initial scan pattern
        movwf PATTERN

SLOOP   movf  PATTERN,w         ; Get scan pattern from mem
        movwf PORTB             ; Set row Low
        movf  KEY_COUNT,w       ; Get Key count
; Now check each column for a zero -----------------------------
        btfss PORTB,5           ; Check column 1
        goto  GOT_IT            ; IF zero THEN found the key!
        incf  KEY_COUNT,f       ; ELSE inc Key
        btfss PORTB,6           ; Check column 2
        goto  GOT_IT            ; IF zero THEN found the key!
        incf  KEY_COUNT,f       ; ELSE inc Key again
        btfss PORTB,7           ; Check column 3
        goto  GOT_IT            ; IF zero THEN found the key!
        incf  KEY_COUNT,f       ; ELSE inc Key again

; Reach here if no closed key ----------------------------------
        rlf   PATTERN,f         ; Shift scan pattern once ->
        btfsc PATTERN,4         ; Check; has the 0 arrived @ RB4?
        goto  SLOOP             ; IF not DO another column

; ELSE no key found anywhere -----------------------------------
        movlw -1                ; Return -1
        goto  S_EXIT

GOT_IT  movf  KEY_COUNT,w       ; Copy Key count into W
S_EXIT  return
```

is, a 0) is found, the next row is tried by shifting the initial test pattern one position. Although the value of the Carry flag is rotated in from the left, the progressive nature of the check means that its value is irrelevant.

There are two ways out of the loop.

- If a 0 is found during the scan, the count in W is the desired value and the subroutine immediately returns after copying the Key count from memory into W.

- If the row pattern shift results in the sample 0 arriving at bit 4, then the subroutine returns h'FF' to tell the caller that no key has been found.

In the real world a subroutine like this would often read in rubbish due to switch bounce and possibly noise induced in the connections between keypad and the electronics. One way of filtering out this unpredictability is shown in the subroutine of Program 11.3. Here the state of the keypad is interrogated using the SCAN_IT subroutine of Program 11.2. By

---

Program 11.3 Noise filtered keypad scanning.

```
; *******************************************************************
; * FUNCTION: Scans 4x3 keypad and returns with a debounced       *
; * FUNCTION: key identifier                                      *
; * ENTRY    : None                                               *
; * EXIT     : Key in W [MEM]=10, [0]=11, [SET]=12                *
; * EXIT     : Return -1 (h'FF') if no key detected               *
; * RESOURCE: COUNT, NEW_KEY, OLD_KEY                             *
; * RESOURCE: Subroutine SCAN_IT                                  *
; *******************************************************************
;
        cblock                    ; Three global variables
         COUNT:1, NEW_KEY:1, OLD_KEY:1
        endc

GET_IT  clrf  COUNT           ; The no-change count zeroed
GLOOP   call  SCAN_IT         ; Raw value returned in W
        movwf NEW_KEY         ; Is new value
        subwf OLD_KEY,w       ; New and old the same?
        btfsc STATUS,Z
        goto  EQUAL           ; IF same go to EQUAL

; Otherwise the readings are different, so: --------------------
        movf  NEW_KEY,w       ; Make old key = new key
        movwf OLD_KEY
        goto  GET_IT          ; and start all over again

; IF readings are the same THEN -------------------------------
EQUAL   incfsz COUNT,f        ; Increment count; IF not
        goto   GLOOP          ; rolled around to 00 repeat

        movf   OLD_KEY,w      ; ELSE that's it!
        return
```

keeping the state of the previous reading in Data memory, any change can be detected. Only if no change occurs over 256 readings will sub-routine GET_IT return with the keypad state. Depending on the quality of the keypad, ambient noise and processor speed, the outcome can be improved at the expense of response time by including a short delay in the loop, or by using a 2-byte stability count to increase the number of readings.

Program 11.4 Scanning the keypad in **C**.

```
unsigned int scan_it(void)
{
unsigned int key, pattern;
key=1; pattern = 0xFE;          /* Initial pattern b'11111110' */
while(key<13)                   /* DO 13 tries                 */
    {
    PORT_B = pattern;           /* Activate one row            */
    if(!COL1) {break;}          /* And try each column in turn */
    key++;                      /* Breaking if a 0 is fount    */
    if(!COL2) {break;}          /* ELSE incrementing Key count */
    key++;
    if(!COL3) {break;}
    key++;
    pattern = pattern << 1;     /* Shift left checks next row  */
    }
if(key==13) {key = 0xFF;}       /* IF 13 was got THEN no key!  */
return key;
}
```

Program 11.4 shows the equivalent coding in CCS **C** to that in Program 11.2, as an example of interaction with parallel ports. This assumes that Port B has already been initialised as follows:

```
#include  <16f627.h>
#use      fast_io(b)           /* Default            */
#byte PORT_B = 6
#bit  COL1  = PORT_B.5         /* Column 1 is RB5    */
#bit  COL2  = PORT_B.6         /* Column 2 is RB6    */
#bit  COL3  = PORT_B.7         /* Column 3 is RB7    */

unsigned int scan_it(void);
int main()
{
set_tris_b(0xF0);
port_b_pullups(TRUE);
```

The CCS compiler handles parallel I/O in several different ways. The #use fast_io(b) statement above leaves it up to the programmer to ex-plicitly setup the appropriate TRIS registers. The use_standard_io(b) alternative allows the programmer to ignore such minutia, but then the compiler will set-up the port configuration *each* time it is used, even if that configuration remains unchanged from the last usage. In a similar manner the function `port_b_pullups(TRUE)` is an alternative to setting the $\overline{\text{RBPU}}$ bit in the Option register.

The logic of the program is very similar to our assembler coding, with a shifting pattern zeroing each row in turn. The only difference is that the loop is executed a fixed number of times using a count, rather than testing bit 4 of the test pattern. This makes the process more transparent, although the latter is more efficient.

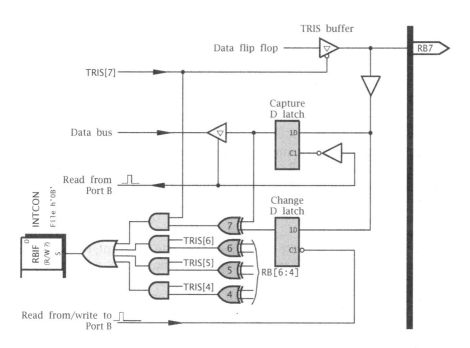

Fig. 11.11 The Port B Change feature.

Interfacing keypads to microcontrollers is such a common task, most PIC MCU families have the ability to detect whenever a variation happens at some Port B inputs. The logic of this Port B Change feature is shown in Fig. 11.11. The top four Port B I/O pins have a second D latch in parallel but in antiphase to the usual input Capture latch. When the CPU reads Port B the Capture latch samples the state of the input pin in the normal way. However, at the same time the Change latch becomes transparent. When the reading action is over, the Change latch freezes and captures the pin state as it is at the time of reading. The outputs of both the Capture and Change latch are Exclusive-ORed together. As we have seen on page 14, an XOR gate detects *differences* between its two inputs. As the Capture latch is now transparent, any subsequent variation at the pin input will cause the output of the associated XOR gate to go to logic 1. Each of the four Port B cells RB[7:4] has a Change feature and the four XOR gates are ORed to give a composite signal which sets the RBIF (Register B Interrupt Flag) in the INTCON of Fig. 7.3 on page 191. If the RBIE (Register B Interrupt Enable) bit is set to 1, then this is a convenient way of awaking a PIC MCU slumbering in its Sleep state. If the GIE (General Interrupt Enable) bit is set as well, an alteration in the top nybble of Port B will cause an interrupt as well. Each XOR gate is ANDed with the appropriate TRIS line, so that only bits that are programmed as an input can contribute to the Change signal.

In the specific case of our keypad of Fig. 11.10, if all the row lines are set to the Low state, then when any switch is pressed a column line will change state. If RBIE has been set to enable the Port B Change facility, then when the RBIF flag sets an interrupt will occur. The keypad may then be scanned in the ISR to determine which key has been closed.

Mid-range[6] 8-pin devices have a similar ability for their GPIO port. For greater flexibility, each GP-pin has an individual selectable Change enable.

Care must be taken in using this facility. For instance, using the lower (non-Change) part of Port B (e.g., bcf PORTB,0) can affect the Change facility by forcing all the latches to resample. Also in older devices, an alteration may occur at the instant the port is being read and may be missed. Neither of these foibles are factors if a keypad is used to awaken a sleeping processor.

Once the PIC MCU has responded to the Change interrupt, the Change signal setting RBIF should be removed by reading Port B again, which equates the state of the two D latches. Only then should RBIF be cleared. Failure to do this initial read will result in this interrupt flag being immediately set again.

As an example, using the keypad to awaken the PIC MCU with the assumption that GIE is zero (no interrupt) should be implemented as:

```
movf   PORTB,w        ; Read Port B to cancel any difference
bcf    INTCON,RBIF    ; Clear the Change interrupt flag
bsf    INTCON,RBIE    ; Enable the Change interrupt enable
sleep                 ; Go to sleep
; zzzzz
call   DELAY          ; On wakening let things settle
movwf  PORTB,w        ; before canceling any difference
bcf    INTCON,RBIF    ; Clear the Change interrupt flag
bcf    INTCON,RBIE    ; Disable Change interrupt facility
```

---

[6]Most PIC12CXXX devices, such as the PIC12C508/9, are in fact low-range 12-bit cores.

Most PIC microcontrollers have relatively few I/O port lines; see Table 11.1. Even the larger footprint devices, such as the PIC16F877 with 33 peripheral pins, may not have enough parallel I/O resources for some projects; especially as several other peripheral devices may need to use the shared I/O pin budget.

As an example, consider a multi-purpose intruder alarm which can monitor up to eight zones—for example, floors in a multi-story building. Each zone can have up to eight movement sensors. A display of eight lamps back at base is to be used to indicate in which zone the intruder is located.

Based on this specification, a budget of 72 (64 input and 8 output) parallel I/O pins will be required. Rather than using one PIC MCU per zone reporting back to a main controller[7] it has been decided to expand the I/O capabilities of a single PIC16F627 device.

One expansion strategy is shown in Fig. 11.12. Here Port B is used to implement an external data bus which connects to the eight zone 3-state buffers and one indicator flip flop array. Each zone's set of sensors are interfaced to this local bus via an octal 3-state buffer. One of eight buffers can be enabled using a 3- to 8-line decoder addressed from Port A. For example, if RA[2:0] were b'111' and RA3 = 0 then Zone 7's buffers are enabled and its eight sensors can be read in at Port B.

To activate the one output lamp array, RA3 should be logic 1 and Port B set to output. Data can then be clocked into the flip flop array by pulsing RA0 Low then High to give a rising edge.

The number of output ports may be expanded in this architecture to eight by using a second 3- to 8-line decoder to select the port enabled when RA3 = 1. However, up to two extra output ports could be added by simply substituting RA0 by RA1 and RA2 to enable these two additional flip flop arrays. For instance, one port could show which sensor(s) within the zone was active and RA4 used to sound a buzzer if any zone was active.

To show how this hardware interacts with the software, consider the subroutine in Program 11.5 below that reads Zone N and if non-zero then lights lamp N; where N is an integer 0–7 in a File called ZONE on entry. We assume that an active sensor gives logic 1 and a lamp illuminates on a logic 0.

Checking the state of the Zone N sensor group is simply a matter of copying the Zone address from ZONE to Port A set up as output. If RA3 is Low then the 3- to 8-line decoder enables the appropriate Zone buffer. After a short delay, introduced to allow data to settle, the Zone data is read in via Port B. For a real system, a delay of several hundreds of milliseconds and a digital smoothing routine, such as the debounce

---

[7]An implementation that is perfectly feasible and cost effective; see SAQ 11.1.

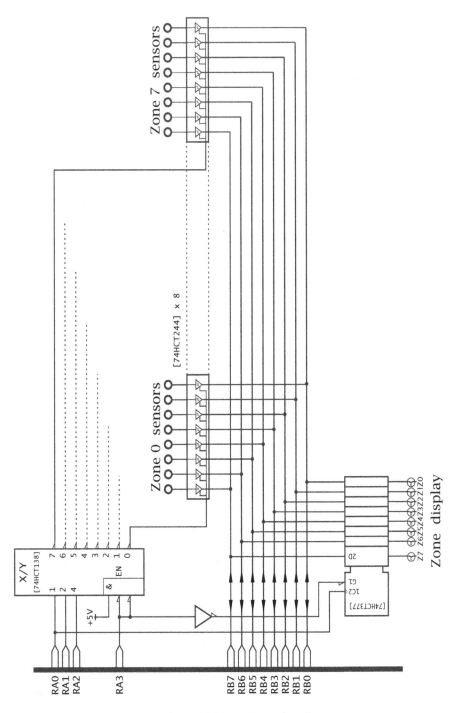

Fig. 11.12 A multi-zone intruder alarm.

routine of Program 11.3, would be needed for reliable data acquisition; assuming that the Zone buffers were geographically distant.

Activating the eight lamps is a little more tricky. In this situation Port B must be configured as all output. The lamps are then actuated by sending the appropriate pattern to Port B, bringing RA3 High to disable the Zone

---

Program 11.5 Interacting with the intruder hardware.

```
;       ************************************************************
;       * FUNCTION: Reads Zone N and activates lamp N             *
;       * ENTRY    : N is in File ZONE as b'00000nnn'             *
;       * EXIT     : Lamp N active if Zone N is non-zero          *
;       * EXIT     : ZONE zero and TEMP unchanged                 *
;       ************************************************************
ZONE_N    bsf     STATUS,RP0    ; Change to Bank 1
          movlw   h'FF'         ; Set Port B to input
          movwf   TRISB
          clrf    TRISA         ; Set Port A to output
          bcf     STATUS,RP0    ; Change to Bank 0
; ------------------------------------------------------------------
          movf    ZONE,w        ; Get N (in File ZONE on entry)
          movwf   PORTA         ; used to select Zone N's buffers
          nop                   ; Delay to allow long lines
          nop                   ; to settle
          movf    PORTB,w       ; Now read data on Port B
          btfsc   STATUS,Z      ; IF not zero THEN an intruder!
          goto    LAMP_OFF      ; otherwise all clear
; Intruder found, now activate lamp -------------------------------
          bsf     STATUS,RP0    ; Change to Bank 1
          clrf    TRISB         ; Port B now set to output
          bcf     STATUS,RP0    ; back to Bank 0
; Convert binary number to unary equivalent to activate lamp ---
          movlw   h'FF'         ; All ones
          movwf   TEMP          ; into TEMP
          bcf     STATUS,C      ; Zero Carry bit
          incf    ZONE,f        ; Map zone range to 1 -- 8
Z_LOOP    rlf     TEMP,f        ; Shift pattern <--
          bsf     STATUS,C      ; Set Carry bit
          decfsz  ZONE,f        ; Decrement Zone number
          goto    Z_LOOP        ; and repeat N times
; TEMP holds the unary lamp activation pattern -----------------
          movf    TEMP,w        ; Get it
LAMP_OUT  bsf     PORTA,3       ; Enable output port
          movwf   PORTB         ; Lamp data
          bsf     PORTA,0       ; Clock it in by pulsing RA0
          bcf     PORTA,0
          return                ; All done
; Go here if no intruder found and turn off all lamps ----------
LAMP_OFF  bsf     STATUS,RP0    ; Change to Bank 1
          clrf    TRISB         ; Port B now set to output
          bcf     STATUS,RP0    ; back to Bank 0
          movlw   h'FF'         ; All ones turns lamps off
          goto    LAMP_OUT
```

---

decoder and then pulsing RA0. This is implemented in Program 11.5 as routine LAMP_OUT. The lamp datum is simply all logic 1s (all lamps off) where no intruder has been detected; that is, where the sensor data has been read as all zeros.

When an intruder has been detected, then lamp N alone must be lit; for instance, b'10111111' for Zone 6. To do this, the binary zone code in ZONE must be converted to the appropriate unary (one of n) code. For instance, Zone 2 b'00000010' maps to b'11111011', Zone 3 b'00000011' maps to b'11110111', etc.

In the program the unary code is built up in File TEMP, which is initially set to b'11111111'. By clearing Carry *before* entering the loop at Z_LOOP but setting it to 1 *within* the loop, a single zero can be shifted left using the Rotate Left File instruction rlf TEMP,f. This gives the sequence b'11111111 ← 11111110 ← 11111101 ← 11111011...01111111'. As this shift progresses, the ZONE datum (mapped to the range 1 – 8 so that at least one shift is implemented) is decremented and the loop exited when this reaches zero. Thus the position of the lone 0 (the initial C = 0) represents the original Zone number. This unary code is then sent out to the lamp port at LAMP_OUT to activate the Zone indicator.

Examples

Example 11.1
A 2N3055 NPN bipolar transistor is to be used to activate the field coils of a small stepper motor. Taking into account the minimum gain of the transistor over the range $+85 \rightarrow -40°C$, it has been calculated that the base current must be at least 10 mA. The transistor is to be controlled from a port pin and its base-emitter voltage can be assumed to be no more than 0.7 V, with a V_{DD} of 5 V. What is the maximum value of the base resistor R_B, and, given this value, what will be the worst-case maximum base current?

Solution
For currents of this magnitude we can assume that the pin voltage will be less than 5 V. The data sheet specifies a minimum voltage of 4.3 V (a drop of 0.7 V) for a I_{OH} of -3 mA, but for currents greater than this we must resort to graphical techniques.

Figure 11.13 shows the graphical relationship of output source current I_{OH} for a High-output voltage state V_{OH}. The grey area is bounded by the minimum situation, which is at $+85°C$ and maximum condition at $-40°C$.

This voltage V_{OH} is also a function of the transistor input base resistor circuit according to the equation $V_{OH} = 0.7 + I_{OH} \times R_B$. This straight line relationship (called a **load line**) is shown on the graph from (0,0.7) drawn

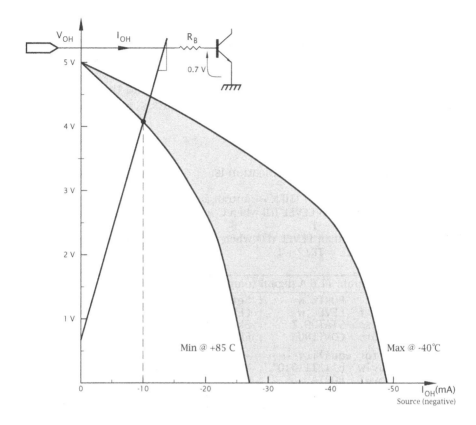

Fig. 11.13 Source current against voltage.

to intersect the minimum locus at a current of −10 mA. This crossover is the only point that satisfies both current–voltage relationships. The slope of the load line $\frac{\Delta V}{\Delta I}$ is the resistance in kΩ (as current is in mA) and measures 280 Ω. Notice that the High-output state has fallen to 4 V (−10, 4.0).

Extending the load line onwards gives the maximum current as the X coordinate of the intersection with the maximum locus, which is approximately 11.5 mA; not much different. If the current requirement had been larger, then the minimum/maximum currents diverge showing a significant temperature sensitivity. For instance, a 20 mA minimum base current requires a base resistor of ≈ 120Ω (assuming a base voltage of 0.8 V) and the maximum base current would be 28 mA.

Example 11.2
A mid-range PIC MCU is to be used as a digital comparator where a parallel-input 8-bit word P is to be compared to a byte datum located in a File named TRIP. Outputs are to indicate Lower-Than, Equivalent,

and Higher-Than. The comparator is to have an hysteresis of ±1 bit. That is, if previous comparisons showed N < TRIP then the trigger level is increased to TRIP + 1 for equality. Similarly, on a downward trajectory the equality level is to be decreased to TRIP − 1.

Datum P is to be input via Port B setup as input and the lower three Port A pins give the active-High comparator outputs <, ==, > at RA2, RA1, RA0, respectively.

Solution

The task list for such a specification is:
1. Subtract P from LEVEL.
2. IF Equal (EQ when Z = 1) THEN == output active.
3. ELSE IF P Higher than LEVEL (HI when C = 0, Borrow) THEN > output active AND LEVEL = TRIP − 1.
4. ELSE IF P Lower than LEVEL (LO when C = 1, No Borrow) THEN < output active AND LEVEL = TRIP + 1.

Program 11.6 A digital comparator with hysteresis.

```
COMP        movf   PORTB,w      ; Get input P
            subwf  LEVEL,w      ; LEVEL - P
            btfss  STATUS,Z     ; Skip if equality
            goto   CONTINUE     ; ELSE IF not THEN try alternative
; This code for equality ---------------------------------------------
            movlw  b'11111010'  ; Make == output logic 1
            movwf  PORTA        ; Other outputs logic 0
            goto   COMP_END     ; and exit
CONTINUE    btfsc  STATUS,C     ; Skip if borrow (P higher than)
            goto   LO           ; ELSE P < LEVEL
; This code if P > LEVEL ---------------------------------------------
HI          movlw  b'11111001'  ; Set > output RA0 to logic 1
            movwf  PORTA        ; Rest to 0
            decf   TRIP,w       ; Copy TRIP-1 to w
            movwf  LEVEL        ; The new comparator level
            goto   COMP_END     ; and exit
; This code when P < LEVEL -------------------------------------------
LO          movlw  b'11111100'  ; Set < output RA2 to Low
            movwf  PORTA        ; Rest to 0
            incf   TRIP,w       ; Copy TRIP+1 to w
            movwf  LEVEL        ; The new comparator level
COMP_END    return
```

(a) Assembly-level subroutine.

```
void compare(unsigned int trip)
{
EQ = HI = LO = 0;
if      (PORTB == LEVEL) {EQ = 1;}
else if (PORTB > LEVEL)  {HI = 1; LEVEL = trip - 1;}
else {LO = 1; LEVEL = trip + 1;}
}
```

(b) CCS **C** function.

The subroutine given in Program 11.6(a) assumes that the main program has setup the port directions accordingly and the fixed value is in TRIP. Initially LEVEL would have been set to the same value as TRIP but would subsequently vary by ±1 as per the specification—the hysteresis band.

Software solutions to traditional hardware functions, such as comparison, have the advantage of greater flexibility, albeit at the price of a lower data throughput. Using low-cost "computing engines" such as the PIC MCU means that relatively simple functions traditionally implemented by dedicated hardware can be replaced by embedded processors.

In this instance, flexibility could be replacing the fixed trip level by a variable datum input via, say, Port C; see SAQ 11.5. Example 12.1 on page 390 shows how an external datum can be serially acquired. Alternatively, an analog signal could represent one or both of the levels in devices with integral A/D converters; see Chapter 14. In all these situations the hysteresis may advantageously be made a fraction of the trip voltage, e.g., $\pm\frac{1}{32}$, rather than a fixed ±1 bit.

In the case of the **C** function equivalent of Fig. 11.6(b), the names EQ, HI and LO have been defined externally as the appropriate bits in Port A. Here we are assuming that `trip` is a variable, acquired elsewhere, and passed to the function. The body of the function does the comparison and actuates the appropriate pin. If required, the global variable LEVEL is altered in order to shift the trigger level. If `trip` is fixed, then it need not be passed to the function and could be a literal.

Example 11.3

The principle of a stepper motor is shown in Fig. 11.14. In essence there are four coils, labeled A, B, C, D, which may be selectively energized either singly or in pairs, to generate a magnetic field in one of eight directions

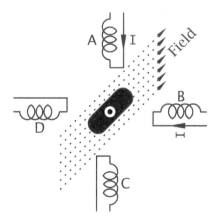

Fig. 11.14 The stepper motor.

in divisions of $45°$.[8] Thus Coil A alone gives a northerly field, A and B together give a north-easterly direction, B alone is east, etc. The rotor follows the field as it changes direction, provided that inertial considerations allow it to keep up during acceleration and de-acceleration.

Write a subroutine originating at h'050' in the Program store to advance the rotor by a passed value of one to 256 steps. Assume that the Port A pins RA[3:0] are connected respectively to coils A, B, C, D. The rate is to be nominally 100 steps per second, based on a 10 ms delay, which is to be written to be largely independent of the crystal frequency. The latter is to be indicated by the programmer by the constant FREQ, which is the multiple of 100 kHz; e.g., d'40' for a 4 MHz crystal.

Solution

Our first step is to devise a table showing energization patterns for the eight possible field directions, as shown in Table 11.2.

Table 11.2: Energization pattern for the eight field directions.

Position	A	B	C	D	Bearing
0	1	0	0	0	↑
1	1	1	0	0	↗
2	0	1	0	0	→
3	0	1	1	0	↘
4	0	0	1	0	↓
5	0	0	1	1	↙
6	0	0	0	1	←
7	1	0	0	1	↖

The coding shown in Program 11.7 comprises three subroutines.

MOTOR

The main subroutine simply modulo-8 increments the position vector by post-ANDing with b'00000111' to give a wrap around from 7 to 0. This vector is then converted to the appropriate energizing pattern and sent out to the motor after a nominal 10 ms delay. The process is repeated until the decrementing STEP datum reaches zero; if initially zero then 256 steps will be actioned.

PATTERN

Returns one of eight energization patterns corresponding to the field vector as listed in Table 11.2. The mechanism of this look-up table coding

---

[8]A real stepper motor repeats the coil set several times around the peripheral motor stator giving a finer mechanical step resolution. Thus, if there are four sets of stator coils the $45°$ electrical step translates to $11.25°$ mechanical.

Program 11.7 Driving a stepper motor.

```
    #define  FREQ  d'40'  ; Programmer gives value in 100k steps
    org      h'50'        ; Code begins at h'050'
; ****************************************************************
; * FUNCTION: Advances stepper motor 1 -- 256 steps            *
; * ENTRY   : Step number in STEP                              *
; * ENTRY   : Current field position in POSITION               *
; * EXIT    : POSITION updated, STEP = -1, W destroyed          *
; * RESOURCE: Subroutine PATTERN, DELAY_10MS                   *
; ****************************************************************
MOTOR    incf    POSITION,w  ; Advance field direction
         andlw   b'0111'     ; Modulo-8
         movwf   POSITION    ; updated
         call    PATTERN     ; Get the energization pattern
         movwf   PORTA       ; Send to stepper motor
         call    DELAY_10MS  ; Hold off 10ms
         decfsz  STEP,f      ; Decrement step count
          goto   MOTOR       ; until zero
         return

; ****************************************************************
; * FUNCTION: Maps an integer 0 -- 7 to field pattern          *
; * ENTRY   : Modulo-8 integer in W                            *
; * EXIT    : Stepper energization pattern in W                *
; ****************************************************************
PATTERN addwf   PCL,f       ; Increment Program Counter
        retlw   b'1000'     ; North
        retlw   b'1100'     ; Northeast
        retlw   b'0100'     ; East
        retlw   b'0110'     ; Southeast
        retlw   b'0010'     ; South
        retlw   b'0011'     ; Southwest
        retlw   b'0001'     ; West
        retlw   b'1001'     ; Northwest

; ****************************************************************
; * FUNCTION: Delays 10 ms delay independent of clock freq     *
; * ENTRY   : FREQ is xtal frequency in multiples of 100kHz    *
; * EXIT    : 10ms delay; DELAY zero, W destroyed              *
; ****************************************************************
DELAY_10MS
        movlw   FREQ        ; The programmer's statement
        movwf   TEMP        ; Gives the PIC MCU frequency
; DO loop (10ms at f = 100kHz xtal -- 1 cycle = 40us) FREQ times
DLOOP1 movlw   d'62'        ; Loop count
        movwf   DELAY
DLOOP2 decf    DELAY,f     ;  62 * 40us
        btfss   STATUS,Z    ;  63 * 40us
         goto   DLOOP2      ; 123 * 40us

        decfsz  TEMP,f      ; Decrement frequency parameter
         goto   DLOOP1      ; and repeat until zero
        return
```

has been described in Program 6.6 on page 162. As this suite of subroutines originates at h'050' in the Program store, the 8-bit addition to the Program Counter will not result in roll-over across page boundaries.

DELAY_10MS
This subroutine gives a nominal 10 ms delay independent of the processor crystal frequency. This is defined by the programmer in the program header as the constant FREQ which denotes the number of multiples of 100 kHz. For instance, for a 8 MHz crystal FREQ is set to 80 using the #define directive, before the program is assembled.

The core of the subroutine is a loop needing a nominal 10 ms execution time at a crystal frequency of 100 kHz—that is, a 40 μs instruction cycle. This loop is transversed FREQ times. Thus our 8 MHz example will have a loop execution of $\frac{10}{80}$ ms but will be executed 80 times to give our required 10 ms delay.

Example 11.4
Enhance the **C** coded keypad encoder function of Program 11.4 by including a debounce routine in the manner of Program 11.3.

Solution
Function get_it() shown in Program 11.8 keeps a count of the calls to scan_it(), each time comparing the returned datum, assigned to the variable new_key, to the previous value in old_key. If there is disagreement then the loop count is zeroed. Only when 254 readings match will the loop exit and the stable value is returned to the caller.

Program 11.8 Debouncing the **C** keypad device driver.

```
unsigned int get_it(void)
{
unsigned int count, old_key, new_key;
count = 0;
while(count<255)
    {
    new_key = scan_it();
    if(new_key == old_key)
        { count++;}
    else
        {
        old_key = new_key;
        count = 0;
        }
    }
return (old_key);
}
```

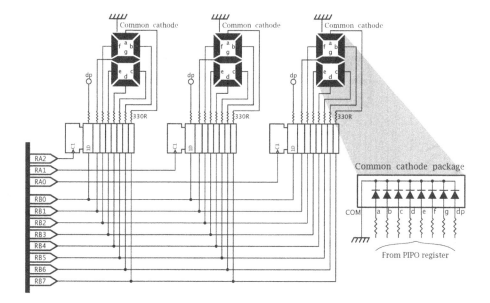

Fig. 11.15 Using port expansion to drive three 7-segment displays.

Example 11.5

Despite the increasing use of liquid-crystal alphanumeric readouts, discrete 7-segment LED displays are commonly used to show multiple numerical digits. Such readouts are particularly effective in low ambient light situations and where large displays are needed.

Assuming each display requires eight lines (seven segments plus decimal point) then a budget of $8 \times n$ parallel lines are required for an n-digit display. The straightforward solution to this problem is shown in Fig. 11.15, where a 3-digit display is driven from three parallel registers on a local bus, in the manner of Fig. 11.12. The principle can be extended as required by using the appropriate number of registers.

The displays shown in the diagram are common cathodes and the appropriate LED is illuminated when the register output is High, with the source current limited by the series resistor. In practice some logic circuitry can sink more current into a Low-state output as compared to sourcing current from a High state, and because of this, common anode displays are often used with the LEDs activated on a Low state. In some larger displays, e.g., 5 cm (2"), several LEDs may be paralleled or in series in each segment. In this situation larger voltages and/or currents may be needed, and suitable drivers used to boost the register outputs.

An alternative approach, shown in Fig. 11.16, is frequently used with LED-based displays. Instead of using a register for each digit, all readouts are connected in *parallel* to the one PIC MCU port. Each readout is enabled in turn for a short time with the appropriate data from the

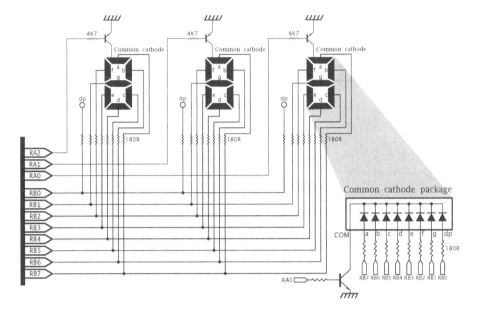

Fig. 11.16 Scanning a multiplexed 3-digit 7-segment array.

output port. Provided that the scan rate is greater than 50 per second (preferably greater than 100) the brain's persistence of vision will trick the onlooker into visualizing the display as flicker free.[9] Of course the current flowing through the segment must be increased to compensate for the mark:space ratio, but LEDs are more efficient when pulsed in this manner and the reduction of series resistance need not be pro-rata.

Discuss the pros and cons of these arrangements, with reference to the tradeoff of software and hardware. Illustrate your answer by displaying the decimal equivalent of the binary byte in File h'20'. For instance; if the contents of BINARY were h'FF' then the display should be 255.

Solution
From the software perspective two main functions can be identified. First, the binary code in a File called BINARY has to be decomposed into three BCD digits; HUNDREDS, TENS and UNITS. Once this is done then each BCD digit ranging from 0 to 9 must be converted to 7-segment code to illuminate the relevant segments to form the appropriate characters. We already have a subroutine to implement the former in Program 6.11 on page 174 and the latter in Program 6.6 on page 162. Based on this code in situ, we have as a task list for software to interact with the hardware of Fig. 11.15:

---

[9]Of course this is how the brain interprets a series of 24 still frames per minute in a movie as a moving image. Each frame is shown twice using a 2-bladed shutter, giving a flicker rate of 48 per second.

1. Convert the binary byte into BCD.
2. DO
   (a) Copy contents of HUNDREDS into W and convert to 7-seg.
   (b) Copy 7-segment code to Port B.
   (c) Pulse _/‾_ RA2.
3. DO
   (a) Copy contents of TENS into W and convert to 7-seg.
   (b) Copy 7-segment code to Port B.
   (c) Pulse _/‾_ RA1.
4. DO
   (a) Copy contents of UNITS into W and convert to 7-seg.
   (b) Copy 7-segment code to Port B.
   (c) Pulse _/‾_ RA0.

The coding implementing this task list is shown in Program 11.9.

The interaction of the software to the hardware of Fig. 11.16 is not so straightforward as there are no registers to dump the data and run! Instead, data has to be continuously sent out in sequence with the appropriate display being enabled. If we use a scan rate of 100 updates each second, then this data should be held for 10 ms before moving on. Thus we have as our new task list:

1. Convert the binary byte into BCD.
2. DO forever:
   (a)
   - Copy contents of HUNDREDS into W and convert to 7-segment code.
   - Copy 7-segment code to Port B.
   - Bring RA2 Low ‾_.
   - Delay 10ms.
   - Bring RA2 High _/‾.
   (b)
   - Copy contents of TENS into W and convert to 7-segment code.
   - Copy 7-segment code to Port B.

Program 11.9 Displaying the decimal equivalent of a binary byte.

```
; Task 1 ------------------------------------------------------
DISPLAY movf  BINARY,w    ; Get binary byte
        call  BIN_2_BCD   ; Convert to 3-digit BCD
; Task 2 ------------------------------------------------------
        movf  HUNDREDS,w  ; Get Hundreds nybble
        call  SVN_SEG     ; Convert to 7-segment code
        movwf PORTB       ; Send out to PortB
        bsf   PORTA,2     ; Clock into register
        bcf   PORTA,2
; Task 3 ------------------------------------------------------
        movf  TENS,w      ; Get Tens nybble
        call  SVN_SEG     ; Convert to 7-segment code
        movwf PORTB       ; Send out to PortB
        bsf   PORTA,1     ; Clock into register
        bcf   PORTA,1
; Task 4 ------------------------------------------------------
        movf  UNITS,w     ; Get Units nybble
        call  SVN_SEG     ; Convert to 7-segment code
        movwf PORTB       ; Send out to PortB
        bsf   PORTA,0     ; Clock into register
        bcf   PORTA,0
```

- Bring RA1 Low ⌐_ .
- Delay 10ms.
- Bring RA1 High _/⌐ .

(c)

- Copy contents of UNITS into W and convert to 7-segment code.
- Copy 7-segment code to Port B.
- Bring RA0 Low ⌐_ .
- Delay 10ms.
- Bring RA0 High _/⌐ .

The coding in Program 11.10 makes use of the 10 ms delay subroutine illustrated in Program 11.7 to regulate the scanning rate. Apart from the length of the enabling pulse, the core of the program is identical to our previous situation. However, the code must run continually to give the impression of a constant display. This illustrates the trade-off between hardware and software. Reducing the hardware has led to greater loading on the software. Indeed, as illustrated here, the entire existence of the PIC MCU will be to service the display! However, in practice the situation can be redeemed somewhat by interrupting the PIC MCU at 10 ms intervals to avoid the need for time-wasting delay routines. The listing on page 423 shows how this can be done, but of course the Timer cannot be used for anything else. Alternatively an external 100 Hz oscillator can be used in its place, but some of the hardware advantages are then lost. With a 10 ms

---

Program 11.10 Displaying a 3-digit decimal number on a scanning readout.

```
; Task 1 ------------------------------------------------------------
DISPLAY movf  BINARY,w      ; Get binary byte
        call  BIN_2_BCD     ; Convert to 3-digit BCD
; Task 2(a) ----------------------------------------------------------
LOOP    movf  HUNDREDS,w    ; Get Hundreds nybble
        call  SVN_SEG       ; Convert to 7-segment code
        movwf PORTB         ; Send out to PortB
        bcf   PORTA,2       ; Enable Hundreds display
        call  DELAY_10MS    ; for 10ms
        bsf   PORTA,2       ; and turn off
; Task 2(b) ----------------------------------------------------------
        movf  TENS,w        ; Get Tens nybble
        call  SVN_SEG       ; Convert to 7-segment code
        movwf PORTB         ; Send out to PortB
        bcf   PORTA,1       ; Enable Tens display
        call  DELAY_10MS    ; for 10ms
        bsf   PORTA,1       ; and turn off
; Task 2(c) ----------------------------------------------------------
        movf  UNITS,w       ; Get Units nybble
        call  SVN_SEG       ; Convert to 7-segment code
        movwf PORTB         ; Send out to PortB
        bcf   PORTA,0       ; Enable Units display
        call  DELAY_10MS    ; for 10ms
        bsf   PORTA,0       ; and turn off
        goto  LOOP          ; DO forever
```

digit rate, up to ten digits may be handled with no additional interface hardware and still have a scan rate no worse than 100 per second.

Another issue that can occur with scanning, is noise introduced by pulsing relatively large currents on a continual basis. This can be a particular problem where analog circuitry is adjacent. Good power-supply decoupling can reduce this problem to some extent.

Self-Assessment Questions

11.1 One problem with the intruder alarm configuration of Fig. 11.12 is the need to cable the Zone ports with eight conductors plus one per zone. An alternative approach would be to replace each zone's 3-state buffer by a PIC MCU. Each MCU would drive a 4-wire common bus back to the main base controller. One wire can be used as a shared handshake line to signal the base that an intruder has been sensed at the zone indicated on the three data wires.

Show how a PIC16F84 could be configured as a local Zone controller paying particular attention to the usage of the single handshake line shared with all zones.

Would it be possible to reduce the number of wires to three? How could a local display be added to show which sensor has been set off?

11.2 A certain PIC MCU running at 20 MHz has its Port C connected to LEDs tied High through a 1 kΩ resistor and with a 300 pF capacitance to ground. All LEDs are off and the programmer attempts to turn on LED 7 and LED 0 as follows:

```
bcf     PORTC,7   ; Turn on LED7
bcf     PORTC,0   ; Turn on LED0
```

However, only LED 0 actually turns on. What is happening?

11.3 Pins RC[1:0] are to be configured as outputs with an initial value of 0 on Power-up reset. The following code is designed to clear both flip flops before changing the port bits to output. On testing, the end result for RC0 is the opposite to the desired outcome. Why is this so and can you modify the code to rectify the situation?

```
bcf     PORTC,0      ; Clear flip flop 0 (see Fig. 11.3(d)
bcf     PORTC,1      ; Clear flip flop 1
bcf     STATUS,RP0   ; Change to Bank 1
movlw   b'11111100'  ; Make RC[1:0] outputs
movwf   TRISC
bcf     STATUS,RP0   ; Change back to Bank 0
```

11.4 A certain system needs to be able to both activate eight LEDs and to read the state of up to eight normally-open (N.O.) push switches. It has been proposed that a single Port B might be able to combine these functions—the former when set to output, the latter when set to input. Can you devise a suitable circuit?

11.5 Extend the digital comparator of Example 11.2 to compare two *external* digital bytes presented to a 28-pin footprint PIC MCU, with byte P being input at Port B and Q at Port C.

11.6 In a low-power wireless data logging system placing the PIC MCU in its Sleep mode will not affect the current consumption of the radio transmitter. It is proposed to use a port pin to supply current to the transmitter and in this way the auxiliary circuitry can be switched on and off as necessary. Discuss.

11.7 The variation of logic 0 output voltage V_{OL} against sink current I_{OL} for the two extremes of the commercial temperature range is shown in Fig. 11.17. Using this graphical relationship determines the maximum value of a series resistor to ensure that a current of no less than 20 mA will flow through an LED connected to +5 V, as shown in the diagram, for any temperature. With this value what will be the current at −40°C? Assume that the conducting voltage across the LED is a constant 2 V.

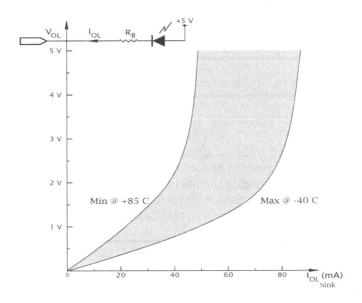

Fig. 11.17 Low-level output voltage against sink current.

CHAPTER 12

One Bit at a Time

Parallel data transmission is fast, with a minimum of software overhead. However, there are many circumstances where its use is inappropriate, either because of the additional hardware cost (for instance, see Fig. 11.12 on page 316) or more commonly where the receivers are geographically distant, with the concomitant cost or non-availability of multiple communication channels and their necessary interface hardware. In such situations data can be sent *one bit at a time* and assembled by the remote device into the original data bytes. In this manner a comparison can be made with the Parallel port on a PC, commonly used for local peripherals, such as a printer, and the Serial or USB ports frequently used with a modem to link into the internet via a single telephone line.

In this chapter we will examine a range of techniques used to serially transmit data, both using bespoke shift register circuits and industrial devices using standard communication protocols. After reading this chapter you will:

- Understand the need for serial transmission.
- Be able to design serial ports and associated software routines to communicate with standard parallel peripheral devices.
- Be capable of interfacing serial peripheral devices using both the SPI™ and I²C protocols.
- Appreciate the need for asynchronous serial communication and be able to write software drivers conforming to this protocol.
- Be able to use the integral universal synchronous asynchronous receiver-/transmitter port (USART) for asynchronous protocols.
- Understand the necessity for buffering long-distance communication circuits.

As an example of serial data communications, consider the smart cards in your wallet. Each card has an embedded microcontroller, typically 8-bit, giving it its intelligence. Cost constraints are severe to give a manufacturing price of under $1, and a large component of this is accounted for by the non-corrosive gold-plated contacts via which the microcontroller is powered and clocked when in contact with the card reader. In order to keep the mechanical precision of the reader low and

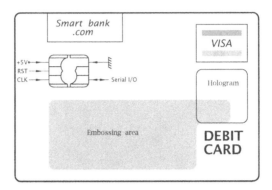

Fig. 12.1 The smart card.

hence reliability high, the number of contacts must be minimized and the pad size maximized.

The standard arrangement shown in Fig. 12.1 uses contacts to provide the two power nodes, Reset and Clock, and *one* line to allow data to be shifted in or out *one* bit at a time. Although this is relatively slow, in comparison to the human–mechanical constraints speed is not an issue. Furthermore, contact between the reader/automatic teller and the central computer, perhaps several thousands of miles/kilometers away, will typically be via a single channel telephone line.

Check the parallel 3-digit 7-segment display interface of Fig. 11.15 on page 325, which uses both Parallel Ports A and B. Although this is a working circuit, most of the parallel port budget of an 18-pin footprint device has been used up. Speed is certainly not a factor here, so a slower mode of data transmission is acceptable.

Consider the serial equivalent shown in Fig. 12.2. Here only two port pins are used. One labeled **SDO** (**Serial Data Output**) outputs the data bit by bit, with the most significant bit first. The other, labeled **SCK** (**Serial ClocK**), is used to clock the three shift registers at the same time, and hence shift the data right one bit at a time, in the manner of Fig. 3.8 on page 62.

Each display has an associated 74HCT164 8-bit shift register;[1] see Fig. 2.22 on page 37. The 74HCT164 has a positive-edge triggered shift input clock C1 and two serial data inputs ANDed together at 1D. One of these data inputs can be used to gate the other input, but in our example they are both connected together to give a single serial input. There is also an active-Low Reset input to clear the register contents, which are held High in the diagram. If desired, another port line can be used to drive R.

---

[1] All data outputs are simultaneously available and thus the 74HCT164 is best described as a serial-in parallel-out (SIPO) register as well as a SISO shift register.

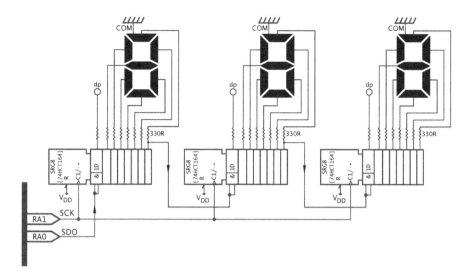

Fig. 12.2 Serial interface to a 3-digit 7-segment display.

To change the display, a total of 24 bits will have to be shifted into the register array. To see how this can be done we will repeat the 7-segment driver routine of Program 11.9 on page 327 which converts a binary byte to an array of BCD digits in HUNDREDS, TENS and UNITS. These are mapped to 7-segment code and then sent out to each digit 8-bits at a time.

To serialize this process we need to design a subroutine to copy each bit of a File, we will call DATA_OUT, to the RA0/SDO pin; beginning with the leftmost bit. At the same time the RA1/SCK pin is pulsed to clock the data. A task list for such a subroutine is:

1. Bring SCK to its Low state.
2. COUNT = 8.
3. WHILE COUNT > 0 DO:
   (a) Copy most-significant bit of DATA_OUT to SDO.
   (b) Shift DATA_OUT left one place.
   (c) Pulse SCK ___/‾___ .
   (d) Decrement COUNT.

Program 12.1 shows two subroutines. The first called DISPLAY is closely akin to Program 11.9, in that it calls the subroutines BIN_2_BCD and then sends the 7-segment coded bytes out to the interface registers. In this instance the Units byte is sent first, as this will eventually be shifted to the far end of the chain, followed by the Tens and finally the Hundreds byte.

The actual serial transmission is handled by the subroutine SPI_WRITE, which implements our task list. The state of bit 7 of the datum placed by the caller in File DATA_OUT is tested and its state used to make the Serial Data Out pin RA0 High or Low. The Serial ClocK pin RA1 is then

Program 12.1 Displaying the decimal equivalent of a binary byte using a serial data stream.

```
SDO       equ    0
SCK       equ    1

DISPLAY bcf     PORTA,SCK   ; Initialize the clock line
        movf    BINARY,w    ; Get binary byte
        call    BIN_2_BCD   ; Convert to 3-digit BCD
        movf    UNITS,w     ; Get Units nybble
        call    SVN_SEG     ; Convert to 7-segment code
        movwf   DATA_OUT    ; Copy into the serial register
        call    SPI_WRITE   ; Shift it out

        movf    TENS,w      ; Get Tens nybble
        call    SVN_SEG     ; Convert to 7-segment code
        movwf   DATA_OUT    ; Copy into the serial register
        call    SPI_WRITE   ; Shift it out

        movf    HUNDREDS,w  ; Get Hundreds nybble
        call    SVN_SEG     ; Convert to 7-segment code
        movwf   DATA_OUT    ; Copy into the serial register
        call    SPI_WRITE   ; Shift it out
        return

; *********************************************************
; * FUNCTION: Clocks out a byte in series, MSB first   *
; * ENTRY   : Datum in DATA_OUT                         *
; * EXIT    : DATA_OUT zero                             *
; *********************************************************
; Task 1 ------------------------------------------------
SPI_WRITE
        bcf     PORTA,SCK   ; Make sure clock starts at Low
; Task 2 ------------------------------------------------
        movlw   8           ; Initialize loop counter to 8
        movwf   COUNT
; Tasks 3(a)&(b) ----------------------------------------
LOOP    bcf     PORTA,SDO   ; Zero data bit
        btfsc   DATA_OUT,7  ; Skip if MSB is 0
        bsf     PORTA,SDO   ; ELSE make data bit 1
        rlf     DATA_OUT,f  ; Shift datum right one place
; Task 3(c) ---------------------------------------------
        bsf     PORTA,SCK   ; Pulse clock
        bcf     PORTA,SCK
; Task 3(d) ---------------------------------------------
        decfsz  COUNT,f     ; Decrement count
        goto    LOOP        ; and repeat until zero
        return
```

toggled once __/‾__ to shift the data into the shift register chain. The data byte is then shifted left and the process repeated in total eight times, to complete the transaction. This takes a maximum of 87 cycles to complete, depending slightly on the data pattern. A complete update of the 3-digit display will take around $120\,\mu s$ with a processor clock of 8 MHz; excluding the time spent in doing the data conversion.

Program 12.2 A **C** implementation of the SPI_WRITE subroutine.

```
void spi_write(datum)
{
int k;
for(k=0;k<8;k++)                    /* DO eight times            */
    {
    if((datum & 0x80)) {SDO = 1;}
    else {SDO = 0;}
    SCK = 1;                        /* Clock the external receiver */
    SCK = 0;
    datum = datum << 1;            /* Shift datum left one place  */
    }
}
```

Program 12.2 shows a possible **C** implementation of our output sub-routine. Function spi_write() accepts a data byte and in a loop of eight copies bit 7 out to SDO, whilst shifting left. The two SPI pins have been previously defined as the appropriate port pin.

Where a long chain of shift registers is being serviced, speed may be improved a little if each register has its own data feed but all clocked with the same SCK pin or sharing the same lines but each with a separate Enable. This latter technique is the method used in Fig. 12.7.

One problem with our shift register technique is that for the period where shifting is in process, the data appearing at the port outputs are not valid—for 23 clock pulses in our example. Of course in this situation the response of the eye to microseconds-long changes in illumination makes this observation spurious. However, this may not always be the case and in such instances the shift register may be buffered from the parallel outputs using an array of D flip flops or latches, which can be loaded after the shifting process has been completed to give a single update.

Rather than employing a separate buffer register, many devices opti-mised for serial data transmission have integral PIPO registers. For ex-ample, the 74HCT595 shown in Fig. 12.3, is a latched shift register with integral 8-bit parallel-in parallel-out (PIPO) register between the shift reg-ister and the outside world. A rising edge _/‾ on the RCK (Register ClocK) pin transfers the serialized data to the parallel outputs. The last stage output of the shift register is made available to allow cascading to any length. In this situation, all RCK pins can be pulsed together to allow the entire chain to update simultaneously.

One example where rippling of data may be undesirable is where a digital datum is to be converted to its analog equivalent. In Fig. 12.3 the conversion is carried out using a National Semiconductor DAC0800. Essentially the analog voltage is a linear function of the 8-bit digital input

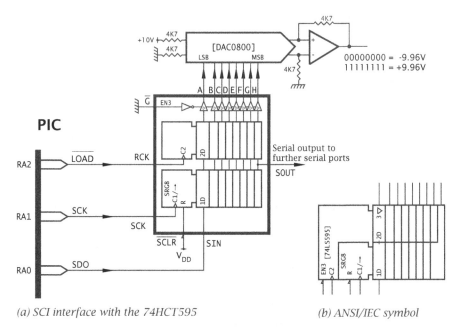

(a) SCI interface with the 74HCT595   *(b) ANSI/IEC symbol*

Fig. 12.3 Serially interfacing to a DAC digital-to-analog converter using a 74HCT595 octal shift register with output register.

varying from −9.96 V for an input of b'00000000' through +9.96 V for b'11111111'; see Fig. 14.17 on page 469.

Using a 74HCT595 registered shift register, the digital input does not change until the new datum is in place and the PIC MCU pulses the RCK Register Clock. This gives clean changes in the data presented to the DAC and corresponding analog output.

Data can be input serially in a similar manner using parallel-in serial-out (PISO) shift registers. The example shown in Fig. 12.4 is a serialized version of the intruder alarm of Fig. 11.12 on page 316 using only three lines to connect to all eight sensor groups; a considerable economy compared to the original 16 lines.

Each sensor group is attached to a 74HCT165 8-bit PISO shift register, with the serial output of the further register feeding the serial input of the next left register. Once the data has been loaded in, it may be shifted into the **SDI (Serial Data In)** Parallel Port A input pin RA1 and assembled bit by bit. In the specific case of the multi-zone intruder alarm, after each eight shifts the assembled byte can be tested for non-zero and the appropriate action taken; see SAQ 12.1.

Also shown in Fig. 12.4 is the single output port used to display the active zone. As both input **SDI** and output **SDO** serial channels share the same shift clock SCK, then shifting data in will simultaneously clock this

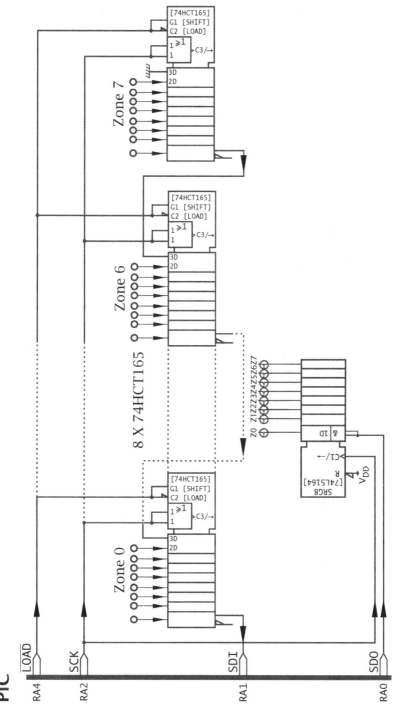

Fig. 12.4 Serially interfacing to the multi-zone intruder alarm.

serial output port. Conversely, sending data to the output port will shift data in from the Zone ports. In this example there is no problem, as microsecond fluctuations in the Zone lamps are of no consequence, and the sequence of operations ends with the output port being accessed with the earmarked data. Where this interaction is undesirable, then either the appropriate datum should be continually presented to SDO at the same time as it is read in at SDI or a latched register, such as the 74HCT595, should be used to staticize the display data. As an alternative, separate serial clock lines could be utilized.

The core serial interface software SPI_READ is the input counterpart of subroutine SPI_WRITE in Program 12.1, and implements the following task list:

1. Bring SCK to its Low state.
2. COUNT = 8.
3. WHILE COUNT>0 DO:
   (a) Pulse SCK _／‾＼_ .
   (b) Shift DATA_IN left once place.
   (c) Copy input state of pin SDI into least-significant bit of DATA_IN.
   (d) Decrement COUNT.

Program 12.3 Input serial byte subroutine.

```
; ******************************************************************
; * FUNCTION: Clocks in a byte in series, MSB first               *
; * ENTRY    : None                                               *
; * EXIT     : Datum in DATA_IN; COUNT = 0                        *
; ******************************************************************
; Task 1: Bring SCK Low -------------------------------------------
SPI_READ
        bcf     PORTA,SCK     ; Make sure clock starts at Low
; Task 2: COUNT=8 -------------------------------------------------
        movlw   8             ; Initialize loop counter to 8
        movwf   COUNT
; Task 3: WHILE COUNT>0 DO: ---------------------------------------
; Task 3 (a): Pulse SCK -------------------------------------------
SER_IN_LOOP
        bsf     PORTA,SCK
        bcf     PORTA,SCK
; Task 3(b): Shift datum left -------------------------------------
        bcf     STATUS,C      ; Zero Carry flag
        rlf     DATA_IN,f     ; Shift it in and datum once left
; Task 3(c): IF SDI is 1 THEN set bit 0 (rightmost bit) -------
        btfsc   PORTA,SDI     ; Skip if SDI == 0
        bsf     DATA_IN,0     ; ELSE set bit0 to 1
; Task 3(d): Decrement COUNT and repeat Task3 WHILE>0 ---------
        decfsz  COUNT,f       ; Decrement count
        goto    SER_IN_LOOP   ; and repeat until zero

        return
```

This task list is similar to that on page 333, except that File DATA_IN is shifted left once and the state of the SDI pin following the clock pulse at pin SCK copied as the new bit 0. After eight clock-shift-test loops the datum in DATA_IN is the parallelized byte assembled from the serial input port, with the first bit ending up in the leftmost significant placeholder in DATA_IN.

The SPI_READ subroutine coded in Program 12.3 is similar to the output subroutine SPI_WRITE of Program 12.1. Indeed they may be combined so that data is shifted out of the specified output data File at the same time as it is shifted in to the specified input data File. This type of scheme is referred to as **full duplex**, as opposed to **half duplex** where only one direction at a time is possible. A serial link where data flow can only be in one fixed direction is known as **simplex**.

The **C** coding of Program 12.4 follows the same coding strategy as the assembly counterpart. Note how Inclusive-OR'ing with b'0000001' using the **C** | operator is used to set bit 0 of the variable DATA_IN. Similarly AND'ing with b'11111110' clears bit 0. Specifically in CCS **C** the non-standard integral functions bset(DATA_IN,0) and bclr(DATA_IN,0) can be used to set or clear any bit in a variable and when single bits are involved, is often more efficient than using logic operators.

Program 12.4 A **C** implementation of a SPI input read.

```
unsigned int spi_read()
{
unsigned int k;
for(k=0;k<8;k++)              /* DO eight times            */
    {
    SCK = 1;                  /* Clock Slave TX bit to SDI */
    SCK = 0;
    DATA_IN = DATA_IN << 1;   /* Shift left one place      */
    if(SDI)
        {DATA_IN = DATA_IN | 0x01;} /* Set bit 0 IF SDI is 1 */
    else
        {DATA_IN = DATA_IN & 0xFE;} /* ELSE make it a 0      */
    }
return data_in;               /* Return complete byte      */
}
```

The serial protocol similar to that described in this example is commonly known as **serial peripheral interface (SPI™)**.[2] Microwire™ is a similar, but not identical, serial protocol.[3] SPI is a sufficiently standardized protocol used by most microcontrollers to allow manufacturers to produce a range of peripheral devices specifically designed to directly inter-

---

[2]SPI™ is a trademark of Motorola, Inc.
[3]Microwire™ is a trademark of National Semiconductor Corporation.

face to this bus without the necessity to add external shift registers. As an example of this genre, the MAX549A of Fig. 12.5 is a dual digital-to-analog converter (DAC) which is powered with a V_{DD} of +2.5 V to +5.5 V. Its operating current is typically 150 μA per DAC at 5 V and either or both DACs can be shut down to reduce the current drain to less than 1 μs in its Standby mode. Data can be clocked in at a rate of up to 12.5 MHz. All this functionality is available in an 8-pin package and should be contrasted with the 20-pin MAX506 of Fig. 14.16 on page 468, designed for direct parallel port connection.

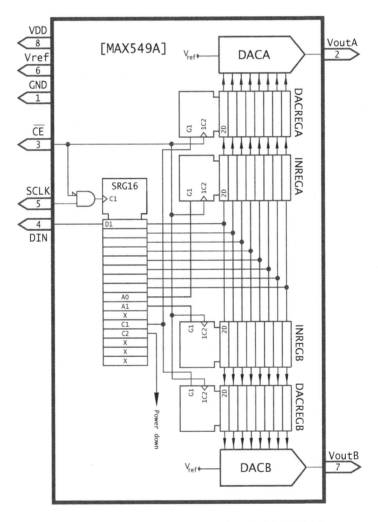

Fig. 12.5 The MAXIM MAX549A SPI dual 8-bit DAC.

The simplified functional model of the MAX549A shown in the diagram shows an integral 16-stage shift register clocked from SCLK and fed data via DIN using the normal SPI protocol. The additional eight locations are used to store four control bits, with the following functionality:

A0
Enables the input PIPO register for channel A and which is clocked on a rising edge at the $\overline{CE}$ pin.

A1
Enables the input PIPO register for channel B and which is clocked on a rising edge at the $\overline{CE}$ pin.

C1
Gates both DAC registers; allowing them to be updated simultaneously by a $\_\_/\overline{}$ on $\overline{CE}$.

C2
When 1 will power down any DAC selected with A0 or/and A1. This disconnects the reference voltage V_{ref} from the DAC's resistor network (see Fig. 14.15 on page 467) and leaves only a residual current of less than $1\,\mu A$ to activate the internal registers, whose contents remain unchanged.

Both DACs have a 2-layer register pipeline isolating them from the shift registers. The first layer is the In registers, which are gated when A0 or A1 as appropriate is 1. The data sitting in the first byte of the shift register can then be clocked in by pulsing $\overline{CE}$ (pin 3) Low. This change will be stored but will not appear at the input of the DAC until the next layer of PIPO registers are clocked. These registers are enabled when C1 is 1 and $\overline{CE}$ is pulsed. This means that one data byte can be sent to, say, DACA and then another to DACB. The DAC registers can then be updated together, resulting in both outputs V_{outA} and V_{outB} changing simultaneously; see Program 12.5. This can even be done when the MAX549A is asleep, as the registers are not affected by this power-down state. From this discussion we see that each transition from the PIC MCU takes two 8-bit transfers | Control | Data | followed by a $\_\_/\overline{}$ on the $\overline{CE}$ pin.

For our example we will send the contents of File h'20' to Channel A and then the contents of File h'21' to Channel B; at that point updating both DAC registers and hence outputting the analog equivalent of File h'20' to pin V_{outA} and File h'21' to pin V_{outB}.

Our implementation will involve the transmission of four bytes of information:

1. Control byte 1: b'XXX00X01'
   No power down, update Channel A, no output change.
2. Data byte 1:
   Contents of File h'20'.
3. Pulse $\overline{}\backslash\_/\overline{}$ $\overline{CE}$.

4. Control byte 2: b'XXX01X10'
   No power down, update Channel B, both outputs change.
5. Data byte 2:
   Contents of File h'21'.
6. Pulse ⎺_/⎺ CE.

The hardware-software interaction is shown in Program 12.5. Four bytes are transmitted using subroutine SPI_WRITE, with the MAX549A's CE being pulsed _/⎺_ after each | Control | Data | byte pair. The final process sets C1 High, which transfers both data bytes to the DAC registers. At the same time the Channel B In register is updated.

Looking at the three pins on the MAX549A would give a waveform similar to that of Fig. 12.6 for the transmission of the first | Control | Data | byte pair. During the transmission CE remains Low, with the data shifting into the MAX549A's integral shift register. After the second byte, i.e., the 16th clock pulse, bringing CE High activates the selected internal registers, executing the instruction.

The diagram shows transitions on the DIN line from the PIC MCU's SDO pin, occurring sometime before the active rising edge on SCK. Sometime

---

Program 12.5 Interacting with the MAX549A dual-channel SPI DAC.

```
CE      equ 2

;  ****************************************************************
;  * FUNCTION: Sends out Channel A & B data in SPI protocol to   *
;  * FUNCTION: MAX549A simultaneously updating outputs           *
;  * RESOURCE: Subroutine SPI_WRITE                              *
;  * ENTRY   : Channel A in File h'20', Channel B in File h'21'  *
;  * EXIT    : Both analog outputs updated                       *
;  ****************************************************************
        movlw   b'00000001'  ; Control byte 1
        movwf   DATA_OUT     ; Put in designated location
        call    SPI_WRITE    ; and send out to MAX549A
        movf    h'20',w      ; Get Channel A data
        movwf   DATA_OUT     ; Put in designated location
        call    SPI_WRITE    ; and send out to MAX549A
        bsf     PORTA,CE     ; Pulse CE
        bcf     PORTA,CE

        movlw   b'00001010'  ; Control byte 2
        movwf   DATA_OUT     ; Put in designated location
        call    SPI_WRITE    ; and send out to MAX549A
        movf    h'21',w      ; Get Channel B data
        movwf   DATA_OUT     ; Put in designated location
        call    SPI_WRITE    ; and send out to MAX549A
        bsf     PORTA,CE     ; Pulse CE
        bcf     PORTA,CE
        return
```

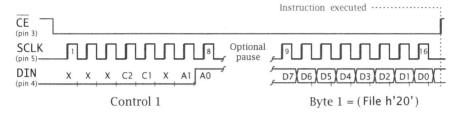

Fig. 12.6 SPI waveforms for the MAX549A.

is a vague term; obviously it must occur no later than a minimum time before $\_\!\!\!/\!\!\!\overline{}$ and be held for a short time after. The MAX549A data sheet gives the minimum set-up time t_{DS} of 30 ns and hold time t_{DH} of 10 ns. Even at a PIC MCU clock rate of 20 MHz an instruction cycle takes 200 ns, so timing will not be violated.

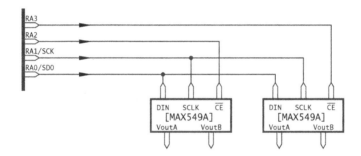

Fig. 12.7 Multiple MAX549As on the one SPI circuit.

By judicious use of the MAX549A's $\overline{CE}$ input, several DACs may be connected to the SCK/SD0 lines, with a serial transmission only being shifted into the device which has its $\overline{CE}$ Low. Figure 12.7 shows two MAX549As sharing the one SPI link, giving four analog output channels in total. Using a 2- to 4-line decoder in conjunction with RA[3:2] would enable up to four MAX549As with a total budget of only four port lines.

Most mid-range and all extended-range PIC MCUs feature an integral synchronous serial port (SSP) which implements, amongst others, the SPI protocol. Three closely related modules are available according to the age of the device. The first of these is known as the Basic SSP (BSSP), which later evolved to the (plain) SSP. Recently introduced devices feature the **Master Synchronous Serial Port** (MSSP). This module adds a few additional options for the SPI clock, but in particular the ability to automatically act as a Master I^2C device, and hence its name.

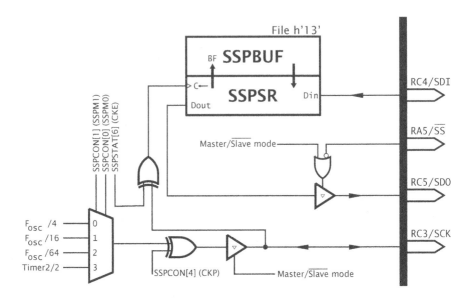

Fig. 12.8 The basic SPI Serial Synchronous Port set to implement SPI. Pinning is shared with Parallel Port C for 28-pin+ devices.

A somewhat simplified representation of the MSSP module set-up for the SPI protocol is shown in Fig. 12.8. The heart of the Master Synchronous Serial Port is the Special Purpose Register File (SPR) **SSPBUF** (**SSP BUFfer**) at File h'13'. A datum byte written into this SPR will automatically be transferred into the **SSP Shift Register** (**SSPSR**) and shifted out of the PIC MCU's dedicated SD0 pin. At the *same* time, eight bits of data will be shifted in from the RC4/SDI pin. When this frantic burst of activity is completed, the new byte is automatically transferred to SSPBUF, whence it can be read. This transfer is signaled by setting the **BF** (**Buffer Full**) flag in the **SSPSTAT** (**SSP STATus**) register, shown in Fig. 12.9. Once SSPBUF is read, BF is automatically cleared.

Apart from parallel ports, in general interface modules are configured and monitored with a set of associated Control and Status registers. In addition, interrupt mask bits and flags are located either in the INTCON register, or more usually in one or two Peripheral Enable and Peripheral Interrupt registers, such as shown in Fig. 7.6 on page 200. Peripheral control, status and interrupt registers are normally set-up as part of the startup initialization routine, where parallel ports are configured as input/output. As such modules invariably multiplex their pins with the parallel ports, even if these latter are not used such shared pins often need to be considered in this initialization phase. Pin I/O settings may be automatically overridden if the peripheral module is enabled or may need to be "manually" set-up by the software. Unfortunately the settings are not always obvious and the data sheet should be consulted for specific information.

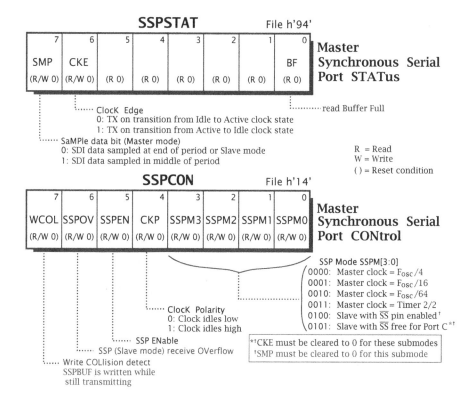

Fig. 12.9 The MSSP module's CONtrol and STATus registers as appropriate for the SPI mode.

Returning to the specific case of the MSSP, Fig. 12.9 shows the **SSPCON** (**SSP CONtrol**) register and **SSPSTAT** (**SSP STATus**) registers set-up for the SPI protocol and I/O pinning. In connecting to the outside world, four pins need to be considered.

RC5/SDO
Bit TRISC[5] must be cleared to 0 to allow this pin to output data.

RC4/SDI
This pin is overridden by the MSSP irrespective of the state of its associated TRISC[4] bit.

RC3/SCK
When in one of the Master modes, bit TRISC[3] must be cleared to allow this pin to output the clock signal. Conversely, when in one of the Slave

modes, TRISC[3] must be set to allow this pin to input the clock signal from an external Master.

RA5/$\overline{SS}$

In Slave mode b'0100' this pin should be configured as an input; i.e., TRISA[5] should be set to 1, in order to allow the external Master to select this device.

On any kind of reset both SSPCON and SSPSTAT registers are cleared and the internal data bit counter is zeroed. In this situation the module is disabled and if the programmer wishes to use the MSSP then the various control bits[4] must be set-up.

SSPEN

Setting SSPCON[5] to 1 enables the Synchronous Serial port. If disabled the associated pins can be used as normal parallel port lines.

SSPM[3:0]

The four SSP Mode switch bits located in SSPCON[3:0] are used to set the communication protocol and various Master/Slave options. The diagram shows the six combinations relevant to the SPI protocol.

The four Master submodes only differ in selecting the one of four internal clocking frequencies. Three of these frequencies are derived from the main PIC MCU oscillator. For example, with a 20 MHz crystal the SCK shift rate can be selected as 5, 1.25 MHz and 312.5 kHz (200, 800 ns and 3.2 μs). The final selection gives the shift rate as half the frequency generated by Timer 2 overflowing; see Fig. 13.8 on page 422. This option is used where very slow shift rates are required.

The Slave options use a clock coming from an outside Master driving the SCK pin.. Optionally the $\overline{SS}$ pin can be used by this external Master device to select one of several Slaves; see Fig. 12.12.

SSPOV

In a Slave mode this status bit indicates that a new byte has been received before the previous byte has been read; that is, a byte or bytes have been lost. This is not automatically cleared when the SSPBUF is read; it must be zeroed manually; that is, in software. The SSP OVerflow status bit does not operate in a Master mode.

WCOL

When software attempts to write a byte into the SSPBUF before the last byte in the SSPSR has been completely shifted out, the action is aborted and the Write COLlision bit is set. This bit can be tested, and if a collision is confirmed, should be cleared in software and the process subsequently tried again.

---

[4]Some of which are in the Status register due to lack of space!

CKP, CKE, SMP

These three bits work in tandem to ensure that the correct clock edges are used to shift data into and out of the remote receivers and transmitters and that incoming data is sampled when it has stabilized.

In order to illustrate the various possibilities, consider the situation when the MSSP is set-up as a Master. As a Master device the MSSP has complete control of the clocking signal at SCK, which is used to clock *both* the remote Slave transmitter and receiver shift registers. The Slave receiver shift registers require an active edge on this clock when the Master data at pin SDO is stable. In addition, Slave transmitter shift registers need to be clocked so that their data bit is stable when the MSSP reads it at its SDI pin. An example of this situation is shown in Fig. 12.12, where the PIC-based Master SPI device can select one of two Slaves, which can both transmit and receive data simultaneously. The Slaves can be other PIC microcontrollers (as shown) or any SPI circuit.

Each byte transmission is broken up into eight clock phases; as shown in Fig. 12.11. In all situations, the next data bit D_n will be presented at the SDO pin shortly (in Industrial devices, not more than 50 ns) after the beginning of each clock phase; see top of diagram. The remote Slave

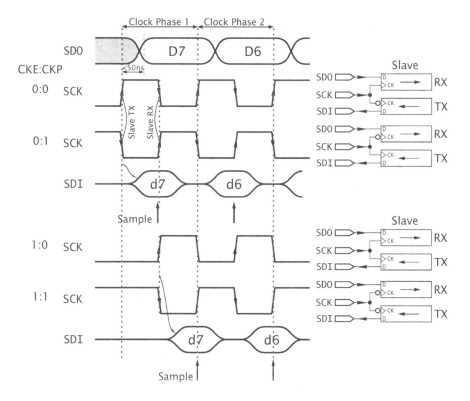

Fig. 12.10 Clocking data in and out to remote Slave devices.

receiver should then be in a position to clock it mid-phase. Similarly, the remote Slave transmitter should present its data bit d_n to pin SDI in time to be sampled by the Master.

Figure 12.10 is split up into two broad situations. The top two SCK waveforms are used when the remote transmitters and receivers have opposite active shift clock edges. As the transmitter is clocked at the beginning of each phase it should be sampled mid-phase by setting SMP = 1.

CKE:CKP = 0:0

When the remote transmitter is clocked by a rising edge _⟋‾ then its data at SDI should be sampled mid-phase. Such data should be present at least 100 ns before this point and be held for at least 100 ns afterwards. The remote Slave receiver clocks in its data from SDO on a falling edge ‾⟍_ on SCK, also mid-phase. In standard SPI terminology, this is described as mode 0,1.

CKE:CKP = 0:1

SPI mode 1,1 is similar, but the remote transmitter is clocked by a ‾⟍_ edge and the remote receiver by a _⟋‾ edge.

Where all remote transmitters and receivers have the same active shift clock edge, then the bottom waveforms are relevant. As the transmitter is clocked mid-phase, its data at SDI should be sampled at the end of the clock phase by making SMP = 0.

CKE:CKP = 1:0

SPI mode 0,0 is used where both Slave transmitters and receivers are clocked together mid-phase on a _⟋‾ edge. By that time the Master data bit D_n will be stable to be clocked into the remote Slave. The remote Slave's data should be ready for sampling by the end of the clock phase.

CKE:CKP = 1:1

SPI mode 1,0 generates a ‾⟍_ edge mid-phase to trigger both Slave transmitters and receivers.

When the MSSP is being used in one of the Slave modes, the clock comes from a remote device. As before, any data previously loaded into the SSPBUF will be clocked out from the SDO pin at the beginning of each clock phase. The CKE and CKP bits still need to be set according to whether the remote sender's data is output on a _⟋‾ or ‾⟍_ edge of its clock and on the active edge of a remote receiver. Also if the remote Master has its first D_n bit presented before or after the first clock pulse. In all such cases the Slave MSSP-configured module should sample such data as it presented at its SDI pin at the end of each clock phase; that is, SMP = 0. The reader should refer to the device data sheet for specific waveforms.

When a PIC MCU is set-up as a Slave SPI device, the $\overline{SS}$ ($\overline{Slave\ Select}$) pin is used by the remote Master to select it for an 8-bit transfer. When $\overline{SS}$ goes High, even in the middle of a transmission, the internal bit counter is reset to zero. Also the SDO pin goes open-circuit, so that another device can take over the line.

BF, SSPIF

When a complete frame of eight bits has been shifted in and been dumped into the SSPBUF Buffer register, BF goes to 1 to indicate that a new datum is ready for collection. This transfer also sets the SSPIF flag in PIR1[3], (see Fig. 7.6 on page 200) and this can be used to initiate an interrupt if the companion SSPIE mask bit in PIE1[3] has been set. If the MSSP has been configured as a Slave and the PIC MCU is sleeping this can be used to waken the device. This is possible, as the SCK pin is clocked by the external Master device and thus the PIC MCU need not be active; that is, the system oscillator can be off.

Reading the newly arrived datum from SSPBUF automatically clears this Buffer Full bit. If not read on time, the datum will be lost and the SSPOV flag will be set to record this. The SSPIF interrupt flag needs to be manually cleared in any interrupt service routine.

Using Figs. 12.8 and 12.9 as a programmer's model we can now deduce the hardware–software interaction in order to action a transmission of a byte and/or receive a new byte:

1. Configure SSP module.
   - Set-up SCK, RC5/SDO as outputs and RC4/SDI, and if appropriate RA5/$\overline{SS}$, as input.
   - Set-up Master/Slave mode with appropriate clock source.
   - Choose active clock edges with CKP:CKE:SMP.
   - Enable the SSP by setting SSPEN.
2. Move datum to SSPBUF to initiate transmission.
3. IF WCOL = 1 THEN reset WCOL and go to item 2.
4. Poll BF for 1.
5. Move RX data from SSPBUF, which also resets BF.

To illustrate this process, consider a subroutine SPI_IN_OUT, which combines the function of SPI_READ and SPI_WRITE; that is, it transmits the datum in File DATA_OUT, whilst at the same time returning the consequently received byte to DATA_IN. Assume that the remote shift registers are all _/‾ triggered; that is, SPI mode 0,0.

The implementation of this subroutine depends on setting up the MSSP during the initialization phase of the main software after a reset. In the following code fragment we are using the $f_{OSC}/4$ clock rate Master mode:

```
        .include  "p16f877.inc"
MAIN    bsf    STATUS,RP0      ; Change to Bank 1
        movlw  b'11010111'     ; RC5/SDO, RC3/SCK outputs
        movwf  TRISC           ; RC4/SDI input
        movlw  b'11000000'     ; Make CKE and SMP = 1
        movwf  SSPSTAT

        .....  .....
        bcf    STATUS,RP0      ; Return to Bank 0
        movlw  b'00100000'     ; Enable SSP, TX clock idles Low
        movwf  SSPCON          ; SPI Master, Fosc/4 rate
```

The coding shown in Program 12.6 follows the task list exactly. Data to be transmitted is moved from the designated File to SSPBUF and status bit WCOL checked to see that it got there. If there is a transmission in progress then the datum is not stored in SSPBUF and WCOL is set. If this subroutine is the only code to access the SSP then this should rarely be the case and in most instances this check is omitted, but its inclusion makes the system more robust.

Once the transmit datum is in situ, the transmit sequence is immediately initiated, as shown in Fig. 12.11, and progresses to its conclusion. When the Buffer Full status flag BF in Bank 1 is set, the received datum can be moved out of SSPBUF to its ordained location. This automatically resets BF.

Program 12.6 Using the SSP for SPI data input and output.

```
;   ****************************************************************
;   * FUNCTION: Transmits and simultaneously receives one byte   *
;   * FUNCTION: from the SSP using the SPI protocol              *
;   * ENTRY   : Data to be transmitted is in DATA_OUT            *
;   * EXIT    : Data received is in DATA_IN                      *
;   ****************************************************************
SPI_IN_OUT
        movf   DATA_OUT,w       ; Get datum for transmission
SSP_IN_OUT_LOOP
        movwf  SSPBUF           ; Put into SSPBUF
        btfss  SSPCON,WCOL      ; Did it make it?
         goto  SPI_IN_OUT_CONT  ; IF so THEN continue
        bcf    SSPCON,WCOL      ; ELSE reset WCOL and try again
        goto   SSP_IN_OUT_LOOP
SPI_IN_OUT_CONT
        bsf    STATUS,RP0       ; Change to Bank1
        btfss  SSPSTAT,BF       ; Check for Buffer Full
         goto  SPI_IN_OUT_CONT  ; IF not then poll again
        bcf    STATUS,RP0       ; Back to Bank 0

        movf   SSPBUF,w         ; ELSE get the received datum
        movwf  DATA_IN          ; Put away
        return
```

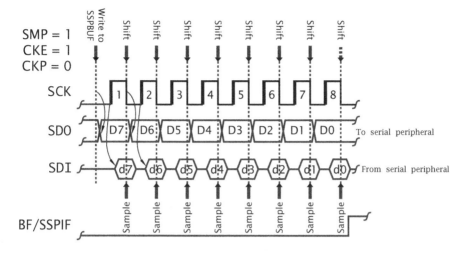

Fig. 12.11 SSP SPI-mode Master waveforms.

Apart from a slight reduction in the code length, the advantage of using this hardware is the increase in speed. The actual transmit/receive takes eight SCK cycles, which in our case is eight instruction cycles. With an f_{OSC} of 20 MHz, the clocking rate is 5 MHz (that is, a bit rate of 5 million bits per second, commonly written as 5 Mbit/s or 5 Mbps), giving a total time of 1.6 μs per byte.

Figure 12.11 shows the SPI mode timing for our subroutine. As we have cleared CKP and set CKE then SCK is idling Low. As soon as SSPBUF is written to, the MSB of the TX datum appears at SDO. In mid-phase the rising edge clocks this data into the remote receiver.

With the remote receiver also clocked at mid-phase there is plenty of time for its data to be presented to the PIC MCU's SDI. This data is then sampled by the PIC MCU at the end of each clock phase.

One use of serial transmission is to connect a number of devices together in one multiprocessor network. For instance, a robot arm may have a MCU controlling each joint, communicating with a master processor. A simple multidrop circuit of one Master and two Slave processors is shown in Fig. 12.12.

In this configuration the Master PIC MCU externally drives the SCK of both Slaves, thus controlling when and how fast transmission occurs across the network. Both Slaves are configured in Mode 0100 so that the $\overline{\text{Slave Select}}$ inputs are enabled. Thus, if the Master wishes to read a datum from Slave 2, the latter's $\overline{\text{SS}}$ is brought Low and the Master clocks the eight bits from Slave 2's SSPBUF/SSPSR, into its own SSPBUF/SSPSR. At

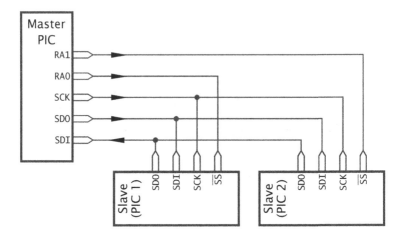

Fig. 12.12 A multidrop SPI communications network.

the same time any data transmitted by the Master will be received by the Slave.

SPI transactions may be coded in **C** either by mimicking the assemble-level code and setting/reading the appropriate registers, or by using built-in functions specific to the task. The key CCS compiler internal functions used for the SSPort in its SPI mode are:

set-up_spi(spi_master|spi_h_to_l|spi_clk_div_4);
This function configures the SSP as an SPI Master, with clock polarity rising edge and a ÷4 clock frequency. These scripts, and others such as spi_slave, spi_sample_at_end and spi_xmit_l_to_h, are part of the included header file; e.g. 16f877a.h. This function also sets the direction of the appropriate Port A and Port C pins.

spi_write(value);
This is used to write out the value from the SSP. It checks that the BF flag is set before returning.

spi_read();
This is virtually identical to spi_write() except that it returns the value read by the SSP. If a value is passed to this function then it will be clocked out of SDO.

spi_data_is_in();
This function returns non-zero if a datum has been received over the SPI connection; that is, if BF is set.

To illustrate this technique, consider our interface to the MAX549A coded in Program 12.5. In order to do this the SSP needs to be configured using code of the form:

```
#include <16f877a.h>
#bit     CE = 5.2         /* Port A, bit 2 to MAX549A's CE    */
void MAX549A(unsigned int channel_A, unsigned int channel_B);
main()
{
set_tris_a(0xFB);         /* CE = RA2 output                  */
set-up_adc(NO_ANALOGS); /* Ports A & E all digital          */
set-up_spi(spi_master|spi_1_to_h|spi_clk_div_4);
```

in which we are assuming that the MAX549A's $\overline{CS}$ is connected to Port A's
RA2 pin; as shown in Fig. 12.7.

The program comprises four spi_write() calls, with CE being pulsed
between | Control | Data | pairs. This function may be called with an
evocation something like MAX549A(data_x, data_y);

Program 12.7 Interfacing to the MAX549A in **C**.

```
void MAX549A(unsigned int channel_A, unsigned int channel_B)
{
spi_write(0x01);          /* Send out Control 1              */
spi_write(channel_A);     /* Send out Data 1                 */
CE=0;                     /* Pulse CE                        */
CE=1;
spi_write(0x0A);          /* Send out Control 2              */
spi_write(channel_B);     /* Send out Data 2                 */
CE=0;                     /* Pulse CE                        */
CE=1;
}
```

Although the SPI protocol is relatively fast, it requires a minimum of
three data lines plus one select line for each duplex Slave device. Apart
from the cost, adding a device to an original design will require some
hardware modification. By increasing the intelligence of the Slave device,
it is possible to send both control, address and data in the one serial
stream. The **inter-integrated circuit** (I^2C™) protocol developed by the
Philips/Signetics Corporation[5] in the early 1980s embodies this concept
and also reduces the interface to only two lines by permitting bidirectional transmission; see Fig. 12.13.

SCL

This is the clock line sychronizing data transfer, serving the same function as SCK in the SPI protocol. However, SCL is bidirectional to allow
more than one Master to take control of the bus at different times.

The original I^2C specification set an upper limit on shift frequency
of 100 kHz; that is, 100 kbit/s, but the specification was augmented in

---

[5]I^2C™ is a trademark of the Philips Corporation.

1993 by a Fast mode with an upper data rate of 400 kbit/s, which is the current de facto standard. In 1998 a compatible High-Speed mode was added with an upper bit rate of 3.4 Mbit/s.

SDA

This I^2C data line allows data flow in either direction. This bidirectionality allows communication from Master to Slave (Master-Write) or from Slave to Master (Master-Read). Furthermore it allows the receiver to signal its status back to the transmitter at the end of each byte.

The I^2C protocol is relatively complex and its full specification can be viewed at the Philips/Signetics Corporation web site.[6] Before looking at the basic protocol, we need to examine the SCL and SDA lines in more detail. When no data is being transmitted, both lines should be High; the **Idle condition**. A device wishing to seize control of an idling bus must bring its SDA output Low. This is known as the **Start condition**. In order for the would-be Master to be able to pull this line Low, all other devices hung on the line must have their SDA pins open circuit and the line as a whole pulled up to the High state through a single external resistor; see Fig. 12.14(a). To implement this, SDA (and also SCL) outputs must be open-collector or open-drain; see Fig. 2.2(b) on page 19. This means that any device hung on the bus is able to pull its line Low by outputting a logic 0.

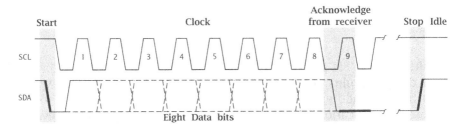

Fig. 12.13 Data transfer on the I^2C bus.

In addition to generating the Start event, the Master is responsible for generating the clock signal and also for sending an address code to the other entities on the bus, to establish communications with one or more Slave devices. Along with this address, a single bit tells the Slave if the data flow is to be from the Master to the Slave (Master-Write) or Slave to Master (Master-Read).

Each packet sent between Master and Slave comprises nine bits. Eight of these are data synchronized by the clock. Changes on SDA *must* only

---

[6]www.semiconductors.com/acrobat/literature/9398/59340011.pdf

occur when the clock line is *Low*. Data is clocked into the receiver on the following SCK rising edge. These bytes may represent address or control information from the Master or data from either Master or Slave. The I^{2}C protocol includes a handshaking (see Fig. 11.2 on page 297) mechanism. During the ninth clock pulse, the transmitting device releases the SDA line and the receiving device on the bus *acknowledges* the data sent by the transmitter. SDA is held Low by the receiver if the datum has been successfully acquired, giving an **ACK** state; as shown in Fig. 12.15. The alternative, where the receiver either signals a problem or that it doesn't want any more data, is called a Not ACKnowledge, or **NACK**. Normally in this latter situation, the transmitter will try again for a number of times before giving up.

Rather less drastic, the Slave device can hold the clock line Low. **Clock stretching** is useful where the Slave device cannot process incoming data fast enough. The Master will attempt to send clock pulses until SCL is released by the Slave.

In any situation it is the responsibility of the Master to terminate the conversion by bringing the SDA line High when the clock line SCK is High; signaling a **Stop** condition. Another conversion can be started, if desired with a different Slave, by the Master sending another Start signal. It is possible for the Master to send out repeated Starts without first stopping. For instance, a Master may wish to send (Master-Write) an internal address to an I^{2}C memory device (see Example 12.3) and then do a Master-Read of the pointed-to data. This requires a change in the conversation direction which is done by doing another Start with a new Slave Address:Direction packet being sent; see Fig. 12.27. The difference between using repeated Starts and a Stop condition is that the latter signals other devices that the Master has relinquished the bus and another device can become a Master on a multi-Master bus system.

In using a PIC MCU to implement the I^{2}C standard in software, a problem arises as port outputs are not open-drain; that is, the logic 1 output state is not open-circuit as required in Fig. 12.14(a). However, it is possible to get around this, simulating the high impedance state by switching the port line output to input. For example, if we wish to use RA2 as the SCL data line then to pulse SCL Low and then off/High we have:

```
bcf    PORTA,2     ; Sometime during set-up make RA2 = 0
....   .....
bsf    STATUS,RP0  ; Change to Bank1
bcf    TRISA,2     ; RA2 is output = 0
nop                ; Short delay
bsf    TRISA,2     ; Float RA1 by making it an input
bcf    STATUS,RP0  ; Return to Bank0
```

where the High state is a consequence of the external pull-up resistor and the high *input* impedance; as shown in Fig. 12.14(b)*ii*.

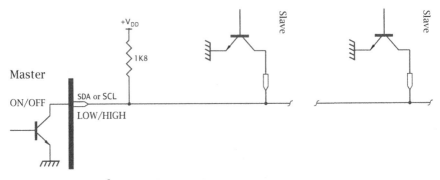

(a) Connection of I^2C devices to the I^2C bus.

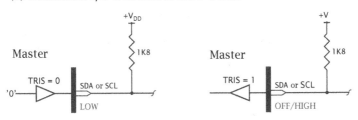

i Output is low   *ii Output is pulled high*

(b) Using the PIC to simulate open collector.

Fig. 12.14 Sharing the SCL and SDA bus lines.

A complete transmission between Master and Slave comprises a packet of several byte/Acknowledge transfers sandwiched between Start and Stop conditions. To some extent the form of this packet depends on the requirements of the Slave device; however, all packets conform to the general sequence Slave address:Control/Command:Data shown in Fig. 12.15.

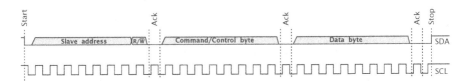

Fig. 12.15 A I^2C packet transmission.

The essence of the I^2C protocol is the requirement that each type of Slave device has an **address**. This address is allocated[7] to the manufacturer of the I^2C peripheral and is factory programmed. To allow

---

[7]By the I^2C-bus committee.

more than one device of the same kind to share the same bus, most I^2C-compatible devices allow up to four bits of this address to be set locally by the designer; usually by connecting Slave address pins to the appropriate logic levels. On receipt of a Start bit, all Slaves on the bus will examine the first seven bits for their personal address. If there is no match then the rest of the conversation is ignored until the next Start bit. Bit 8 is a direction bit; $R/\overline{W}$ is Low if the Master is to be the transmitter; that is, to Write to the Slave, and High if the Master wishes to Read from the Slave.

Not all 7-bit addresses are valid. All addresses matching b'0000XXX' or b'1111XXX' are reserved for special situations; leaving 224 valid addresses in total. For instance, the address b'0000000' indicates a General Call broadcast to *all* Slaves on the bus, rather than to one specific device. Along with the introduction of a Fast mode, the I^2C protocol was extended to permit a 10-bit address. This is signaled by the reserved address b'11110XXX', where the three XXX bits are added to the following 7-bit address packet.

After the address byte(s), the next byte is usually treated by the addressed Slave as a Command/Control word, passing configuration information. For instance, a I^2C memory may require the internal address where the data is to be written to; see Example 12.3. Bytes following this are usually pure data or a mixture of data and control bytes.

In order to illustrate these concepts we will use the Maxim MAX518 DAC, shown in Fig. 12.16, as our exemplar. This is the I^2C counterpart to the SPI protocol MAX549, with a 2-layer register pipeline, two channels and a power-down feature.

The MAX518 has a 7-bit Slave address of the form **01011AD1AD0** where **AD1** and **AD0** should match the logic state of pins 5 and 6, respectively. If we assume that both pins are connected to GND then the Address byte sent out by the Master will be $\boxed{01011\,|\,00\,|\,0}$. $R/\overline{W}$ is 0, as this device can only be written to.

The Command byte is of the form 000 RST PD XX A0, with three active control bits:

A0
This enables the input PIPO register for Channel 0 if 0 and Channel 1 if 1.

PD
When 1 this control bit will power down both DAC channels, reducing the supply current to typically $4\,\mu A$. The contents of the internal registers remain unchanged and data may be shifted in and registers updated in this condition. The state information is only executed whenever a Stop bit is sent by the Master, at which point the last transmitted value of PD is acted upon.

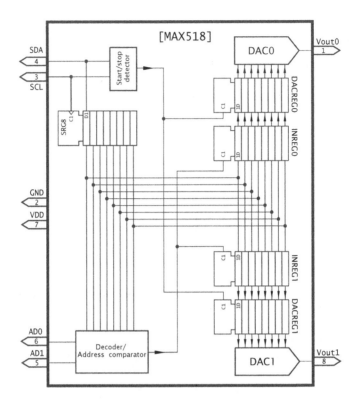

Fig. 12.16 The MAXIM MAX518 I²C dual digital to analog converter.

RST
All internal registers are cleared irrespective of the following data byte which may be treated as a dummy byte. Analog outputs go to zero after the Stop condition.

In all cases the Stop condition updates the analog outputs according to the commands and data byte. If there have been several Command:Data byte pairs since the last Stop then the most recent command and data are reflected in the state and output of the device.

In order to interface to the MAX518, we will need to design subroutines to send out a Start condition, a Stop condition and a Master-Write byte. To design the device driver we need to look more closely at the time relationship between Clock and Data signals, which generally are more tightly defined than in the SPI protocol.

The MAX518 and most current I²C-compatible devices are designed to the Fast mode specification and the figures used in Fig. 12.17 relate to this 400 kHz clocking rate. Of particular note is the requirement that the clock SCL should be held High not less than 0.6 μs ($t_{HD;STA}$) after the active ⎴⎍ of SDA to signal a Start condition. Similarly, a Stop condition requires that

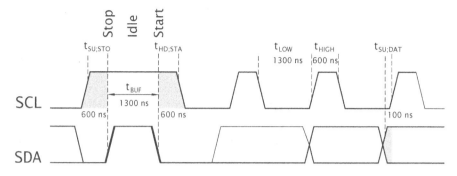

Fig. 12.17 Minimum timing relationships for the Fast I²C mode.

the clock be set-up High at least $0.6\,\mu s$ ($t_{SU;STO}$) before the active $\_\!\!/\ulcorner$ of SDA. A minimum of $1.3\,\mu s$ is required with the bus free (t_{BUF}) in the Idle state between a Stop and a following Start condition. These requirements allow time for the Slave devices to detect these synchronizing events without ambiguity.

During a data byte transmission the clock should be Low (t_{LOW}) no less than $1.3\,\mu s$ and High (t_{HIGH}) no less than $0.6\,\mu s$ within the $2.5\,\mu s$ overall duration limitation imposed by the 400 kHz clock rate. Data changes only when the clock is Low, and any change should be complete no less than 100 ns ($t_{SU;DAT}$) before the clock goes High.

Not shown in the diagram is the maximum rise and fall times, which should not exceed 300 ns with a maximum bus capacitance of 400 pF. To keep within this transition restriction, the pull-up resistors of Fig. 12.14 should not be more than $1.8\,k\Omega$ with this value of capacitance. With short bus runs and few Slave devices this value of resistance can be increased by up to a factor of ten, to reduce energy dissipation.

In implementing the I²C timings, a PIC MCU with a crystal above 3.2 MHz, with an execution time of less than $1.25\,\mu s$, may need to insert short delays between actions. For example, a 20 MHz crystal driven PIC MCU implementing the instruction pair:

```
    bcf  TRISA,SCL
; Drag Clock Low by making pin an output to logic 0
    bsf  TRISA,SCL
; Float clock into the High state by making pin an input
```

would give High and Low durations of only $0.2\,\mu s$. Short delays are conveniently implemented using nop (No OPeration) instructions; each taking one instruction cycle ($f_{OSC}/4$). For instance, to give a nominally 400 kHz clock at 20 MHz we have:

```
        bcf   PORTA,SCL    ; Clock Low
        nop                ; 0.2us
        nop                ; 0.4us
        nop                ; 0.6us
        nop                ; 0.82us
        nop                ; 1.0us
        nop                ; 1.2us
        bsf   PORTA,SCL    ; Clock High
        nop                ; 1.6us
        nop                ; 1.8us
        nop                ; 2.0us
        nop                ; 2.2us
        nop                ; 2.4us
        nop                ; 2.6us
```

Of course slower clock speeds require less nops, but rather than tailor our subroutines for one particular crystal we will use the assembler macro called Delay_600, coded in Program 12.8, that will expand to the appropriate number of nops to give a nominal 600 ns (0.6 μs) delay, depending on the value of the constant XTAL defined by the programmer at the head of the source file. For example, to alter the coding of Program 12.9 to suit a 12 MHz crystal system then the one line #define XTAL 20 should be altered to #define XTAL 12 and the program reassembled.

Program 12.8 A crystal frequency-independent short delay macro.

```
Delay_600 macro                  ; Delays by nominally 0.6us
            if (XTAL <= 6)
            nop                  ; One nop if XTAL is less than 6MHz
            endif
            if ((XTAL > 6) && (XTAL <= 13))
             nop                 ; Two nops delays if
             nop                 ; XTAL is between 6 & 13MHz
            endif
            if (XTAL > 13)
             nop                 ; Three nop delays if
             nop                 ; XTAL is above 13MHz
             nop
             endif
            endm
```

The coding of Program 12.8 makes use of the conditional assembler directive if – endif. This is similar to the C language statement if(true){do this;} of page 264 in that all instructions down to the following endif are implemented if the argument of the if directive is true. For example, if((XTAL>6)&&(XTAL<=13)) states that if the constant XTAL is greater than 6 AND less than or equal to 13 then insert two nop instructions. At 13 MHz this will be approximately 600 ns. In practice, extra delays will be introduced by instructions toggling the bus lines and executing housekeeping tasks. Thus some fine-tuning can be undertaken if maximum speed is a criterion.

Based on the macro of Program 12.8 and the following initialization code (which for convenience uses the File Select Register to indirectly access the direction bits for the SCL and SDA pins):

```
        include "p16f84a.inc"
        #define XTAL 20

SCL equ  0
SDA equ  1

MAIN movlw h'86'     ; Set-up the File Select Register
     movwf FSR       ; to point to TRISA (File h'85')
     bcf   PORTA,SCL ; Preset Clock & Data pins to 0
     bcf   PORTA,SDA ; so that line can be dragged Low
     bsf   INDF,0    ; Float Clock line High (TRISA[0])
     bsf   INDF,1    ; & Idle the Data line  (TRISA[1])
```

which is assuming that we are using Port A bits 0 and 1 of a 20 MHz PIC16F84A to implement our SCL and SDA lines, we can code the three subroutines outlined in Program 12.9 to allow us to communicate with the I^2C MAX518.

START
This subroutine releases both the SCL and SDA lines which are then pulled High to ensure the bus is in its Idle state for the minimum duration 1.3 μs t_{BUF}. Bringing SDA Low gives the characteristic Start $\neg\!\_$, which is followed by a 0.6 μs delay to implement $t_{HD;STA}$ (HolD; STArt; see Fig. 12.17) before the subroutine exits with both SCL and SDA Low.

STOP
The Stop condition is implemented by ensuring that both SCL and SDA lines are Low (which should be the case after an Acknowledge condition) and then releasing the SCL line which is then pulled High. After a 0.6 μs delay to implement $t_{SU;STO}$ (Set-Up; STOp), SDA is released to give the characteristic Stop $\_\!\sqrt{}$. The subroutine exits with both lines released idling in preparation for the next Start condition.

I2C_OUT
This subroutine clocks out the eight bits placed in DATA_OUT by the caller, MSB first, and then checks that the Slave has Acknowledged the transaction.

The first part of this process is implemented by repetitively shifting the datum in DATA_OUT and inspecting the Carry flag. SDA is set to mirror C and the SCL line toggled to accord with the t_{LOW} and t_{HIGH} parameters illustrated in Fig. 12.17.

Program 12.9 Low-level I²C subroutines.

```
;       **************************************************************
;       * FUNCTION: Outputs the Start condition                     *
;       * ENTRY   : FSR points to the I2C port's TRIS register      *
;       * EXIT    : Start condition and SCL, SDA pins Low           *
;       **************************************************************
START       bsf     INDF,SDA    ; Ensure that we start with the
            bsf     INDF,SCL    ; Data and Clock lines pulled High
            Delay_600           ; 1.3us delay in Idle state
            Delay_600
            bcf     INDF,SDA    ; Low-going edge on Data line
            Delay_600           ; Wait for Slave to detect this
            bcf     INDF,SCL    ; Exit with the Clock line Low
            return

;       **************************************************************
;       * FUNCTION: Outputs the Stop condition                      *
;       * ENTRY   : FSR points to the I2C port's TRIS register      *
;       * EXIT    : Stop condition and SCL, SDA pins High (Idle)    *
;       **************************************************************
STOP        bcf     INDF,SCL    ; Make sure that Clock line is Low
            bcf     INDF,SDA    ; and the Data line is Low
            bsf     INDF,SCL    ; Bring Clock line High
            Delay_600           ; for a minimum of 0.6us
            bsf     INDF,SDA    ; Rising edge on Data signals Stop
            return              ; including the return time

;       **************************************************************
;       * FUNCTION: Transmits byte to Slave and monitors Acknowledge*
;       * ENTRY   : 8-bit data to be TXed is in DATA_OUT            *
;       * RESOURCE: START and STOP subroutines                      *
;       * EXIT    : Byte transmitted. ERROR is 01 IF no Ack received*
;       * EXIT    : from Slave ELSE 00. SCL Low                     *
;       **************************************************************
I2C_OUT     bcf     INDF,SCL    ; Make sure that Clock line is Low
            clrf    ERR         ; Start with no error
            movlw   8           ; Loop counter = 8
            movwf   COUNT
I2C_OUT_LOOP
            bcf     INDF,SDA    ; Start with Data bit Low
            rlf     DATA_OUT,f  ; Shift data left once into Carry
            btfsc   STATUS,C    ; Is C 0 or 1
             bsf    INDF,SDA    ; IF the latter THEN make Data High
            Delay_600           ; Delay plus xtra instructions OK
            Delay_600
            bsf     INDF,SCL    ; Bring Clock pin High
            Delay_600           ; for at least 0.6us
            bcf     INDF,SCL    ; Bring Clock Low
            decfsz  COUNT,f     ; Decrement loop count
             goto   I2C_OUT_LOOP ; and repeat eight times
; Now check Acknowledge from Slave ----------------------------
            bsf     INDF,SDA    ; Release Data line
            Delay_600           ; Keep Clock line Low
            Delay_600           ; long enough for Slave to respond
            bsf     INDF,SCL    ; Bring Clock line High
            btfsc   INDF,SDA    ; Check if Data is Low from Slave
             incf   ERR,f       ; IF not THEN ERROR1
            bcf     INDF,SCL    ; Now finish ACK by bringing Ck Low
            return
```

Once the loop count reaches zero, the Data line is released with SCL Low for the duration t_{LOW}. SCL is then released High and the state of SDA, which should have been dragged Low by the Slave, checked. If not Low, the No ACKnowledge (NACK) situation is returned with ERR = h'01'; otherwise it will be zero.

Our use of errors here is very rudimentary. For instance, errors can also occur if some other device has locked either line Low; that, is the bus is busy.

We have not coded a Master-Receive I^2C counterpart to subroutine I2C_OUT, as the MAX518 only demands a Master-Transmit data interchange. However, Program 12.18 gives the I2C_IN mirror.

As our example we will send the contents of File h'40' to the MAX518 Channel 0 and then the contents of File h'41' to Channel 1; at that point updating both DAC registers and hence simultaneously outputting the analog equivalent of File h'40' to pin V_{out0} and File h'41' to pin V_{out1}. We assume that both AD0 and AD1 pins are connected to Ground.

Our implementation will involve the transmission of a group of five packets of information, sandwiched between a Stop and a Start condition.

1. Start.
2. Address byte: b'01011000'
   Slave address b'01011(00)', Write.
3. Command byte 1: b'00000XX0'
   No ReSeT, no Power Down, Channel 0.
4. Data byte 1:
   Contents of File h'40'.
5. Command byte 2: b'00000XX1'
   No ReSeT, no Power Down, Channel 1.
6. Data byte 2:
   Contents of File h'41'.
7. Stop and update both DAC registers.

The listing of Program 12.10 follows our itemization exactly. On return from each call to I2C_OUT the Error datum is tested for zero. If not zero then the process is restarted. Repeated Starts are allowed by the I^2C protocol. However, if there was a hardware fault with the bus or Slave then this process would continue indefinitely. Thus, for robustness a time-out mechanism should be implemented to prevent hang-ups.

All Serial Synchronous Port modules support the use of a PIC MCU as an I^2C bus device. Earlier versions only implement Slave protocols; however, the MSSP module allows automatic use as a Master device on a Multi-Master bus, hence the name. A **Multi-Master** I^2C bus allows for more than one device to take over as a Master, but of course not simultaneously. Avoiding a bus collision situation complicates matters, and the MSSP Master I^2C operation is correspondingly complex and beyond

Program 12.10 Interacting with the MAX518 dual-channel I²C DAC.

```
ANALOG   call    START       ; Start a transmission packet
; Address byte ------------------------------------------------
         movlw   b'01011000'; Slave address Master-Write
         movwf   DATA_OUT    ; Copied to pass location
         call    I2C_OUT     ; Send it out
         movf    ERR,f       ; Check for an error
         btfsc   STATUS,Z    ; IF Zero THEN continue
         goto    ANALOG      ; ELSE try again
; Command byte 1 ----------------------------------------------
         movlw   b'00000000'; No ReSeT, No Power Down, Channel0
         movwf   DATA_OUT    ; Copied to pass location
         call    I2C_OUT     ; Send it out
         movf    ERR,f       ; Check for an error
         btfsc   STATUS,Z    ; IF Zero THEN continue
         goto    ANALOG      ; ELSE try again
; Data byte 1 ------------------------------------------------
         movf    h'40',w     ; Channel0's datum from memory
         movwf   DATA_OUT    ; Copied to pass location
         call    I2C_OUT     ; Send it out
         movf    ERR,f       ; Check for an error
         btfsc   STATUS,Z    ; IF Zero THEN continue
         goto    ANALOG      ; ELSE try again
; Command byte 2 ----------------------------------------------
         movlw   b'00000001'; No ReSeT, No Power Down, Channel1
         movwf   DATA_OUT    ; Copied to pass location
         call    I2C_OUT     ; Send it out
         movf    ERR,f       ; Check for an error
         btfsc   STATUS,Z    ; IF Zero THEN continue
         goto    ANALOG      ; ELSE try again
; Data byte 2 ------------------------------------------------
         movf    h'41',w     ; Channel1's datum from memory
         movwf   DATA_OUT    ; Copied to pass location
         call    I2C_OUT     ; Send it out
         movf    ERR,f       ; Check for an error
         btfsc   STATUS,Z    ; IF Zero THEN continue
         goto    ANALOG      ; ELSE try again

         call    STOP
```

the scope of this text. Details are given in the Microchip application note AN7578, *Use of the SSP Module in the I²C Multi-Master Environment*. Here we will confine ourselves to the use of the MSSP module as an I²C Slave device.

Figure 12.18 shows a block diagram of a SSP configured as an I²C Slave. Typically pin RC4 is used as the bidirectional I²C SDA data channel and RC3 implements the SCL clock line. Both pins need to be set as inputs for the I²C Slave protocol.

Internally, data I/O is via the SSPSR shift register, which is used both for Slave transmission or reception.

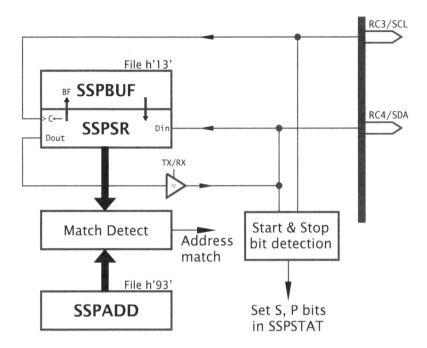

Fig. 12.18 Block diagram of a MSSP module set-up as an I²C Slave device.

Transmission

Where the Slave is sending data to a remote Master (Master-Read) the datum placed in the SSPBUF buffer register will automatically be transferred into the SSPSR (if empty) whence it is shifted out of SDA in eight clock pulses. If the SSPSR is full, this transfer does not happen and a Write-collision error is set.

Reception

If the Slave expects to read a packet from the remote Master, then the data is shifted in via SDA and when collected is transferred into the SSPBUF register. The MSSP module then automatically ACKowledges the safe reception of the datum during the ninth clock pulse, unless an overflow error has occurred. This happens where the previously received byte has not been read from the SSPBUF register in time.

Once a Start condition is sensed, all Slaves on the bus shift in the first packet from the Master, looking for a match with the pattern set-up in software in the **SSP ADDress register SSPADD**. If the top seven bits match (bit 0 is R/$\overline{\text{W}}$) then it is ACKnowledged and the Slave is now ready to converse with the remote Master. Both the BF and SSPIF flags will be set to signal an I²C event. As we have seen in Fig. 12.15, bit 0 of this first address packet tells the Slave to either receive or transmit as directed until the next Start or Stop condition.

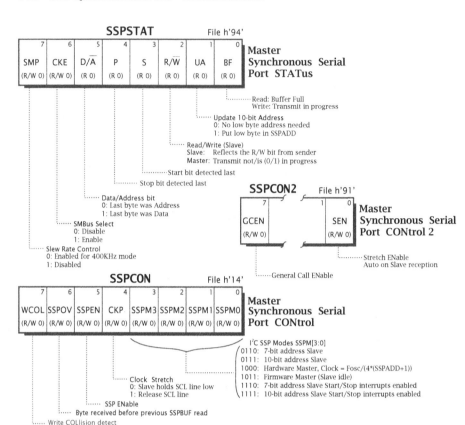

Fig. 12.19 The Master SSP module's CONtrol and STATus registers as appropriate for the I²C Slave modes.

As in the case of the SPI mode, the Control and Status registers need to be set-up to configure and monitor the operation of the MSSP module. Figure 12.18 shows the situation applicable for the four possible I²C Slave modes. These should be compared to those shown for the SPI modes of Fig. 12.9. The same SSPSTAT and SSPCON registers are used and indeed, rather confusingly, some of the bit names have been retained, such as CKE, even though their function is very different. As compared to the older SSP modules, the MSSP has a second control register named **SSP-CON2**. Apart from bits 7 and 0, SSPCON2 deals solely with the Master I²C environment.

SSPEN
Setting SSPCON[5] enables the Synchronous Serial port. On any type of reset, the MSSP is disabled and pins RC3 and RC4 may be used for normal Port C I/O duties.

SSPM[3:0]

Four combinations of these SSP Mode switches are relevant to this discussion. For simplicity, we will assume a 7-bit Slave address mode. Where a 10-bit address mode is used, the software must first place the top address byte b'11110A_9A_8' in SSPADD for comparison and then follow it with the bottom 8-bit address byte b'$A_7A_6A_5A_4A_3A_2A_1A_0$'. Modes b'0110' and b'1110' differ only in that the SSP interrupt Flag SSPIF is optionally set when a Start or Stop situation occurs.

BF, SSPIF

Buffer Full generally indicates that something is still happening with data in the SSPBUF register. SSPIF is the interrupt flag associated with the MSSP module, and is set whenever an I^2C event has occurred.

Slave-Receive

When a complete frame is received from the Master and has been dumped into the SSPBUF register, BF is set to indicate that a new datum is ready for collection and an ACKnowledge is sent back during clock pulse 9. SSPIF in PIR[3] is also set to potentially generate an interrupt. When this byte is read by the software, BF is automatically cleared (it is a read-only flag), but SSPIF has to be manually reset in the normal interrupt flag manner.

Should a complete new byte be shifted in before the previous datum is read; that is, if BF is still 1, then it will not be transferred to SSPBUF and the SSPOV flag will be set to show an OVerflow condition has occurred. In this case an ACKnowledge is Not sent; that is, NACK.

Slave-Transmit

While a byte is being shifted out to the Master, BF is set to show that a transmit is in progress. If a new byte is written into SSPBUF, it will not be transferred into SSPSR and the WCOL flag is set to indicate a Write COLlision has occurred.

SSPOV

In a Slave-Receive situation, failure to read the SSPBUF register before the next byte has arrived is signaled by setting this flag.

A NACK is sent to the Master to indicate an overflow. The NACK state can be used by the Slave deliberately to inform the Master that it should try again later. This condition is cleared by reading the datum from the SSPBUF register to clear BF and manually zeroing SSPOV.

WCOL

An attempt to write to the SSPBUF while a transmission is in progress sets the WCOL flag to indicate the Write COLlision has occurred. This flag has to be manually cleared.

S, P

These flags indicate that a Start or stoP bit, respectively, was detected last. They are normally the inverse of each other except that they are both zero after any Reset or when the module is enabled — SSPEN → 1. A set P indicates that the bus is free; which is useful if a listening device wishes to take over as a Master.

D/$\overline{A}$, R/$\overline{W}$, UA

These flags all relate to the packet(s) following the Start condition, which contains information regarding the Slave address and direction of subsequent data packets.

D/$\overline{A}$ indicates whether the byte sitting in the SSPBUF is Data or Address.

R/$\overline{W}$ informs the software whether the message is going to be Master-Read (R/$\overline{W}$ = 1 or Master-Write (R/$\overline{W}$ = 0). Effectively, it is bit 0 of the (first) address packet.

UA is used only in the 10-bit address modes. In this case the seven MSBs of the first address byte b'11110$A_9A_8$0' are matched first. the LSB represents R/$\overline{W}$ and is 0 to indicate that the next address packet is to be written from the Master. UA is then automatically set to 1, to indicate that the software must now place the lower address byte into SSPADD for match detection. When this is done, UA is automatically cleared.

GCEN

When the General Call Enable bit is 1, the SSPIF interrupt flag will be set when the General Call address b'00000000' is received; irrespective of an address match. This indicates that the Master wishes to make a general broadcast to all Slaves. This facility is not available in earlier SSP modules.

CKP, SEN

When CKP is cleared, the Slave device holds the SCL pin Low, so that the Master device is unable to send clock pulses. When CKP is set to 1, the SCL pin is released and the Master can clock a new packet.

Although CPK can be toggled anytime by the software (that is, manually), clock stretching can be performed automatically.

SEN = 0

When the Stretch ENable bit is clear (the default Reset state) or in older SSP modules, the CKP bit will be cleared automatically by the module at the end of each packet it sends to the Master. The Slave must set CKP after a new packet is copied into SSPBUF to release the clock and allow the next packet to be transferred. Clock stretching for such Slave-Write transmissions is not optional, and is always implemented irrespective of the SEN bit or in modules without SEN.

SEN = 1

In the latest version of the MSSP[8] when SEN is set, automatic clock stretching is enabled *both* for Slave transmission and reception. This

---

[8]For instance, in the PIC16F87X group, SEN does not affect Slave operation, but in the PIC16F87XA variants it has the function described here.

latter is useful if the Master is sending packets at too fast a rate for the Slave to process.

CKE
Only in the MSSP module can this be set to alter the electrical characteristics of SDA and SCL to conform to the SME bus standard.

Using the MSSP module as a Slave is a sequential process that requires the software to react to each possible I^2C event. Although the SSPIF flag in the PIR1 register may be polled, our exemplar software will use an interrupt-driven approach.

Before looking at these events and showing example code, the MSSP module needs to be initialised in the start-up code. Typical set-up code for a PIC16F877A device acting as a 100 kHz 7-bit address Slave h'06' would be something like:

```
       include "p16f877a.inc"

SETUP movlw b'00110110'  ; Enable MSSP, no clock stretch, CKP=1
      movwf SSPCON       ; 7-bit Slave mode 0110

      bsf   STATUS,RP0   ; Now move into Bank 1

      bsf   SSPSTAT,SMP  ; Slew rate for 100kHz
      bsf   SSPCON2,SEN  ; SEN=1 for auto clock stretch on RX
      movlw h'0C'        ; Address 06 shifted left one place
      movwf SSPADD       ; for matching

      bsf   PIE1,SSPIE   ; Enable the SSP interrupt (SSPIE)
      bsf   INTCON,PEIE  ; Enable the Peripheral module group
      bsf   INTCON,GIE   ; Globally turn on enabled interrupts

      bcf   STATUS,RP0   ; Back to Bank 0
```

where:

1. The three MSSP Status/Control registers are set-up appropriately for our specification.
2. The Slave address (h'06' aligned with the top seven bits of the Address packet) is copied into the SSPADD for matching.
3. The SSP interrupt enabled by setting to 1 the specific SSPIE as well as the general Peripheral group mask bit PEIE and Global mask GIE; see Fig. 7.5 on page 199.

Once the MSSP and interrupt system are initialized, the interrupt service routine (ISR) can be written to recognise the various I^2C events. Any legitimate event will transfer processing from the backgound process to the ISR. Even if the Slave is asleep, a I^2C state will set SSPIF and awaken it in the usual way. These events are:

1: Master-Write: Last packet was an Address

S = 1   Start condition occurred last.
R/W̄ = 0   Master-Write conversation pending.
D/Ā = 0   This packet is an Address.
BF = 1   Buffer is full.

It is important to read the SSPBUF register to clear the BF flag, even if the address byte sent by the Master is going to be discarded. If this is not done, the next byte sent by the Master will cause an SSP OVerflow (SSPOV → 1) and the MSSP will NACK the Slave.

If SEN is 1 then CKP will automatically be cleared and the clock stretched. When appropriate, CKP must be set to 1 to allow the Master to proceed.

2: Master-Write: Last packet was Data

After the Address packet, the Master will send one or more Data packets. The Slave must read each one to avoid SSP OVerflow and ensure an ACK at the end of each packet. Also CKP must be manipulated as described in State 1 if SEN is active. The Status settings differ only in that D/Ā is 1.

S = 1   Start condition occurred last.
R/W̄ = 0   Master-Write conversation pending.
D/Ā = 1   This packet is Data.
BF = 1   Buffer is full.

3: Master-Read: Last packet was an Address

The bus device begins a Master-Read with an Address packet with R/W̄ = 1 to indicate that the Slave is expected to start sending Data packets back. Once a Slave recognizes its personal address, or a General Broadcast if GCEN is active, then the key SSPSTAT settings will be:

S = 1   Start condition occurred last.
R/W̄ = 1   Master-Read conversation pending.
D/Ā = 0   This packet is an Address.
BF = 0   Buffer is free for transmission.

Notice that the BF bit is clear in this case. This is because in a Master-Read situation BF is only 1 to show that SSPBUF is ready to be loaded with a Data byte to be sent to the Master. Thus SSPBUF does not need to be read, as in State 1.

Once its Address has been recognized, the Slave can send the first byte to the Master by copying it into SSPBUF and then setting the CKP bit to release the SCL pin. Clock stretching is always automatically asserted on a Master-Read packet, independent of the setting of any SEN bit.

4: Master-Read: Last packet was Data

This is similar to State 3, with the same treatment of the CKP clock stretching switch. A new data byte should not be copied into SSPBUF whilst BF is 1 otherwise the WCOL bit will set to show a Write COLlision.

The SSPSTAT settings are identical to State 3 except that D/Ā is 1 to show that a Data packet was sent last:

S = 1   Start condition occurred last.
R/$\overline{W}$ = 1   Master-Read conversation pending.
D/$\overline{A}$ = 1   This packet is Data.
BF = 0   Buffer is free for transmission.

5: Master-Read: Master sent a NACK

This occurs typically when the Master does not wish to receive any more bytes from the Slave. The NACK signals the end of the message, and when sensed by the Slave MSSP resets the I^{2}C logic.

S = 1   Start condition occurred last.
R/$\overline{W}$ = 0   R/$\overline{W}$ bit cleared by Slave logic.
D/$\overline{A}$ = 1   This packet is Data.
BF = 0   Buffer is free for transmission.

The NACK event is identified because the R/$\overline{W}$ bit is cleared but BF is 0. This is a conflicting state, as it appears to say that a Data packet has been received from the Master but the buffer is empty!

For our example, let us assume that the Slave PIC16F877A set-up to respond to address h'06' is monitoring eight temperature transducers using its Analog-to-Digital Convertor (ADC) module; see Chapter 14. The Master wishes to read any one of these digitized channels via its I^{2}C port by first of all initiating a Master-Write sending the channel number N to the Slave and then launching a Master-Read of the Slave, which then sends the designated datum. We assume the subroutine GET_ANALOG of Program 14.1 on page 461 is in situ.

A suitable task list would be:

1. Master sends a Start pulse followed by a call to Slave 06 requesting that it read the next packet (Master-Write).
2. Master sends one Data packet, specifying a channel number 0–7.
3. Master sends a Start pulse (repeat Start) for Slave 06, this time requesting that it write the next packet (Master-Read).
4. Slave stretches clock while its ADC digitizes its specified analog input.
5. Slave sends the requested datum.
6. Master NACKs Slave to signal end of transmission.

For clarity, our foreground software is shown as two separate routines. Overall, we are assuming that the temporary system locations used to store context variables, such as _status, are in common Bank-independent locations; which in the PIC16F877A are from File h'60' through File h'7F'. Program variables are taken to be in Bank 0.

Program 12.11 gives the ISR framework software, with the context being saved on entry and restored on exit; in the manner discussed on page 194. Before the context is restored, the CKP switch bit is set to 1 to lift any clock stretching and the SSPIF is cleared in the normal way.

The meat of this sandwich simply checks if the SSPIF interrupt flag is set and if not exits. In a real-world situation, interrupts may come from

Program 12.11 The I²C temperature acquisition ISR.

```
;************************************************************
; * FUNCTION    : ISR to send digitized channel N via I2C link*
; * ENTRY       : An I2C event has occurred              *
; * ENVIRONMENT: Uses subroutines GET_ANALOG, I2C_HANDLER    *
;************************************************************
; First save the context ------------------------------------
ISR         movwf  _work     ; Put away W
            swapf  STATUS,w  ; and the Status register
            movwf  _status
; ==========================================================
; Check the SSPIF flag is set? ------------------------------
            bcf    STATUS,RP0 ; Change to Bank 0
            bcf    STATUS,RP1
            btfss  PIR1,SSPIF ; Is this a MSSP interrupt?
            goto   ISR_EXIT   ; IF not THEN exit
            bsf    STATUS,RP1 ; ELSE change to Bank 1
            movf   SSPSTAT,w  ; Get the settings from SSPSTAT
            bcf    STATUS,RP1 ; and return to Bank 0
            andlw  b'00101101'; Zero all but S, D/A, R/W and BF
            movwf  I2C_STATUS ; and copy into a Temporary File
            clrf   I2C_ERROR  ; Zero Error flag
            call   I2C_HANDLER; Now respond to the I2C event
; ==========================================================
; Restore the context ---------------------------------------
ISR_EXIT    bcf    PIR1,SSPIF ; Clear interrupt flag
            bsf    SSPCON,CKP ; Release clock line
            swapf  _status,w  ; Untwist the original Status
            swapf  _work,f    ; Set the original W reg back
            swapf  _work,w    ; leaving STATUS unchanged
            retfie            ; and return to background
```

several sources, and this part of the code would be elaborated to check for each of the applicable interrupt flags.

If SSPIF is 1 then a copy of the state of the SSPSTAT Status register in Bank 1 is made in the more convenient Bank 0 location we have named I2C_STATUS, with the non-critical bits first cleared. The program variable named I2C_ERROR is also zeroed. If any problem is encountered then this will be changed to indicate an error situation for the background program.

Once intialization is complete, the actual I²C interpretation software is called as the subroutine shown in Program 12.12. This software is structured as a series of five separate cases; each one corresponding to one of the I²C states listed on page 369. This is done by using the xorlw instruction to check for no differences (see page 125) between the copy of the SSPSTAT Status pattern and the listed I²C state byte. Where a match is found, an action is taken corresponding to our task list, or in some cases just housekeeping to enable the MSSP to continue in the correct

Program 12.12 The I²C temperature acquisition handler subroutine.

```
; ****************************************************************
; * FUNCTION   : Interprets I2C state & responds appropriately*
; * ENTRY      : Copy of SSPCON in I2C_STATUS                 *
; * EXIT       : Appropriate action taken                     *
; * EXIT       : I2C_ERROR = -1 IF state not recognized       *
; ****************************************************************
I2C_HANDLER; Are we in State 1? (Address packet, Master-Write)
          movf   I2C_STATUS,w; Get copy of SSPSTAT status
          xorlw  b'00001001' ; Check for S=1, D/A=0, R/W=0, BF=1
          btfss STATUS,Z    ; Equal?
           goto STATE2      ; IF not THEN try for State 2
          movf   SSPBUF,w    ; ELSE do a dummy read to clear BF
; Are we in State 2? (Data packet, Master-Write) -------------
STATE2    movf   I2C_STATUS,w; Get copy of SSPSTAT status
          xorlw  b'00101001' ; Check for S=1, D/A=1, R/W=0, BF=1
          btfss STATUS,Z    ; Equal?
           goto STATE3      ; IF not THEN try for State 3
          movf   SSPBUF,w    ; ELSE read in Channel number
          call   GET_ANALOG  ; Digitize Channel N' analog input
          movwf TEMP         ; and copy outcome into TEMPerature
; Are we in State 3? (Address packet, Master-Read) -----------
STATE3    movf   I2C_STATUS,w; Get copy of SSPSTAT status
          xorlw  b'00001100' ; Check for S=1, D/A=0, R/W=1, BF=0
          btfss STATUS,Z    ; Equal?
           goto STATE4      ; IF not THEN try for State 4
          movf   TEMP,w      ; ELSE read in Temperature N
          movwf SSPBUF       ; Put in BUFfer reg for TX to Master
; Are we in State 4? (Data packet, Master-Read) --------------
STATE4    movf   I2C_STATUS,w; Get copy of SSPSTAT status
          xorlw  b'00101100' ; Check for S=1, D/A=1, R/W=1, BF=0
          btfss STATUS,Z    ; Equal?
           goto STATE5      ; IF not THEN try for State 5
; Datum has been sent, so do nothing!!! ---------------------
; Are we in State 5? (Master NACKs Slave) -------------------
STATE5    movf   I2C_STATUS,w; Get copy of SSPSTAT status
          xorlw  b'00101000' ; Check for S=1, D/A=1, R/W=0, BF=0
          btfss STATUS,Z    ; Equal?
           decf I2C_ERROR,f ; IF not THEN signal an error
          return
```

state. For instance, nothing needs to be done for I²C State 5, where the Master NACKs the Slave, as the MSSP will be automatically reset. If no state match is found, I2C_ERROR is decremented to signal an error state.

Another example of the use of the SSP module is given in the Microchip application note AN734, *Using the PIC Microcontroller SSP for Slave I²C Communications.*

As for the SPI protocol, many **C** compilers targeted to the PIC MCU have built-in functions to implement the I²C protocol and avoid bit banging user-defined functions.

To illustrate the technique, consider Program 12.13 which replicates the assembly-level coding of Programs 12.9 and 12.10 using the CCS compiler.

Program 12.13 Interfacing to the MAX518 in **C**.

```
#include <16F84a.h>
/* PortA, bit0 is the Master SCL, bit1 is the Master SDA,
   fast protocol                                            */
#use i2c(master, scl=PIN_A0, sda=PIN_A1, fast)
#byte DATA_X = 0x20
#byte DATA_Y = 0x21

void MAX518(unsigned int channel_0, unsigned int channel_1);

main()
{
/*      Various code lines                                  */
MAX518(DATA_X, DATA_Y); /* Send out the two data bytes      */
/*      More code                                           */
}

void MAX518(unsigned int channel_0, unsigned int channel_1)
{
i2c_start();             /* Start condition                 */
i2c_write(0x58);         /* Send out Slave address; Master-Write*/
i2c_write(0);            /* Send out Command 1              */
i2c_write(channel_0);    /* Send out datum to channel 0     */
i2c_write(0x01);         /* Send out Command 2              */
i2c_write(channel_1);    /* Send out datum to Channel 1     */
/* Updates both channels                                    */
i2c_stop();              /* Stop condition                  */
}
```

---

i2c_start();
Generates the Master Start condition.

i2c_stop();
Generates the Master Stop condition.

i2c_read();
Reads a byte over the bus. If an optional parameter of 0 is used then will NACK the received data. In Master mode, will generate a clock.

i2c_write(value);
Sends a single byte over the bus. In Master mode will generate a clock.

#use i2c(master, scl=PIN_A0, sda=PIN_A1, fast)
This is a directive by which the programmer informs the compiler which pins are used for the I²C lines, the fast or standard protocols and Master or Slave mode. At the time of writing the Master functions of the MSSP are not supported, and these functions are implemented in software. Slave functions can be implemented using the MSSP module by using the FORCE_HW option for the #use i2c() directive.

The key characteristic of the various serial protocols discussed up to now is that a clock signal is transmitted by the Master, which allows the Slave to receive or transmit data in perfect synchronization. An alternative approach is to send data under the assumption that the transmitter and receivers are running at approximately the same frequency. This **asychronous** protocol has been in use for data communications systems for over a century to send alphanumeric data over telegraph, telephone, and radio links to implement the Telex system.

One of the features of early computer development in the 1940s/1950s was the extensive use of existing technology. An essential adjunct of any computer-oriented installation is a data terminal. At that time the communications industry made considerable use of the teletypewriter (TTY).[9] Serial data were converted between serial and parallel formats in the terminal itself as well as providing keyboarding and printing functions.

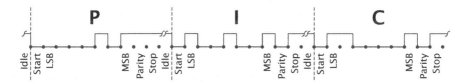

Fig. 12.20 Transmitting the message string "PIC" in the asynchronous serial mode, with odd one's parity and a minimum of one stop bit.

Until the early 1980s, TTYs were electromechanical machines, driven by a synchronous electric motor. This meant that synchronization between remote terminals could only be guaranteed for short periods. To get around this problem, each word transmitted was proceeded by one Start bit and followed by one or more Stop bits. A typical example is shown in Fig. 12.20. While the line is idling, a logic 1 (break level) is transmitted. A logic 0 signals the start of a word. After the word has been sent, a logic 1 terminates the sequence. Electromechanical terminals typically print ten characters per second, and require a minimum

---

[9]Literally a "typewriter from afar"; Greek, tele = far.

of two Stop bits. For 8-bit words, this requires a transmission rate of 110 bits per second, or 110 **baud**.[10]

The first purely electronic terminals required only one Stop bit, and could print at 30 characters per second, giving a rate of 300 baud. Traditionally communication channels use multiples of 300; e.g., 1200, 2400, 4800, 9600, ... PC Serial ports can run up to 115,200 baud. However, this $300 \times n$ series is not mandatory, as long as receiver and transmitter are running at the same nominal rate.

Typically a receiver, on detecting an incoming datum, will try and sample each bit at approximately the mid-point. This means that a frequency drift of ± 0.5 bit time can be tolerated in the space of 10 bits. Thus the receiver and transmitter local sample clocks must be within $\pm 5\%$. The two will be resynchronized at the start of each datum.

Although not the most efficient of techniques, the asynchronous protocol outlined here has the major advantage of being an international standard. There are several variants; for instance, the word can typically be from 5 to 9 bits long. In our example the word length is 8 bits with the eighth bit being used to provide a limited error-checking capability. This **parity bit** is set in Fig. 12.20 so that the number of 1s in the word is always odd. This can be checked at the receiver (see SAQ 5.17 on page 145) to detect a single bit error.

The original teleprinter code developed by Emile Baudot in 1875 is only 5 bits long.[11] Here the string "PIC" is coded as 10110 00110 01110. Although limited in capability, its key advantage over Morse code (Samuel Morse, 1840) was its fixed length (compare with ·– – ·· ·· – ·–·) which considerably simplifies the design of the transmitter and receiver. However, Morse code is more efficient, as the number of bits is approximately inversely proportionally to a letter's statistical frequency of use.

The 7-bit ASCII code of Table 1.1 on page 5, first adopted in 1963, was the first code specifically developed for computer communication systems. In 8-bit systems the extra bit is usually utilized to add a selection of 128 accented, mathematical and graphic symbols, rather than for parity. However, parity can be accommodated by using a 9-bit word format.

For our example we have adopted a format of one Start, eight data bits with no parity and one Stop bit. Using a bit-banging approach, as we have already done for our SPI and I²C protocols, is straightforward provided that we have an accurate $\frac{1}{2}$-bit delay. For instance, for a 9600 baud

---

[10]Strictly speaking the baud rate is a measure of information flow. For a simple baseband system this is equal to the bit rate. However, this equality is not always true. For instance, a telephone modem can use a di-bit modulation scheme where groups of bits two at a time give a carrier tone phase shift of 0°, 90°, 180° and 270° for the patterns 00, 01, 10, 11, respectively. In this case the baud rate is twice the bit rate.

[11]Actually the first binary coded alphanumeric code was devised by Francis Bacon in around 1600. It too was a 5-bit code.

link this would be $(1 \div 9600) \times 0.5 = 52\,\mu s$. As the delay is so short we can use an in-line approach using a macro, in the same manner as in Program 12.8 on page 360, rather than the subroutine approach of Program 6.9 on page 171. If our crystal is 20 MHz, then an instruction cycle is $\frac{1}{5}\,\mu s$; so the code below needs to take $52 \times 5 = 260$ cycles.

```
Baud_delay macro
           movlw  K          ; 1~            ; K is the delay cons
BAUD_LOOP  addlw  -1         ; K~            ; Decrement
           btfss  STATUS,Z ; ((K-1)+2)~  ; until zero
           goto   BAUD_LOOP; 2(K-1)~
           endm
```

which gives a total of $4K$ cycles delay, where each cycle is 4/XTAL microseconds. Thus making K = d'65' gives us our $52\,\mu s$ delay. Altering K will give us differing crystal/baud rates, although some padding out with nop instructions may be necessary for low baud rates.

With our delay macro in situ, the basic input/output subroutines of Program 12.14 are similar to our bit-banging SPI subroutines. The PUTCHAR subroutine simply brings the TX pin Low for two Baud_delay periods and then assigns the pin eight times, corresponding to the contents of DATA_OUT, least-significant bit first—the opposite order to SPI/I²C. Finally TX is held High for the same period to give the Stop/Idle condition.

The input GETCHAR counterpart is more complex. After an Idle state a Low-going voltage at pin RX will be treated as a Start bit. However, if the data stream is subsequentially sampled at intervals of one-bit periods (two evocations of Baud_delay) then as this is just at the transition point of the transmitter, any drift in the two clock rates may cause errors. To avoid this, a half-bit period is evoked and then the state of RX is checked to ensure that the Start bit is still present. If it is, then subsequent samples are taken at two Baud_delay periods, which is approximately at the bit center point. Better noise rejection could be obtained by sampling at a higher rate (over sampling) and then taking a majority decision regarding the logic state of the incoming voltage.

After the eight data bits have been shifted into DATA_IN, the Stop bit is checked for 1. If Stop is 0 then a **Framing error** has occurred. This is signaled by returning a value of -1 in ERR. Other more elaborate schemes may return a variety of error types. For instance, where parity is used then a Parity error can be returned.

As an example, if we wish to transmit the 3-character string "PIC", then the following code fragment would implement our task. For convenience the assembler allows the programmer to represent characters in delimited single quotes to represent their ASCII equivalent, as described on page 239.

Program 12.14 Asynchronous formatted input and output subroutines.

```
;       ************************************************************
;       * FUNCTION: Transmits one 8-bit byte in asynchronous format *
;       * RESOURCE: Macro Baud_delay giving a 0.5 bit delay; COUNT  *
;       * ENTRY   : 8-bit datum in DATA_OUT, XTAL & BAUD predefined  *
;       * EXIT    : Contents of DATA_OUT 00h, byte TXed              *
;       ************************************************************
PUTCHAR         movlw   8               ; Eight data bits
                movwf   COUNT
                bcf     PORTA,TX        ; Start bit
                Baud_delay              ; 2x0.5 bit delay
                Baud_delay
; Now shift out data, LSB first ---------------------------------
PUTCHAR_LOOP    rrf     DATA_OUT,f      ; Rotate right into Carry
                btfss   STATUS,C        ; Test Carry bit
                 goto   ITS_A_0         ; IF 0 THEN output a 0
                bsf     PORTA,TX        ; ELSE output a 1
                goto    PUTCHAR_NEXT    ; and continue
ITS_A_0         bcf     PORTA,TX        ; Output a 0
PUTCHAR_NEXT
                Baud_delay              ; One-bit duration
                Baud_delay
                decfsz  COUNT,f         ; Repeat eight times
                 goto   PUTCHAR_LOOP
                bsf     PORTA,TX        ; Stop bit
                Baud_delay
                Baud_delay
                return
;       ************************************************************
;       * FUNCTION: Receives one 8-bit byte in asynchronous format  *
;       * RESOURCE: Macro BAUD_DELAY giving a 0.5 bit delay; COUNT  *
;       * ENTRY   : XTAL & BAUD predefined                          *
;       * EXIT    : DATA_IN holds the received byte.                *
;       * EXIT    : Err is 00 if no Framing error ELSE -1           *
;       ************************************************************
GETCHAR         movlw   8               ; Eight data bits
                movwf   COUNT
                clrf    ERR             ; Zero Error byte
GETCHAR_START
                btfsc   PORTA,RX        ; Poll for 0
                 goto   GETCHAR_START
                Baud_delay              ; Hang around for 0.5 bit time
                btfsc   PORTA,RX        ; Check; is it still Low?
                 goto   GETCHAR_START
                Baud_delay              ; IF yes THEN hang around
                Baud_delay
GETCHAR_LOOP    bcf     STATUS,C        ; Clear Carry
                rrf     DATA_IN,f       ; Shift 0 into datum
                btfsc   PORTA,RX        ; Check; is input High?
                 bsf    DATA_IN,7       ; IF yes THEN set bit in datum
                Baud_delay
                Baud_delay
                decfsz  COUNT,f         ; Do eight times
                 goto   GETCHAR_LOOP
                btfss   PORTA,RX        ; Look for a Stop bit (High)
                 decf   ERR,f           ; IF Low THEN signal an error
                return
```

```
movlw   'P'        ; Same as movlw h'50' (ASCII for P)
movwf   DATA_OUT   ; Put in store
call    PUTCHAR    ; Send it out
movlw   'I'        ; Same as movlw h'49'(ASCII for I)
movwf   DATA_OUT   ; Put in store
call    PUTCHAR    ; Send it out
movlw   'C'        ; Same as movlw h'43' (ASCII for C)
movwf   DATA_OUT   ; Put in store
call    PUTCHAR    ; Send it out
```

Handling serial communications this way is only really satisfactory for very simple situations. For example, if the RX pin is not continually monitored a transmission can be missed or synchronization lost. Also it is difficult to implement a full-duplex link. In addition the procedure is software intensive, with most of the processing power being wasted in delay loops. The situation can be improved somewhat by using an internal timer to generate the baud delay and by using interrupt-driven techniques. However, the majority of 28+pin PIC MCUs have an integral communications port to automatically deal with asynchronous transmission.

One of the first applications of the then new LSI fabrication techniques in the late 1960s was the implementation of a dedicated hardware asynchronous serial port known as the **universal asynchronous receiver transmitter**. The UART[12] was already in production by the time microprocessors were developed. Most PCs, even in the 1970s, had a serial port implemented by a UART, as do most current systems. As well as dealing with shifting, error checking, and interrupt handling, most UARTs also have an integral baud-rate generator which can be set-up in software to give the correct bit frequency.

The basic structure of a UART is shown in Fig. 12.21. Any given UART circuit will have three core sections. A Transmit shift register will serialize a datum with appended Start and Stop bits to be shuffled out via a TX pin. Associated with TXREG is a Buffer register holding data for onward transmission. A Status register will hold a flag (TBUF in our diagram) indicating when this buffer is free for more data.

The Receive shift register strips the Start and Stop bits off an incoming frame at a RX pin, transferring the parallelized datum when complete into one or more Buffer registers. At the same time, a flag (RBUF in our case) will be set to allow the software to determine that a new datum is ready for collection. This needs to be read before a new frame has been assembled, otherwise an overrun condition will occur and data will be lost.

The transmission and reception of a frame is not locked in step; that is, they can overlap, but the baud rates are usually the same.

Real devices are more sophisticated in that various options, such as the number of bits in a datum, and error condition detection are sup-

---

[12]Sometimes known as the asynchronous communication interface adapter, or ACIA.

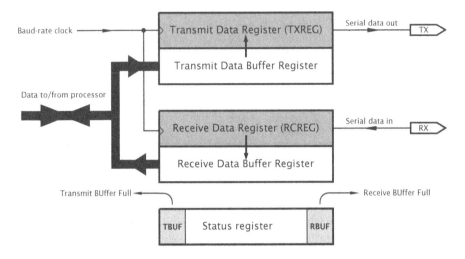

Fig. 12.21 The generic UART.

ported. Thus the associated Control and Status registers are of necessity rather more comprehensive, but the PIC MCU Serial Communication Interface (SCI) module, more usually known as the **USART (Universal Synchronous-Asynchronous Receiver Transmitter** module) shown in Fig. 12.22 is clearly based on a generic UART architecture. This module is actually a dual-purpose SCI in that it supports both the asynchronous protocol described here if the **SYNC** bit in TXSTA[4] is 0 (the default reset value) and a synchronous mode (SYNC = 1) which does not use a Start and Stop delimited frame. This latter protocol uses pin RC6/CK as a clock; output if operating as a transmitter and clock input if a receiver. Pin RC7/DT then operates as a I/O data line as appropriate. Synchronous data can be sent either one byte at a time or as a continuous burst. Because of this synchronous function the module is known as a USART rather than a UART. Here we will concentrate on the asynchronous mode, and the two Status registers shown in Fig. 12.22 are shown as appropriate to this mode of usage.

The core of the USART is the Transmit and Receive shift registers and their associated Buffer and Status registers. To enable the overall USART the **Serial Port ENable (SPEN)** bit in the **ReCeive STAtus register** (RCSTA[7]) must be set. *Both* RC6 and RC7 pins used for the transmission and reception lines TX and RX, respectively, have to be set-up as inputs.[13]

Transmission
The transmitter logic is enabled when the **TranSmit ENable (TXEN)** bit in the **TranSmit STAtus register** (TXSTA[5]) is set. To send a character the

---

[13]Small-footprint PIC MCUs typically use pins RB1 and RB2, respectively. In this case the latter has to be set-up as an output.

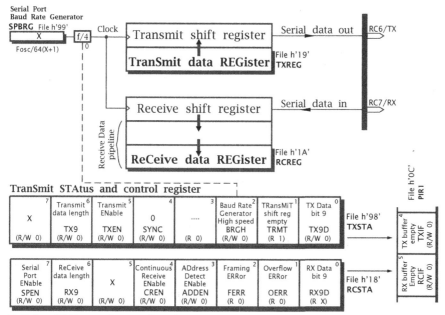

Fig. 12.22 The Serial Peripheral Interface module configured for asynchronous communication.

datum must be moved to the **TranSmit data REGister** (**TXREG**), whence it will be transferred to the **Transmit shift register** and shifted out of pin TX. If a 9-bit format is required the **TX9** bit in TXSTA[6] must be set to 1 and the ninth bit placed in bit 0 of the same register *before* moving the lower eight bits into TXREG. If the Transmit shift register is not empty; that is, it is in the process of shifting out a previous datum, then the new datum will remain in the TXREG buffer register awaiting the completion of transmission before being transferred.

Bit 1 of the TranSmit STAtus register reflects the state of the Transmit shift register whilst the **TranSmit Interrupt Flag** (**TXIF**) in the Peripheral Interrupt Register 1 (PIR1) is automatically set when the TXREG buffer is empty and ready for reloading. If an interrupt on TX buffer is empty is required, the corresponding TXIE mask bit in the Peripheral Enable Register 1 (PIE1[4]) must be set; see Fig. 7.5 on page 199. TXIF is automatically cleared whenever a datum is written into the TXREG, so it doesn't have to be manually cleared in the polling routine or associated ISR.

Reception

Once a Start bit is detected at pin RX then the succeeding eight or nine bits are shifted into the 2-deep **ReCeive data REGister** (**RCREG**) pipeline, irrespective of what is going on at the transmitter section. When a datum has been received, it is automatically stored in the top RCREG buffer whence it moves to the lower buffer, provided that no datum is still waiting to be read. The **ReCeiver Interrupt Flag (RCIF)** is automatically set whenever a datum is waiting for collection and this can be used to generate an interrupt if the RCIE mask bit is set; as well as the GIE and PEIE global masks. RCIF is automatically cleared whenever a datum is read. If a datum is waiting in the top buffer, then it moves down and RCIF is immediately set again, showing that there is another datum ready for collection.

If a third character has been received and the 2-deep Receive pipeline is full then the **Overflow ERRor (OERR)** bit at RCSTA[1] will be set and this newly received datum will be lost. The RCREG can still be read twice to retrieve the two buffered bytes. However, to clear OERR the receive logic must be reset by clearing the **Continuous Receive ENable (CREN)** bit in RCSTA[4] and then setting it again.

The **Framing ERRor (FERR)** bit in RCSTA[2] indicates when a Start bit has been detected but at the end of the shift no Stop bit is found. Both FERR (and any ninth received bit) are double-buffered in the same way as the received data, and so should be checked first *before* the datum in RCREG is read, as this will move data down the pipeline and therefore change these auxiliary bits.

All versions of the USART module support the use of an 8- or 9-bit datum individually selectable for both transmission and reception. In the latter case the **RX9** bit in RCSTA[6] must be set to 0. Typically this extra bit will be used to add a parity error-checking capability. However, another use of a 9-bit datum is to implement a network of asynchronous devices. Here the ninth bit can be used to indicate that the other eight bits represent data or a Slave address. A simple network using this principle is shown in Fig. 12.23. With an 8-bit address, up to 255 Slaves may be addressed and one address reserved for a general broadcast facility.

To facilitate networking in this manner, newer versions of the USART module can be configured so that the arrival of a set ninth bit sets the RCIF interrupt flag automatically. This is enabled by setting the **ADDEN** bit in RCSTA[3] to 1. With both ADDEN and RX9 set to 1 any frame with its MSB = 0 will be ignored and the datum will not be placed in the Receive Data pipeline. If the MSB is 1 then the datum will be copied out of the Receive Shift Register into the Receive pipeline and RCIF set. The Slave can then read this address from RCREG, with this process clearing RCIF. If a match is found in software the Slave can clear ADDEN and any subsequent data frames will be captured in the normal way. The Slave can still monitor the ninth bit in RX9D and thus the Master can terminate its conversation by sending a datum with the ninth bit set to 1.

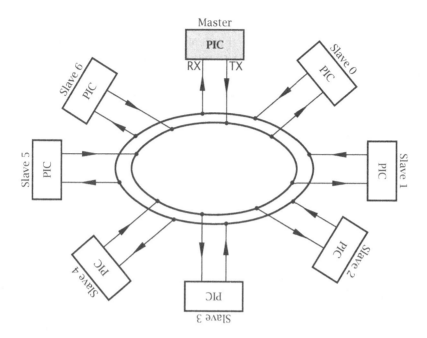

Fig. 12.23 A local area network (LAN) using an asynchronous serial protocol.

Serial Port Baud-Rate Generator, SPBRG

This is basically a programmable 8-bit counter followed by a switchable frequency ÷4 flip flop chain, which can be set-up to give the appropriate sampling and shifting rates for the desired baud rate. Based on the PIC MCU's crystal frequency XTAL we have:

Baud rate $= \frac{XTAL \times 10^6}{64 \times (X+1)}$ Low-speed: SPBRG = 0

Baud rate $= \frac{XTAL \times 10^6}{16 \times (X+1)}$ High-speed: SPBRG = 1

where X is an 8-bit datum written into SPBRG. Using the low-speed option, X has the value $\frac{XTAL \times 10^6}{64 \times BAUD} - 1$. For instance, if we require a baud rate of 9600 on a 20 MHz device, then X = 31 will give a actual baud rate of 9766, an error of +1.73%. At 20 MHz the maximum baud rate is 312,500 whilst the lowest rate is 1221. A baud rate of 1.25 Mbaud is obtainable at 20 MHz in the high-speed mode with X = 1.

Actually, the SPBRG produces a frequency 16 times the base baud rate, to enable the USART to take three samples around bit midpoints and adopt a majority decision. This increases reliability in a noisy environment.

To illustrate how to use the USART, we will repeat our GETCHAR and PUTCHAR subroutines using hardware. First, in the main program we have to set-up the Serial Port Baud Rate Generator and both Transmit and

Receive Status/Control registers. Assuming the programmer has defined the constants XTAL and BAUD, then we can let the assembler evaluate the arithmetic to give us the value of X to put in the SPBRG. With this in mind, the initialization code would look something like:

```
        include "p16f877a.inc"

        #define BAUD  d'4800'  ; For example 4800 baud rate
        #define XTAL  d'8'     ; 8MHz crystal
        #define X     ((XTAL*d'1000000')/(d'64'*BAUD))-1

START   bsf     STATUS,RP0     ; Change to Bank1
        movlw   X              ; Move X to Baud Rate Generator
        movwf   SPBRG
        movlw   b'00100000'    ; 8 data bits, TX enabled
        movwf   TXSTA          ; Low speed SPBRG mode
        bcf     STATUS,RP0     ; Back again to Bank0
        movlw   b'10010000'    ; USART enabled, 8 data bits
        movwf   RCSTA          ; Receiver enabled
```

Note that for the PIC16F87X group, Microchip specifies that both RX and TX pins be configured as inputs for correct operation. As this is the default Reset state, bits TRISC[7:6] have not been explicitly altered in the listing. As observed in footnote 13, other family members may require different settings for the RX and TX pins.

With the USART enabled, the subroutines are coded in Program 12.15. PUTCHAR is simply a matter of polling TXIF, waiting for it to go to 1 and then copying the datum to the TranSmitter REGister.

The input GETCHAR is a little more complex if some error checking is to be incorporated. The subroutine polls the state of RCIF, which goes to 1 whenever there is data to be read. Also returned is the variable ERR which is h'00' if there is no problem, -1 if a Framing error occurred, -2 if a Overflow situation is sensed and -3 if both errors occurred. In these latter situations OERR is zeroed by resetting the receiver logic. After the error conditions have been checked, the data is read from the ReCeive REGister. Error checking is always done *first* before reading the datum, to avoid altering these Status flags.

Some systems may not wish the processor to hang up waiting for a character which is a long time in coming. In such cases an alternative input subroutine, perhaps called getch, could return an ERR of +1 if the return was empty handed. A better approach would be to generate an interrupt each time an incoming character is sensed rather than using a polling technique.

Asynchronous links can be used in **C** as a standard input/output channel. In the specific case of the CCS **C** compiler, the #use rs232 directive tells the compiler which pins are to be used for RX and TX and the baud rate. The normal **C** I/O functions, such as printf(), use these pins

Program 12.15 The USART-based I/O subroutines.

```
; *****************************************************************
; * FUNCTION: Transmits one 8-bit byte in asynchronous format *
; * RESOURCE: PIC MCU USART                                   *
; * ENTRY   : 8-bit datum in DATA_OUT                         *
; * EXIT    : Contents of DATA_OUT unchanged, byte TXed       *
; *****************************************************************
PUTCHAR btfss    PIR1,TXIF    ; Check, is TX buffer full?
        goto     PUTCHAR      ; IF not THEN try again
        movf     DATA_OUT,w   ; ELSE get datum
        movwf    TXREG        ; and copy to USART TX register
        return


; *****************************************************************
; * FUNCTION: Receives one 8-bit byte in asynchronous format  *
; * RESOURCE: PIC MCU USART                                   *
; * ENTRY   : None                                            *
; * EXIT    : DATA_IN holds the received byte.                *
; * EXIT    : ERR is 00 if no error. Framing ERRor only = -1  *
; * EXIT    : ERR = -2 if Overflow ERRor and -3 if both types *
; *****************************************************************
GETCHAR clrf     ERR          ; Zero flag byte
        btfss    PIR1,RCIF    ; Check, is there a char ready?
        goto     GETCHAR      ; IF not THEN try again
; Error return -------------------------------------------------
        btfss    RCSTA,FERR   ; Was there a Framing error?
        goto     CHECK_OERR   ; IF not THEN check for Overflow
        decf     ERR,f        ; ELSE record a Framing error

CHECK_OERR
        btfsc    RCSTA,OERR   ; Check for Overflow ERRor
        goto     GET_EXIT     ; IF none THEN complete
        decf     ERR,f        ; Otherwise register error
        decf     ERR,f
        bcf      RCSTA,CREN   ; and reset the logic
        bsf      RCSTA,CREN
GET_EXIT
        movf     RCREG,w      ; Get datum
        movwf    DATA_IN      ; and put away
        return
```

as their link to the standard channel. Multiple coexisting asynchronous channels can be implemented with this compiler.

As an example, Program 12.16 shows a **C** implementation of an asynchronous 9600-baud duplex link to a terminal; such as shown in Fig. 12.25. A switch is connected to pin RB0, and when the operator sends the character G (for Go) to the PIC MCU it continually monitors this switch. When the switch closes, bringing the pin Low, the terminal is to alert the operator by printing the message "Switch 1 is now closed". The standard

---

Program 12.16 Using a duplex asynchronous channel in **C**.

```
#include  <16f877a.h>
#use delay (clock = 20000000)/* Tell compiler 20MHz xtal    */
/* Tell compiler baud rate & which pins to use for TX  & RX  */
#use rs232(baud=9600, xmit=PIN_A1, rcv=PIN_A2)
#bit SWITCH1 = 6.0                /* Switch connected to RB0       */

main()
{
while(TRUE)
    {
    if(getch() == 'G')
        {
        while (SWITCH1) {;}   /* Do nothing while Switch is hi*/
        printf("Switch 1 is now closed \n");
        }
    }
}
```

C I/O functions printf() for output and getch() are used for output and input, respectively.

As pins RA1 and RA2 are specified as the TX and RX pins, respectively, the compiler will generate a software UART implementation, such as that used in Program 12.14. For this reason the compiler needs to know the crystal frequency so that it can generate the appropriate baud delays. If pins RC6 and RC7 are used with this processor, the compiler will automatically use the USART rather than software for serial interface. In our example, the software UART needed 146 instructions against 74 instructions for the USART module.

There is more to setting up a communication link than establishing a suitable protocol. PIC microcontrollers have normal logic voltage and current levels which are not intended for connections greater than 30 cm (1′). Although with care,[14] distances considerably in excess of this can be employed, in situations with relatively fast bit rates different signaling techniques have to be used.

In the era of electromechanical TTYs the 20 mA loop de facto standard was in common use. This uses zero and 20 mA current to signal logic 0 and logic 1, respectively. Use of current means that line attenuation is not a problem (as current out must equal current in) and this level of current was sufficient to directly activate the receiver solenoid relay.

Current sources are realized by using high voltages in series with a large resistance. The latter gives long time constants, which, while adequate in the era of 110 baud rates, did not transfer well to the intro-

---

[14]Or sometimes ignorance!

duction of electronic terminals, UARTs and modems. **RS-232**[15] was introduced in 1969 as the standard interface for connecting an item of Data Terminal Equipment (DTE), such as a terminal, to approved Data Circuit terminating Equipment (DCE), typically a modem. Thus, not only did it define signaling levels, as shown in Fig. 12.24(a), but also various control and handshake lines, some of which are shown in Figs. 12.24(d) and 12.25. For instance, the modem would signal back to the DTE that a telephone link had been opened with the remote DTE by activating the Clear To Send (CTS) handshake signal. Two data lines plus an optional ground line are needed for a full duplex transmission circuit.

The RS-232 standard has a specified range of 15 m (50′) at a maximum rate of 20 kbaud, which it achieves by mapping logic 0 (often called a **space**) to typically +12 V and logic 1 (often called a **mark**) to typically −12 V. The receiver can distinguish levels down to ±3 V. The **RS-423 standard** (1978) in Fig. 12.24(b) is similar but can manage 1.2 km (6000′) at up to 80 kbaud and 10 Mbaud at 12 m (40′) with up to ten receivers.

Both RS-232 and RS-423 are **unbalanced** (or single-ended) standards, where the receiver measures the potential between signal line and local ground reference. Even though the transmitter and receiver grounds are usually connected through the transmission line return, the impedance over a long distance may support a significant difference in the two ground potentials, which will degrade noise immunity. Furthermore, any noise induced from outside will affect signal lines differently from the ground return, due to their dissimilar electrical characteristics, hence the term unbalanced.

The **RS-422** (1978) and **RS-485** (1983) standards are described as **balanced**. Here each signal link comprises *two* conductors, normally twisted around each other, known as a **twisted pair**. The logic level, is represented as the *difference* of potential across the conductors, not the difference from ground. Calling the conductors A and B, then logic 0 is represented as A < B and logic 1 by A > B. A difference of more than ±200 mV at the receiver is sufficient to establish the logic level and the transmitter will typically generate a $\Delta V = \pm 5$ V. As the A and B conductors have the same characteristics and are tightly wound together, they represent similar targets for induced noise. As the same noise voltage appears in *both* conductors and the receiver only distinguishes *differences*, rejecting common-mode voltages up to ±7 V, then the noise immunity of these balanced links is clearly superior to unbalanced schemes. Commercial twisted-pair cables, used in Local Area Networks (LANs), often carry three or four pairs of conductors, each link having a different twist pitch to reduce cross-talk between links. PC USB ports also use a balanced signal path.

---

[15]Defined in the United States of America as the Electronics Industries Association EIA 232-E standard and in Europe as the V24 interface, by the CCITT.

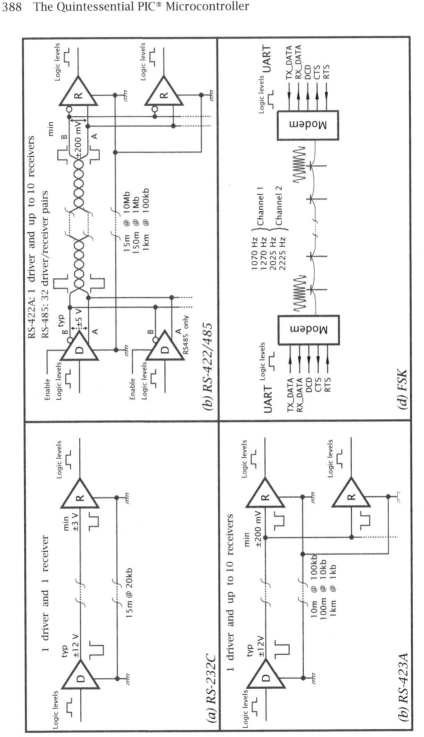

Fig. 12.24 Some signaling configurations.

The main difference between the RS-422 and RS-485 standards is the provision in the latter case for multiple transmitters as well as receivers to implement multi-drop LANs. As only one transmitter can be active at any one time, an RS-485 transmitter buffer must have an enable input, to select the Master device. The single RS-422 transmitter has no need to be disabled.

RS-232 was originally designed for DTE-modem interconnection, although its use is now much more varied; see Fig. 12.25. Figure 12.24(d) shows a simple Frequency Shift Keying (FSK) full duplex system with the mark/space of one channel being represented by the tones 1070/1270 Hz and the other by 2025/2225 Hz; frequencies which fit well inside the normal telephone link bandwidth of 300–3400 Hz. Handshake lines DCD (Data Carrier Detect), CTS (Clear TO Send) and RTS (Ready To Send) are used to control the sequence of modem operations prior to and terminating the communication of data.

Many modem schemes currently use Phase Shift Keying (PSK), where typically at least eight different phases in 45° steps of a single tone are used to encode 3-binary bit code groups (tri-bits) in any one time slot. In this way the baud rate may be increased with the same signaling rate, albeit at the expense of noise immunity, as witnessed by the steady increase in PC-based home telephone Internet data rates in recent years up to 56 kbaud.

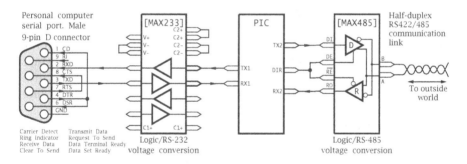

Fig. 12.25 Communicating with a PC via an RS-232 link and the outside world.

As an example, Fig. 12.25 shows the connection between a PIC MCU and the serial port of a PC—or any device with an asynchronous RS-232 port. The Maxim MAX233 dual RS-232 transceiver translates between +12 V and 0 V (logic 0) and between −12 V and +5 V (logic 1). If handshake lines are not being used, as is usual in simple links, the PC can be "fooled" into treating the interface as ready to accept data by linking, as shown in the diagram. For instance, the serial port UART's RTS is looped back to CTS. The MAX233 has two transmit and two receive buffers in all and thus can be used to buffer some additional handshake lines if required.

In Fig. 12.25 the same PIC MCU is shown driving a half-duplex RS-485 link using a Maxim MAX485 voltage converter. Each buffer has a separate Enable of the opposite logic polarity. The PIC MCU can activate the appropriate buffer depending on the communication direction. Alternatively the MAX485 can be used to implement a full duplex channel using two separate links.

The RS-485 link need not use the asynchronous protocol. Any suitable synchronous protocol can be buffered to RS-485, but of course a separate buffered clock channel will be needed.

Examples

Example 12.1

In Example 11.2 we designed a subroutine to compare a *fixed* number TRIP with the byte read in from Port B. In some cases it may be necessary to have the software adapt to changing circumstances, altering the trigger value by reading updates from outside. Rather than using up another eight port lines, it is proposed that the update be fed in from an outside agency in series at pin RA4, with RA3 being used as the clock line. With the assumption that each data bit is stable when the Clock line is High, write a subroutine to read in a new value into memory at TRIP.

Solution

One solution is shown in Program 12.17. The Clock line is monitored for a High state, during which time the Data is stable. By mirroring the state of the Data line into the Carry flag the datum is rotated bit by bit into memory. After each shift the loop is not completed until the Clock line again goes Low.

This is similar to subroutine SPI_READ in Program 12.3, except that the clock is generated from outside; that is, the PIC MCU is acting as a Slave. This causes problems in a real system, where the PIC MCU Slave must be able to tell the Master when it wants a new byte. This could be done by using another port line as a Clear To Send handshake, which will interrupt the Master and initiate the conversion. Of course the Master could be another PIC MCU and if so we have an economical way of connecting two PIC MCUs together. If PIC MCUs with integral serial ports are used, then interrupts can be automatically generated and this is a frequently used method of implementing multi-processor networks.

Example 12.2

Design and code the I2C_IN counterpart of the I2C_OUT subroutine of Program 12.9. You may assume that the same variables are available and that the received datum is in DATA_IN on exit.

Program 12.17 Updating Program 11.6's trip value.

```
;    **********************************************************
;    FUNCTION: Shifts in value for TRIP which is subsequently  *
;    FUNCTION: used as one operand for subroutine COMP         *
;    ENTRY   : Data bit changes at RA4 when at RA3 is Low      *
;    EXIT    : COUNT is 00, datum is in TRIP                   *
;    **********************************************************
SER_TRIP    movlw   8               ; Bit loop count
            movwf   COUNT
SER_TRIP_LOOP1
            btfss   PORTA,3         ; Wait for Clock to go High
            goto    SER_TRIP_LOOP1
            bcf     STATUS,C        ; Carry = 0
            btfsc   PORTA,4         ; Is Data line High?
            bsf     STATUS,C        ; IF yes THEN Carry = 1
            rlf     TRIP,f          ; Shift bit in
SER_TRIP_LOOP2
            btfsc   PORTA,3         ; Wait for Clock to go Low
            goto    SER_TRIP_LOOP2
            decfsz  COUNT,f
            goto    SER_TRIP_LOOP1
            return
```

Solution

The I2C_IN subroutine of Program 12.18 shifts the datum in File DATA_IN eight times through the Carry flag, which mirrors the state of pin SDA. At the same time the Clock line SCL is toggled according to the I^{2}C time and protocol specification as in our I2C_OUT subroutine of Program 12.9. In this protocol the Master signals back to the Slave to stop sending data by letting the SDA line float High in the Acknowledge slot in the ninth clock pulse; see Fig. 12.13. The normal Low state in this slot is called ACK, whilst the deviant High Acknowledge state is called NACK (No AC-Knowledge). To cope with both these situations our I2C_IN optionally generates either situation depending on the state of the variable ACKNO set by the caller. If File ACKNO is zero on entry, then a normal Low ACK is sent in this slot. Any non-zero value in this variable causes a High NACK to be sent back to the Slave. The Slave then terminates its transmission and listens for the next Stop/Start condition.

Example 12.3

Many MCU-based products require storage of data in non-volatile memory for retrieval after the system has been powered down. A typical example is the total distance traveled by a car from new, which should be held independently of the state of the car battery. Such data is typically held in Electrically-Erasable Programmable Read-Only Memory (EEPROM), as

Program 12.18 Reading in a byte using the I²C protocol.

```
;*****************************************************************
; * FUNCTION: Reads in byte from Slave with optional ACK/NACK *
; * ENTRY   : ACKNO = 00 for ACK ELSE NACK                    *
; * RESOURCE: START and STOP subroutines, Delay_600 macro     *
; * EXIT    : DATA_IN holds datum sent from Slave             *
; * EXIT    : ACK or NACK sent to Slave, SCL Low              *
;*****************************************************************
I2C_IN    bcf    INDF,SCL    ; Make sure that Clock line is Low
          movlw  8           ; Loop count = 8
          movwf  COUNT

I2C_IN_LOOP
          bcf    INDF,SCL    ; Clock Low
          Delay_600          ; For minimum period
          Delay_600
          bsf    INDF,SCL    ; Clock High
          bcf    STATUS,C    ; Carry = 0
          btfsc  INDF,SDA    ; Check state of incoming bit?
          bsf    STATUS,C    ; IF 1 THEN make Carry = 1
          rlf    DATA_IN,f   ; and rotate it into the datum
          decfsz COUNT,f     ; Decrement loop count
          goto   I2C_IN_LOOP ; and repeat eight times

; Now determine if Acknowledge is to be sent ------------------
          bcf    INDF,SCL    ; Clock Low
          bsf    INDF,SDA    ; Data output float (NACK)
          movf   ACKNO,f     ; Test the caller's wish
          btfsc  STATUS,Z    ; IF non-zero THEN leave as NACK
          bcf    INDF,SDA    ; ELSE bring Low to signal ACK
          Delay_600          ; Keep Clock Low
          Delay_600
          bsf    INDF,SCL    ; Now High state
          Delay_600
          bcf    INDF,SCL    ; Leave with Clock Low
          return
```

detailed on page 28. Although many PIC microcontrollers have an integral EEPROM data module, as described in Chapter 15, capacity is limited to 256 bytes[16] at most. For larger capacities an external EEPROM memory is required. Most of these devices use an SPI or I²C interface; specifically the I²C 24LCXXX shown in Fig. 12.26. The 24LCXXX 8-pin serial EEPROMs vary from the 1 kbit 24LC01B to the 512 kbit 24LC512, organized as bytes; i.e., 128 byte to 64 kbyte.

The 24XXX serial EEPROMs have the following features.

• 400 kHz I²C compatible (V_{DD} = 5 V), 100 kHz at V_{DD} = 2.5 V.
• Write protection (ROM mode) using the WP pin.

---

[16]Some extended-range devices have a 512-byte integral EEPROM module.

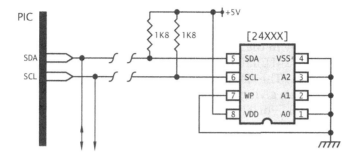

Fig. 12.26 The 24XXX series of I^{2}C serial EEPROMs.

- 2 ms typical Write cycle time.
- 1,000,000 minimum Write cycle endurance per byte cell.
- 3 mA Write, 1 mA Read and 100 μA standby current.
- Internal generation of high programming voltage.

Using a 24LC01B serial EEPROM, show how you could increment a number in the bottom three locations, which represents the total distance in either miles or kilometers depending on the market. You may assume that the PIC MCU is interrupted on each mile/kilometer and that your software is part of the interrupt handler. You have the resources of the subroutines of Program 12.9 and 12.18.

Solution

Before writing code to implement our specification we need to look more closely at the protocol used by the 24XXX serial EEPROMs in communicating with the Master PIC MCU. This is encapsulated in the signals shown in Fig. 12.27.

In all cases the Master initiates a data transfer by sending a Start condition followed by a Command byte. The Control byte contains the I^{2}C Slave address 1010; the chip select address A2 A1 A0 and the R/$\overline{\text{W}}$ bit in the order

| 1 | 0 | 1 | 0 | A2 | A1 | A0 | R/$\overline{\text{W}}$ |

. Although the chip select address is shown as part of the Command byte and the three corresponding pins are shown in Fig. 12.26, newer versions of the smaller EEPROMS do not implement this feature. This is because if EEPROM capacity needs to be expanded then it is more efficient to replace the device by a pin-identical larger version. For example replacing a 24LC01B by a 24LC08B gives an eightfold increase with no hardware alteration. Larger EEPROMS, such as the 24LC256 do implement chip select address pins as the method of expansion, as additional devices will need to be hung on the bus in this situation. Eight 24LC512s will give a capacity of 512 kbyte of non-volatile memory.

This is normally followed by the address in the EEPROM that data is to be written into or read out of. In the specific case of the 24LC01B the data is arranged as 128 cells, each comprising a byte that can be individually written to or read from. This means that a 7-bit address will fit comfortably in the 8-bit address byte. This scheme will cope with devices up to the'24LC02B but beyond this addresses greater than 8 bit wide are needed. This is done by using the three Chip select bits in the Command byte, giving an address width of 11 bits and a capacity of 2 kbytes (16 kbits). For EEPROMs larger than the 24LC16, two Address bytes are used following the Command byte.

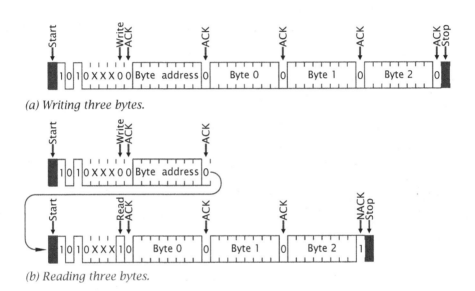

(a) Writing three bytes.

(b) Reading three bytes.

Fig. 12.27 EEPROM Read and Write waveforms.

The process of sending the byte address to the EEPROM is implemented as a Write action in Fig. 12.27(a). This is actioned by setting the R/$\overline{W}$ bit Low in the Command byte. Where a data byte is to be written into the addressed location, this byte comes immediately after the Address byte and then is followed by a Stop condition. If more than one data byte is transmitted before the Stop then this data is stored in a small on-board buffer and the actual programming will not occur until the Stop condition. The 24LC01B can store eight bytes at a time in a single page, with the *lower three* address bits being incremented on each data byte sent. If this address rolls over, earlier addressed data will be overwritten. The size of this page depends on the device; for example, the 24LC256 uses a 64-byte page. In Fig. 12.27(a) three bytes are shown being written into the 24LC01B. As these locations are to be targeted in the bottom three locations, h'00-01-02', then roll-over will not occur.

Program 12.19 Incrementing the non-volatile odometer count.

```
EXTRA_MILE                          ; Get the 3 bytes at h'00:01:02'
            call    START           ; Start a transmission packet
; Command byte 1 to initialize address ----------------------
            movlw   b'10100000';    Slave address Master-Write
            movwf   DATA_OUT        ; Copied to pass location
            call    I2C_OUT         ; Send it out
            movf    ERR,f           ; Check for an Acknowledge error
            btfsc   STATUS,Z        ; IF Zero THEN continue
             goto   EXTRA_MILE      ; ELSE try again
; Address 00 ------------------------------------------------
            clrf    DATA_OUT        ; Pass location
            call    I2C_OUT         ; Send it out

; Command byte 2 to change over to Read ----------------------
            call    START
            movlw   b'10100001';    Slave address Master-Read
            movwf   DATA_OUT        ; Copied to pass location
            call    I2C_OUT         ; Send it out

; Now read in three bytes -----------------------------------
            clrf    ACKNO           ; Enable Acknowledge
            call    I2C_IN          ; Read the High byte in h'00'
            movf    DATA_IN,w       ; Get byte
            movwf   MSB             ; and put in memory
            call    I2C_IN          ; Read the Middle byte in h'01'
            movf    DATA_IN,w       ; Get byte
            movwf   NSB             ; and put in memory
            incf    ACKNO,f         ; Signal a NACK
            call    I2C_IN          ; Read the Low byte in h'02'
            movf    DATA_IN,w       ; Get byte
            movwf   LSB             ; and put in memory
            call    STOP            ; End of Read process

; Now increment 3-byte array --------------------------------
            incf    LSB,f           ; Add one
            btfss   STATUS,Z        ; Is it now zero
             goto   PUT_BACK        ; IF not THEN continue
            incfsz  NSB,f           ; ELSE increment middle byte
             goto   PUT_BACK        ; IF not zero THEN continue
            incf    MSB,f

; Now write back the three bytes ----------------------------
PUT_BACK    call    START           ; Start the Write process
            movlw   b'10100000';    Write state
            movwf   DATA_OUT
            call    I2C_OUT
            clrf    DATA_OUT        ; Address h'00'
            call    I2C_OUT

            movf    MSB,w           ; Get the new High byte
            movwf   DATA_OUT
            call    I2C_OUT
            movf    NSB,w           ; Get the new Middle byte
            movwf   DATA_OUT
            call    I2C_OUT
            movf    LSB,w           ; Get the new Low byte
            movwf   DATA_OUT
            call    I2C_OUT

            call    STOP
```

As soon as the Stop condition is received, the 24LC01B will commence programming the targeted cells with the buffered data. This process takes typically 2–5 ms across the family. If the Master attempts to initiate a process during this time then the EEPROM will not Acknowledge following the Start-Control byte and this can be used as a busy indicator. This polling is shown when the first Control byte is sent out in Program 12.19.

The opposite process of reading bytes from the EEPROM, shown in Fig. 12.27(b), is slightly more involved. As in the previous case an opening address has to be written into the device. After this occurs a repeat Start condition is sent, with the following Control byte having its $R/\overline{W}$ bit High to indicate Master-Read. The Slave EEPROM then transmits the byte at the specified location to the Master, which Acknowledges receipt and the process continues indefinitely with the address incrementing until the Master does not send an Acknowledge. The Slave then releases the bus and the Master is free to issue a Stop condition. If the initial writing of the first address is omitted, then one beyond the last used address is the first location read from.

The software listed in Program 12.19 follows the process outlined in Fig. 12.27 exactly. Once the initial address h'00' has been sent, the Master PIC MCU goes into a listen mode and three sequential bytes are read from memory; terminated by the Master returning a NACK condition followed by Stop. With the triple-byte distance count in locations MSB:NSB:LSB the array is incremented in the usual way. Finally address h'00' is again written out to the EEPROM followed by the three updated bytes and the process terminated by the Master transmitting Stop.

Example 12.4
It is possible to combine some of the attributes of synchronous I^2C and asynchronous signaling to send data asynchronously in both directions half-duplex along a single link. One example of this is the **1-Wire™**[17] interface outlined in Fig. 12.28.

In Fig. 12.28(a) a Maxim Integrated Products/Dallas Semiconductor DS18S20 digital thermometer is shown driven from a single port line with the MCU acting as a 1-Wire Master.

The DS18S20 has the following features.

- Measures temperature from $-55°C$ to $+125°C$ in $0.5°C$ steps as a signed 16-bit datum.
- $\pm0.5\%$ accuracy in the range of $-10°C$ to $+85°C$.
- Converts temperature in 750 ms maximum.
- Zero standby current.
- May be powered from the data line; supply range +3 V to +5.5 V.
- Multidrop capability.

---

[17] 1-Wire™ is a trademark of Maxim Integrated Products/Dallas Semiconductor.

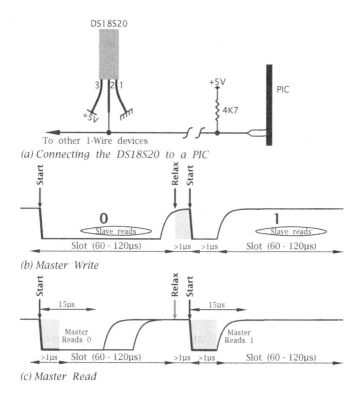

(a) Connecting the DS18S20 to a PIC

(b) Master Write

(c) Master Read

Fig. 12.28 Interfacing the DS18S20 1-Wire digital thermometer.

The various DS18S20 functions, such as Convert (h'44'), Read temperature (h'BE'), are initiated by the Master sending the appropriate data as 8-bit codes, each byte comprising a Start condition (⌐‿) and eight Write slots; as shown in Fig. 12.28(b). As in the I^2C case, the data line DQ is pulled into the High state with a pull-up resistor and the Master simulates the logic 1 state by changing its port line from the Low state to input (see Fig. 12.14(b)). In this state the Master can listen to data sent by the Slave, as shown in Fig. 12.28(c).

For our example we are required to write two subroutines that will respectively write a byte to a 1-Wire Slave and read a byte from the Slave.

Solution

From Fig. 12.28(b) we see that writing a bit to a Slave involves the following tasks:

1. The Master starts the process by forcing the data line Low for at least $1\,\mu s$.
2. The Master either keeps the line Low (Write 0) or releases the line (Write 1) for 60–120 μs.

3. The Slave reads the line state between 15–45 µs later.
4. The Master releases the line (if Write 0) for at least 1 µs to relax the system.

The subroutines of Program 12.20 assume that the port line driving DQ has been set-up as described on page 355 for the I²C bus to give the two states as hard Low and open circuit, pulled up to the High state. Also we assume that we have the delay macro Delay_us in situ which gives a $K\mu s$ delay, where K is the parameter passed to the macro.

```
Delay_us macro K                    ; K is the number of us delay
         local DELAY_US_LOOP
         movlw ((K*XTAL)/d'16')+1; 4˜ (4/XTAL us) per loop: 1˜
DELAY_US_LOOP
         addlw -1                   ; Decrement count : N˜
         btfss STATUS,Z             ; to zero          : N + 1˜
          goto DELAY_US_LOOP        ;                  : 2(N-1)˜
         endm
```

The additional plus one in the formula for loop count ensures that values for K round up.

Both subroutines begin by driving DQ Low for a minimum of 1 µS, defining the Start condition. Writing a single bit to DQ occurs in a slot which has a duration of 60–120 µs, and commences with DQ either Low or released to be pulled High, defining a Write-0 or Write-1 condition. The Slave samples the state of the data line sometime after 15 µs into the slot. Although the duration of the slot is not critical, care needs to be taken as a Low-state duration of between 480 and 960 µs is interpreted by the Slave as a Reset command (see SAQ 12.3).

Eight Write slots are used with a minimum 1 µs relax period interval to transmit the byte, each slot's state following the bit rotated into the Carry flag of the datum byte DATA_OUT. After eight shift/output cycles the process terminates.

Reading from a Slave involves the following tasks:

1. The Master starts the process, forcing the data line Low for at least 1 µs.
2. The Master then listens to data placed on the line by the Slave which is valid for up to 15 µs after the Start edge.
3. The Slave releases the line after 15 µs which should be pulled High by the end of the 60 µs slot.
4. The Master waits a minimum of 1 µs before starting the next slot.

The input subroutine READ_1W follows this task list, sampling the data line sometime before 15 µs into the slot, at which time the Slave's data should have settled to the appropriate voltage level. Each bit is used to appropriately set the Carry flag, which is then shifted into DATA_IN. After eight sample/shift loops, DATA_IN has the received byte datum.

Unlike the I^2C bus, the 1-Wire architecture is designed for a single Master. However, 1-Wire Slaves have device addresses comprising a 64-bit unique code as part of an internal ROM. The first eight bits are a 1-Wire family code — the DS18S20 code is h'10'. The following 48 bits are a unique serial number and the last eight bits are an error-checking byte.

Program 12.20 Reading and writing on a 1-Wire system.

```
;
;  ****************************************************************
;  * FUNCTION: Writes a byte datum to a 1-Wire Slave             *
;  * RESOURCE: macro Delay_us giving N microsecond delay         *
;  * ENTRY    : Datum is in DATA_OUT                             *
;  * EXIT     : DATA_OUT is zero, W, STATUS altered              *
;  ****************************************************************
WRITE_1W  movlw     8            ; Loop count
          movwf     COUNT
W_LOOP    bcf       INDF,DAT     ; Low edge signals Start
          Delay_us  1            ; for 1us
          rrf       DATA_OUT,f   ; LSB first shift into Carry
          btfsc     STATUS,C     ; Was it a 1?
           bsf      INDF,DAT     ; IF it was THEN output High
          Delay_us  d'60'        ; Hold for 60us
          bsf       INDF,DAT     ; Release line to go High
          Delay_us  1            ; Relax for 1us
          decfsz    COUNT,f      ; Repeat eight times
          goto      W_LOOP
          return

;  ****************************************************************
;  * FUNCTION: Reads a byte datum from a 1-Wire Slave            *
;  * RESOURCE: macro Delay_us giving N microsecond delay         *
;  * ENTRY    : None                                            *
;  * EXIT     : Datum is in DATA_IN, W, STATUS altered           *
;  ****************************************************************
READ_1W   movlw     8            ; Loop count
          movwf     COUNT
R_LOOP    bcf       INDF,DAT     ; Low edge signals Start
          Delay_us  1            ; for 1us
          bsf       INDF,DAT     ; Release line
          Delay_us  8            ; Wait 8us for Slave to O/P data
          bcf       STATUS,C     ; Clear Carry
          btfsc     INDF,DAT     ; Check input state
           bsf      STATUS,C     ; IF High THEN set Carry
          rrf       DATA_IN,f    ; Shift bit in -> LSB
          Delay_us  d'48'        ; Wait to end of slot
          decfsz    COUNT,f      ; Repeat eight times
          goto      R_LOOP
          return
```

Self-Assessment Questions

12.1 Rewrite Program 11.5 on page 317 but based on the SPI hardware of Fig. 12.4. *Hint*: Rather than shifting in whole bytes it may be more efficient to simply shift in and test on a bit-by-bit basis.

12.2 Show how you could connect four MAX518 ADCs (see Fig 12.16) on the one I^2C circuit and how channel 1 on the third ADC could be written to.

12.3 Communications along a 1-Wire link begins with a Reset operation where the Master pulls the line Low for 480–960 μs after which the line is released. The Slave then responds by dragging the line Low after no more than 60 μs delay. This Low state persists for a further 60-240 μs after which the Slave releases this line. Design a subroutine that will do this procedure when called. Assume the resources of Program 12.20 are available to you.

12.4 Parity is a technique whereby the number of digits in a word is always either even or odd. This is accomplished by adding an extra bit which is calculated by the transmission software to be 0 or 1 to meet this overall criterion. For instance, for odd parity of an 8-bit word b'01101111' we have b'1 01101111'. The receiver will check that all nine received bits have an odd count. If one bit (or any odd number) has been corrupted by noise, then a **parity error** is said to have occurred.

Based on the PIC MCU USART, write software to set the asynchronous protocol to 9 bit word and calculate the odd one's parity bit of DATA_OUT which should be placed in TX9D of the TXSTA register prior to the loading of the data into TXREG and transmission.

12.5 Rewrite the subroutine GETCHAR of Program 12.14 as an interrupt service routine called GETCH. Compare the two approaches.

12.6 A certain data logger is to sample temperature once every 15 minutes. The power supply current consumption is reduced by using a Low-voltage part at a V_{DD} of 3 V and a crystal of 32.780 MHz. Under these conditions the current consumption with the Timer 1 running is a maximum of 70 μA (45 μA typical). A I^2C EEPROM is to be used to store the data as it is read but is only powered on at sample time—by using a spare port line as the EEPROM's power supply. The logger is to be left submerged at the bottom of a lake for 6 months before being recovered. Can you choose an appropriate 24LCXXX EEPROM and estimate the required capacity of the 3 V battery in mA-hours?

CHAPTER 13

Time Is of the Essence

Of crucial importance in many systems are time-related functions. These may manifest themselves in the measurement of duration, event counting, or control of an external physical event for known periods. An example of the former would be the time between pulses generated by the teeth on a flywheel to measure engine speed for a car dashboard management system; see Fig. 3.8 on page 62.

Where *time is of the essence* these functions are often best implemented by using hardware counters to time events, rather than software delay routines. In this chapter we will look at the various timer modules which are available to the mid-range PIC MCU family. After completion you should:

- Know how a Watchdog timer improves the robustness of a MCU-based system and how to use the integral PIC microcontroller's device.
- Be able to use the basic 8-bit Timer 0 module as both a counter and timer.
- Understand the function of the 16-bit Timer 1 module and its interaction with the Capture/Compare/PWM (CCP) modules.
- Be able to use the 8-bit Timer 2 module together with the CCP modules to generate a pulse-width modulated output.

Many MCU-based systems are hosted in an electrically hostile environment with noise induced outside both through logic lines and the power supply. Our example of an auto dashboard manager is typical of this situation, with induction from the high-voltage ignition sparks and alternator sourced ripple in the battery supply. No matter what precautions in shielding and filtering are taken, it is inevitable that on occasion the MCU will jump out of its proper location in Program memory and "run amok" with potentially serious consequences on the controlled system.[1] In some cases this is little more serious than requiring a manual reset.[2] However, this is not possible in many situations; for instance, in a pacemaker implanted in the patient's body or a Martian space probe.

---

[1] The same can happen as a result of software bugs.
[2] As in a Window's® PC.

One solution to this problem is to use an oscillator/binary counter which resets the processor when the count overflows.[3] If the software is arranged to clear this counter on a regular basis so that overflow never occurs, then the MCU never resets. If something happens and the MCU jumps out of its normal loop then eventually, without this constant clearing, the counter will overflow and the MCU will be reset to its starting point. This circuit is given the name **Watchdog timer**, as it enhances the system security.

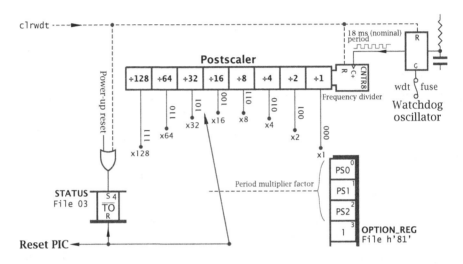

Fig. 13.1 The integral PIC MCU Watchdog timer with Postscaler assigned.

Rather than rely on external Watchdog timers, all PIC MCUs, even the early low-range family, have an integral module, as illustrated in Fig. 13.1. A built-in oscillator *independent* of the processor main clock runs continually if the Watchdog timer is enabled, with a nominal period of 18 ms. The integral capacitor/resistor timing network is not tightly specified and as a consequence the period will vary with device process, temperature and supply voltage from a minimum of 7 ms (lowest temperature −40°C, highest V_{DD} 6 V) up to a maximum of 33 ms (highest temperature +85°C, lowest V_{DD} 2 V); see Fig. 15.8 on page 500.

The Watchdog oscillator is followed by an 8-bit **Postscaler** counter. This can be set to give Watchdog counter overflow time-out periods rising in powers of two up to nominally $18 \times 128 \approx 2.3$ s; 0.9 s minimum, 4.2 s maximum. The actual value is selected by the programmer using the **PS[2:0]** (Pre-/PostScaler rate Select) bits in the **Option register (OPTION_REG)**; see Fig. 13.2. The Watchdog oscillator and Postscaler counter

---

[3]Other approaches typically are based on a retriggerable monostable.

(which we will call the Watchdog timer chain) are both reset with the **clrwdt** (**CLeaR WatchDoG Timer**) instruction, and issuing this at regular intervals is the mechanism whereby the programmer prevents time-out.

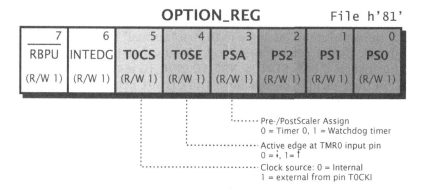

Fig. 13.2 The Option register.

The Watchdog timer Postscaler is a *shared* resource with Timer 0 (see Fig. 13.3) and cannot be used simultaneously for both functions. The **PSA** (**Pre-/PostScaler Assignment**) bit is used to assign this resource either to the Watchdog timer or to Timer 0. On reset the Postscaler defaults to the Watchdog timer and the time-out period is set to ×128.

Microchip regard the Watchdog timer as a system resource, rather than as a peripheral interface module. For this reason the user must enable the Watchdog timer via the **WDTE** configuration fuse when blasting in code to the Program store; as described in Fig. 10.6 on page 281. For instance:

```
__config _HS_OSC & _WDT_ON & _PWRTE_OFF & _CP_OFF
```

or the equivalent that is appropriate to your **C** compiler; for instance, in the CCS compiler:

```
#fuses HS, WDT, NOPUT, NOPROTECT
```

The default is to activate the Watchdog facility. The script for no Watchdog is _WDT_OFF and NOWDT, respectively.

Given a Watchdog-enabled PIC, a Power-on reset will initialize the Watchdog timer chain and will deactivate (set to 1) the $\overline{\text{TO}}$ (**Time Out**) bit in STATUS[4]; see Fig. 4.6 on page 78. After the ordained time the Watch-dog counter chain will overflow and the $\overline{\text{TO}}$ bit will be activated (cleared to 0) as listed in Table 10.4 on page 289. Most microcontroller software runs as an endless task, with various routines being actioned inside this loop. Ensuring that in normal operation a clrwdt instruction will always be executed no matter what sequence of events occur will ensure that time-out will never happen. If something untoward does happen and it times out, then the PIC MCU will automatically reset and start all over again at the Reset vector h'000'. However, this type of reset will not alter the setting of the $\overline{\text{TO}}$ flag, which can be checked if necessary by the software to distinguish from a "normal" restart. The $\overline{\text{TO}}$ flag is read-only; that is, cannot be set to 1 (deactivated) by a normal instruction; for instance, bsf STATUS,NOT_TO (NOT_TO is the Microchip name for $\overline{\text{TO}}$ in their Include header file). The clrwdt instruction will deactivate $\overline{\text{TO}}$ (and also the $\overline{\text{PD}}$ flag activated by the sleep instruction) and of course also restart the Watchdog timer chain.

As an example, consider a system counting cans of beans moving along a conveyer belt in the manner shown in Fig. 13.4, keeping a tally in a File called BEAN_COUNT. On Power-on this tally is to be zeroed. If due to a glitch the PIC MCU's Watchdog times out and the PIC MCU resets, then this tally should be left unchanged. To do this we can use code to check the state of $\overline{\text{TO}}$ and take the appropriate action; for instance:

```
        __config _WDT_ON        ; Enable the Watchdog timer
        org     h'000'          ; The Reset vector
MAIN    btfss   STATUS,NOT_TO ; Was this a Watchdog reset?
        clrf    BEAN_COUNT      ; IF not THEN zero tally
; more initializing code -------------
        clrwdt                  ; Set NOT_TO and reset Watchdog
```

Strictly the clrwdt instruction doesn't need to be executed for a Power-on reset initialization. However, the set-up process will add to the time it takes to get into the normal main code and the normal loop clrwdt; thus an extra clrwdt at the end of any initializing code is useful insurance.

As the Watchdog oscillator is completely independent of the main system clock, it continues to run even if the processor is placed in its Sleep state. To facilitate this, the sleep instruction clears the Watchdog chain and deactivates the $\overline{\text{TO}}$ flag. In addition it also activates the $\overline{\text{PD}}$ (Power Down) flag in STATUS[3] to show that the processor is asleep. In this way a complete Watchdog timer period will elapse after a sleep instruction before time-out. If the Watchdog times out while the PIC MCU is asleep, then it will waken and continue with the following instruction.[4] Normally this will be a clrwdt instruction.

If necessary, the software can determine that the time-out occurred during a Sleep state by examining the $\overline{\text{TO}}$ and $\overline{\text{PD}}$ flags. In the case of the low-range PIC MCU family, Watchdog time-out was the only way of awakening a sleeping device, short of a Manual reset. Mid- and extended-range devices can also be awakened by an external interrupt, and where applicable the Timer 1 and Analog modules, both of which have the option

---

[4]After the normal 1024 clock cycle delay as described on page 277.

of a separate local clock oscillator. Where the Watchdog is enabled and used in the Sleep mode, the designer should be aware that the current consumption (PIC16F87X with $V_{DD} = 3$ V, $-40°$C to $+85°$C) rises from typically 0.9 μA (5 μA maximum) to 7.5 μA (30 μA maximum). If long-term battery operation is required (for instance, see SAQ 12.6 on page 400) this presents serious problems. Running the processor continually at the lower frequency of 32.768 kHz and 3 V power supply with the Watchdog disabled takes typically 20 μA (35 μA maximum) in comparison.

The low-range family features a basic 8-bit counter/Prescaler which was originally called a Real-Time Clock/Counter. Although this term was still used in the early 14-bit PIC16CXXX data sheets, the introduction of additional timers led to the more consistent term **Timer 0** (**TMR0**). However, the term RTCC is still to be found as a relic in older textbooks and software. For instance, the CCS **C** compiler sets the Timer 0 clock source to external Low-going edge with a Prescale value of 4 by calling `setup_counters(rtcc_ext_l_to_h, rtcc_div_4);`.[5]

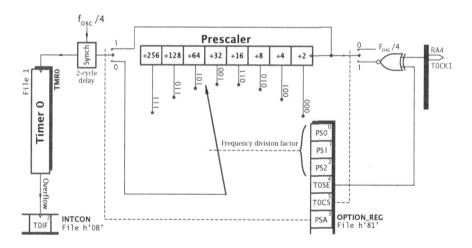

Fig. 13.3 Simplified equivalent circuit for Timer 0.

From Fig. 13.3 we see that Timer 0 chain comprises a primary 8-bit counter located at **File 1** in series with an optional 8-bit Prescaler counter. This gives eight selectable clock rates into the primary counter, as selected by the three PS[2:0] bits in OPTION_REG. This Timer 0 Prescaler is

---

[5]The same function can be used to set up the Watchdog Postscaler; for instance, `setup_counters(WDT_288MS);`.

actually the same circuit as the Watchdog's Postscaler[6] and PSA in OP-TION_REG[3] must be set to 0 to assign the Prescaler to Timer 0.

The Prescaler is assigned to the Watchdog timer by default on a Power-on reset, and in this situation the primary counter is either clocked by the internal instruction cycle clock $f_{OSC}/4$ (that is, at a rate XTAL/4) or from an external source via the **Timer 0 Clock Input RA4/T0CKI** (or **GP5/T0CKI**) pin. The **Timer 0 Clock Select T0SC** bit at OPTION_REG[5] is used to select the internal/external mode. When clocked from outside, the active edge is set using the **Timer 0 Set Edge T0SE** bit at OPTION_REG[4].

An event at the T0CKI pin will appear to be random with respect to the internal clock cycle. In order to allow Timer 0 to be read from or written to in the normal way without interference from such an action, a synchronization stage is necessary. This is done using a 2-stage shift register before the Timer 0 primary counter's clock input. This causes a delay of two instruction cycles; 1 μs with an 8 MHz crystal. Where the primary counter is directly connected to the internal clock, this will cause a 2-cycle delay before anything happens after a datum is written into Timer 0 at File 1.

When Timer 0 overflows (11111111 → 00000000) the **Timer 0 Interrupt Flag T0IF** is set. This event can be sensed by polling INTCON[2] or if the **Timer 0 Interrupt Enable** mask bit **T0IE** in INTCON[5] is set, an interrupt will automatically be generated; see Fig. 7.3 on page 191.

An external clock signal going directly into Timer 0 should be in its High state for at least $2t_{OSC} + 20$ ns and Low for at least the same time. Thus for a 8 MHz crystal ($t_{OSC} = 125$ ns), T_{high} should be at least 270 ns and the same for T_{low}; a maximum counting frequency of 1.8 MHz. When the Prescaler is used, this minimum total period of $4t_{OSC} + 40$ ns can be divided by the Prescaler ratio. The Prescaler input waveform is subject only to a minimum 10 ns pulse width. Thus with a $\div 256$ setting, a nominally 50 MHz signal at T0CKI is acceptable.

The Prescaler is not readable; so the timer is not strictly a 16-bit counter. Reading Timer 0 does not affect the Prescaler but any instruction writing to it (e.g., clrf 1, movwf 1) will both *clear* the Prescaler and the clock synchronizer as well as Timer 0 itself.

As the Prescaler is assigned to the Watchdog on reset, it can be subsequently changed over by clearing **PSA**. However, it is possible that this change-over could cause a Watchdog reset even if it is disabled. Microchip therefore recommend that the change-over be preceded by a clrwdt instruction; as shown in the following code snippet which shows the Prescaler dividing the clock input by four from the T0CKI pin, and incrementing on a ‾_ edge.

---

[6]The PIC18XXXX extended-range family uses physically separate scalers for the Watchdog timer and Timer 0.

```
clrwdt              ; Clears Prescaler and wdt
bsf     STATUS,RP0  ; Change-over to Bank1
movlw   b'11110001' ; External clock on Low-going edge
movwf   OPTION_REG  ; 1:4 Timer0 Prescaler assigned to Timer0
bcf     STATUS,RP0  ; Back to Bank0
```

It is also possible to change the Prescaler over from Timer 0 to the Watchdog timer "on the fly." In the same manner the clrwdt should be executed before altering OPTION_REG to avoid a spurious Watchdog reset.

Timer 0 is mainly used either to *count* external events or to measure the *time* between external events. It can also be used to time software toggling port pins for precisely known durations, without tying up the processor in time-wasting delay routines.

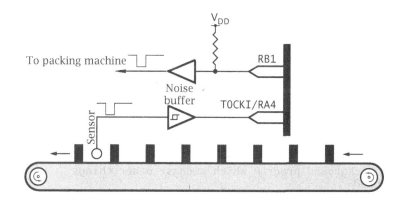

Fig. 13.4 Counting cans of beans on a conveyer belt.

We will illustrate the usage of Timer 0 as an *event counter* and stop clock with two examples. The first, which also illustrates the Watchdog timer, is to tally cans of baked beans traveling along a conveyer belt, as shown in Fig. 13.4. Each 24 cans passing the sensor should generate a pulse to a packing machine, so that the box can be replaced by a new empty container. This pulse need only be a few microseconds in duration. A double-byte count should also to be kept of the number of boxes packed since the last Power-on/Manual reset. This will be uploaded to the central plant computer at the end of the shift for inventory control.

Our first consideration is the set-up and initialization code. This code, shown below, begins by checking the $\overline{\text{TO}}$ flag at the Reset vector. If zero then the bulk of the initialization code is omitted, as reset was due to a Watchdog time-out. If this was not the case then port pin TOCKI is set-up as an input and RB1 is set-up as an output to activate the packing machine.

```
        include  "p16F877a.inc"
        __config _WDT_ON      ; Enable Watchdog

        cblock h'20'
        _work:1, _status:1
        COUNT:2
        endc

        org     0             ; Reset vector
        btfss   STATUS,NOT_TO ; Check if a Watchdog reset
         goto   MAIN_LOOP     ; IF yes THEN no initialization
        goto    MAIN          ; ELSE a fresh start

        org     4             ; Interrupt vector
        goto    ISR           ; Foreground program

MAIN bsf     PORTB,1         ; Idle state of the Packing pulse
     bsf     STATUS,RP0      ; Change to Bank1 and
     bsf     TRISA,4         ; make sure that TOCK1 is I/P pin
     bcf     TRISB,1         ; & RB1/Packing machine an O/P pin
     movlw b'00101111'       ; Timer source external falling edge
     movwf OPTION_REG        ; Prescaler assigned to wdt
     bcf     STATUS,RP0      ; Back to Bank0

     movlw -d'24'            ; Initialize TMR0 to -24 (h'E8')
     movwf TMR0
     clrf  COUNT+1           ; Clear the 2-byte score count
     clrf  COUNT
     bsf   INTCON,T0IE       ; Enable Timer0 interrupt
     bsf   INTCON,GIE        ; Enable all interrupts

; The background program which amongst other things
MAIN_LOOP
     clrwdt                  ; Regularly resets the wdt
     ...   .....             ; More background code
```

Also in Bank 1, the Option register is set up to assign the Prescaler to the Watchdog and extend its time-out period by ×128. The Timer 0 counter chain is set to be clocked from T0CKI on a negative-going edge. Finally, back in Bank 0, Timer 0 itself is set to h'E8' (i.e., −24 decimal) so that 24 can pulses will cause it to overflow and cause an interrupt. Both INTCON flags T0IE and GIE are then set to enable the interrupt.

The main background program commences with a clrwdt instruction. Provided that the background endless loop is no longer than $7 \times 128 = 0.8961$ s, the minimum Watchdog period, then time-out will not occur.

With the initialization code in situ, all that remains is to implement the interrupt service routine (ISR) that will be automatically entered after each batch of 24 cans; that is, when Timer 0 counts up 24 input pulses and overflows back to zero. When this occurs, Timer 0 will set T0IF and the PIC MCU will jump to the Interrupt vector at h'004'. In our initialization code we have placed a goto ISR at this point, and so named the routine in Program 13.1.

Program 13.1 The bean counter interrupt service routine.

```
; **********************************************************
; * The ISR to issue a Packing-machine pulse and          *
; * re-initialize Timer0 to -24.  Also keeps a grand score *
; * total in COUNT:2 for background analysis               *
; **********************************************************
; First save context in usual way
ISR     movwf   _work       ; Put away W
        swapf   STATUS,w    ; and the Status register
        movwf   _status


; =============================================================
; The core code
        btfss   INTCON,TOIF ; Was it a can?
        goto    ISR_EXIT    ; IF no THEN false alarm

        bcf     PORTB,1     ; Pulse packing machine
        movlw   -d'24'      ; Re-initialize Timer0
        movwf   TMR0
        incf    COUNT+1,f   ; Add one to score count
        btfsc   STATUS,Z
        incf    COUNT,f
        bcf     INTCON,TOIF ; Reset interrupt flag
        bsf     PORTB,1     ; End packing machine pulse
; =============================================================

ISR_EXIT swapf  _status,w   ; Untwist the original Status reg
        movwf   STATUS
        swapf   _work,f     ; Get the original W reg back
        swapf   _work,w     ; leaving STATUS unchanged
        retfie              ; and return from interrupt
```

The ISR itself is sandwiched between the normal context switching wrapper described in Program 7.2 on page 202. The core simply implements the following task list in no particular order:

- Toggle RB1 to signal the packing machine.
- Reset Timer 0 to -24.
- Increment the double-byte score count.
- Reset the Timer 0 interrupt flag TOIF.

In Program 13.1 TOIF is tested on entry and if not set the ISR is exited. If there are other sources of interrupt then the switch would be to another part of the ISR, as shown in the listing on page 196.

For an alternative approach using Hardware interrupts see Program 7.2 on page 202.

Our second example illustrates the use of Timer 0 as a *clock* to measure time between events. The events in question are R-points peaks in the ECG waveform illustrated in Fig. 7.1 on page 186. there a peak detector interrupts the MCU, which keeps a 2-byte count from a 10 kHz external oscillator. In this manner the period between events can be determined on each event in increments of $100\,\mu s$, which we call here **jiffies**. For our example we will modify the specification to eliminate this oscillator and use Timer 0 to keep a 1 ms 2-byte Jiffy tally.

For this task we need to use the main PIC MCU oscillator as the clock source which together with the Prescaler will cause Timer 0 to overflow (counts 256) once per millisecond ($1000\mu s$). If we choose a 4.096 MHz crystal we have:

$$\text{Time-out} = 1000\,\mu s = \frac{4}{4.096} \times 256 \times \text{Prescaler ratio}$$

which gives a required Prescaler ratio of 1:4.

With these requirements in mind we have for our initialization background software:

```
        org     0           ; Reset vector
        goto    MAIN        ; Background program

        org     4           ; Interrupt vector
        goto    ISR         ; Foreground program

MAIN    clrwdt              ; Reset Watchdog timer
        bsf     STATUS,RP0  ; Change to bank1
        movlw   b'00000001' ; INT on -ve edge, internal clock
        movwf   OPTION_REG  ; Prescale div4 assigned to Timer0
        bcf     STATUS,RP0  ; Back to Bank0

        clrf    NEW         ; Zero the New flag
        bsf     INTCON,T0IE ; Enable Timer0 interrupt
        bsf     INTCON,INTE ; Enable Hardware interrupt
        bsf     INTCON,GIE  ; Enable all interrupts
```

As well as enabling the Timer 0 interrupt, INTE is set to enable Hardware interrupts from the INT pin, which is going to signal an ECG peak. Neither the double-byte Jiffy count nor Timer 0 need be cleared as the first reading of the series will always be erroneous—because the patient's heartbeat is not synchronized to the PIC MCU reset! However, File NEW, which is set to non-zero each time an ECG peak is detected, is cleared.

The core of the ISR shown in Program 13.2 implements the following task list when an interrupt is received:

1. IF Timer 0 interrupt.
   - Increment 2-byte Jiffy count.
   - Reset Timer 0 interrupt flag T0IF.
   - Return from interrupt.
2. ELSE a Hardware interrupt from peak picker.
   - Copy Jiffy count into general-purpose Files.
   - Zero Timer 0.
   - Set New indicator.
   - Reset Hardware interrupt flag.
   - Return from interrupt.

Program 13.2 Measuring the ECG waveform period to a resolution of 1 ms.

```
; **********************************************************
; * The ISR to increment the 2-byte COUNT IF TMR0 interrupts  *
; * Copies COUNT:2 to DATA:2 if an INT interrupt and sets NEW *
; * to show background prog that new data is available        *
; **********************************************************
;
; First save context in usual way
ISR       movwf   _work        ; Put away W
          swapf   STATUS,w     ; and the Status register
          movwf   _status
; =============================================================
; The core code
          btfss   INTCON,T0IF  ; Was it a heartbeat?
          goto    HEART_BEAT   ; IF yes THEN go to it

          incf    COUNT+1,f    ; Record one more 1ms jiffy
          btfsc   STATUS,Z
          incf    COUNT,f      ; Overflow to upper byte
          bcf     INTCON,T0IF  ; Clear interrupt flag
          goto    ISR_EXIT

HEART_BEAT                     ; Land here if ECG peak
          movf    COUNT+1,w    ; Get new period count LSB
          movwf   DATUM+1      ; Copy into data area
          movf    COUNT,w      ; Get MSB
          movwf   DATUM
          clrf    COUNT+1      ; Zero Jiffy count
          clrf    COUNT
          btfsc   INTCON,INTF  ; Reset interrupt flag
          incf    NEW,f        ; Tell world there is new data
; =============================================================
ISR_EXIT  swapf   _status,w    ; Untwist the original Status reg
          movwf   STATUS
          swapf   _work,f      ; Get the original W reg back
          swapf   _work,w      ; leaving STATUS unchanged
          retfie               ; and return from interrupt
```

Both bytes in COUNT:COUNT+1 are copied into the two Files called DATUM:DATUM+1 when a Hardware interrupt is received and the Jiffy count and Timer 0 are then zeroed ready for the next event. When the background program polls File NEW and finds a non-zero datum, then it knows that a fresh count is ready for collection. It then, for instance, could send it to a serial EEPROM as in Example 12.3 on page 391 or down a serial link to a PC for subsequent processing and display.

Most mid- and extended-range PIC MCUs have at least two additional timer/counters and associated circuitry with the following properties.

Timer 1
This 16-bit counter has its own optional dedicated oscillator and programmable Prescaler. Its state can be sampled by an external event and
it can control the state of a pin when it reaches a predefined value.

Timer 2
This 8-bit counter has both programmable Pre- and Postscaler functions.
Its count length can be set by the programmer and it may be used to generate a pulse-width modulated output with no on-going software overhead.

Capture/Compare/PWM
Both timers can be used in conjunction with additional logic called Cap-
ture/Compare/Pulse Width Modulation (CCP) to implement the Timer 1
sample instant (Capture), the Timer 1 roll-over value (Compare), and the
automatic PWM generation from Timer 2.

Timer 1 comprises a primary 16-bit counter implemented as a pair
of Files with the lower byte named **TMR1L** and with **TMR1H** holding the
high byte. When this counter overflows, then the Interrupt Flag **TMR1IF**
in the Peripheral Interrupt Register 1 PIR1[0] is set.

The source of the counting pulses may be external to the device; either
events at the **T1CKI** (**Timer 1 ClocK Input**) or from a dedicated Timer 1
oscillator. Alternatively, the internal system oscillator $f_{OSC}/4$ may be selected as the counting source. In all cases, the counting pulse train can
be divided down with a Prescaler counter. External counting pulses can
be optionally synchronized to the system oscillator. Gatable versions of
the module can externally disable this pulse train via the $\overline{\text{T1G}}$ (**Timer 1
gate**) pin.

The **Timer 1 CONtrol register TMR1CON**, shown in exploded form in
Fig. 13.5, is used to select the various features of the Timer. All bits in
this Control register are 0 on Power-on/Brown-out resets, which initially
disables Timer 1 and its external oscillator, with a Prescaler value of 1:1
and system clock used as the source.

TMR1ON
Setting T1CON[0] to 1 enables the Timer. In some cases[7] the Timer 1
related pins are then automatically set as input, overriding any TRIS settings and in others[8] the program must ensure that such pins are explicitly
set-up as input.

TMR1CS, T1OSCEN
Timer 1 can be configured to measure time from the internal system clock
if the **TiMeR 1 Clock Select** switch bit in T1CON[1] is 1 or else use an
external source of counting pulses.

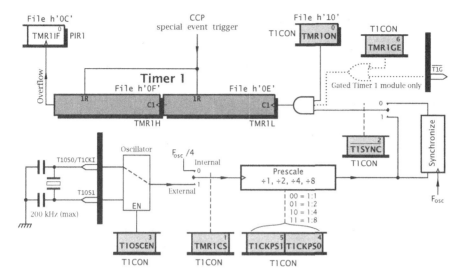

Fig. 13.5 Functional equivalent circuit for Timer 1.

External events can be a _／‾ rising edge on the T1CKI pin, or else if the **Timer 1 OSCillator Enable** switch bit in T1CON[3] is 1 a "private" oscillator separate from the main PIC MCU oscillator. This avoids having to pick the main crystal to suit the timer, as we did in our Timer 0 ECG peak picker of page 410. This external oscillator is timed with a crystal across the T1OS0/T1CKI and T1OSC1 pins, with a maximum value of 200 kHz. Typically a 32.768 kHz (2^{15}) watch crystal is used.

T1CKPS[1:0]
Whatever the source of counting pulses, the primary 16-bit counter may be incremented either directly or on every second, fourth, or eighth event. This is controlled by the setting of T1CON[5:4], as listed in the diagram.

Timer 1 will overflow and set the TMR1IF interrupt flag after $2^{16} =$ 65, 536 counter events from zero. This in turn can be used to interrupt the processor if the paired TMR1IE mask bit in the Peripheral Interrupt Enable 1 register is set (see Fig. 7.5 on page 199), or else polled. In either case the program should manually clear the TMR1IF flag once overflow has been detected.

For instance, if a 32.768 kHz watch crystal is used, then Timer 1 will overflow in two seconds if T1CKPS[1 : 0] = 00 and every 16 seconds if T1CKPS[1 : 0] = 11.

T1SYNC
Output from the programmable Prescaler is by default synchronized to the system clock, giving a 2-instruction cycle delay. However, unlike Timer 0 this synchronization shift register can be bypassed with T1CON[2] set to 1. The asynchronous mode allows Timer 1 to be used with an ex-

ternal count source when the PIC MCU is asleep. As the synchronizer shift register is clocked from the system clock f_{OSC}, which is switched off when in the Sleep mode, a bypass is necessary in this situation. The asynchronous mode also needs to be used when the external count pulse rate is faster than the system clock $4 \times t_{OSC}$, as the synchronizer will then miss some events.

Apart from these cases, $\overline{T1SYNC}$ should be 0, as the lack of synchronization can lead to an unpredictable outcome if software attempts to write to Timer 1 at the same time as the random external event tries to increment the timer. If the Timer 1 state is to be updated in the asynchronous mode, then it should be disabled and thus stopped by clearing TMR1ON during this process. For instance, to change the state of Timer 1 to h'8000':

```
movlw  h'80'         ; New high byte
bcf    T1CON,TMR1ON  ; Stop the timer
movwf  TMR1H         ; Set Timer1 to 8000
clrf   TMR1L
bsf    T1CON,TMR1ON  ; Restart the timer
```

Altering the state of Timer 1 will always clear the Prescale counter.

If Timer 1 is in the synchronized mode, then its value may be altered "on the fly," but care has to be taken when both bytes are to be changed, as in some cases after altering the High byte the Low byte may overflow before the software updates its value and unintentionally alter the high byte. To avoid this, the low byte should first be cleared; for instance to alter the state of Timer 1 on the fly to h'9FFF' we have:

```
movlw  h'9F'    ; Value destined for high byte
clrf   TMR1L    ; Ensure Low byte will not overflow
movwf  TMR1H    ; Update High byte of count
movlw  h'FF'    ; Value for Low byte
movwf  TMR1L    ; Final Low byte in place
```

Timer 1 can be read at any time even if in the asynchronous mode. However, as only one byte can be read at a time[9] it is possible that the timer may have overflowed from lower to higher byte in between the two reads; for instance:

```
; Assume Timer1 is at state h'80FF'
    movf   TMR1L,w  ; Read low byte = h'FF'
    movwf  TEMPL    ; Store away
; «« Timer 1 now increments to state h'8100' »»
    movf   TMR1H,w  ; Get high byte = h'81'
    movwf  TEMPH    ; Store away giving overall h'81FF'!!
```

[9]In the PIC18XXXX family reading one of the bytes of Timer 1/3 automatically makes a copy of the other byte in a temporary register, effectively giving a single 16-bit time sample.

erroneously reads the state as h'81FF' instead of h'80FF'. This is even more likely to occur if another peripheral device interrupts between the two reads.

A predictable reading can be obtained by either stopping Timer 1 before taking the readings or by reading the high byte first and then checking after the low byte has been read that the high byte has not changed. As an example of the latter technique we have:

```
T1_GET movf   TMR1H,w   ; Get High byte
       movwf  TEMPH     ; Store away
       movf   TMR1L,w   ; Get Low byte
       movwf  TEMPL     ; Store away
; Now check to see if High byte has changed
       movf   TMR1H,w   ; Get High byte again
       subwf  TEMPH,w   ; Check to see if same
       btfss  STATUS,Z  ; Skip if difference is zero
       goto   T1_GET    ; ELSE try again
```

In either case, where an accurate reading is required, it is advisable to disable interrupts by clearing the GIE Global mask bit, until both bytes have been successfully read. An alternative method of reading the two bytes in one cycle is discussed on page 418.

When the internal system clock is selected (TMR1CS = 0), synchronization is not necessary. In this case the state of $\overline{T1SYNC}$ is ignored.

TMR1GE, T1GINV

Some newer devices have a Gate control version of the Timer 1 module; that of the PIC12F675 is shown in Fig. 13.5. With T1CON[6] set to 1 counting can be externally halted by bringing pin $\overline{T1G}$ to its High state.

Other versions of the Gateable Timer 1 module use switch bit T1CON[7] to select which level of voltage on $\overline{T1G}$ is active. This version of the module (e.g., PIC16F684) optionally allows Comparator 2 (see Fig. 14.6 on page 443) to switch this stream on and off, rather than a voltage on $\overline{T1G}$. This allows the timing of an analog event to be measured.

For our example assume that we require a low-power temperature logger that will read the sensor and transmit its value back to base once every 15 min. It is proposed that Timer 1 be used to action this process and that the Timer 1 oscillator with a 32.768 kHz watch crystal is to give the timebase.

As the maximum possible overflow time is only 16 s (see page 413) we need to keep a count of overflows to record 900 s in total. Setting the overflow period to 4 s gives us a Jiffy count requirement of $\frac{900}{4} = 225$ to record our 15 min. Thus our set-up and main skeleton software would be something like that shown in Program 13.3. Here Timer 1 is set up to use the external oscillator with a Prescaler ratio of 1:2, giving our 4 s Jiffy.

Program 13.3 Generating a 15-min data logger timebase.

```
            include  "p16f877a.inc"
            __config _WDT_OFF & _CP_OFF

            cblock  h'20'
            JIFFY:1
            endc

MAIN        movlw   b'00011111'  ; Timer on, external clock, asynch
            movwf   T1CON        ; Extern osc enabled, PS ratio 1:2

            clrf    JIFFY        ; Zero Jiffy count

            bsf     STATUS,RP0   ; To Bank1
            bsf     PIE1,TMR1IE  ; Enable the Timer1 interrupt
            bcf     STATUS,RP0   ; Back to Bank0

DOZE        sleep                ; Slumber & wait for Timer1 interrupt
            bcf     PIR1,TMR1IF  ; Zero the interrupt flag
            incf    JIFFY,f      ; Record one more Jiffy
            movlw   d'225'       ; Check, 225 Jiffies = 15 minutes?
            subwf   JIFFY,w
            btfss   STATUS,Z
             goto   DOZE         ; IF not THEN go back to sleep
            clrf    JIFFY        ; ELSE reset Jiffy count
            call    SAMPLE       ; Sample temperature and transmit
            goto    DOZE         ; and go back to sleep
```

In order to reduce power consumption the PIC MCU is to be in its Sleep mode, and will be woken up every four seconds. To facilitate this, the TMR1IE mask bit in PIE1[0] is set to 1. Once the PIC MCU is awake, the TMR1IF interrupt flag is cleared and one is added onto the Jiffy count. This is tested for 225 and if equal, then it is zeroed and the subroutine to transmit temperature to base is called.

It should be noted that an enabled, Timer 1 adds something of the order of $20\,\mu A$ current drain, which is a consideration that is especially important if it is intended to use Timer 1 to waken the processor from a low-current Sleep state which typically only uses $0.9\,\mu A$ (all figures for the PIC16F87X).

Associated with Timer 1 (and as we shall see later, Timer 2) are one (e.g., PIC16F62X) or two (e.g., PIC16F87X) **Capture/Compare/Pulse-Width Modulation** (**CCP**) modules. Each CCP module essentially comprises a 16-bit register to match the double-byte Timer 1 and a 16-bit digital comparator to check for equality between the Timer state and the contents of the CCP register. As module CCP1 and CCP2 are virtually identical, sharing the same Timer 1 time-base,[10] but with separate input/output pins **CP1** and **CP2**, we will focus on CCP1. Any differences will be noted as appropriate.

A CCP module has three main functions.

- When configured in a **Capture** mode, an outside event on the associated CCP pin causes the state of Timer 1 to be copied into the CCP register. This can be used to derive the time or duration of this event, to a resolution down to 12.5 ns.
- Configured in a **Compare** mode; when the state Timer 1 equals that in the CCP register, the state of the associated CCP pin is changed or Timer 1 is reset. This can be used to generate a precisely timed event in hardware with a 200 ns resolution.
- When configured in a **PWM** mode, a CCP module, in conjunction with Timer 2, can generate by hardware a pulse-width modulated output with a variable period and duty cycle of up to 10-bit resolution (0.1%).

In all cases involving Timer 1, a synchronized clock must be used to guarantee correct operation; that is, $\overline{T1SYNCH} = 0$.

Each CCP module has an associated Control register used to set the mode. In all cases the appropriate CCP pin needs to be explicitly set-up as an input or output as appropriate.

The Capture mode is illustrated in Fig. 13.6. The various submodes are:

0000

On a Power-on/Brown-out reset all bits are zeroed. This turns off the associated CCP module and clears the Prescaler. One way of avoiding spurious interrupts when changing mode is to turn off the module before making the change.

0100

On a ‾_ edge at the CCP1 pin both Timer 1 bytes are simultaneously copied into the CCPR1H:L pair of Files. At the same time the CCP1 Interrupt Flag **CCP1IF** is set in PIR1[2] and if the **CCP1IE** mask bit in PIE1[2] is set, an interrupt will be generated.

The CCP2 module is the same, but there is no room for the CCP2 Interrupt Flag **CCP2IF** and its associated **CCP2IE** mask bit in the PIR1/PIE1 registers. Instead PIR2[0] and PIE2[0] are used, and in many mid-range processors is the only occupant of these registers.

0101

The time capture described above is triggered on a _／‾ edge at the relevant CCP pin.

0110

Capture is actioned after four rising edges at the CCP pin.

0111

Capture occurs after 16 rising edges at the CCP pin.

---

[10]In the PIC18XXXX extended-range family, Timer 1 and Timer 3 can be used to give each CCP module its own time-base.

Once a defined event has taken place, the processor can read this frozen value—that is, the time—either in an ISR or when the appropriate CCPIF flag is polled as a 1. If Timer 1 is reset after each capture, then the sampled datum is the time since the last event. Alternatively, as Timer 1 continues to increment, its captured value can be subtracted from the previous reading to give the difference. As the mode may be altered on the fly, the time between rising and falling edge on CCP1 can be measured by toggling CCP1M[0] between captures. This may cause the CCP1IF interrupt flag to be set. To prevent false interrupts, CCP1IE should be cleared before the change-over and CCP1IF after the change-over. Alternatively, the CCP1 module can be used to capture the rising edge and CCP2 the falling edge; see Example 13.3.

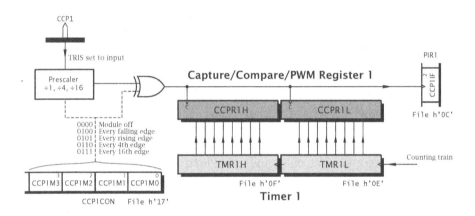

Fig. 13.6 Capturing the time of an event.

Although it seems perverse, if the pin CCP pin is set as an output, then under program control a capture can take place by altering the state of this pin from inside. This capture can be used to time an internal event, or even as a non-orthodox way of doing a simultaneous 2-byte read of Timer 1. The outcome can later be read from the CCP Files without the problems associated with two time-spaced single File reads.

As our example, consider that we wish to measure the period of our ECG signal with the peak detector of Fig. 7.1 on page 186 connected to pin CCP1. If we assume Timer 1 is synchronously clocked by its own 32.768 kHz watch crystal, our set up code is something like this:

```
movlw    b'00001011'  ; Timer on, external clock, synched
movwf    T1CON        ; Oscillator enabled, PS ratio 1:1

movlw    b'00000100'  ; Capture mode, event = falling edge
movwf    CCP1CON

clrf     NEW          ; Zero NEW flag

bsf      STATUS,RP0   ; To Bank1
bsf      PIE1,CCP1IE  ; Enable the CCP1 interrupt
bcf      STATUS,RP0   ; Back to Bank0
bcf      PIR1,CCP1IF  ; Ensure that interrupt flag is zero
bsf      INTCON,PEIE  ; Enable Timer/CCP interrupts
bsf      INTCON,GIE   ; Global interrupts enabled
```

The ISR simply reads the contents of the CCP register and stores it away in two temporary locations, setting File NEW to indicate to background program that a new time datum exists. Timer 1 is then cleared ready for the next event.

With a crystal of 32.768 kHz the time resolution of the captured datum is 30.5 μs with our 1:1 Prescale setting. Timer 1 will overflow in 2 s, which is sufficient to record a heart rate down to 30 beats per minute.

Program 13.4 Capturing the instant of time an ECG R-point occurs.

```
; ******************************************************************
; First save context in usual way
ISR      movwf    _work        ; Put away W
         swapf    STATUS,w     ; and the Status register
         movwf    _status

; =================================================================
; The core code
         btfss    PIR1,CCP1IF  ; Was it a CCP1 interrupt?
         goto     ISR_EXIT     ; IF no THEN false alarm

         incf     NEW,f        ; Signal a new capture
         bcf      PIR1,CCP1IF  ; Reset interrupt flag
         movf     CCPR1L,w     ; Get captured low byte
         movwf    TEMP+1       ; Store away
         movf     CCPR1H,w     ; Get captured high byte
         movwf    TEMP         ; Store away
         clrf     TMR1L        ; Zero Timer1
         clrf     TMR1H

; =================================================================

ISR_EXIT swapf    _status,w    ; Untwist the original Status reg
         movwf    STATUS
         swapf    _work,f      ; Get the original W reg back
         swapf    _work,w      ; leaving STATUS unchanged
         retfie                ; and return from interrupt
```

A more robust software system would also enable the Timer 1 over-flow interrupt. If this occurs it indicates that the subsequent captured data will be invalid—although time-outs can be counted and thus extend the validity of the captured time. However, in our system it is more likely to be used to set off an alarm!

Modes 1000–1011 listed in Fig. 13.7 give four **Compare modes**. Here a digital equality comparator detects when the 16-bit Timer 1 datum equals the setting in the 16-bit **CCPR1H:L** (**CCP Register 1**). When an equality match occurs, the CCP1IF interrupt flag in PIR1[2] will be set and this can cause an interrupt if the corresponding **CCP1IE** mask bit in PIE1[2] is set.

Besides setting CCP1IF, depending on the setting of the CCP1M[3:0] mode bits, one of four actions are possible on Timer 1 matching CCPR1:

1000: Set Output Pin on Match
Pin CCP1 is forced High. The CCP latch can only be cleared by switching the CCP module to Mode 0000; that is, by turning it off.

1001: Clear Output Pin on Match
Pin CCP1 is forced into its Low state. The CCP latch is set when the CCP module is turned off.

1010: Generate Software Interrupt on Match
Pin CCP1 unchanged, but CCP1IF still set.

1011: Trigger Special Event on Match
Timer 1 is cleared. With CCP2 (this is the only functional difference be-tween CCP1 and CCP2) an ADC module conversion can be optionally initiated; see Fig. 14.11 on page 454.

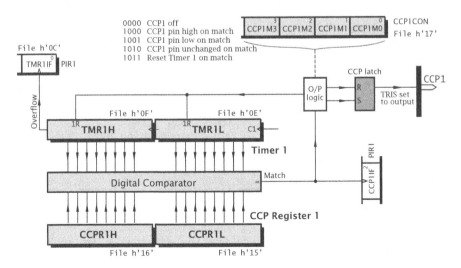

Fig. 13.7 The CCP1 module set to Compare mode.

In modes 1000 and 1001 the parallel port bits shared with CCP1 and CCP2 should be set-up as outputs. Whilst the CCP module is off (as it is after any sort of reset) these pins will reflect the state of those port bits. An example is turning off the CCP module. Because clearing CCPCON to mode 0000 is the only way of relaxing the CCP latch to its pre-match value, it is advisable to set each associated port bit to this value to avoid spurious pulses. For instance, if mode 1001 is being used, then the port bit should be set to 1 in the initialization code to ensure a High-state reset value. Typically for a 28+ footprint device, CCP1 is shared with parallel port pin RC2 and CCP2 is shared with RC1.

As an example, consider that we wish to set up Timer 1 as configured in the last example to overflow every 10 seconds. To do this we need to set the timer to time-out after 16 s (Prescaler ratio 1:8) and then shorten the cycle. This is implemented by loading the CCPR1 register with the fraction $\frac{10}{16}\left(2^{16} \times \frac{10}{16}\right)$; which translates to h'A000'. Whenever Timer 1 reaches this value it will automatically be reset and an interrupt will occur if the CCP1IE mask bit (and global PEIE and GIE masks) are set.

Initialization code for this is:

```
movlw   h'A0'        ; Set up CCPR1 to h'A000'
movwf   CCPR1H
clrf    CCPR1L
movlw   b'00001011'  ; CCP Compare mode 1011. Special event
movwf   CCP1CON
movlw   b'00111011'  ; Timer1 on (1), external clock (1)
movwf   T1CON        ; Synched (0), oscillator (1) 1:8 (111)
bsf     STATUS,RP0   ; Change to Bank 1
bsf     PIE1,CCP1IE  ; Enable CCP1 interrupts
bcf     STATUS,RP0   ; Change back to Bank 0
bsf     INTCON,PEIE  ; Enable Timer/CCP interrupts
bsf     INTCON,GIE   ; Enable all interrupts
```

The PIC MCU will then automatically be interrupted every 10 seconds.

Because the CCP1 pin is not changed by Compare mode 1011, this pin can be used as a normal parallel port input/output independently of the CCP1 module.

Timer 2 is an 8-bit counter with both a programmable Prescaler and Postscaler; as shown in Fig. 13.8. Input to this counter is always a derivative of the system clock. Unlike the two previous timers, output is not taken from the counter chain but from the **Timer 2 Comparator**. This compares the state of Timer 2 with that in the **Period Register (PR2)**. On equality an output pulse is generated which resets Timer 2 at the *next* count pulse. This may optionally be used to determine the SSP port's SPI clock rate, as listed in Fig. 12.9 on page 345. As determined by the Postscaler, any integer number from 1 to 16 of these reset events will set the Timer 2 Interrupt Flag **TMR2IF** in PIR1[1]. If the Timer 2 mask bit **TMR2IE** is also set, an interrupt is generated.

The value of the Pre- and Postscaler ratio and actuation of Timer 2 is set-up using the T2CON Control register as listed below. All bits are cleared on a reset, turning Timer 2 off with 1:1 Pre- and Postscaler ratios.

TMR2ON
Setting T2CON[2] to 1 enables Timer 2.

T2CKPS[1:0]
Timer 2 can be incremented either directly at the instruction cycle rate $f_{OSC}/4$ or frequency divided by four or 16. The three settings of T2CON[1:0] are listed in the diagram.

TOUTPS[3:0]
The number of Timer 2 periods activating the TMR2IF interrupt flag can be set to between one and 16 with T2CON[5:2]. This 4-bit code n maps to $n + 1$ events; from b'0000' = 1:1 to b'1111' = 1:16.

The advantage of this architecture is that time-out can be fine tuned without using a CCP module, by setting the Period Register to an appropriate value. The delay until TMR2IF is set is given as:

$$4/f_{OSC} \times \text{Prescale} \times (\text{PR2} + 1) \times \text{Postscale}$$

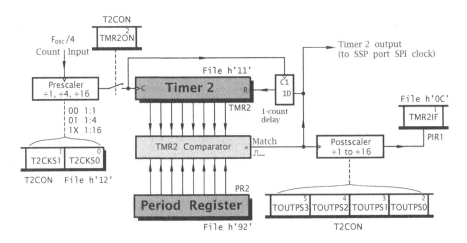

Fig. 13.8 A simplified equivalent circuit for Timer 2.

For our example, consider the need for an interrupt 100 times per second as part of a digital real-time clock. Assuming a 4 MHz crystal, choosing a Prescaler ratio of 1:4 gives a clocking period for Timer 2 of

$4\,\mu$s. If the Period Register is set to 249 then the Timer 2 comparator output period is $250 \times 4 = 1\,$ms. Thus setting the Postscaler to 1:10 (1001) will give the required 10 ms (100 Hz) interrupt rate. By varying the Postscaler from 1 to 16 we can have a corresponding interrupt rate from 1 to 16 ms. For fine adjustments a unit change in PR2 alters the rate in $4 \times$ Postscale μs steps.

Set-up code for this example is:

```
movlw   b'01001101' ; Postscale 1:10 (1001), Timer2 on (1)
movwf   T2CON       ; Prescale 1:4 (01)
bsf     STATUS,RP0  ; Change to Bank1
movlw   d'249'      ; Set up period register to 249
movwf   PR2
bsf     PIE1,TMR2IE ; Enable Timer2 interrupts
bcf     STATUS,RP0  ; Change back to Bank0
bsf     INTCON,PEIE ; Enable all Timer/CCP interrupts
bsf     INTCON,GIE  ; Global enable
```

The setup_timer_2(mode,period,postscale); function is the CCS C equivalent to initialize Timer 2.

```
setup_timer_2(T2_DIV_BY_4,249,10);
enable_interrupts(INT_TIMER2);
enable_interrupts(GLOBAL);
```

Timer 2 can be read from using the get_timer2() function and written to with the set_timer2() function.

One of the more common applications of MCU-based systems is the control of power circuits, such as heating, lighting and electric motor speed control. One approach to this problem would be to use a digital-to-analog converter, such as that discussed in Fig. 12.16 on page 358, driving a power amplifier. Such linear control is expensive and inefficient due to the large current:voltage products that must be handled by the power amplifier. A rather more efficient and cost effective approach rapidly switches the load on and off at a reasonably fast rate. A power switch, such as a thyristor, dissipates relatively little power, as when the switch is off no current flows and when the switch is on the voltage across the switch is small—ideally zero.

An example of such waveforms is shown in Fig. 13.9. The average amplitude is simply $A \times N$, where N is the duty cycle fraction of the repeat period. If we vary N from 0 to 100% then the average power will vary in a like fashion—all without the benefit of analog circuitry. This digital-to-analog conversion technique is known as **pulse width modulation** (**PWM**).

The thermal or mechanical inertia of most high-power loads is such that even with a relatively low repetition rate (typically no lower than 100 Hz) the "bumps" will be smoothed. Low switching rates are more efficient, as each switching action dissipates energy. If PWM is used

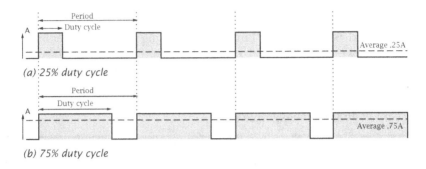

(a) 25% duty cycle

(b) 75% duty cycle

Fig. 13.9 Pulse width modulation.

for more conventional digital-to-analog conversion, such as for audio applications, then a low-pass filter may be utilized to reduce the high-frequency harmonics. In such cases a sampling rate of at least ten times the maximum analog signal should be used to space out the harmonics (see Fig. 14.3 on page 440) and reduce the necessary filtering burden.

Generating a PWM waveform is conveniently implemented using a counter and digital equality comparator. The output pin is driven from a latch, which is always set as the counter rolls over. The latch is reset when the counter state equals a number representing the duty cycle. The larger is the Duty number the longer a proportion of the cycle will the pin remain in its High state.

As a simple example, consider a 3-bit count with a Duty number of b'011':

Set High			Reset Low				
$000 \rightarrow$	$001 \rightarrow$	$010 \rightarrow$	$\boxed{011} \rightarrow$	$100 \rightarrow$	$101 \rightarrow$	$110 \rightarrow$	111
			Match				

In this instance the pin will remain in its High state for three counts, giving a duty cycle of $\frac{3}{8}$, or 37.5%. By changing this number, the average power can be altered with a resolution of $\frac{1}{8}$ from zero up to 87.5%.

PIC microcontrollers implement this scheme using the CCP modules. In the PWM mode, Timer 2 is used to implement the timebase counter and the Duty number is fed in via a double-buffered register to a 10-bit PWM comparator. In Fig. 13.10 the CCP1 module is shown, with pin CCP1 being used as the PWM output. If the CCP2 module is implemented, then it too can be used in this manner but with pin CCP2 being used to generate the waveform. Any CCP pin used as a PWM output must be set-up as an output, with associated TRIS bit logic 0. Although both CCP modules can operate in parallel with different Duty numbers they share the same Timer 2 and thus have the same period.

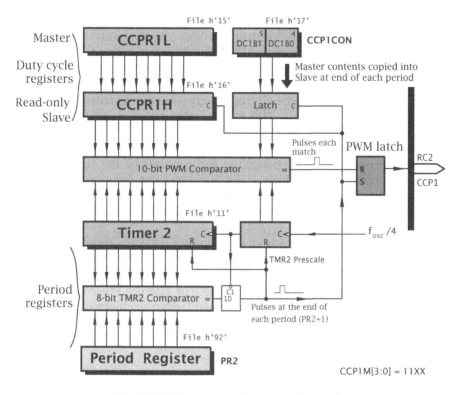

Fig. 13.10 Timer 2 and the PWM CCP mode.

Period

The timebase is set using Timer 2 in the manner outlined in Fig. 13.8. The roll-over period is a function of the main instruction cycle time $4 \times t_{OSC}$, the Prescaler ratio, and the contents of the Period register PR2. Recalling that Timer 2 resets on the clock pulse *after* equality with PR2 is reached, the total repetitive period is given as:

$$(4 \times t_{OSC}) \times \text{Prescaler ratio} \times (\text{PR2} + 1)$$

For instance for a 16 MHz crystal, Prescaler ratio 1:16 and PR2 contents of h'63' = d'99' we have:

$$\text{Period} = \left(4 \times \frac{1}{16}\right) \times 16 \times (99 + 1) = 400\,\mu s$$

Each time Timer 2 overshoots the Period number, three things happen.
1. Timer 2 is reset to zero; unless PR2 is zero.
2. The PWM latch is set and pin CCP1 goes to its High state.
3. The 10-bit content of the Master register is copied into the Slave register and presents the next Duty number to the 10-bit PWM digital comparator.

Duty Cycle

The Duty number is presented to the 10-bit PWM equality comparator in a 2-deep 10-bit wide pipeline. The outer word is located in the 8-bit CCPR1L together with the two lowest bits held in CCP1CON[5:4], which together are labeled in the diagram as the Master register. The contents of the Master register can be altered at any time by the software as two separate movwf instructions. This word is only moved down the pipeline to be presented as the Duty number to the comparator at the end of each period. This reduces the possibility of a mid-period glitch, due to the random nature of any changes in the contents of the Master register in relation to the timebase. The Slave register comprises CCPR1H companded with a 2-bit internal latch. While in this mode the CCPR1H register is read-only. This prevents direct access by software to the Duty number.

The core Timer 2 register is only eight bits wide. In order to extend Timer 2 to 10 bits, to match the Duty number, two lower bits are added. These extra two bits either originate from the Prescaler counter which is dividing down the system clock to Timer 2 or else if a Prescaler ratio of 1:1 is chosen, the 2-bit count defining the quadrature clocks of Fig. 4.4 on page 76. In either case, the result is to give a maximum 10-bit (1:1024) resolution in the Duty cycle, with a counting rate of ×4 of that of the 8-bit Timer 2 core.

When this 10-bit count equals the Duty number, the PWM latch is reset, and the CCP1 pin drops ti its Low state. It stays low until the next period begins, then Timer 2 rolls over and the cycle repeats ad infinitum. In all cases the datum in in CCPR1L must be smaller than that in PR2, otherwise the PWM latch will never reset! If PR2 is h'FF' then the resolution of the system is a full 10 bits. Smaller values of Timer 2 period data will reduce this resolution. For instance, if PR2 = h'3F' then the resolution is reduced to 8 bits; six in Timer 2 proper and two extension bits.

For our example, let us assume the situation described previously where our timebase period is $400\,\mu s$ (2.5 kHz) for a 16 MHz crystal with Prescaler ratio of 1:16 and a PR2 value of h'63'. If we wish to generate a 25% duty cycle, as in Fig. 13.9(a), the the set-up code would be something like:

```
bsf    STATUS,RP0  ; Change to Bank1
movlw  h'63'       ; Set up Timer2 Period register to d'99'
movwf  PR2
bcf    TRISC,2     ; Make CCP1 an output
bcf    STATUS,RP0  ; Change back to Bank0
movlw  h'19'       ; Set-up Master to 1/4 full scale (h'63/4')
movwf  CCPR1L      ; That is, b'0001 1001'
movlw  b'00001100'; CCP1 module PCM mode (1100)
movwf  CCP1CON     ; with CCP1CON[5:4] (00)
movlw  b'00000110'; Timer2 Prescale 1:16 (10)
movwf  T2CON       ; Timer2 on (1). Start waveform
```

The Timer 2 Postscaler does not affect the PWM generation but still sets the TMR2IF in the normal way. The CCP1IF is not altered in this mode.

Many high-power applications require two or four switching wave-forms to control the load in a bridge configuration. Devices, such as the PIC16F684 with an Enhanced CCP module, are specifically designed to control these bridge circuits. They also include a dead-band period to avoid possible current short circuits due to delays in the electronic switches simultaneously turning on and off.

Examples

Example 13.1
Show how you could use Timer 0 to generate a PWM version of a digital byte in File DATUM using pin RA0 as the output. Assuming an 8 MHz crystal, calculate the PWM duration.

Solution
Timer 0 will give a time-out related to the number loaded into the timer register at the beginning of the period. If we load in the 2's complement of the byte (the negative value) then the duration will be proportional to this value—the larger it is, the longer the timer has to count before overflowing. Conversely loading in the value of DATUM will give a time-out duration inversely proportional to the value. By alternately loading the 2's complement of DATUM and making the pin High followed by DATUM itself making the pin Low will give us a total period approximately the same as a total Timer 0 time-out as if sequentially counting through all 256 states.

The coding of Program 13.5 sets up Timer 0 to count at a 2 MHz rate ($f_{OSC}/4$) with no Prescaler. Thus the total PWM rate is $\frac{2}{256}$ or 7.8125 KHz. When Timer 0 overflows it generates an interrupt. The ISR checks the state of PORTA[0] and if 0 changes its state and then calculates the 2's complement of the data byte (invert plus one). However, there is a 2-cycle delay in Timer 0 responding when its state is changed due to the clock synchronizer circuit, and so another two is added to compensate for this extra delay. If the port pin was already 1 then it is zeroed and the datum itself plus the compensatory two is written into Timer 0.

Adding the compensation will cause problems at either extremes of the mark:space ratio. Why is this so, and what action could you take to ameliorate it?

Program 13.5 Pulse-Width Modulation using Timer 0.

```
MAIN      bsf     STATUS,RP0   ; Change into Bank 1
          clrf    OPTION_REG   ; Internal clock, Prescale disabled
          bcf     TRISA,0      ; Make RA0 the PWM output
          bcf     STATUS,RP0   ; Change back to Bank 0
          bsf     INTCON,T0IE  ; Enable Timer 0 interrupt
          bsf     INTCON,GIE   ; Enable all interrupts

; <<<< More background code >>>>

; ***************************************************************
; * The ISR to generate a PWM waveform at RA0 by altering       *
; * time out proportional to DATUM and 2's complement DATUM     *
; * Digital byte is DATUM. PORTA[0] holds current PWM state O/P *
; ***************************************************************
; First save context in usual way
ISR       movwf   _work        ; Put away W
          swapf   STATUS,w     ; and the Status register
          movwf   _status

; ==============================================================
; The core code
          btfss   INTCON,T0IF  ; Has Timer0 overflowed?
          goto    ISR_EXIT     ; IF no THEN false alarm

          bcf     INTCON,T0IF  ; Reset interrupt flag
          movf    DATUM,w      ; Get datum
          btfsc   PORTA,0      ; Is current output Low?
          goto    MAKE_LO      ; IF not THEN bring it Low
MAKE_HI   bsf     PORTA,0      ; ELSE output goes High
          xorlw   b'11111111'  ; and compute 2's complement
          addlw   1            ; Invert +1
          goto    SET_UP       ; and set Timer0 up

MAKE_LO   bcf     PORTA,0      ; Bring pin to Low state

SET_UP    addlw   2            ; Compensation for synch delay
          movwf   TMR0         ; Initialize Timer

; ==============================================================

ISR_EXIT  swapf   _status,w    ; Untwist the original Status reg
          movwf   STATUS
          swapf   _work,f      ; Get the original W reg back
          swapf   _work,w      ; leaving STATUS unchanged
          retfie               ; and return from interrupt
```

Example 13.2

A certain tachometer is to register engine speed in the range 0–12,000 rpm (revolutions per minute). The engine generates one pulse per revolution and it is intended that a PIC16F877 be used to count the number of pulses each second and calculate the equivalent rpm. Using two of the three available timers, can you design a suitable hardware–software configuration?

Solution

A speed of 12,000 rpm translates to a maximum pulse count of 200 rps (revolutions per second). Thus we propose to use Timer 0 as the pulse counter driven from pin T0CKI with no Prescaler.

Timer 1 in conjunction with CCP1 set to a Compare mode, will give a 1 s time-out with its own oscillator and 32.768 kHz watch crystal together with a count spanning h'0000–7FFF'. However, to simplify the mathematical relationship rpm = rps × 60 it is proposed to shorten the timebase by the factor $\frac{60}{64}$ to implement the equivalent relationship $\frac{rps \times 60}{64} \times 64$. This can be done by reducing the count span to h'7FFF' × $\frac{60}{64}$ = h'77FF'. The final ×64 can easily be implemented by either shifting left six times (<<6) or more efficiently placing the rps count as the high byte of the double-byte rpm datum and shifting right twice; i.e.,

$$rpm = (rps \times 256) >> 2$$

Overall this is considerably more efficient than using a 1 s timebase and multiplying by 60.

One possible solution is shown in Program 13.6. Here the initialization code implements the following task list:

- Set Timer 0 to count ⌐‿ events at T0CKI.
- Set CCP1 to Compare mode 1011 to reset Timer 1 on equality.
- Enable an interrupt from this event.
- Configure CCPR1H:L to set the timebase to $\frac{60}{64}$ s.

The ISR itself simply copies the rps reading from Timer 0 extended to a double byte, zeros the timer and then converts the reading to rpm as described above. After the two Shift-Right operation >>2, the top two bits of RPM are cleared to remove erroneous carry-ins. The resulting 14-bit datum in RPM:RPM+1 is the required outcome, which can then perhaps be used by the background program to activate the display or maybe transmit the data to a computer over a serial link.

To make the system more robust, the Timer 0 interrupt flag T0IF should be checked as part of in the ISR to signal overflow and thus to activate an overspeed warning indicator.

Program 13.6 Tachometer software.

```
MAIN movlw   h'77'              ; Setting h'77FF' to give
     movwf   CCPR1H             ; a time base of 60/64 seconds
     movlw   h'FF'
     movwf   CCPR1L

     bsf     STATUS,RP0         ; To Bank1
     movlw   b'00111000'        ; Timer0 external negative edge
     movwf   OPTION_REG         ; No Prescaler
     bsf     PIE1,CCP1IF        ; Enable interrupts from CCP1
     bcf     STATUS,RP0         ; Back to bank0

     movlw   b'00001011'        ; CCP Compare mode 1011
     movwf   CCP1CON            ; resets Timer1 on equality

     movlw   b'00001011'        ; Timer1 PS1:1, oscillator synched
     movwf   T1CON              ; and enabled

     bsf     INTCON,PEIE        ; Enable Timer/CCP interrupts
     bsf     INTCON,GIE         ; Global enable mask bit on

; <<<< More background code >>>>

; ****************************************************************
;
; First save context in usual way
ISR     movwf   _work          ; Put away W
        swapf   STATUS,w       ; and the Status register
        movwf   _status
; ================================================================
; The core code ------------------------------------------------
        btfss   PIR1,CCP1IF    ; Did CCP register Timer1 reset?
        goto    ISR_EXIT       ; IF no THEN false alarm

        movf    TMR0,w         ; Get totalized pulse count
        clrf    TMR0           ; Zero pulse count
        movwf   RPM            ; Save totalized count away

; Now multiply by 64 -------------------------------------------
        clrf    RPM+1          ; Clear lower byte
        rrf     RPM,f          ; RPM as MSB; i.e., X256
        rrf     RPM+1,f        ; >>2 to convert rps to rpm
        rrf     RPM,f
        rrf     RPM+1,f
        bcf     RPM,7          ; Zero top two bits
        bcf     RPM,6

        bcf     PIR1,CCP1IF    ; Reset interrupt flag
; ================================================================
ISR_EXIT swapf   _status,w     ; Untwist the original Status reg
         movwf   STATUS
         swapf   _work,f       ; Get the original W reg back
         swapf   _work,w       ; leaving STATUS unchanged
         retfie                ; and return from interrupt
```

Example 13.3

A PIC16F877 is to be used to measure the duration of an event. This duration is the time a signal is in its High state, as shown in Fig. 13.11. You can assume that the main crystal is 8 MHz and the duration of the event is guaranteed to be no more than 100 ms.

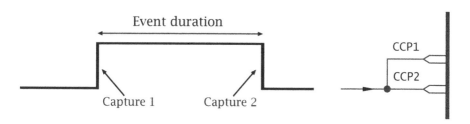

Fig. 13.11 An event manifesting itself as a pulse duration.

Solution

One way of tackling this problem is to feed the signal shown in the diagram into both pins CCP1 and CCP2 in parallel. Using one CCP module to capture the rising edge and the other to capture the falling edge gives the duration as the difference in the two captured values. In Program 13.7 Timer 1 is zeroed on a rising edge and thus the second captured Timer 1 state is our duration. If we use a Prescaler ratio of 1:4 and the system machine cycle clock. then we have as our counting rate 500 kHz; i.e., the system resolution is 2 μs. The overall maximum duration that can be measured in this way is $2^{16} \times 2 = 131,077\mu s$, which is large enough not to overflow for a maximum duration of 100 ms.

The ISR in Program 13.7 simply tests each CCP interrupt flag in turn and goes to the appropriate routine. If CCP1 has signaled a ⎍ event then the timer is zeroed to restart the count. Timer 1 has been configured to increment at a 500 kHz rate and when the next ⎍ occurs the CCP2 module captures the state of this timebase and places it in the 16-bit CCPR2 register. The ISR then copies it into the two Files TIME:TIME+1 and this is the period in 2 μs ticks.

Actually resetting Timer 1 on the first event introduces some inaccuracy into the process, as the clearing event takes some time. In our application this is of little consequence but it may cause some problems in shorter high-resolution situations. In this case Timer 1 can be left to run continually and the two captured 16-bit data subtracted to give the required difference at relative leisure.

Program 13.7 Measuring the duration of a pulse.

```
MAIN    movlw   b'00000101'     ; CCP1 module captures rising edge
        movwf   CCP1CON
        movlw   b'00000100'     ; CCP2 module captures falling edge
        movwf   CCP2CON

        bsf     STATUS,RP0      ; To Bank1
        bsf     PIE1,CCP1IE     ; Enable interrupts from CCP1
        bsf     PIE2,CCP2IE     ; Enable interrupts from CCP2
        bcf     STATUS,RP0      ; Back to bank0

        movlw   b'00100001'     ; Timer1 enabled (1), int osc (0)
        movwf   T1CON           ; Synched (0), Prescale 1:4 (10)

        clrf    NEW             ; Clear the New flag

        bsf     INTCON,PEIE     ; Enable Timer/CCP interrupts
        bsf     INTCON,GIE      ; Global enable mask bit on

; <<<< More background code >>>>

; ****************************************************************
; First save context in usual way
ISR     movwf   _work           ; Put away W
        swapf   STATUS,w        ; and the Status register
        movwf   _status
; ================================================================
; The core code
        btfsc   PIR1,CCP1IF     ; A CCP1 rising edge capture?
        goto    CAPTURE1        ; IF yes THEN go to it!
        btfss   PIR2,CCP2IF     ; A CCP2 falling edge capture?
        goto    ISR_EXIT        ; IF not THEN false alarm!
CAPTURE2
        movf    CCPR2L,w        ; Get low byte of captured time
        movwf   TIME+1          ; and put away
        movf    CCPR2H,w        ; Get high byte of captured time
        movwf   TIME            ; and put away
        bcf     PIR2,CCP2IF     ; Clear CCP2 interrupt flag
        incf    NEW,f           ; Tell the world: A new time datum
        goto    ISR_EXIT

CAPTURE1
        clrf    TMR1L           ; Zero time count
        clrf    TMR1H
        bcf     PIR1,CCP1IF     ; Reset CCP1 interrupt flag
; ================================================================
ISR_EXIT swapf  _status,w       ; Untwist the original Status reg
        movwf   STATUS
        swapf   _work,f         ; Get the original W reg back
        swapf   _work,w         ; leaving STATUS unchanged
        retfie                  ; and return from interrupt
```

Self-Assessment Questions

13.1 Using Timer 1 and CCP1, design a system to generate a continuous square wave with a total period of 20 ms from pin CCP1. You may assume that the main crystal is 8 MHz. *Hint*: Remember that the state of the CCP pin will only change when a match occurs, so the Compare mode will have to be changed on the fly every 10 ms.

13.2 The echo sounding hardware shown in Fig. 7.9 on page 211 uses an external 1.72 kHz oscillator to interrupt the PIC MCU once per 5.813 ms; that is, once every time sound travels 1 cm through air. Assuming that a 20 MHz PIC MCU is used, show how Timer 2 could be used to generate this interrupt rate to an accuracy better than 0.1%.

13.3 The mid-range PIC MCU family has only one hardware input pin; namely INT. Suggest some way to use Timer 0 to simulate another hardware interrupt with pin T0CKI.

13.4 As part of a software implementation of an asynchronous serial channel running at 300 baud, a delay of 3.33 ms is to be generated. Assuming that a 8 MHz PIC MCU is the host processor, show how you could use a timer to generate an interrupt each baud period. Extend your routine to enable baud rates up to 19,200 in doubling geometric progression.

13.5 Show how you would use Timer 1 with its separate integral oscillator with a 32.768 kHz watch crystal, to keep the central heating real time clock array HOURS : MINUTES : SECONDS of Example 7.3 on page 206 up to date.

13.6 The CCS **C** compiler has integral functions dealing with the Timer and CCP modules. For instance, Timer 1 can be written to using set_timer1(datum); and read from using get_timer1();. The function setup_timer_1(mode); is used to initialize the timer. Similarly setup_ccp1(mode); initializes the CCP1CON register. Mode scripts for Timer 1 and the CCP Compare configuration are:

```
T1_DISABLED          T1_INTERNAL   T1_EXTERNAL
T1_EXTERNAL_SYNCH    T1_CLK_OUT    T1_DIV_BY_1
T1_DIV_BY_2          T1_DIV_BY_4   T1_DIV_BY_8
CCP_COMPARE_RESET_TIMER
```

Where separate modes can be separated by the Inclusive-OR | operator.

Show how you would code your solution to SAQ 13.5 in **C**. In CCS **C** a function can be turned into a CCP1 interrupt service routine by preceding it by the directive #INT_CPP1; see Program 9.6 on page 265 for details.   You an also assume that the reserved variable CCP_1 represents the 16-bit CCPR1H:L register.

13.7 Pulse-width modulation can be used to control the speed of a DC. motor by altering the average winding current. However, starting up such a motor is problematic, as the current flow is much greater than normal until the motor reaches normal running speed and the consequent back emf reduces winding current. It is proposed that to avoid damage to the current driving transistors, a PWM technique be used to gradually increase the duty cycle from zero to full over a period of several seconds. Show how you could do this assuming a PIC MCU with a 4 MHz crystal and CCP module.

13.8 Light-controlled pedestrian crossings (Pelican crossings) in the UK follow the listed sequence of operations once one of the cross-request switches are closed.
   1. Green light only (standby).
   2. Amber light only for 3 s.
   3. Red light plus buzzer for 15 s.
   4. Flashing amber light—five flashes each comprising 3 s on and 3 s off.
   5. Return to standby.
   Using a suitable PIC microcontroller with a Timer 1 module, design the software to control the lights and buzzer. Although the lights are on both sides of the road, you may assume that they are connected in parallel and are activated by a High state on the relevant parallel port pin. The two input switches, CROSS_REQUEST0 and CROSS_REQUEST1 give 0 when closed. The buzzer is activated by a Low state on the connected parallel port pin.

CHAPTER 14

Take the Rough with the Smooth

Given that digital microcontrollers are in the business of monitoring and controlling the real environment—which is commonly analog in nature—we need to consider the interaction between the analog and the digital worlds. In some cases all that is required is a **comparison** of two analog voltage levels. However, for more sophisticated situations, analog input signals need conversion to a digital equivalent; that is, **analog-to-digital conversion (ADC)**. Thereafter the digital patterns can be processed in the normal way. Conversely, if the outcome is to be in the form of an analog signal, then a **digital-to-analog conversion (DAC)** stage will be necessary.

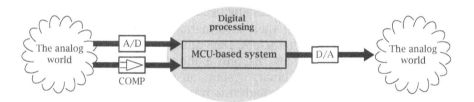

Fig. 14.1 Analog world—digital processing.

Of these various processes, illustrated in Fig. 14.1, A/D conversion is by far the more complex. Many PIC microcontrollers feature integral multi-channel A/D facilities. However, analog outputs frequently require external circuitry to implement the D/A process.

In this chapter we will look at the properties of analog and digital signals and the conversion between them as relevant to the PIC MCU. After completion you will:

- Understand the quantization relationship between analog and digital signals.
- Be aware of the need to sample an analog signal at least twice the highest frequency component.
- Appreciate how the successive approximation technique can convert an analog voltage to a binary equivalent.

- Understand the operation and be able to configure the Analog Comparator, Voltage Reference and ADC modules.
- Know how to configure I/O pins as either analog or digital.
- Be able to write assembly-level programs to acquire analog data using polling, interrupt-driven, and Sleep techniques, and to interrogate the state of the analog comparators.
- Be able to code high-level C programs to set-up and interact with the various analog modules.
- Know how to parallel interface to a proprietary DAC.

The information content of an **analog signal** lies in the continuously changeable worth of some constituent parameters, such as amplitude, frequency, or phase. Although this definition implies that an analog variable is a continuum in the range $\pm\infty$, in practice its range is restrained to an upper and lower limit. Thus a mercury thermometer may have a continuous range between, say, $-10°C$ and $+180°C$. Below the bottom number mercury disappears into the bulb. Above the highest number and the top of the tube is blown off!

Theoretically the quantum nature of matter sets a lower bound to the smooth continuous nature of things. However, in practice noise levels and the limited accuracy of the device generating the signal sets an upper limit to the resolution that processing needs to take account of.

Digital signals represent their information content in the form of arrangements of discrete characters. Depending on the number and type of symbols making up the patterns, only a finite totality of value portrayals are possible. Thus in a binary system, an n-digit pattern can at the most represent 2^n levels. Although this *rough* grainy view of the world seems inferior to the infinity of levels that can be *smoothly* represented by an analog equivalent, the quantizing grid can be tailored to fit the accuracy of the task to be undertaken. For instance, a telephone speech circuit will tolerate a resolution of around 1%. This can use an 8-bit depiction, which gives up to 256 discrete values with a corresponding $\approx 0.5\%$ resolution. A music compact disk uses a 16-bit scheme, giving a one part in 65,636 grid—around 0.0015% resolution.

From this discussion it can be seen that any process involving interconversion between the analog and digital domains will involve transition through the **quantization** state. Therefore we need to look at how this affects the information content of the associated signals.

As an example, consider the situation shown in Fig. 14.2, where an input range is represented as a 3-bit code. In essence the process of quantizing a signal is the comparison of the analog value with a fixed number of levels—eight in this case. The nearest level is then taken as expressing the original as its digital equivalent. Thus in Fig. 14.2 an input voltage of 0.4285 of full scale is 0.0536 above quantum level 3. Its quantized value will then be taken as level 3 and coded as b'011' in our 3-bit scheme of things.

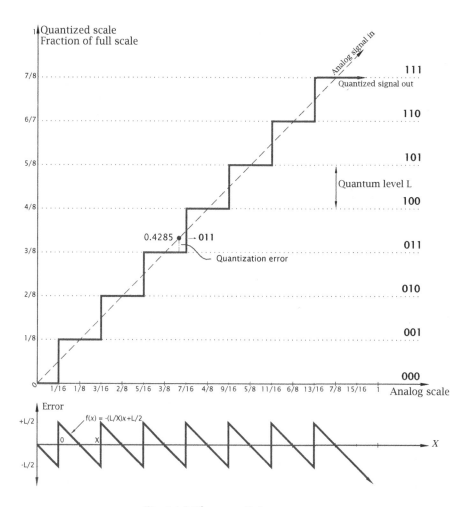

Fig. 14.2 The quantizing process.

The residual error of −0.0536 will remain as quantizing noise, and can never be eradicated; see also Fig. 14.3(d). The distribution of quantization error is given at the bottom of Fig. 14.2, and is affected only by the number of levels. This can simply be calculated by evaluating the average of the error function squared. The square root of this is then the root mean square (rms) of the noise.

$$\mathcal{F}(x) = -\frac{L}{X}x + \frac{L}{2}$$

The mean square is:

$$\frac{1}{X}\int_0^X \mathcal{F}(x)^2\,dx \;=\; \frac{1}{X}\int_0^X \left[\frac{L^2}{X^2}x^2 - \frac{L^2}{X}x + \frac{L^2}{4}\right]dx$$

$$= \; \frac{1}{X}\left|\frac{L^2}{3X^2}x^3 - \frac{L^2}{2X}x^2 + \frac{L^2}{4}x\right|_0^X = \frac{L^2}{12}$$

Thus the rms noise value of $\frac{L}{\sqrt{12}} = \frac{L}{2\sqrt{3}}$, where L is the quantum level.

A fundamental measure of a system's merit is the signal-to-noise ratio. Taking the signal to be a sinusoidal wave of peak to peak amplitude $2^n L$, we have an rms signal of $\frac{\left(\frac{2^n L}{2}\right)}{\sqrt{2}}$; that is, $\frac{\text{peak}}{\sqrt{2}}$. Thus for a binary system with n binary bits, we have a signal-to-noise ratio of:

$$\frac{\left(\frac{2^n L}{2\sqrt{2}}\right)}{\left(\frac{L}{\sqrt{12}}\right)} \;=\; \frac{2^n\sqrt{12}}{2\sqrt{2}} = 1.22 \times 2^n$$

In decibels we have:

$$\text{S/N} = 20\log 1.22 \times 2^n = (6.02n + 1.77)\,\text{dB}$$

The dynamic range of a quantized system is given by the ratio of its full scale ($2^n L$) to its resolution, L. This is just 2^n, or in dB, $20\log 2^n = 20n\log 2 = 6.02n$. The percentage resolution given in Table 14.1 is of course just another way of expressing the same thing.

Table 14.1: Quantization parameters.

Binary bits n	Quantum levels (2^n)	% resolution	Resolution dynamic range	S/N ratio (dB)
4	16	16.25	24.1 dB	26.9 dB
8	256	0.391	48.2 dB	49.9 dB
10	1024	0.097	60.2 dB	61.9 dB
12	4096	0.024	72.2 dB	74.0 dB
16	65,536	0.0015	96.3 dB	98.1 dB
20	1,048,576	0.00009	120.4 dB	122.2 dB

The exponential nature of these quality parameters with respect to the number of binary-word bits is clearly seen in Table 14.1. However, the implementation complexity and thus price also follows this relationship. For example, a 20-bit conversion of 1 V full scale would have to deal with quantum levels less than $1\,\mu$V apart. Pulse-code modulated telephonic links use eight bits, but the quantum levels are unequally spaced, being closer at the lower amplitude levels. This reduces quantization hiss where conversations are held in hushed tones! Linear 8-bit conversions are suitable for most general purposes, having a resolution of better than $\pm\frac{1}{4}\%$. Actually video looks quite acceptable at a 4-bit resolution, and music can just about be heard using a single bit—i.e., positive or negative!

S/N ratios presented in Table 14.1 are theoretical upper limits, as errors in the electronic circuitry converting between representations and aliasing (discussed below) will add distortion to the transformation.

The analog world treats time as a continuum, whereas digital systems sample signals at discrete intervals. Shannon's sampling theorem[1] states that provided this interval does not exceed half that of the highest signal frequency, then no information is lost. The reason for this theoretical twice highest frequency sampling limit, called the Nyquist rate, can be seen by examining the spectrum of a train of amplitude modulated pulses. Ideal impulses (pulses with zero width and unit area) are characterized in the frequency domain as a series of equal-amplitude harmonics at the repetition rate, extending to infinity. Real pulses have a similar spectrum but the harmonic amplitudes fall with increasing frequency.

If we modulate this pulse train by a baseband signal $A \sin \omega_f t$, then in the frequency domain this is equivalent to multiplying the harmonic spectrum (the pulse) by $A \sin \omega_f t$, giving sum and different components thus:

$$A \sin \omega_f t \times B \sin \omega_h t = \frac{AB}{2}(\sin(\omega_h + \omega_f)t + \sin(\omega_h - \omega_f)t)$$

for each of the harmonic frequencies ω_h.

More complex baseband signals can be considered to be a band-limited (f_m) collection of individual sinusoids, and on the basis of this analysis, each of these pulse harmonics will sport an upper (sum) and lower (difference) sideband. We can see from the geometry of Fig. 14.3(b) that the harmonics (multiples of the sampling rate) must be spaced at least $2 \times f_m$ apart, if the sidebands are not to overlap.

A low-pass filter can be used, as shown in Fig. 14.3(d), to recover the baseband from the pulse train. Realizable filters will pass some of the harmonic bands, albeit in an attenuated form. A close examination of the frequency domain of Fig. 14.3(d) shows a vestige of the first lower sideband appearing in the pass band. However, most of the distortion in the reconstituted analog signal is due to the quantizing error resulting from the crude 3-bit digitization. Such a system will have a S/N ratio of around 20 dB.

In order to reduce the demands of the recovery filter, a sampling frequency somewhat above the Nyquist limit is normally used. This introduces a guard band between sidebands. For instance, the pulse code telephone network has an analog input band limited to 3.4 kHz, but is sampled at 8 kHz. Similarly the audio compact disk uses a sampling rate of 44.1 kHz, for an upper music frequency of 20 kHz.

---

[1] Shannon, C.E.; *Communication in the Presence of Noise*, Proc. IRE, vol. 37, Jan. 1949, pp. 10–21.

A more graphic illustration of the effects of sampling at below the Nyquist rate is shown in Fig. 14.4. Here the sampling rate is only 0.75 of the baseband frequency. When the samples are reconstituted by filtering, the resulting pulse train, the outcome—shown in Fig. 14.4(b)—bears no simple relationship to the original. This spurious signal is known as an **alias**. Where an input analog signal has frequency components *above* half the sampling rate, maybe due to noise, then this will appear as distortion in the reconstituted signal. For this reason analog signals are usually

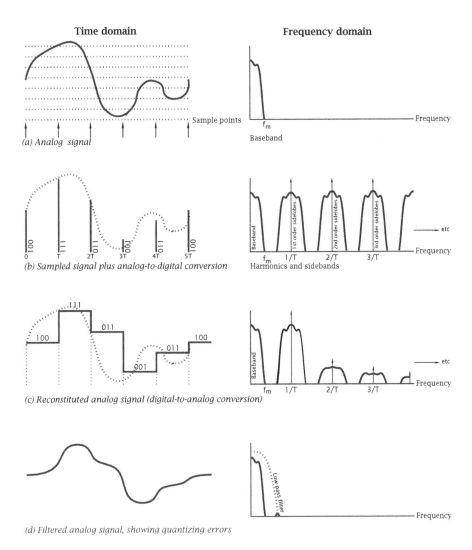

Fig. 14.3 The analog–digital process.

(a) *Sampling below the Nyquist rate*

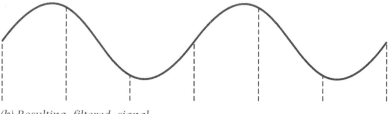

(b) *Resulting filtered signal*

Fig. 14.4 Illustrating aliasing.

low-pass filtered at the input of an A/D converter. This process is known as anti-aliasing filtering.

In dealing with analog inputs many situations simply need to make a true:false decision on whether a voltage is above or below a reference value V_{ref}. For instance, the signal shown in Fig. 14.5 (see also Fig. 14.20) represents the current during the discharge of an EKG diphasic defibrillator, as generated using a Hall effect current to voltage sensor. When nothing is happening the baseline voltage is 2.6 V. When the defibrillator begins its discharge, this voltage rapidly rises to a peak of 3.6 V over a few tens of microseconds. If the MCU is to sample the voltage over the next several tens of milliseconds, say, to calculate the total shock energy, then to begin this process it needs to know when this voltage rises above a threshold. In the diagram this is shown as 3.4 V. It could of course rapidly sample the analog signal using its integral Analog-to-Digital module (if it has one), as described later on page 454, but this continuous sample-and-check routine would use most of the processing capability of the processor. It would be much more software efficient to be able to automatically generate an interrupt in hardware when the input voltage V_{defb} rises above this threshold. The resulting ISR could then begin sampling the signal and performing the real-time analysis.

In Fig. 14.5 the analog signal V_{defb} is used as the input to the non-inverting (+) terminal of an **analog comparator**. The inverting termi-

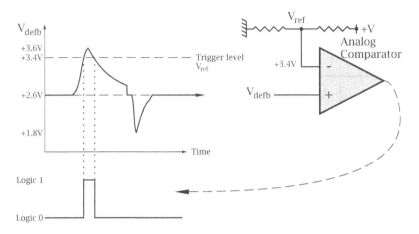

Fig. 14.5 Using an analog comparator to determine the start of the EKG defibrillator discharge.

nal (−) is connected to a fixed reference V_{ref} of 3.4 V. Whenever V_{defib} rises above V_{ref}, the comparator's output voltage changes from logic 0 to logic 1, and conversely when $V_{defb} < V_{ref}$ the output goes back to logic 0.

An analog comparator is basically a high-gain analog differential amplifier with no negative feedback. With a very large open-loop gain the amplifier will saturate at either near its positive or negative power supply if the difference between inputs is more than an exceedingly small value. An ordinary operational amplifier can act as an analog comparator, but circuits specifically designed for this purpose give standard logic levels at their output and have a snap action whenever slowly changing noisy inputs cross the differential threshold.

All three of our exemplar PIC microcontrollers feature a **Comparator module**. The PIC12F675 8-pin device has one analog Comparator. However, the dual configuration available in the PIC16F87XA group, shown in Fig. 14.6, is more typical for larger footprint devices.

The **COMparator CONtrol CMCON** register, usually at File h'9C', is used to set-up one of the eight configurations listed in the diagram, as commanded by the three **CM[2:0] Comparator Mode** bits in CMCON[2:0]. In the specific case of the PIC16F87XA devices these bits reset to mode b'111', which effectively completely removes the Comparator module from view. Most other devices reset to mode b'000' which, while also disabling the Comparators, rather subtly configures the associated pins as analog inputs.

It is a universal rule in PIC microcontrollers with analog modules, that all potential analog inputs pins (usually Port A, E or GP) *always come out of a Power-up reset as analog inputs*. This Power-on reset requirement is to prevent physical damage to the input digital buffers (see Fig. 11.7 on

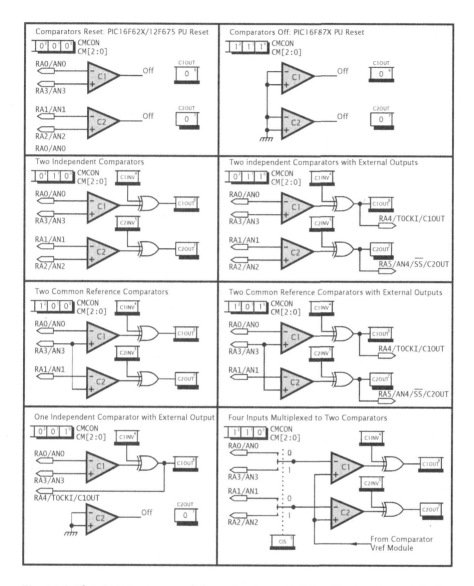

Fig. 14.6 The Comparator module as implemented for the PIC16F87X device group.

page 306) if an analog input voltage, say 2.6 V, were present at a pin on powering up. If that pin was set to be a digital input, expecting a voltage around 0 V or V_{DD}, then an intermediate voltage could cause several transistors to conduct at the same time, possibly causing thermal damage. As analog voltages are not well defined, even where a pin is configured as analog, an external resistor is often used to limit current flow if the analog voltage exceeds V_{DD} or goes negative, as shown in Fig. 14.20.

In the specific case of the PIC16F87XA group, to preserve compatibility with the slightly older PIC16F87X which didn't have a Comparator module, the default is to completely remove this facility. However, all devices in this group have an Analog-to-Digital module, which on reset configures all associated pins as analog and so complies with this rule. Mode b'111' has the lowest power consumption, so it should be chosen if there are no analog input signals and the module is not to be used, especially when in the Sleep state.

The six active modes basically allow either completely independent operation for one or two Comparators, or both non-inverting inputs can be combined to be used as a common reference input. Outputs of any active Comparator may be read at any time from bit 6 **C1OUT** and bit 7 **C2OUT** of CMCON. Each output has an associated programmable invertor control in CMCON[4] and CMCON[5], respectively, labeled **C1INV** and **C2INV**. When $V_{in+} > V_{in-}$ and the associated INVersion bit is 0 then the Comparator output will read as 1, otherwise as 0.[2] As described on page 415, in some versions of Timer 1 the output of Comparator 2 may be used to gate the counting pulse train and thus measure the duration of time an analog signal is above a threshold voltage. In modes b'011' and b'101' the Comparator outputs can also be read externally via pins RA4/C1OUT and RA5/C2OUT—or equivalent shared pins in other devices. In this case pins RA4 and RA5 should be set-up as outputs using TRISA[5:4] in the usual way. Any parallel port pins to be used as analog inputs should similarly be set-up via the appropriate TRIS resister as inputs.

When there is a *change* in an active Comparator output, the **CoMparator Interrupt Flag CMIF** (in PIR2[6] for the PIC16F87XA) will be set and will generate an interrupt if the associated **CoMparator Interrupt Enable mask CMIE** (in PIE2[6] for the PIC16F87XA) and global mask bits GIE and PEIE have been set to 1. As each Comparator does not have its own interrupt flag, the software needs to maintain information regarding the status of the output bits C1OUT and C2OUT to determine which Comparator actually changed. This can be updated as part of the ISR. The act of reading CMCON will end the Change mismatch—in the same manner as the Port B Change interrupt described on page 313. Only then can the CMIF flag can be cleared, in the normal manner. If the Comparator mode is to be changed "on the fly" then interrupts should be disabled beforehand. After a delay of not less than 10 μs after the mode change, to allow voltage levels to stabilize, CMCON should be read again to clear

[2]There is a small uncertainty range in this difference signal of $\pm 10\,mV$ maximum ($\pm 5\,mV$ typical) due to Comparator offset voltages.

any Change mismatch and CMIF cleared afterwards before re-enabling the interrupt system.

As the Comparator module does not depend on the system oscillator, an active Comparator can be used to waken a sleeping PIC MCU when an external voltage crosses a V_{ref} threshold and sets CMIF. After wakening, the PIC MCU should cancel the Change mismatch and clear CMIF following the sleep instruction or in the ISR if the Comparator interrupt is enabled.

It should be noted that an active Comparator uses considerably more current than the base Sleep value. For instance, the PIC12F629/75 group has a typical quiescent current at 5 V of 2.9 nA (995 nA maximum). The Comparator module uses a current of typically $11.5\,\mu A$ ($16\,\mu A$ maximum). Thus if Comparators are not being used during the Sleep duration, they should then be disabled.

Mode b'110' allows one of two voltages to be sampled for each of the comparators individually, as switched via the **Comparator Input Switch CIS** bit in CMCON[3]—which is zeroed on a Power-on reset. In addition, both non-inverting inputs are commoned to an internal voltage reference source, generated from the **Comparator Voltage Reference** module.

All family members with a Comparator module have a separate but related **Comparator Voltage Reference CVR** module. As can be seen from Fig. 14.7, this is essentially a resistor chain with an analog multiplexer gating through one of 16 different voltages, as selected with the **Comparator Voltage Reference** bits **CVR[3:0]** in the **Comparator Voltage Reference CONtrol** register **CVRCON[3:0]**. The CVR module is enabled when the **Comparator Voltage ENable CVREN** in CVRCON[7] is set to 1. This also connects the chain of typically $2\,k\Omega$ resistors to the V_{DD} supply voltage.

Two voltage ranges are available, as set with the **Comparator Voltage Reference Range CVRR** bit in CVRCON[5], which switches in or out an extra 8R resistor at the bottom of the chain. Denoting the 4-bit value of CVR[3:0] as n, these are:

CVRR	Value	Minimum	in 16 steps of	Maximum
0 (reset)	$V_{DD} \times (0.25 + n/32)$	$0.25 \times V_{DD}$	$V_{DD}/32$	$0.71875 \times V_{DD}$
1	$V_{DD} \times n/24$	0 V	$V_{DD}/24$	$0.625\ \ \times V_{DD}$

where n ranges from 0 through 15.

The accuracy of this module is given as $\frac{1}{2}$ of a single step, but in reality the absolute value is directly proportional to the supply voltage; a quantity that is not normally tightly regulated. In addition, V_{DD} can change as the power supply or battery drifts with temperature or load current. Any noise on this line will also be coupled into this reference voltage, although this can be reduced somewhat by capacitive decoupling at the V_{DD} pin and judicious routing of the power supply lines. With these points in mind, where an accurate voltage level is required, an external precision voltage

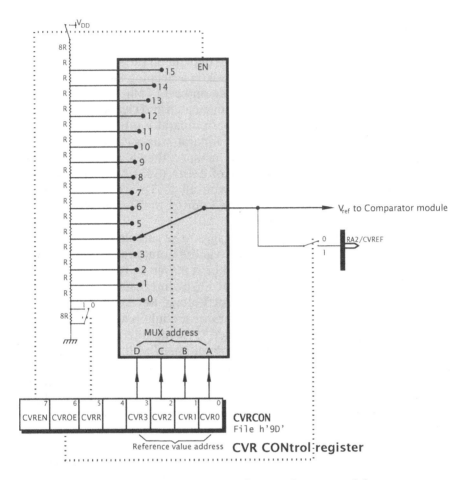

Fig. 14.7 The Comparator Voltage Reference module.

reference device is often used; for instance, with Comparator module mode b'100' connected to pin RA3; see Fig. 14.20.

As our example, assume that V_{DD} is 5 V and we are going to generate our threshold voltage of 3.4 V for Fig. 14.5. We will need to use the high range; that is, CVRR = 0, and calculate a value for CVR[3:0]:

$$5 \times (0.25 + n/32) = 3.4$$
$$0.25 + n/32 = 3.4/5$$
$$n = (3.4/5 - 0.25) \times 32 = 13.76$$

giving $n = 14$ as our closest approximation. Making CVR[3:0] = b'1110' gives an actual V_{ref} of 3.4375 V.

Some family members have an additional control bit to switch the reference voltage to a port pin, so that it can be accessed externally. When the **CVR Output Enable bit CVROE** in CVRCON[6] is set to 1, the analog voltage V_{ref} is gated through to the appropriate pin. Due to the relatively high value of resistance, which also depends on the selected tap, Microchip recommend that this external reference voltage be buffered— typically with an operational amplifier. If necessary, the amplifier gain can be altered to give finer control over the V_{ref} value and filtering can be added to reduce high-frequency noise. Used in this way, the Voltage Reference module can be used as a simple 4-bit digital-to-analog converter.

The code to set-up the Comparator and CVREF modules for our defibrillator example using Comparator 1 with RA3 as the analog input is then:

```
include  "p16f877a.inc"
bsf     STATUS,RP0   ; Change to Bank 1
movlw   b'00001110'  ; Comparator mode 110
movwf   CMCON        ; Switch to RA3 (CIS = 1)

movlw   b'10001110'  ; CVREF module on (1), not external (0)
movwf   CVRCON       ; Hi range (0), CVR[3:0] = 1110

bsf     PIE2,CMIE    ; Enable Comparator interrupts

call    DELAY_10US   ; Allow 10us for voltages to settle
movf    CMCON,f      ; Read CMCON to clear any Change state

bcf     STATUS,RP0   ; Back to Bank 0
bcf     PIR2,CMIF    ; Zero the Comparator interrupt flag
bsf     INTCON,PEIE  ; Enable Peripheral interrupt group
bsf     INTCON,GIE   ; & Globally enable interrupt system
```

Notice especially that before enabling the interrupt system a delay of $10\,\mu s$ is executed to allow internal analog voltages to attain equilibrium. Reading the CMCON register then clears any Change situation, after which the Comparator Interrupt Flag CMIF is cleared. The general interrupt system can then be enabled by setting the PEIE and GIE mask bits in the usual way in the INTCON register.

Some device data sheets, for instance, the PIC12F675, label this module the **Voltage Reference** module. Here the associated Control register is labeled **VRCON** and the various Control bits similarly have their C prefix removed; e.g., **VREN** instead of CVREN.

In many situations more information on the analog signal is needed than a bang-bang comparison with a reference voltage. For instance, in the waveform shown in Fig. 14.5 the deviation of the voltage squared from the baseline, integrated with time, would be required to measure power. In such a situation the incoming signal would have to be sampled and converted from an analog amplitude to a digitized equivalent.

The mapping function from an analog input quantity to its digital equivalent can be expressed as:

$$V_{in} \mapsto V_{ref} \sum_{i=1}^{n} k_i \times 2^{-i}$$

where k_i is the ith binary coefficient having a Boolean value of 0 or 1 and $V_{in} \leq V_{ref}$, where V_{ref} is a fixed analog reference voltage. Thus V_{in} is expressed as a binary fraction of V_{ref} and the Boolean coefficients k_{-i} are the required binary digits.

To see how we might implement this in practice, consider the following mechanical successive approximation analogy. Suppose we have an unknown weight W (analogous to V_{in}), a balance scale (equivalent to an analog comparator) and a set of precision known weights 1, 2, 4, and 8 gm (analogous to a V_{ref} of 15 gm). A systematic technique based on the task list might be:

1. Place the 8 g weight on the pan. IF too heavy THEN remove ($k_1 = 0$) ELSE leave ($k_1 = 1$).
2. Place the 4 g weight on the pan. IF too heavy THEN remove ($k_2 = 0$) ELSE leave ($k_2 = 1$).
3. Place the 2 g weight on the pan. IF too heavy THEN remove ($k_3 = 0$) ELSE leave ($k_3 = 1$).
4. Place the 1 g weight on the pan. IF too heavy THEN remove ($k_4 = 0$) ELSE leave ($k_4 = 1$).

This technique will yield the nearest lower value as the sum of the weights left on the pan. For instance, if W were 6.2 g then we would have a weight assemblage of $4 + 2$ g or b'0110' for a 4-bit system.

The electronic equivalent to this **successive approximation** technique uses a network of precision resistors or capacitors configured to allow consecutive halving of a fixed voltage V_{ref} to be switched in to an analog comparator, which acts as the balance scale.

Most MCUs use a network of capacitors valued in powers of two to subdivide the analog reference voltage, such as shown in Fig. 14.8. Small capacitance values are easily fabricated on a silicon integrated circuit and although the exact value will vary somewhat between different batches of ICs, within the one device all capacitor values will closely match and track with changes in temperature and supply voltage. Multiples of this base value can be fabricated by paralleling unit devices—typically FET gate-source capacitance.

Before the conversion process gets underway, the network has to be primed with the unknown analog input voltage V_{in}, as shown in Fig. 14.8(a). The dynamics of this **sampling** acquisition process involves charging up this capacitance network through both internal and external resistance allowing for the settling time of the internal analog switches. If we take the 10-bit ADC module of Fig. 14.11 as an example, then the parallel capacitor network with nominal unit value of 0.12 pF appears at the AN pin

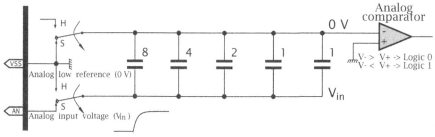

(a) The sample process

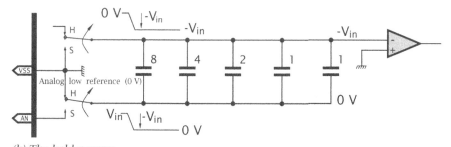

(b) The hold process

Fig. 14.8 Initializing the 8-4-2-1 capacitor network for a 4-bit convertor.

as a $120\,\text{pF}$ ($120 \times 2^{-12}\text{F}$) capacitor. Internal resistance is of the order of $7.5\,\text{k}\Omega$, but is rather temperature and supply voltage dependent. Externally the maximum recommended value is $2.5\,\text{k}\Omega$ in order to keep any ohmic voltage offset due to the pin leakage current of $\pm\frac{1}{2}\,\mu\text{A}$ less than one quantum level (least-significant bit).

The time constant τ (CR) with the values given here is $120 \times 10^{-12} \times \times10^4 = 1.2\,\tau\text{s}$ for a total resistance of $7.5 + 2.5 = 10\,\text{k}\Omega$. In order to get within 0.05% of the final voltage; that is, $\frac{1}{2}$ of a 10-bit quantum level, takes approximately $8 \times \tau \approx 10\,\mu\text{s}$. The data sheet gives the maximum switch settling time of $10\,\mu\text{s}$, although examples use a value of $2\,\mu\text{s}$. Taking a worse-case scenario, a sampling time of $20\,\mu\text{s}$ should ensure stability before the conversion.

Our analysis assumed that the input voltage on the capacitor network needed to be changed by the full range since the last sample. This is likely to be the case if sampling one of several different analog channels or a long time has elapsed since the last sample, allowing charge to leak away. A smaller external source resistance will reduce the time constant. Of course, to evaluate the maximum rate of samples that can be taken, the actual conversion time must be added to this acquisition time.

During the sample (S) period the top capacitor electrodes are held to $0\,\text{V}$ and bottom electrodes are charged to V_{in}. The change-over to the hold (H) position, shown in Fig. 14.8(b), grounds the bottom electrodes and allows the top electrodes to float. The voltage across a capacitor can only change if charge is transferred across electrodes, $\Delta Q = C\Delta V$. Thus the change in voltage $\Delta V = -V_{in}$ at the bottom electrodes is matched at the top floating electrodes, which now become $0 - V_{in}$, as charge cannot flow in or out of the floating top electrodes. Thus at the start of the conversion process the inverting input of the analog comparator is $-V_{in}$.

A 4-bit version of the successive approximation network at the heart of the ADC module is shown in a simplified form in Fig. 14.9. The step-by-step process is sequenced by a shift register (SRG, see Fig. 2.22 on page 37) when the programmer sets the GO/DONE bit in the ADC Control register. As the Control shift register is clocked, a single 1 moves down to activate each step in the sequence:

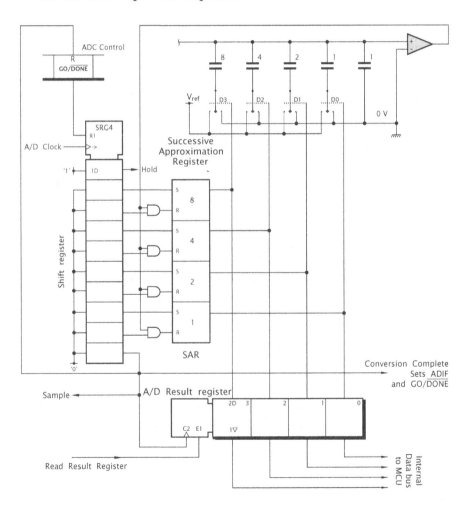

Fig. 14.9 Simplified view of a 4-bit successive approximation A/D converter.

Hold	bit 3	bit 2	bit 1	bit 0	Complete/Sample

The capacitor network is switched to Hold and each capacitor, beginning with the largest value, is switched to V_{ref} in turn. The outcome of the comparator then determines the state of the corresponding bit in the Successive Approximation Register (SAR). The process is detailed in Fig. 14.10. After four set-try-reset actions, the outcome in the SAR is transferred to the Analog-to-Digital RESult register. The $GO/\overline{DONE}$ flag is now cleared to indicate the End Of Conversion and the Analog/Digital Interrupt Flag **ADIF** set. Finally, the analog input is again switched back into the capacitor network (Sample) which then charges up ready for the next conversion after a suitable period.

The total conversion time is approximately six times the clock period t_{AD} of the sequencer shift register—one period for each bit plus one each for the Hold and Ready/sample slots. In the case of a 10-bit module, this will be approximately 12 times the clock period. For the PIC MCU modules, the minimum clocking period is $1.6\,\mu s$ ($\approx 600\,kHz$) for all but the older $2\,\mu s$ PIC16C71/711 devices. There is no specified lower clocking frequency, but as charge slowly leaks away from the network capacitors, a t_{AD} of more than nominally $20\,\mu s$ (50 kHz) should be avoided. From Fig. 14.11 we see that the ADC clock can be derived from one of four sources. The first three of these are fractions of the system clock rate and the fourth is a stand-alone CR oscillator with a nominal t_{AD} of $4\,\mu s$.

The conversion process, where each successive half-fraction of V_{ref} is added to and conditionally taken away from the initial value is illustrated in Fig. 14.10. As we have seen in Fig. 14.8, at the end of the acquisition period the top plates of the capacitor array are at $-V_{in}$. As an example, let us assume that V_{in} is $0.4285V_{ref}$.

1. The process begins by switching in V_{ref} into the lower plate of the largest capacitor, as controlled by the SAR_8 latch in Fig. 14.9. This causes an injection of charge $\Delta Q = C_{total}V_{ref}$, which is identical across both the 8-unit capacitor C_1 and the rest of the capacitors which also have a parallel value of 8 units in Fig. 14.10. Thus the voltage at node N rises by $V_{ref}/2$ to $-0.485 + 0.5 = +0.07125V_{ref}$. In general $\Delta V_N = V_{ref}C_k/C_{total}$. The comparator output is now logic 0 and thus the SAQ_8 latch is consequently cleared, reversing the $V_{ref}/2$ step.

2. SAQ_4 switches V_{ref} into the next highest capacitor giving a $V_{ref}/4$ step at N ($\frac{4}{16}$). The resulting voltage of $-0.485 + 0.25 = -0.178V_{ref}$ giving a comparator output of logic 1 and SAR_4 remains set with the node voltage staying at $-0.1785V_{ref}$.

3. SAQ_2 switches V_{ref} into the second lowest capacitor giving a $V_{ref}/8$ step at N ($\frac{2}{16}$). The resulting voltage of $-0.1785+0.125 = -0.0535V_{ref}$ giving a comparator output of logic 1 and SAR_2 remains set with the node voltage staying at $-0.0535V_{ref}$.

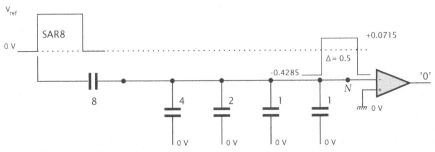

(a) The most significant bit

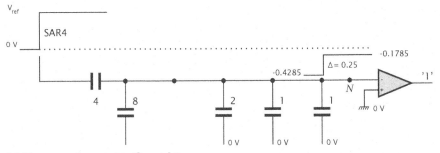

(b) The second most significant bit

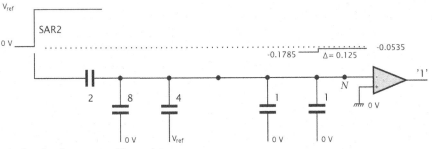

(c) The third most significant bit

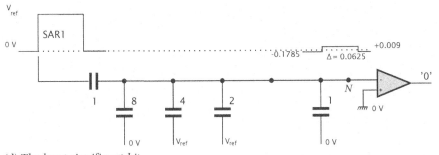

(d) The least significant bit

Fig. 14.10 The successive approximation process.

4. SAQ_1 switches V_{ref} into the lowest capacitor giving a $V_{ref}/16$ step at N ($\frac{1}{16}$). The resulting voltage of $-0.0535 + 0.0625 = +0.009V_{ref}$ giving a comparator output of logic 0 and SAR_1 is cleared reversing the $V_{ref}/16$ step.

The state of the SAR of b'0110' or $0.375V_{ref}$, represents the best 4-bit fit to $V_{in} = 0.4285V_{ref}$. The residue $0.0535V_{ref}$ is the quantizing error.

Most MCUs use an 8- or 10-bit capacitor array. In principle the technique can readily be extended to higher resolutions, but in practice the difficulty in matching ever greater capacitors and internal logic noise means the majority of processors are limited to 12-bit resolution. External high-speed successive-approximation devices with 12+ bit resolution, usually using a resistor ladder network, are readily available, but are relatively expensive.

Matching of the array capacitors, offsets, and resistance of internal switches, leakage currents, and analog comparator non-linearities all contribute to errors in the conversion process. It is beyond the scope of this text to analyze the various measures of error but the device data sheet lists sources and values of these component errors in terms of the least significant bit. For instance, in the PIC12F675 data sheet the 10-bit ADC module is listed as having a total absolute error of $\pm 1\,LSB$. This guarantees that the transfer is monotonic; that is, the binary code will never move in the reverse direction for any change ΔV_{in} of input voltage. This error figure is for $V_{ref} = V_{DD}$; if V_{ref} is lower than V_{DD} then accuracy deteriorates, although values down to 2 V will give acceptable results in many cases.

Both the PIC12F675 and the PIC16F87X exemplar group have an integral 10-bit ADC module. Earlier devices, such as the PIC16F73, use an 8-bit version of this module, which is very similar in architecture and operating process to the 10-bit module shown in Fig. 14.11 and so will not be discussed here. All PIC MCU ADC modules use a capacitor network with characteristics previously described. From the user's perspective the details of the conversion process are less important than the system aspects in integrating this module into software.

In all relevant family members the single analog-to-digital converter is fronted with an analog multiplexer. This allows the software to select up to eight separate analog voltages *one at a time*. Two Control registers allow the program to select any one channel for sampling and to determine the source of the sequence clock. In addition the appropriate pins may be set-up as either analog (by default on a Power-on reset) or digital, with some control over the source of reference voltage. The conversion is initiated via the GO/$\overline{DONE}$ bit, which also indicates when the process is complete and the 10-bit outcome can then be read from the two Result registers.[3]

---

[3] 8-bit modules only require one AD Result register.

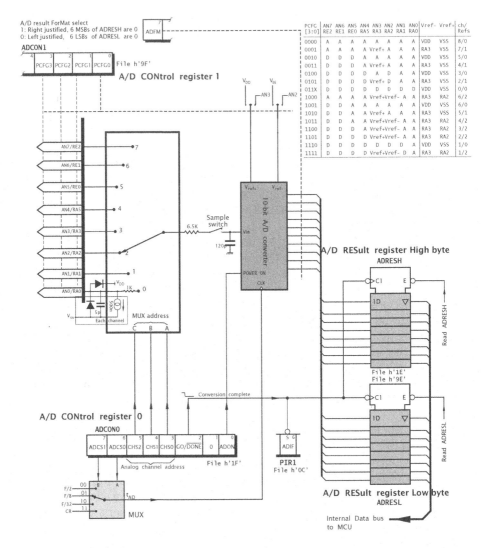

Fig. 14.11 The PIC16F87X 10-bit 8-channel analog-to-digital conversion module.

Our description of the ADC module can be split into the initial set-up and the conversion process.

Initialization
In setting up your module you need to consider the following points:
1. How do I enable the module?
2. How am I going to clock the module?
3. Which channels am I going to use?
4. Do I only need an 8-bit outcome?

All these options are set up using **AD Control register 0, ADCON0** and **AD CONtrol register 1, ADCON1**.[4]

ADON

On a Power-on reset the ADC module is disabled. Setting ADCON[0] to 1 turns the module on. An enabled module typically uses $220\,\mu A$ (PIC16F87X) even when idling, so it should be disabled when not in use where power consumption is a consideration. The $GO/\overline{DONE}$ switch bit should not be set to 1 in the same instruction as the module is enabled to avoid starting a conversion at the same time as the module is being started up.

ADCS[1:0] A/D Clock Select

The module needs a clock signal in order to time the set and test sequence of Fig. 14.10. If the clock rate is too fast, changes in switching voltages will not have time to settle. The data sheet specifies the upper frequency in terms of the A/D clock period t_{AD}, as $1.6\,\mu s$ ($3\,\mu s$ for low-voltage situations) or approximately 600 kHz. For instance, a 5 MHz crystal with ADCS[1:0] = 01 gives a t_{AD} of $1.6\,\mu s$ ($\frac{5}{8}$ MHz) using a $\div 8$ ratio. Table 14.2 shows suggested settings for five typical crystal values.

Table 14.2: ADC clocking frequency versus device crystal frequency.

ADC clock source t_{AD}		PIC MCU crystal frequency				
	ADSC1:0	20 MHz	8 MHz	4 MHz	1 MHz	333 kHz
$f_{OSC}/2$	00	—	—	—	$2\,\mu s$	$6\,\mu s$
$f_{OSC}/8$	01	—	—	$2\,\mu s$	$8\,\mu s$	—
$f_{OSC}/32$	10	$1.6\,\mu s$	$4\,\mu s$	$8\,\mu s$	—	—
CR[1]	11	$2\text{-}6\,\mu s$	$2\text{-}6\,\mu s$	$2\text{-}6\,\mu s$	$2\text{-}6\,\mu s$	$2\text{-}6\,\mu s$
CR[2]	11	$3\text{-}9\,\mu s$	$3\text{-}9\,\mu s$	$3\text{-}9\,\mu s$	$3\text{-}9\,\mu s$	$3\text{-}9\,\mu s$

Note 1: Standard devices; average $4\,\mu s$.
Note 2: Extended-range and low-voltage devices; average $6\,\mu s$.

To allow operation in a low-speed system clock environment; for instance, when a 32.768 kHz watch crystal is used, a separate internal Capacitor-Resistor (CR) oscillator is provided. As this stand-alone oscillator is separate from the system clock, a conversion can be completed while the PIC MCU is in its Sleep state. In this situation, the End Of Conversion interrupt can be used to waken the processor. Doing a conversion with the system clock turned off makes sense,

---

[4]The PIC12F675 has a slightly different arrangement of Control registers with an ANSEL register replacing ADCON1 and a different distribution of Control bits, but the principles are the same.

as this gives a quiet environment with little digital noise. If the separate CR clock is used with a system clock of greater than 1 MHz, then Microchip recommends using a Sleep conversion, as the lack of synchronization between the two clock rates increases noise induced into the analog circuitry.

Unusually, the PIC12F675 has three AD Clock Select bits, giving a frequency division option of ÷64. This is useful to cope with a 20 MHz system clock to give the 3 μs minimum specified for the larger voltage range for this device.

CHS[2:0] CHannel Select

Devices with ADC modules can select to digitize the voltage in one of several possible analog input pins. This ranges from four channels shared with Port GP for the diminutive 8-pin PIC12F675 through to eight shared with Port A and E for the 40-pin PIC16F874/7.

On a Power-on reset, all such shared port pins default to analog inputs; see page 442. As can be seen from Fig. 14.12, an I/O pin configured as an analog input simply disables the digital input buffer—compare with Fig. 11.3 on page 300. No other circuitry is affected. From this we can make the following deductions.

- A port pin configured as analog will read as logic 0 due to the disabled digital input buffer.

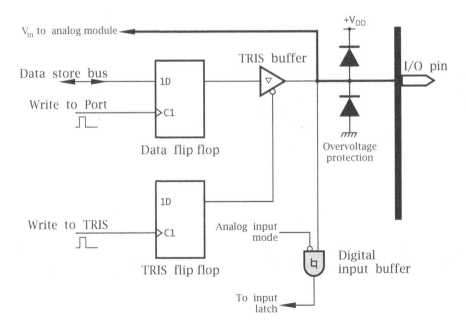

Fig. 14.12 Configuring the analog inputs for Port A and Port E.

- The TRIS buffer is not affected and thus the appropriate TRIS bits should be 1; that is, the direction of the port pins configured as analog should be set to input to prevent contention between the analog V_{in} and the digital state of the Data flip flop.
- The ADC can read an analog voltage at the pin even if that pin has not been configured as analog. However, the still active digital input buffer may consume an excessive current outside of the device's specification.

PCFG[3:0] Port ConFiGuration

If the number of analog channels required for a particular application is less than the maximum available, some unused channels can be reclaimed for use by the associated parallel port. This is accomplished using the appropriate bit pattern in ADCON1[3:0]. The actual choices, number, and location of these bits are device specific, but those patterns applicable to the PIC16F87X group are shown in Fig. 14.11. For instance, if you only require a single analog channel for your project, the pattern b'1110' will leave pin RA0/AN0 as analog and pins RA5, RA[3:1] and RE[2:0] are available for other purposes.

In the situation where no analog activity is required, the ADCON1 register still needs to be set-up; this time to pattern b'0110' or b'0111' to configure all pins as digital.[5] Failure to do this is one of the more common errors, as most newer devices have analog module(s) and, as described on page 442, all relevant pins always default to analog on a Power-on reset. Port pins set-up as analog always read through a parallel port as 0. As observed, any pins used for analog purposes should have their associated TRIS bits set to 1; that is, set-up as inputs.

As we have seen in Fig. 14.10, the successive approximation process essentially comprises a series of tries of ever-halving fractions of a fixed reference voltage. The precision of this process depends on the quality of this reference. One measure of this is the worth of the least significant bit (LSB); that is, the quatization increment. In the case of a 10-bit module this value is $V_{ref}/1024$, or better than 0.1% of this reference.

This reference voltage can be set-up to be internal, using the power supply voltage, say 5 V. For instance, the pattern b'1110' sets pin RA0 to be an analog input and V_{DD} will be the reference used. In this situation the digitization will essentially give the fraction of the supply this analog voltage is.

Using the supply voltage is rather noise prone and its value can vary somewhat. Where more precision or different voltages are required, analog pins can be used as the source of external reference

---

[5]In some devices (not the PIC16F87XA) the Analog Comparator module's Control register also needs to be changed from its default value.

voltages. All ADC modules will allow for at least one external voltage. In the case of the PIC16F87X, one or two external references may be used. For instance, pattern b'0101' would set-up pins RA[1:0] as two analog channels and use RA3 for an external precision voltage reference V_{ref+}; see Fig. 14.20.[6] V_{ref+} can be anywhere between $V_{DD} - -2.5$ V and $V_{DD} + 0.3$ V, with a 2 V minimum.

In some cases it can be an advantage to use a different lower bound than V_{SS} (0 V or ground). Some ADC modules, e.g., the PIC16F87X, allow for a separate lower reference voltage V_{ref-}. For our example, pattern b'1101' also specifies two analog channels with V_{ref+} on pin RA3. However, this time pin RA2 is used for V_{ref-}. This should not be higher than 2 V nor below -0.3 V. Overall, the full range $V_{ref+} - V_{ref-}$ should never be less than 2 V.

ADFM A/D ForMat
Our example ADC module needs two Files to hold the 10-bit outcome. As the total capacity of ADRESH:ADRESL is 16 bits, there are two ways of aligning these ten bits.

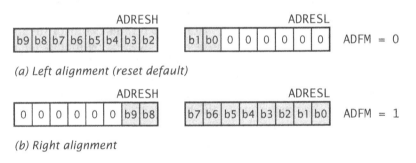

Fig. 14.13 Aligning the 10-bit digital outcome in a 16-bit field.

Many applications only require 8-bit resolution and processing. Where this is the case, the bottom two bits of the outcome word can be thrown away. From Fig. 14.13(a) we see that this is facilitated by left alignment. The content of ADRESL is simply ignored.

Where a full 10-bit word is necessary, setting ADCON1[7] to 1 will right align the datum. As can be seen from Fig. 14.13(b) the outcome is a 10-bit datum extended to a 16-bit format by padding with leading zeros. Normal 16-bit arithmetic and other processing algorithms can then be used.

---

[6]This could be derived from the Voltage Reference module if implemented (see Fig. 14.7) although this has many of the disadvantages of using the raw supply voltage.

Conversion Process

After the module has been configured, from the user's perspective digitizing a selected analog channel is relatively straightforward. Assuming first that interrupts are not being used, the following steps can be identified (including, for completeness, the initialization process) and is visualized with the timeline in Fig. 14.14.

1. Configure ADC module.
   - Set up port pins as analog/voltage reference (ADCON1).
   - Select ADC conversion clock source (ADCON0).
   - Select ADC input channel (ADCON0).
   - Turn on ADC module (ADCON0).
2. Wait for the required acquisition time, typically $20\,\mu s$.
3. Start conversion by setting the GO/$\overline{\text{DONE}}$ bit.
4. Wait for ADC conversion to complete by polling the GO/$\overline{\text{DONE}}$ bit for logic 0.
5. Read the ADRES registers.
6. For next conversion go to step 1 or step 2 as required.

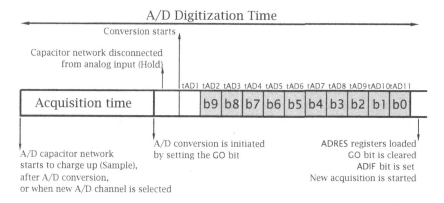

Fig. 14.14 Timeline for the conversion process.

As an example, consider that we wish to continually read each of the eight analog channels of a PIC16F874/7 in turn while outputting the most significant eight digitized bits to Port B and the channel number to the lower three bits of Port D. The main crystal is 20 MHz and the power supply is to be used as the reference voltage.

The listing of Program 14.1 assumes that the ADC module has been initialized at reset with start-up code of the form:

```
include "p16f877.inc"

bsf      STATUS,RP0 ; Change over to Bank 1

clrf     ADCON1     ; Shared port inputs are all analog
                    ; left aligned for 8-bit resolution
clrf     TRISB      ; Port B is all output
movlw    b'11111000'; Low 3 bits of Port D are Output
movwf    TRISD

bcf      STATUS,RP0 ; Back to Bank 0

movlw    b'10000001'; AD clock/32 (10), Ch0 (000)
movwf    ADCON0     ; No conversion (0), ADC turned on (1)
```

which configures the module to enable all eight analog channels with internal reference voltages and for a left-aligned outcome. ADCON1 is initialized to $\boxed{10}\,\boxed{000}\,\boxed{0}\,\boxed{0}\,\boxed{1}$ to select the ADC module clock source as $f_{OSC}/32$ (b'10'); i.e., $\frac{20}{32} = 625\,\text{kHz}$ or $1.6\,\mu s$, channel zero (the initial value is irrelevant) and with the module turned on. With an initial zero value of GO/$\overline{\text{DONE}}$ no conversion is actioned.

With the module initialized, the main software of Program 14.1 spends all its time in a loop reading the digitized equivalent of each channel in turn from ADRESH and copying it in turn to Port B. Before the digitization, the Channel counter is sent to Port D as a modulo-3 number.

The acquisition itself is implemented using the GET_ANALOG subroutine, to which is passed the desired channel number in the rightmost three bits of the Working register. This is copied into a temporary location TEMP, where it is logic shifted three places to the left to align the channel number with the CHSn bits in ADCON0[5:3]. After clearing the CHS[2:0] bits, the shifted Channel number can then be added into ADCON0 to set CHS[2:0] to the appropriate channel.

After the channel number has been set up, a delay subroutine is called to allow for switch delay and stabilization. As we require 8-bit resolution, only a $6\tau \approx 7\,\mu s$ delay is needed to charge up to within 0.25% of the final value, on top of the worst-case switch time of $10\,\mu s$; see page 449. Then the GO/$\overline{\text{DONE}}$ bit in ADCON0 is set to initiate a conversion.[7] The completion of the process can then be monitored by polling GO/$\overline{\text{DONE}}$ until this goes to 0. At this point the contents of ADRESH are the 8-bit outcome of the conversion.

Each actual conversion takes around $13 \times 1.6 \approx 21\,\mu s$, giving a total channel time of $17 + 21 = 38\,\mu s$. Thus a 8-channel scan takes around $38 \times 8 \approx 300\,\mu s$ to complete. That is, around 3300 scans per second.

Rather than polling for completion, the end of conversion can be used to generate an interrupt. In particular if a conversion is to be done in the

---

[7]A conversion may be aborted at any time by clearing GO/$\overline{\text{DONE}}$.

Program 14.1 Scanning an 8-channel data acquisition system.

```
MAIN    clrf    CHANNEL         ; Use a GPF to hold the Channel count
MAIN_LOOP
        movf    CHANNEL,w       ; Get the Channel number
        andlw   b'00000111'     ; Zero the top five bits
        movwf   PORTD           ; Copy to Port D

        call    GET_ANALOG      ; Digitize it; returned in W
        movwf   PORTB           ; and copy to Port B

        incf    CHANNEL,f       ; Advance to next channel
        goto    MAIN_LOOP       ; and DO forever
; ******************************************************************
; * FUNCTION: Analog/digital conversion at channel n               *
; * RESOURCE: Subroutine DELAY_17US, byte TEMP                     *
; * ENTRY   : Channel number in W                                  *
; * EXIT    : Digitized analog value in W                          *
; ******************************************************************
GET_ANALOG
        movwf   TEMP            ; Copy of Channel number in TEMP
        bcf     STATUS,C        ; Shift channel number left 3 places
        rlf     TEMP,f          ; to align with ADCON0[5:3]
        rlf     TEMP,f          ; that is, the CHS[2:0] bits
        rlf     TEMP,w          ; and copy into W
        bcf     ADCON0,CHS0     ; Zero channel bits
        bcf     ADCON0,CHS1
        bcf     ADCON0,CHS2
        addwf   ADCON0,f        ; Moves Channel number to ADCON0[5:3]
        call    DELAY_17US      ; Wait 17us to stabilize
        bsf     ADCON0,GO       ; Start conversion
GET_ANALOG_LOOP                 ; Takes around 20us to finish
        btfsc   ADCON0,GO       ; Check for End Of Conversion
        goto    GET_ANALOG_LOOP

        movf    ADRESH,w        ; Fetch byte when GO/NOT_DONE zero
        return

; ******************************************************************
; * FUNCTION: Delays 17us at 20MHz (85 cycles)                     *
; * ENTRY   : None                                                 *
; * RESOURCE: None                                                 *
; * EXIT    : W is zero                                            *
; ******************************************************************
DELAY_17US
        movlw   d'20'           ; Delay constant
DELAY_17US_LOOP
        addlw   -1              ; Decrement
        btfss   STATUS,Z        ; Until zero
        goto    DELAY_17US_LOOP
        return
```

Sleep mode then this interrupt can be used to waken the device. The ADC module can operate when the PIC MCU is in its Sleep state as it has the option of its own private oscillator to sequence the conversion even if the system oscillator is disabled. The main advantage of a conversion while asleep is the electrically quiet environment when the system oscillator is off. Against this is the considerably longer conversion time, as when the PIC MCU is wakened, there will be the normal 1024-cycle delay to restart the system oscillator; see page 277.

This personal oscillator may be used even where the PIC MCU is not put to sleep. However, as there is no synchronization between the system and local oscillators, clock feedthrough noise becomes a problem, especially with system clock rates above 1 MHz.

The following task list outlines the Sleep state conversion process.

- The ADC clock source must be set to CR, ADCS1:0 = 11.
- The ADIF flag must be cleared to prevent an immediate interrupt.
- The ADIE and PEIE mask bits must be set to enable the ADC interrupt to awaken the processor.
- The GIE mask bit must be 0 unless the programmer wishes the processor to jump to an ISR when it awakens.
- The GO/$\overline{\text{DONE}}$ bit in the ADCON0 register must be cleared to initialize the conversion, followed immediately by the sleep instruction.
- On wakening, the ADRESH:L registers hold the digitized value.

For instance, consider a Sleep state version of the GET_ANALOG subroutine of Program 14.1. This time the initialization code must set up the interrupt system as specified in the task list, to ensure that when the AD Interrupt Flag ADIF is set at the end of the conversion (at the same time as the GO/$\overline{\text{DONE}}$ flag goes to 0) the PIC MCU is woken up.

```
include "p16f877.inc"

bsf     STATUS,RP0  ; Change over to Bank 1
clrf    ADCON1      ; Shared port inputs are analog, 8-bit res

clrf    TRISB       ; Port B is all output
movlw   b'11111000'; Low 3 bits of Port D are Output
movwf   TRISD

bsf     PIE1,ADIE   ; Enable AD interrupts
bcf     STATUS,RP0  ; Back to Bank 0

movlw   b'11000001'; CR AD clock (11), Ch0 (000)
movwf   ADCON0      ; No conversion (0), ADC turned on (1)

bcf     PIR1,ADIF   ; Zero the AD interrupt flag
bsf     INTCON,PEIE; & enable the Peripheral interrupt group
```

Apart from the initialization of the interrupt system, the only change is to the setting of ADCON0[7:6], which is made b'11' to select the internal ADC oscillator as the clock.

Program 14.2 Scanning an 8-channel data acquisition system.

```
; ****************************************************************
; * FUNCTION: A/D conversion at channel n while asleep         *
; * RESOURCE: Subroutine DELAY_17US, byte TEMP                 *
; * ENTRY    : Channel number in W                             *
; * EXIT     : Digitized 8-bit analog value in W               *
; ****************************************************************
;
GET_ANALOG
        movwf   TEMP            ; Copy of Channel number in TEMP
        bcf     STATUS,C        ; Shift channel number left 3 places
        rlf     TEMP,f
        rlf     TEMP,f
        rlf     TEMP,w          ; and copy into W
        bcf     ADCON0,CHS0     ; Zero channel bits
        bcf     ADCON0,CHS1
        bcf     ADCON0,CHS2
        addwf   ADCON0,f        ; Moves Channel number to ADCON0[5:3]
        call    DELAY_17US      ; Wait 17us to stabilize

        bcf     INTCON,GIE      ; Disable all interrupts
        bcf     PIR1,ADIF       ; Ensure the AD Int flag is 0 before
        bsf     ADCON0,GO       ; starting the conversion

        sleep                   ; Doze in quiet while converting

        bsf     INTCON,GIE      ; Re-enable interrupts (optional)
        movf    ADRESH,w        ; Fetch byte when awake
        return
```

The Sleep version of GET_ANALOG shown in Program 14.2 is virtually identical to the original version, with a the following changes.

1. GIE may need to be cleared if other devices can request an interrupt.
2. Before the conversion is started, the ADIF flag is cleared to ensure that the Sleep state is not prematurely terminated.
3. A sleep instruction follows the setting of the GO/$\overline{\text{DONE}}$ switch. Where the local clock option is selected, an extra t_{AD} period is automatically inserted to ensure that conversion only begins after the sleep instruction has been executed.
4. There is no need to poll the GO/$\overline{\text{DONE}}$ status flag, as the PIC MCU will only restart after the conversion has completed and will then execute the following instruction. In our example the GIE mask bit has been cleared, and it should then be set again to 1 if there is to be interrupt activity from other sources. If GIE is permanently left at 1 then the processor will automatically jump to an ISR after it awakens.

For our final example we are going to code a 20 MHz PIC16F874 in CCS **C** to act as a magnitude comparator in the manner of Example 11.2 on page 319. Here we want to measure up the parallel-input 8-bit word N at Port B against an analog input at Channel 1. Outputs at RC[2:0] are to represent Analog Lower Than N (b'001'), Equivalent (b'010') and Higher Than N (b'100') respectively. The comparator is to have a hysteresis of $\Delta = \pm 1$ bit; called delta in our program. That is, if a previous comparison showed Analog < N then the new trigger level is $N + 1$. Similarly, on a downward trajectory the trigger level is decreased to $N - 1$.

The function compare() of Program 14.3 assumes that initialization code of the form:

```
#include <16f874.h>
#byte PORT_B = 0x06
#byte PORT_C = 0x07
#device ADC=8              /* Configure for an 8-bit outcome  */
/* Declare function to which is send delta (+1 or -1)
and which returns updated value +1 or -1                     */
unsigned int compare(unsigned int delta);

int main()
{
unsigned int hysteresis = 0;
set_tris_c(0xF8);
setup_adc(ADC_CLOCK_DIV_32);
setup_adc_ports(RA0_RA1_RA3_ANALOG);
set_adc_channel(1);
```

has already been executed.

The key internal functions used here are:

setup_adc(ADC_CLOCK_DIV_32);
This function configures bits ADCS[1:0] in ADCON0[7:6] to select the module's clock source; here the processor oscillator/32. The script ADC_CLOCK_INTERNAL may be used to select the internal CR oscillator.

setup_adc_ports(RA0_RA1_RA3_ANALOG);
This configures bits PCFG[3:0] in ADCON1[3:0] to select which port pins are analog, which are digital, and if external reference voltages are to be used. The script RA0_RA1_RA3_ANALOG indicates that port lines RA3 and RA[1:0] are to be analog with internal reference voltages, with the rest being digital — PCFG[3:0] = b'0100'; see Fig. 14.11. The equivalent script using an external V_{ref+} at RA3 is RA0_RA1_ANALOG_RA3_REF. Scripts appropriate to any particular device are stored in the corresponding header file; in this case 16f874.h. All devices with an ADC module have scripts ALL_ANALOG and NO_ANALOGS.

set_adc_channel(n);
This is used to set up the channel number bits CHS[2:0] in ADCON0[5:3].

read_adc();
This activates GO/$\overline{\text{DONE}}$ in ADCON0[2] and returns with the digitized value from ADRESH:L when GO/$\overline{\text{DONE}}$ goes to 0.

#device ADC=8
This directive configures the ADC module to left align the 10-bit outcome (see Fig. 14.13) and is used by the function read_adc() to return an 8-bit int, which it gets from ADRESH. The directive device ADC=10 returns a long int from ADRESH:L.

Program 14.3 A digital/analog comparator with hysteresis.

```
unsigned int compare(unsigned int delta)
{
unsigned int analog;
analog = read_adc();
if(analog > PORT_B + delta) {PORT_C = 0x04; delta = 0xff;}
if(analog == PORT_B) {PORT_C = 0x02;}
else {PORT_C = 0x01; delta = 1;}
return delta;
}
```

The function `compare()` in Program 14.3 expects the value of the hysteresis, called `delta`, which here is either +1 or −1 (h'FF'). After the ADC module is read, the digitized value `analog` is compared with the contents of Port B plus `delta` and the three Port C bits (`RC[2:0]`) set to their appropriate state.

At the same time as the comparison is resolved, `delta` will be updated to reflect the outcome (i.e., +1 if `analog` < (PORT_B + delta), −1 if `analog` > (PORT_B + delta)). The value `delta` is returned by the function to allow the caller function to update its variable; called, say, `hysteresis`. Thus to activate the comparator outputs and also update `hysteresis` at the same time, the caller might have a statement such as `hysteresis = compare(hysteresis);`. An alternative would be to define the variable `hysteresis` before the main function `main()` making it global; that is, known to all functions. In this situation its value need not be passed by the caller back and forth to any appropriate function.

Conversion from a digital quantity to an analog equivalent is somewhat simpler than the converse and not so commonly required. Perhaps for these reasons digital-to-analog converters (DACs) are not often found as an integral function in most MCU families.

We have already seen that one way of providing this mapping is to vary the mark:space ratio of a pulse train of constant repetitive duration, as shown in Fig. 13.9 on page 424. Here a small digital number gives a skinny pulse, which when smoothed out by a low-pass filter (which gives the average or d.c. value) translates to a low voltage. Conversely, a large digital number leads to a correspondingly large mark:space ratio, which in turn, after smoothing, yields a higher voltage.

PWM conversion can be very accurate and is simple to implement. However, extensive filtering is required to remove harmonics of the pulse rate and this makes the conversion slow to respond to changes in the digital input. Normally PWM is used to control heavy loads, such as motors or heaters, where the inertia of these devices inherently provides the smoothing action. Furthermore, the pulsed nature of the signal is ideally suited to power control, activating thyristor firing circuits.

Another way is to switch in a tapping on a chain of resistors, each adding one least significant bit increment to the grand total. This is the principle used in the Comparator Voltage Reference module of Fig. 14.7. However, rather a lot of resistors are needed; e.g., 1024 for 10-bit resolution.

Many commercial DAC devices are available which can be controlled externally. Two examples were given in Figs. 12.3 and 12.5 on pages 336 and 340, where the MCU transferred digital data in series. Here for completeness, we will look at an example where parallel data transfer is used.

The majority of proprietary devices are based on an R-2R ladder network, such as that shown in Fig. 14.15(a). Voltage appearing at any bit switch node emerges at the output node in an attenuated form. As our analysis will show, each move to the left attenuates this voltage b_i by 50%, which is the binary weighting relationship:

$$V = \sum_{i=0}^{N+1} b_i \times 2^i$$

for an N-bit word.

In Fig. 14.15(b), at node A looking to the left we see a resistance of R (2R//2R) and the voltage is attenuated by two. As we move to the right the process is repeated with each voltage divided by two. Thus, at node B the voltage $b_0/2$ is further divided by two as is voltage b_1, giving $V_B = b_0/4 + b_1/2$. As the network is symmetrical the resistance looking right at any mode is also 2R. This means that as seen from *any* digital switch, the total resistance is 2R + 2R//2R = 3R. This is important as the characteristics of a transistor switch, such as resistance, are dependent on current, and keeping this the same reduces error.

For clarity our analysis has been for three bits. This can be extended by simply moving the leftmost terminating resistor over and inserting the requisite number of sections. This does not affect the resistance as seen left of the mode, and therefore does not change the conditions of the rightmost sections. An inspection of our analysis shows that nowhere does the absolute value of resistance appear. In fact the accuracy of the analysis depends only on the R:2R ratio. While it is relatively easy to fabricate accurate ratioed resistors on a silicon die, this is certainly not the case for absolute values. For this reason R:2R networks are the standard technique used for most integrated circuit DACs.

The Maxim MAX506 of Fig. 14.16 is an example of a commercial D/A converter (DAC). This 20-pin footprint device contains four separate DACs sharing a common external V_{ref}. Digital data is presented to the D[7:0] pins and one of four latch registers selected with the A[1:0] address inputs. Once this is done, the datum byte is loaded into the selected register n and appears at the corresponding output VOUTn.

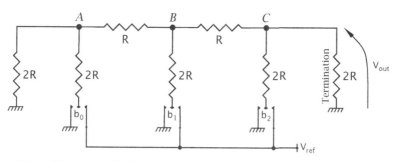

(a) A 3-bit R-2R ladder network

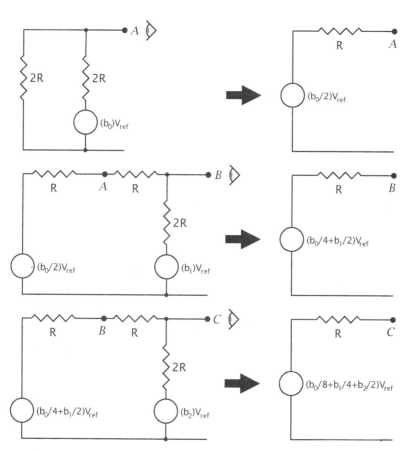

(b) Reducing the circuit

Fig. 14.15 R-2R digital-to-analog conversion.

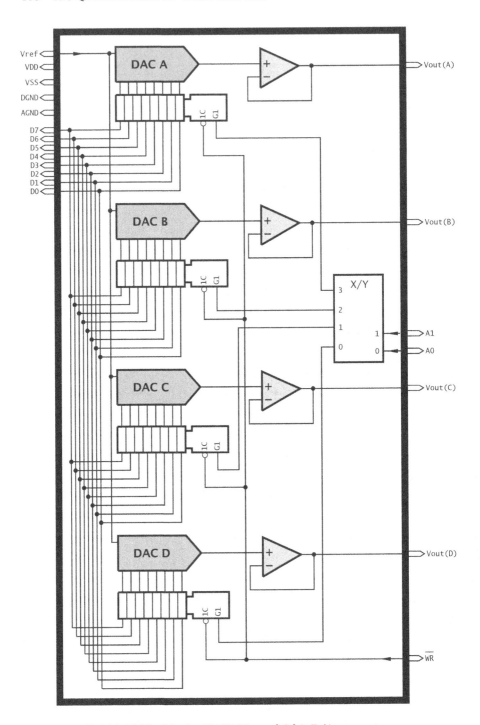

Fig. 14.16 The Maxim MAX506 quad 8-bit D/A converter.

This output analog voltage ranges from zero (Analog GrouND) for a digital input of h'00' through to V_{ref} for a digital input of h'FF'. Where V_{SS} is connected to ground, then V_{ref} can be anything between 0 V and V_{DD} (+5 V). However, V_{SS} can be as low as −5 V and in this situation V_{ref} can be anywhere in the range ±5 V. If V_{ref} is negative for dual supplies then the output voltage will also be negative. In either case, effectively the output can be treated as the product $D \times V_{ref}$ where D is the digital input byte scaled to the range 0–1 (h'00–FF').

The MAX505 is a 24-pin variant which permits separate reference voltages to be used for each of the four DAC channels. In addition, the MAX505's DAC latches are isolated from the converter ladder circuits by a further layer of latches all clocked at the *same* time with a $\overline{LDAC}$ (Load DAC) control signal. This double buffering permits the programmer to update all four DACs simultaneously after their individual latches have been set up.

As an example, consider that a MAX506 quad DAC has its Address selected via RA[1:0] and RA2 drives the $\overline{WR}$ input to latch in the addressed data from Port B. We need to generate the continuous staircase sawtooth waveform shown in Fig. 14.17 from DACD. A suitable software routine would be something like the following listing:

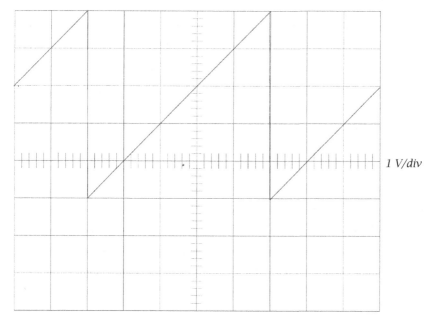

TIME BASE 0.1 ms/div

Fig. 14.17 Generating a continuous sawtooth using a MAX506 DAC.

```
        movlw   b'0111'     ; DACD is channel 3 (b'11'), WR = 1
        movwf   PORTA       ; To MAX506 WR, A1:0
LOOP    movwf   PORTB       ; Datum to MAX506's D7:0
        bcf     PORTA,2     ; WR = 0; Latch datum in
        bsf     PORTA,2     ; WR = 1; by pulsing WR
        addlw   1           ; Increment staircase count
        goto    LOOP        ; and repeat forever
```

where we are assuming that Port B and Port A[2:0] have been set up as outputs.

The typical DAC staircase output waveform shown in the oscillogram in Fig. 14.17 is based on a 12 MHz crystal clocked PIC MCU. With a loop cycle count of six cycles gives a sawtooth duration of $(256 \times 6)/3 \approx 0.5$ ms, at $2\,\mu s$ per step.

Examples

Example 14.1
The analog input channel voltage range for most ADC modules[8] is limited to the positive range $0 - V_{ref+}$, where V_{ref+} can either be the internal V_{DD} voltage or an external voltage at RA3 in the range $3 - V_{DD}$. Many situations require a digitized mapping from bipolar analog signals. Design a simple resistive network to translate a bipolar voltage range of ± 10 V to a unipolar range of $0 - 5$ V, assuming V_{ref+} is +5 V. Extend the design to give an anti-aliasing filter, assuming a sampling rate of 5000 per second.

Solution
One possibility is shown in Fig. 14.18. The value of the three resistors must be such that the input voltage V_{in} range of ± 10 V will be shifted so that the midpoint of 0 V gives half-scale ($V_{ref+}/2 = 2.5$ V) at the input pin AN. The range at this pin must also be attenuated by a factor of 4. A more general way of expressing this is given by the relationship $V_{in} = \pm G \times V_{ref+}$.

1. When V_{in} is zero, the voltage at the summing node is half-scale, which maps to b'10000000'. To do this, R_1 paralleled with R_2 must have the same resistance as R_3, i.e.,
   $$R_3 = R_1//R_2$$

2. The attenuation of the network is a function of the potential divider between R_1 and $R_2//R_3$. This gives us the value of G as:
   $$2G = (R_1 + (R_2//R_3))/(R_2//R_3)$$

Where in our instance $G = 2$.

After some manipulation we have:

$$R_1 = (G - 1) \times R_2$$
$$R_2 = G \times R_3$$

---

[8]The PIC16C77X devices can be configured to accept bipolar input analog voltages.

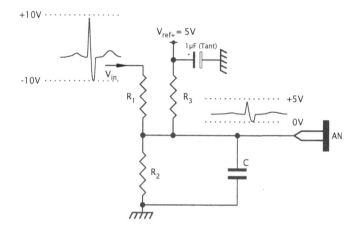

Fig. 14.18 A level-shifting resistor network.

Of course we have three unknowns and only two equations, so we have to start off by choosing a value for one of them. If we pick a value of $5\,k\Omega$ for R_3, then we have $R_2 = 2 \times 5 = 10\,k\Omega$ and $R_1 = 10\,k\Omega$.

The resistance looking out from the pin is all three resistors in parallel; which in our case is $2.4\,k\Omega$. This meets the maximum to keep within a LSB leakage error for a 10-bit conversion. For 8-bit resolution, the resistor values could be increased by a factor of four.

A small capacitor at the summing node can be used to implement a simple first-order low-pass filter to attenuate high frequencies from external sources, such as the MCU's system clock, and act as an anti-aliasing filter, as described in Fig. 14.4. With a sampling rate of 5000 per second, then ideally the filter break frequency should be no more than $2.5\,kHz$— half the sampling frequency. As this filter has an attenuation of only $6\,dB$ per octave, choosing a break frequency $\frac{1}{2\pi CR}$ of $1\,kHz$ provides a generous margin. We then have:

$$\frac{1}{2\pi CR} = 1000$$
$$C = \frac{10^{-6}}{4.8 \times \pi}$$
$$C \approx 66\,nF$$

To further reduce noise, the filter capacitor should have good high-frequency characteristics, e.g., polyester (capacitors become inductors at high frequencies) and together with the resistors, be physically as close as possible to the pin and not adjacent to any digital lines. It is always good practice to decouple the reference voltage and power supply with a low-value Tantalum electrolytical capacitor or/and a $0.1\,\mu F$ ceramic capacitor to reduce switching noise from the MCU and other devices taking power from the same source. Using a separate supply and ground connection to the power supply to the PIC MCU should also reduce noise from this source.

Example 14.2

As part of a smart biomedical monitor, the peak analog value of an electrocardiogram (EKG) signal is to be determined anew for each cycle. This R-point (see Fig. 7.1 on page 186) maximum value is to be output from Port B and RA5 is to be pulsed High whenever this value is being updated. Assuming that a PIC16F87X is used to implement the intelligence, and the EKG signal (conditioned as shown in Fig. 14.18) is connected to Channel 1 RA1, devise a possible strategy. Timer 0 is being used to interrupt the processor at nominally 2000 times per second; see Program 13.2 on page 411. Design a suitable ISR to implement your strategy.

Solution

As in any biomedical parameter the EKG signal will vary from cycle to cycle in gain, shape, and period. Even if this were not so, imperfections in the data acquisition system, notably the skin electrodes, can cause slow baseline (d.c.) drift. Thus the threshold at which the signal is to be tracked to its peak R-value must be reset at some sensible fraction of its previous peak during the period following the last update.

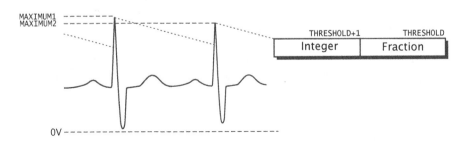

Fig. 14.19 EKG detection strategy.

One possibility is shown in Fig. 14.19. Here the threshold is slowly decremented after the peak to ensure that a following peak of lower amplitude is not missed. On the basis of a lowest EKG rate of 40 beats per minute (period 1.5 s), if we reduce the threshold by $\frac{1}{64}$ of a bit every sample then the maximum reduction would be a count of ≈ 47 at a sample rate of 2000 per second. To do this the threshold value THRESHOLD in Program 14.4 is stored as a double-byte number of form integer:fraction and $\frac{1}{64}$ of an integer (i.e., fraction = b'00000100') subtracted in each sample where the peak value MAXIMUM is not updated. This droop rate can be altered by changing the subtracted fraction.

The task list implemented by this listing is:
1. DO a conversion to get ANALOG.
2. IF (ANALOG > THRESHOLD)
   - MAXIMUM = ANALOG.
   - THRESHOLD = ANALOG.
   - PORTB = ANALOG.
   - RA5 = 1.
3. ELSE
   - Reduce THRESHOLD by $\frac{1}{64}$.
   - RA5 = 0.

In updating THRESHOLD (where ANALOG > THRESHOLD) the integer byte takes the new value of MAXIMUM whilst the fractional byte is zeroed. Treating this byte pair as a 16-bit word, this effectively equates the threshold as MAXIMUM × 256 or THRESHOLD = MAXIMUM << 8, where MAXIMUM has been shifted left eight places. We are assuming that THRESHOLD has been zeroed in the background program during the initialization phase and that we are doing an 8-bit conversion.

If the digitized analog sample is less than the threshold trip value then h'04' = b'00000100' is subtracted from the lower byte at THRESHOLD+1 and if this produces a borrow, then the upper byte at THRESHOLD is decremented. This subtract $\frac{1}{64}$ routine is skipped if the threshold has reached zero, thus preventing underflow.

Program 14.4 uses the subroutine GET_ANALOG of Program 14.1 and its associated 17 μs delay subroutine. However, as there is a considerable period between calls, the 17 μs delay can be reduced to 10 μs if required.

Program 14.5 gives the C coded version implementing our task list. The #int_rtcc directive tells the compiler to treat the following function as a Real-Time Counter Clock (Timer 0) ISR. In function ecg_isr(), the variables threshold and maximum are declared static. This means that their value will be retained after the function has exited and will be available next time on entry. The default way of treating C function variables is to hold their value only for the duration of the function. An alternative way of dealing with this problem is to declare such variables outside any function, in which case they will be global and retain their value indefinitely.

The threshold variable is defined as a long int and the CCS compiler will then treat this datum as a 16-bit variable as required. The definition in equating threshold to zero will only initialize it once when the program begins its run, as the variable is static. Again this is not the normal behavior of the default auto variable.

In equating threshold to the new maximum value, the latter is multiplied by 256 by shifting left eight times. A good compiler will automatically change a N*256 to N<<8; or even better just take the upper byte of

Program 14.4 EKG peak picking.

```
; *****************************************************************
; * FUNCTION: ISR to update the EKG parameters                    *
; * ENTRY   : On a Timer0 interrupt                               *
; * EXIT    : Update MAXIMUM and THRESHOLD:THRESHOLD+1            *
; * RESOURCE: GET_ANALOG subroutine gets 8-bit digitized data     *
; *****************************************************************
; First save context
EKG_ISR movwf   _work           ; Put away W
        swapf   STATUS,w        ; and the Status register
        movwf   _status

; ================================================================
        btfss   INTCON,T0IF     ; Was this a Timer0 interrupt?
         goto   EKG_EXIT        ; IF not THEN exit

        bcf     INTCON,T0IF     ; ELSE clear flag
        movlw   1               ; Initiate a conversion of
        call    GET_ANALOG      ; Channel 1

        movwf   TEMP            ; Save digitized byte
        subwf   THRESHOLD,w     ; THRESHOLD - ANALOG
        btfsc   STATUS,C        ; IF no Borrow THEN
         goto   BELOW           ; don't update MAXIMUM
        movf    TEMP,w          ; ELSE get digitized value
        movwf   MAXIMUM         ; which is the new MAXIMUM
        movwf   PORTB           ; made visible to outside
        bsf     PORTA,5         ; which is signaled
        movwf   THRESHOLD       ; Now update double-byte
        clrf    THRESHOLD+1     ; threshold
        goto    EKG_EXIT        ; and finish

; Land here if the input is below the threshold
BELOW   bcf     PORTA,5         ; Signal no update
; Now reduce the threshold by 0.5 unless it is zero
        movf    THRESHOLD,f     ; Is integer threshold zero?
        btfsc   STATUS,Z        ; Skip if not
         goto   EKG_EXIT        ; IF it is THEN leave alone

        movlw   h'04'           ; 1/64 = b'00000100'
        subwf   THRESHOLD+1,f   ; Take away from fraction byte
        btfss   STATUS,C        ; Skip if no borrow
        decf    THRESHOLD,f     ; ELSE decrement integer thresh
; ================================================================

EKG_EXIT swapf  _status,w       ; Untwist the original Status reg
        movwf   STATUS
        swapf   _work,f         ; Get the original W reg back
        swapf   _work,w         ; leaving STATUS unchanged
        retfie                  ; and return from interrupt
```

Program 14.5 An implementation of the EKG peak picker in **C**.

```
#bit  RA5     = 5.5          /* Pin RA5 is bit 5 of Port A */
#byte PORT_B = 6             /* Port B is File 6           */

#int_rtcc
ecg_isr()
{
unsigned int analog;
static unsigned long int threshold = 0;
static unsigned int maximum;
analog = read_adc();

if(analog > (threshold>>8))
    {
    maximum = analog;          /* New maximum value     */
    PORT_B  = analog;          /* Show the outside world */
    threshold = maximum << 8;  /* New 2-byte threshold  */
    RA5 = 1;                   /* Tell outside world    */
    }
else
    {
    threshold = threshold - 0x0004; /* Reduce by 1/64    */
    RA5 = 0;                        /* Signal no update  */
    }
}
```

the pair as the outcome. This double-byte form allows for the reduction by $\frac{1}{64}$ of a bit h'0004' to give the specified falling trip level.

Example 14.3
A microcontroller is to be used to calculate a measure of power discharged by the diphasic defibrillator of Fig. 14.5. When the MCU detects the beginning of the discharge, 256 samples are to be taken at a nominal rate of 20,000 per second, with the sum of the squares of the deviation from the baseline voltage being an analog measure of the power—assuming that the resistance of the patient's chest/electrodes remaining constant whilst all this is going on!

A 4.096 V voltage reference device is to act as an external reference voltage for a ADC module, giving a 16 mV resolution for an 8-bit conversion. After the process begins, pin RA4 is to be pulsed as a trigger for a storage oscilloscope, which allows the waveform to be captured for archiving purposes. When the process has been completed, the top byte of the power summation is to be output via Port B for display.

Show how you might use a 20 MHz PIC16F87XA device to implement the logic of the measurement system. You can assume that the voltage reference device can be biased as for a Zener diode. In practice an optional potentiometer can be used to trim the voltage slightly for more accurate results.

Solution

Figure 14.20 shows a suitable hardware configuration, from which we can estimate the peripheral budget. The signal itself ranging between +1.8 and +3.6 V (see Fig. 14.5) is connected to Analog channel 0 at pin RA0/AN0. The 10 kΩ resistor protects the analog input against overvoltage as well as implementing an anti-aliasing filter, with the 3.3 nF capacitor giving a 450 kHz nominal breakpoint. As the actual defibrillator uses very large voltages (of the order of 25 kV) the two 1N4004 diodes act as additional protection against high-voltage spikes, supplementing the internal diodes shown in Fig. 14.12.

The external 4.096 V reference voltage is directly connected to pin RA3, which for the ADC module configuration mode b'0101' is used as the

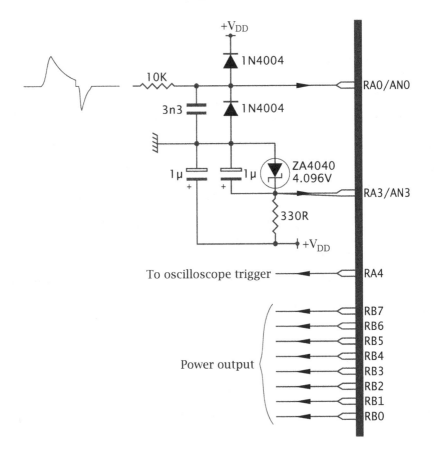

Fig. 14.20 Measuring the discharge power for an EKG defibrillator.

external reference. Both the V_{DD} and V_{ref+} voltages are decoupled with 1 μF tantalum capacitors to reduce noise at this point.

An internal analog comparator is used to detect the initial rise of the discharge voltage as described in Fig. 14.5. If Comparator module mode b'1110' is used then the CVREF module can be used to generate an internal reference voltage as described on page 447. With the CIS bit cleared (see Fig. 14.6) Comparator 1 can share Analog channel 0 with the ADC module.

Finally, both RA4 and all of Port B must be configured as digital outputs. The former is going to be used to generate a synchronization pulse, and the latter to output the end result of the analysis.

```
        include "p16f877a.inc"

        org   0             ; Reset vector
        goto  SET_UP        ; On reset go to background routine
        org   4             ; Interrupt vector
        goto  EKG_ISR       ; On Comparator change go to ISR

SET_UP  bsf   STATUS,RP0    ; Change to Bank 1
        movlw b'00000110'   ; Comparator mode 110 CIS = 0
        movwf CMCON
        call  DELAY_17US     ; Allow 17us for voltages to settle
        movf  CMCON,f        ; Read CMCON to clear Change state
        bsf   PIE2,CMIE      ; Enable Comparator interrupts

        movlw b'10001110'   ; CVREF module on (1), not extern (0)
        movwf CVRCON        ; Hi range (0), CVR[3:0] = 1110

        movlw b'00000101'   ; RA0/1 analog inputs
        movwf ADCON1        ; RA3 is Vref+ input

        movlw b'101111'     ; Make RA4 an output
        movwf TRISA
        clrf  TRISB         ; PortB is all output

        bcf   STATUS,RP0    ; Back to Bank 0

        movlw b'10000001'   ; Clock /32 turn on ADC module
        movwf ADCON0

        bcf   PIR2,CMIF     ; Zero the Comparator interrupt flag
        bsf   INTCON,PEIE   ; Enable Peripheral interrupt group
        bsf   INTCON,GIE    ; & Globally enable interrupt system
```

Based on our analysis, the initialization code is shown above. The modules are set-up as follows:

1. The Analog comparator module is turned on in Mode b'110' with CIS = 0. For convenience, the 17 μs delay subroutine DELAY_17US used by GET_ANALOG is also employed to allow the module to settle, rather than a separate 10 μs delay. After this, the CMCON register is read to clear any Change condition. This allows the CMIF flag to be subsequently zeroed and interrupts enabled.

2. The CVREF module is enabled and set to tapping b'1110' in the high range to give a 3.4375 V reference.

3. The ADC module is enabled and set to mode b'0101' to configure pin RA0 as an analog channel (and also RA1) and use RA3 for the external V_{ref+}. Alignment is set-up to facilitate an 8-bit outcome. The ADC clock is sourced as the system 20 MHz frequency divided by 32. This gives a 625 kHz rate, as shown in Table 14.2.

4. PORTA[4] is set-up as an output. Other Port A pins are left as inputs, as required for AN0, AN1 and AN3. All of Port B is configured as output.

The actual software itself is shown in Program 14.6. The Main routine itself simply sleeps until the Comparator module changes state and generates an interrupt. When control returns to the background routine, the top byte of the triple-byte Power accumulator is copied to Port B and the process repeated for the next run.

After saving the context in the normal way, the foreground routine first confirms the source of the interrupt and then clears a loop counter and the three bytes used to store the grand sum of the 256 squares of the sampled voltage. Pin RA4 is then pulsed to tell the outside world that the discharge is beginning.

The GET_ANALOG subroutine listed in Program 14.1 is used to acquire an 8-bit digitized sample. The difference between the baseline voltage of 2.6 V (see Fig. 14.5) is then determined. If this is negative; that is, a Borrow is generated after the subtraction showing that the input voltage is below 2.6 V, then the difference byte is 2's complement inverted (see page 9) to switch from negative to positive. This modulus voltage is then squared using the SQR routine of Program 8.3 on page 232. The two global return bytes SUM:SUM+1 are then added to the triple-byte total POWER:POWER+1:POWER+2 array.

This is repeated 256 times with an extra loop delay of 470 μs to give an approximate 500 μs total delay necessary to give the specified 2 kHz sampling rate. When this has been completed, taking a total time of around 128 ms, the Comparator module is read to clear the difference condition. This is done at the end of the process, rather than at the beginning, as the input voltage will fall back through the 3.4375 V Comparator threshold part way through the process and trigger another change! The CMIF flag is then cleared and the context, restored.

Of course this is rather rudimentary. For instance, the baseline voltage may vary with time, so a learning run prior to an analysis may be necessary. If fairly stable, this value can be burnt into non-volatile memory as described in the next chapter. The use of a fixed number of samples can be restrictive, and additional loops can be implemented until the voltage difference drops below a certain threshold.

Program 14.6 Gauging the defibrillator discharge power.

```
MAIN        sleep               ; Idle
            nop
            movf   POWER,w      ; Get top byte of power
            movwf  PORTB        ; and output
            goto   MAIN
;  ****************************************************************
;  * FUNCTION: ISR to begin the defibrillator analysis          *
;  * ENTRY   : On a Comparator module interrupt                 *
;  * EXIT    : Update POWER:3                                    *
;  * RESOURCE: GET_ANALOG subroutine gets 8-bit digitized data  *
;  * RESOURCE: SQUARE subroutine does 8x8 multiplication         *
;  ****************************************************************
;
; First save context
EKG_ISR     movwf  _work        ; Put away W
            swapf  STATUS,w     ; and the Status register
            movwf  _status
; ===============================================================
            btfss  PIR2,CMIF    ; Was this a Comparator interrupt?
            goto   EKG_EXIT     ; IF not THEN exit
            bcf    PIR2,CMIF    ; ELSE clear the interrupt
            clrf   POWER        ; Zero the 3-byte grand total
            clrf   POWER+1
            clrf   POWER+2      ; LSB
            clrf   COUNT        ; Prepare to do loop 256 times
            bcf    PORTA,4      ; Pulse pin RA4
            bsf    PORTA,4      ; to generate a synch pulse
            bcf    PORTA,4
ACQUIRE     clrw                ; Analog channel 0 (W is h'00')
            call   GET_ANALOG   ; Do a conversion
            addlw  -BASELINE    ; Difference from baseline voltage
            btfsc  STATUS,C     ; IF Borrow (C==0) THEN skip
            goto   EKG_CONT     ; as difference is positive
            xorlw  b'11111111'  ; ELSE invert
            addlw  1            ; plus one makes -ve diff positive
EKG_CONT    call   SQR          ; Square it
            movf   SUM+1,w      ; Get LSB of squared voltage
            addwf  POWER+2,f    ; Add it to the low byte of Power
            btfss  STATUS,C     ; Check for a Carry
            goto   NEXT_BYTE    ; IF not THEN add next byte
            movlw  1            ; Increment the high byte of Power
            addwf  POWER+1,f
            btfsc  STATUS,C     ; Check; did this generate a Carry?
            incf   POWER,f      ; IF yes THEN increment upper byte
NEXT_BYTE   movf   SUM,w        ; Get MSB of squared voltage
            addwf  POWER+1,f    ; Add it to the high byte of Power
            btfsc  STATUS,C     ; Check for a Carry; IF yes
            incf   POWER,f      ; THEN increment the Upper byte
            call   DELAY_470US  ; Hang around until the next sample
            incfsz COUNT,f      ; Increment the loop count and do
            goto   ACQUIRE      ; another acquisition if not zero
; ===============================================================
EKG_EXIT    bsf    STATUS,RP0   ; First clear the Comparator
            movf   CMCON,f      ; Change situation
            bcf    STATUS,RP0   ; by reading it in Bank 1
            bcf    PIR2,CMIF    ; and clear the interrupt flag
            swapf  _status,w    ; Untwist the original Status reg
            movwf  STATUS
            swapf  _work,f      ; Get the original W reg back
            swapf  _work,w      ; leaving STATUS unchanged
            retfie              ; and return from interrupt
```

Example 14.4

Using **C** coding show how a 10-bit digitized reading from Channel 3 of a PIC16F874 can be acquired with the processor in its Sleep state.

Solution

The CCS compiler uses the sleep() function to put the MCU to sleep; this simply translates to the sleep instruction. A Sleep conversion cannot be implemented using the read_adc() function of Program 14.3 as no processing is done in the Sleep state. Instead we need to set and clear individual interrupt related bits before going to sleep in the manner outlined in the assembly-level Program 14.2. On wakening the state of ADRESH:L registers can then be read "manually" and combined to give the 10-bit outcome.

Coding for this specification is shown in Program 14.7. Here the PEIE, ADIF and GO/$\overline{\text{DONE}}$ bits are defined using the #bit directive. This time the script ADC_CLOCK_INTERNAL is used with the setup_adc() internal function to select the internal CR clock for the DAC module, as necessary for the Sleep conversion.

Program 14.7 Sleep conversion in **C**.

```
#include <16f874.h>
#device ADC=10          /* Configure for a 10-bit outcome       */
#use delay(clock=8000000) /* Tell compiler it's an 8MHz clock */
#bit ADIF = 0x0C.6      /* The A/D interrupt flag in PIR1[6]    */
#bit PEIE = 0x0B.6      /* The group interrupt flag in INTCON[6] */
#bit GO  =  0x1F.2      /* The Go/NOT_DONE bit in ADCON0[2]     */

#byte ADRESH = 0x1e   /* The Result registers                 */
#byte ADRESL = 0x9E

int main()
{
unsigned long int result; /* 16-bit digitized outcome          */
set_tris_a(0x0E);
setup_adc(ADC_CLOCK_INTERNAL);
setup_adc_ports(RA0_RA1_RA3_ANALOG);
set_adc_channel(3);
delay_us(17);                   /* Allow voltages to stabilize   */
disable_interrupts(GLOBAL);/* Disable all ints (GIE & PEIE=1) */
ADIF = 0;
enable_interrupts(INT_AD);
PEIE = 1;                  /* Enable the auxiliary group interrupts */
/*    Code                                                     */
GO = 1;
sleep();
/* When awake read each byte with high byte x 256              */
result = (long int)ADRESH * 256 + ADRESL;
}
```

The internal function disable_interrupts(GLOBAL) clears *both* GIE and PEIE mask bits. The complementary enable_interrupts(GLOBAL) sets both bits, but we need to enable the PEIE only and leave GIE cleared. This is implemented by the "bit-twiddling" statement PEIE=1;. Similarly, clearing the ADIF flag is directly actioned by ADIF=0;. Before calling sleep() the statement GO=1; manually starts the conversion. After sleep() the ADRESH register is read and cast to a long int to ensure that the compiler treats it as a 16-bit object. Multiplying by 256 tells an intelligent compiler to treats it as the top byte of a 16-bit object. Adding ADRESL puts this in the low byte of the 2-byte outcome.

Self-Assessment Questions

14.1 In Example 14.2 the decay of the threshold level was linear. Although this is fairly effective for situations where the nominal period is known a priori and does not vary greatly, an exponential decay would be better suited where this is not the case. To generate this type of relationship a fixed *percentage* of the value at each sample point should be subtracted to give the new outcome rather than a constant. Show how you could modify Programs 14.4 and 14.5 to decrement at a rate of approximately 0.025% ($\approx \frac{1}{4096}$) on each sample and determine the time constant in terms of the number of samples.

14.2 Real-world analog signals are noisy. In practice this means that some form of filtering or smoothing is frequently required. In any circumstance, noise coming in from outside should not have any appreciable frequency components above half the sampling rate since such noise will be frequency shifted back into the baseband, as shown in Fig. 14.4. Such low-pass filtering must be applied to the signal *before* the A/D conversion, as shown in Fig. 14.20.

Although this external **anti-alias** filter must by definition be implemented using hardware circuitry (such as a CR network), noise within the passband can be smoothed out using software filtering routines. One simple approach to digital filtering is to take multiple readings and average them to give a composite outcome. For example, 16 readings summed and shifted right four times ($\div 16$) would reduce random noise by a factor of $\sqrt{16} = 4$.

Another approach well known to statisticians, is to take a moving average; for example, of a stock price over a month interval. A comparatively efficient algorithm of this type is a 3-point average:

$$\text{Array[i]} = \frac{S_n}{4} + \frac{S_{n-1}}{2} + \frac{S_{n-2}}{4}$$

where S_n is the nth sample from the analog module.

Show how you could modify the GET_ANALOG subroutine to re-member the last samples from the two previous calls and return the smoothed value.

14.3 It has been proposed that as part of the EKG monitor of Example 14.2 that a MAX506 DAC be used to introduce an automatic gain control (AVC) function preceding the PIC MCU's analog input. The aim of the AVC is to keep the peak of the analog input between $\frac{3}{4}$ and $\frac{7}{8}$ full scale. How might you go about implementing this subsystem? *Hint*: Recall that each channel of a MAX506 is the product of its digital input and V_{ref} and that the latter can vary between $0\,V$ and V_{DD}.

14.4 An input analog sinusoid signal, conditioned as shown in Fig. 14.18, is to be full-wave rectified; that is, voltages that were originally neg-ative are to have their sign changed. Design a routine to do this assuming that the 8-bit digitized input voltage is available at ADRESH and the processed output is to be presented via Port B to a DAC.

14.5 Figure 14.21 is based on Fig. 10 of Microchip's application note AN546 *Using the Analog-to-Digital (A/D) Converter* as a means of provid-ing an external voltage reference source for power-sensitive applica-tions. How do you think the circuit works and what factors govern the choice of current limiting resistor?

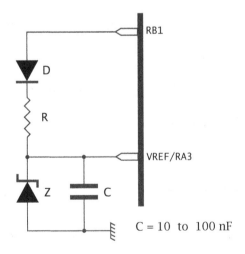

Fig. 14.21 A controllable external voltage circuit.

CHAPTER 15

To Have and to Hold

Many mid- and enhanced-range PIC MCUs feature a small EEPROM scratch-pad memory that can be controlled and accessed indirectly via special-purpose registers (SPRs) in the same manner as other peripheral devices. An integral non-volatile scratchpad enables the programmer to read and modify static data, such as the odometer tally in a car, which needs to be retained in the absence of a power supply; see Example 12.3 on page 391. Although this facility can be implemented using an external EEPROM memory, such as the 24XXX of Fig. 12.26 on page 393, where only a modest amount of non-volatile data needs to be kept, integral EEPROM storage increases reliability and reduces cost, size, and power requirements.

Our objective here is to examine these non-volatile storage facilities. After reading this chapter you will:

- Be familiar with the characteristics of the EEPROM Data module.
- Know how to both read and write data to the EEPROM Data module.
- Understand how the main Flash EEPROM Program memory can be used in some devices to store and retrieve non-volatile data.
- Be able to contrast the EEPROM Data module and Flash Program memory as a location for non-volatile data.

The now obsolete PIC16C84, introduced in 1994, was the first PIC microcontroller to use EEPROM technology for its main Program store. As we have seen in Fig. 2.13 on page 28, Electrically Erasable PROM is similar to EPROM but does not require UV radiation to erase data. Although EEPROM technology is more expensive than EPROM, its use in implementing the Program store was convenient in prototyping and educational or hobbyist applications. Along with this innovation, an EEPROM Data peripheral module was featured which enabled the programmer to store up to 64 bytes of non-volatile data independently of the normal Data store.

The PIC16C84 and its analogous Flash EEPROM Program store successor, the PIC16F84, remained the only EEPROM family member until the introduction of the PIC16F87X in 1998. As a matter of policy, mid-range introductions from 2000 were nearly always Flash EEPROM parts or available in addition to PIC16CXXX EPROM counterparts. All our exemplar parts use Flash EEPROM Program stores and have an integral EEPROM Data module.

Before examining the details, it is instructive to look at an application requiring the use of non-volatile storage. A good example of this is the smart card of Fig. 12.1 on page 332. Here we need to store, amongst other things, the card account number, PIN number, start and expiry dates. Some of this data, such as the account number, is essentially fixed. Security data may need to be altered occasionally by the user from a terminal. If the card is used as a cash card, its credit will need to be charged via an ATM and discharged when payments are made. The size and cost sensitivities of a smart card processor is such that *integral* EEPROM data storage is vital.

Figure 15.1 shows the logic organization of the PIC16F62X EEPROM Data module.[1] The memory matrix is not part of the normal Data and Program stores but is indirectly accessed via four SPRs, which address the target byte, collect/hold data, and control the Read and Write processes.

EEPROM Matrix

The mid-range[2] EEPROM Data module architecture supports up to 256 byte-sized cells. The PIC16F627/8, PIC16F873/4 and PIC12F629/75 devices only implement the bottom 128 locations. The PIC16F648 and PIC16F876/7 implement all 256 memory cells. Key features are:

- 1,000,000 minimum (10^7 typical) Erase/Write cycle endurance for each cell at 5 V and 25°C.[3]
- Maximum Erase/Write cycle time 8 ms; 4 ms typical.
- Data retention greater than 40 years (100 years for the PIC16F62X group).

EEPROM ADdRess Register EEADR

The 8-bit EEADR register can address up to a maximum of 256 (2^8) bytes of EEPROM data. Where less than maximum capacity is implemented, unused upper address bits must be 0 to ensure that the address is within the physical address space. In the PIC16F627/8, allowable addresses are in the range h'00 – 7F'.

EEPROM DATA Register EEDATA

The EEDATA register either holds the 8-bit datum read out of the addressed cell or the byte the programmer wishes to write to the target EEPROM cell.

EEPROM CONtrol Register 1 EECON1

The EEPROM Data module has two modes of operation, with EECON1 controlling and monitoring the Read and Write actions; see Fig. 15.2.

---

[1]The earlier PIC16F84 has a similar architecture but with only 64 bytes of storage and with the EEIF flag in the EECON1 register and the EEIE mask in the INTCON register.

[2]The enhanced-range core EEPROM Data module is virtually identical, although extended versions exist to give larger capacities.

[3]Compare with 10,000 to 100,000 for Flash Program memory.

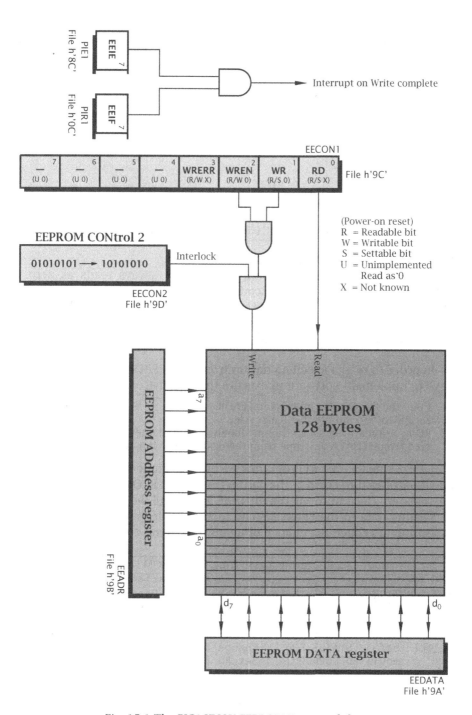

Fig. 15.1 The PIC16F62X EEPROM Data module.

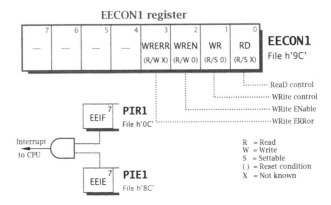

Fig. 15.2 The PIC16F62X EECON1 register.

EEPROM CONtrol Register 2 EECON2

This register is not physically implemented; it always reads as zero. Rather the action of writing the code pattern b'01010101' followed immediately by b'10101010' *with no interruption* is used to unlock the Write cycle. This arcane incantation is deliberately designed to convolute the process, as security against accidential alterations in the data.

In order to read a specified datum from the EEPROM Data module we have to implement software to execute the task list:

1. Copy the target cell's address to EEADR.
2. Set RD to 1 to initiate the Read cycle.
3. RD is automatically cleared immediately and the target 8-bit datum can be read from EEDATA any time from the next instruction cycle, as convenient.

Subroutine EE_GET in Program 15.1 directly implements this process and illustrates the return of the datum from the EEPROM cell to the Working register. The datum will remain in EEDATA until the register is reused.

Writing data to the EEPROM Data module is deliberately made more Byzantine to reduce the chance of a spurious Write corrupting the data

---

Program 15.1 Retrieving a byte from the EEPROM Data module.

```
;  ****************************************************************
;  * FUNCTION: Gets one byte from the EEPROM Data module         *
;  * ENTRY    : Address in EEADR                                 *
;  * EXIT     : Datum in W and in EEDATA. System in Bank0        *
;  ****************************************************************
EE_GET  bsf     STATUS,RP0   ; Change to Bank1
        movlw   b'00000001'  ; Set RD for Read cycle and
        movwf   EECON1       ; read datum into EEDATA
        bcf     STATUS,RP0   ; Back to Bank0
        movf    EEDATA,w     ; Copy datum into W
        return               ; for return
```

due to a software bug or processor malfunction because of, say, a power glitch. The task list to write a datum to a specified cell is:

1. Copy the target cell address to EEADR.
2. Set WREN in EECON1[2] to enable the Write process.
3. Disable all interrupts.
4. Send h'55' to EECON2.
5. Send h'AA' to EECON2.
6. Set WR to initiate the Write cycle.
7. Clear WREN.
8. Enable interrupts.
9. Wait until WR returns to zero, signaling the completion of the Write cycle, and exit.

The Write cycle will not initiate if the interlock sequence in items 4 – 6 is not exactly followed without interference. For instance, in an interrupt-driven system an interruption during the interlock sequence will abort the Write cycle. Thus in this situation interrupts should be disabled by clearing GIE until the Write cycle has been initiated.

If desired, the completion of the Write cycle can be used to interrupt the processor. This is enabled by setting the EEIE mask bit in PIE1[7]. When the interrupt flag EEIF, located in PIR1[7], is set on completion of the Write action, then the interrupt is generated in the normal way. EEIF should be cleared in the ISR.

It is possible that the processor is reset; for instance, by a Watchdog overflow, before the Write cycle is complete. In this situation, the EEPROM datum may be corrupt. The **WRERR** flag in EECON1[3] will be set if the Write operation has been prematurely terminated with a Reset action. If this is not the case, when the cycle is complete the datum may be read back and verified to give extra security. The WREN bit may be cleared at this point to help prevent an accidental Write. Doing this before the Write is complete will not affect the operation.

Program 15.2 implements this task list. Both EEDATA and EEADR are set up by the caller program with the byte datum and address. The subroutine is not exited until the Write cycle has completed, typically 4 ms. This ensures that these SPRs will not be altered during the cycle, which may possibly give an erroneous outcome.

In order to illustrate these concepts we will repeat Example 12.3 on page 391, which incremented a non-volatile triple-byte odometer total in external serial EEPROM, but this time using the internal EEPROM Data module. We will assume that the odometer count is located at EEPROM cells h'10 – 12'.

The coding shown in Program 15.3 makes use of the two subroutines EE_GET and EE_PUT to read and subsequently write the three odometer bytes from/to the EEPROM Data module. The address of the first (highest) byte is copied into EEADR at the beginning of the subroutine and is

```
           Program 15.2 Putting a byte into the EEPROM Data module.
;   *************************************************************
;   * FUNCTION: Writes one byte into the EEPROM Data module     *
;   * ENTRY   : Datum byte in EEDATA, module address in EEADR    *
;   * EXIT    : Interrupts disabled for 9 instructions          *
;   * EXIT    : System in Bank0                                 *
;   *************************************************************
;
EE_PUT   bsf     STATUS,RP0  ; Change to Bank1
         bsf     EECON1,WREN ; Enable for Write cycle
         bcf     INTCON,GIE  ; Disable interrupts

         movlw   h'55'       ; Now do the interlock
         movwf   EECON2
         movlw   h'AA'
         movwf   EECON2

         bsf     EECON1,WR   ; Initiate the Write cycle
         bcf     EECON1,WREN ; Optionally disable any other Writes
         bsf     INTCON,GIE  ; Re-enable interrupts
EE_EXIT  btfsc   EECON1,WR   ; Check, has the Write completed?
         goto    EE_EXIT     ; IF not THEN retry
         bcf     STATUS,RP0  ; Go back to Bank0
         return              ; and return when cycle has finished
```

subsequently incremented and decremented in situ to point to the appropriate datum.

After the 3-byte odometer state has been fetched and copied into memory it is incremented in exactly the same manner as in Program 12.19 on page 395. The augmented array is then written back into EEPROM in the opposite sense as it was read, with EEADR being decremented. The EE_PUT subroutine checks that the Write cycle has been completed before returning and thus timing need not be checked by the calling program.

As well as altering data under program control it is possible to initialize the state of the EEPROM Data module at the same time as the executable program is being externally blasted into the Program memory; as illustrated in Fig. 10.6(a) on page 281. The area of Program memory beyond the user Program store belongs to the special Test/Configuration memory space h'2000 – 30FF' and can be accessed only during external programming. In Fig. 10.6(b) & (c) we observed that the Configuration fuse word is located at h'2007'. The EEPROM Data module also lies in this space, located at h'2100 – 21FF'. For instance, to store the value of sine every 10° between 0° and 90° as part of the program source code we have:

```
     org h'2100'       ; The EEPROM Data module
SINE de  0,      h'2C', h'57', h'7F', h'A4', h'C4'
     de  h'DD', h'F0', h'FB', h'FF'
```

Program 15.3 Incrementing the non-volatile odometer count in Data EEPROM.

```
;  ****************************************************************
;  FUNCTION: Adds one onto the triple-precision odometer total*
;  RESOURCE: Subroutines EE_GET and EE_PUT               *
;  ENTRY   : Current total in EEPROM module at h'10:11:12'   *
;  EXIT    : Incremented total back in EEPROM module        *
;  EXIT    : also available in RAM at LSB:NSB:MSB           *
;  ****************************************************************
EXTRA_MILE
         bsf    STATUS,RP0; Change into Bank 1
         movlw  h'10'     ; Address of high-byte odometer total
         movwf  EEADR     ; Copy into EEPROM address register
         call   EE_GET    ; Read byte from EEPROM module
         movwf  MSB       ; and put into File MSB
         bsf    STATUS,RP1; Back to Bank 1
         incf   EEADR,f   ; Address of middle byte odometer
         call   EE_GET    ; Read byte from EEPROM module
         movwf  NSB       ; and put into File NSB
         bsf    STATUS,RP1; Back again to Bank 1
         incf   EEADR,f   ; Address of low byte odometer
         call   EE_GET    ; Read byte from EEPROM module
         movwf  LSB       ; and put into File LSB

; Now increment 3-byte array
         incfsz LSB,f     ; Add one; IF not Zero THEN continue
          goto  PUT_BACK  ; IF not THEN continue
         incfsz NSB,f     ; Increment middle byte
          goto  PUT_BACK  ; IF not zero THEN continue
         incf   MSB,f

; Put the augmented odometer count back in Data EEPROM
PUT_BACK movf   LSB,w     ; Get new odometer low byte
         bsf    STATUS,RP1; Change to Bank 1
         movwf  EEDATA    ; Put in EE Data register
         call   EE_PUT    ; Write to EEPROM cell h'12'
         movf   NSB,w     ; Get new odometer middle byte
         bsf    STATUS,RP1; Back to Bank 1
         movwf  EEDATA    ; Put in EE Data register
         decf   EEADR,f   ; Address of middle byte
         call   EE_PUT    ; Write to EEPROM cell h'11'
         movf   MSB,w     ; Get new odometer low byte
         bsf    STATUS,RP1; Again go to Bank 1
         movwf  EEDATA    ; Put in Data register
         decf   EEADR,f   ; Address of high byte
         call   EE_PUT    ; Write to EEPROM cell h'10'
         return
```

where the assembler directive **de** (Data EEPROM) specifies the comma-delimited list of data. After the PIC MCU has been programmed, the contents of the EEPROM Data module will look like Fig. 15.3.

Any data programmed in this way can be subsequently read by the program. For instance, to read sin(50) the contents of EEPROM Data module location h'05' ($\frac{50}{10}$) is read; giving from our diagram h'C4' or decimal 196 ($\frac{196}{256} = 0.76525$).

Although it is possible to initialize the Program store in a similar manner using the **dw** (Data Word) directive, as shown in Program 15.5, this is

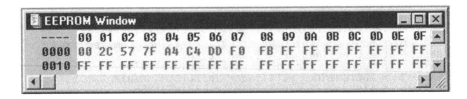

Fig. 15.3 The first 32 bytes of EEPROM holding the sine look-up table.

of little use, as the Harvard architecture's separation of Data and Program store memory spaces means that there is no way an instruction can read this data. Instructions can only access the Data store. However, some newer Flash EEPROM PIC microcontrollers allow the program to read and write such data in a similar indirect manner to that used for the EEPROM Data module; in particular our PIC16F87X group. Details differ to some extent from the original device and the later A-version. Initially we will use the former as our exemplar.

Both the PIC16F873/4 devices have a 4 Kbyte Flash Program memory and a 128-byte EEPROM Data module, whilst the PIC16F876/7 have 8 Kbyte and 256-byte capacities, respectively, but otherwise have the same characteristics.

Key EEPROM properties for the PIC16F87XA group are:

- 100,000 minimum (10^6 maximum) EEPROM Data module Erase/Write cycle endurance per cell.
- 10,000 minimum (10^5 maximum) Flash EEPROM Program store Erase-/Write cycle endurance.
- Maximum Write/Erase time 8 ms (typical 4 ms) for both the Data module and Flash memory.

Of particular note is the endurance limit of 10,000[4] Write cycles for the Flash EEPROM. Whilst this is entirely satisfactory when changing the device's program, it is a limitation for some non-volatile data storage situations. Thus Flash Program memory storage is more applicable to constant data, such as the sine lookup table, rather than for information that requires frequent update, such as the odometer.

Flash EEPROM has a smaller geometry than normal EEPROM. Whilst this speeds up its operation, charges which eventually trap in the floating gate insulation have a disproportional effect on the storage mechanism and leads to earlier deterioration.

Figure 15.4 shows the PIC16F87X EEPROM Data module with the Flash Program store superimposed. This form of representation is used as the EEDATA and EEADR registers are common to both EEPROM arrays. Of course, the Flash Program store is larger both in the number of cells (8 Kbytes against 256 bytes) and in cell width (14 bits against 8 bits). Thus

---

[4]Early Flash memories had a minimum endurance of only 100 Write cycles.

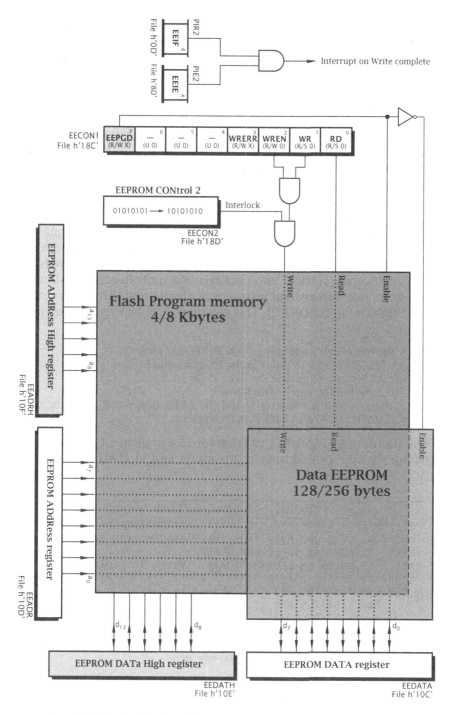

Fig. 15.4 The PIC16F87X Flash and Data EEPROM storage system.

both Data and Address registers have the high-end extensions **EEDATH** and **EEADRH**, respectively, to cope with this additional capacity.

As we shall see, the process of reading from and writing to either array is similar. The target module is chosen using the **EEPGD** (EEProGram/Data) control bit control bit in EECON1[7]. Apart from this additional bit and the relocation of the EEIF and EEIE bits to the PIR2 and PIE2 registers, respectively, the EECON1 register of Fig. 15.4 is identical to the basic PIC16F62X version shown in Fig. 15.2. The virtual EECON2 interlock register remains the same.

Reading and writing to the EEPROM Data module is identical to that used for the more basic PIC16F62X devices. The only change necessary to subroutines EE_GET and EE_PUT relates to the Bank 2 location of EEDATA and EEADR and Bank 3 for EECON1 and EECON2.

The process of reading from the Flash Program store is similar to that of the EEPROM Data module but using double Data and Address registers. However, we are interacting with the same Program store from which code is being fetched into the execution unit. In consideration of this dualism, two dummy nop instructions should follow the instruction setting the RD bit in EECON1[0]. This gives our Read task list for the Flash Program store as:

1. Copy the target cell's address to EEADRH:EEADR on entry.
2. Set the EEPGD bit to 1 to point to the Program store.
3. Set the RD bit to 1 to initiate the Read cycle.
4. Execute two dummy nop instructions.
5. RD is then cleared automatically and the 14-bit target datum can be read from EEDATH:EEDATA in Bank 2 as convenient.

The subroutine FLASH_GET of Program 15.4 implements this process assuming that the cell address is already in situ in EEADRH:EEADR on entry.

Program 15.4 Reading a word from the Flash Program store.

```
; ******************************************************************
; * FUNCTION: Gets one byte from the Flash Program store          *
; * ENTRY    : Address in EEADRH:EEADR                            *
; * EXIT     : Datum in EEDATH:EEDATA. System in Bank0            *
; ******************************************************************
FLASH_GET bsf    STATUS,RP1       ; Change to Bank3
          bsf    STATUS,RP0
          movlw  b'10000000'      ; Point to Program memory
          movwf  EECON1           ; by setting EEPGD in EECON1[7]
          bsf    EECON1,RD        ; Set RD for Read cycle
          nop                     ; Dummy nops
          nop
          bcf    STATUS,RP1       ; Return to Bank0
          bcf    STATUS,RP0
          return
```

For our example we will design a subroutine that will return the square of an integer between 0 and 100 in EEDATH:EEDATA. We could of course calculate this by multiplication, but for the purposes of this study we will implement this exercise as a look-up table located in Flash Program store. As this is a table of constants we can load the data into Flash memory at the same time as the rest of the program code.

In Program 15.5 the table is located at h'300' in the Program store. The directive **dw** (Data Word) is similar to de directive but each datum in the comma separated list is 14-bits. For convenience the **radix** directive is used to specify constants which by default are treated as decimal.

Directly following the table is the executable code. In this manner Program 15.5 is comparable to a **C++ class** where a program object comprises both data members and member functions (subroutines).

Program 15.5 Squaring an integer.

```
        radix   dec     ; Default number base is decimal
        __config _CPD_OFF & _WRT_ENABLE_OFF
        org     h'300'  ; Table starts @ h'300' in Program store
;       *****************************************************************
;       * FUNCTION: Generates the square of an integer                 *
;       * RESOURCE: Subroutine FLASH_GET                               *
;       * ENTRY    : Integer in W range 0 -- 100                       *
;       * EXIT     : 14-bit square in SQRH:SQRL. In Bank0              *
;       *****************************************************************
TABLE_OF_SQUARES        ; Table of constants expressed in decimal
   dw 0,1,4,9,16,25,36,49,64,81,100,121,144,169,196,225
   dw 256,289,324,361,400,441,484,529,576,625,696,729,784,841
   dw 900,961,1024,1089,1156,1225,1296,1369,1444,1521,1600,1681
   dw 1764,1849,1936,2025,2116,2209,2304,2401,2500,2601,2704
   dw 2809,2916,3025,3136,3249,3364,3481,3600,3721,3844,3969
   dw 4049,4225,4356,4489,4624,4761,4900,5041,5184,5329,5476
   dw 5625,5776,5929,6084,6241,6400,6561,6724,6889,7056,7225
   dw 7396,7569,7744,7921,8100,8281,8464,8649,8836,9025,9216
   dw 9409,9604,9801,10000

SQUARE  bsf     STATUS,RP1  ; Move to Bank2
        bcf     STATUS,RP0
        movwf   EEADR       ; Build up the address, low byte
        movlw   3
        movwf   EEADRH      ; High byte
        call    FLASH_GET   ; Get table entry n in h'3nn'
        bsf     STATUS,RP1  ; Move back to Bank2 from Bank0
        bcf     STATUS,RP0
        movf    EEDATA,w    ; Get lower byte of square
        bcf     STATUS,RP1  ; Go to Bank0
        movwf   SQRL        ; Copy to SQRL in Bank0
        bsf     STATUS,RP1  ; Back to Bank2
        movf    EEDATH,w    ; Get high byte of square
        bcf     STATUS,RP1  ; Return to Bank0
        movwf   SQRH        ; and copy to SQRH in Bank0
        return
```

The subroutine itself builds up the table element nn address by placing the integer nn passed in W in EEADR and the constant h'03' in EEADRH. This gives the double-byte address as h'3nn'. After this is done, the subroutine FLASH_GET retrieves the 14-bit datum from the table. The subroutine then moves both bytes from EEDATH:EEDATA and returns the datum in the two Files SQRH:SQRL. As these Files are in Bank 0, each byte copied from EEPROM SPRs in Bank 2 needs switching to Bank 0 when the byte has reached the Working register. As the PIC16F874/7 have 16 GPR Files reflected across the four banks, it would have been helpful to place SQRH:SQRL in this common area.

When the program has been burnt into Flash memory by the external programmer the Program store in the area around h'300' will look like Fig. 15.5.

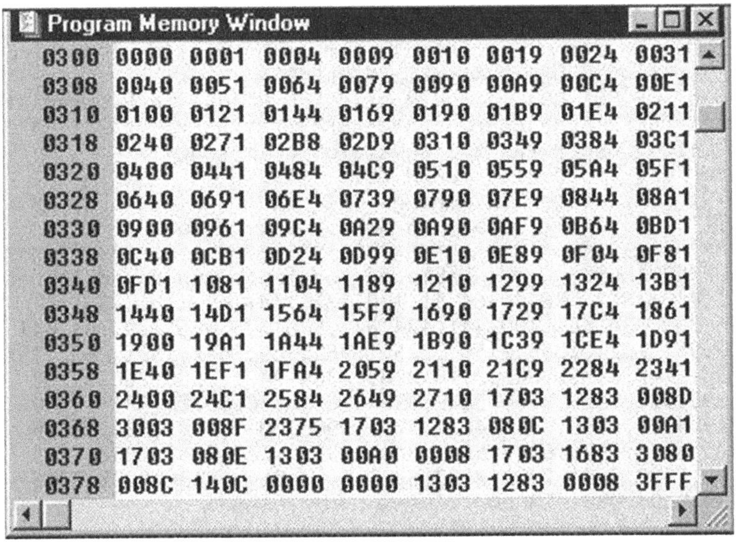
Program Memory Window

0300	0000	0001	0004	0009	0010	0019	0024	0031
0308	0040	0051	0064	0079	0090	00A9	00C4	00E1
0310	0100	0121	0144	0169	0190	01B9	01E4	0211
0318	0240	0271	02B8	02D9	0310	0349	0384	03C1
0320	0400	0441	0484	04C9	0510	0559	05A4	05F1
0328	0640	0691	06E4	0739	0790	07E9	0844	08A1
0330	0900	0961	09C4	0A29	0A90	0AF9	0B64	0BD1
0338	0C40	0CB1	0D24	0D99	0E10	0E89	0F04	0F81
0340	0FD1	1081	1104	1189	1210	1299	1324	13B1
0348	1440	14D1	1564	15F9	1690	1729	17C4	1861
0350	1900	19A1	1A44	1AE9	1B90	1C39	1CE4	1D91
0358	1E40	1EF1	1FA4	2059	2110	21C9	2284	2341
0360	2400	24C1	2584	2649	2710	1703	1283	008D
0368	3003	008F	2375	1703	1283	080C	1303	00A1
0370	1703	080E	1303	00A0	0008	1703	1683	3080
0378	008C	140C	0000	0000	1303	1283	0008	3FFF

Fig. 15.5 View of the Flash Program module showing the look-up table and subroutine SQUARE.

Although the table has been placed here at an even 256-byte boundary for convenience, in practice it can be located anywhere in memory. In the general case the table offset nn will need to be added to the complete 14-bit address TABLE. SAQ 15.2 discusses how this address arithmetic could be done.

For the original PIC16F87X group the Write cycle also is virtually identical to its EEPROM Data module counterpart but with the addition of a double-nop relaxation phase. However, after this point processing will cease for the approximately 4 ms it takes to erase and then blast the datum into the target Program store location. Processing then continues as normal. This gives us our Flash Write cycle task list:

1. Copy the target cell address to EEADRH:EEADR.
2. Set WREN in EECON1[2] to enable the Write process.
3. Disable all interrupts if used.
4. Send h'55' to EECON2.
5. Send h'AA' to EECON2.
6. Set WR to initiate the Write cycle.
7. Execute two dummy nop instructions.
8. Clear WREN.
9. Enable interrupts.
10. It is not necessary to wait until WR returns to zero, signaling the completion of the Write cycle, as processing will cease during this time and will only start up again on conclusion.

The subroutine FLASH_PUT in Program 15.6 assumes that the cell address is in EEADRH:EEADR and 14-bit datum is in EEDATH:EEDATA on entry.

All EEPROM-based Program store devices have code-protection fuses in their Configuration word. The primary function of code protection is to prevent the external Programmer reading code from the Program store, to give a measure of security against unauthorized peeking at the code. In the specific case of the original PIC16F87X devices, two code bits **CP[1:0]** (duplicated as bits 13:12 and 5:4) in the Configuration word in the

---

Program 15.6 Writing to Flash Program memory.

```
;   ****************************************************************
;   * FUNCTION: Writes one byte into the Flash Program store      *
;   * ENTRY   : Datum byte in EEDATH:EEDATA                        *
;   * ENTRY   : Datum address in EEADRH:EEADR                      *
;   * EXIT    : Interrupts disabled for the duration              *
;   * EXIT    : System in Bank0                                    *
;   ****************************************************************
FLASH_PUT  bsf    STATUS,RP0   ; Go to Bank 3
           bsf    STATUS,RP1
           bsf    EECON1,EEPGD; Target the Flash Program store
           bsf    EECON1,WREN  ; Enable for Write cycle
           bcf    INTCON,GIE   ; Disable all interrupts

           movlw  h'55'        ; Now do the interlock
           movwf  EECON2
           movlw  h'AA'
           movwf  EECON2
           bsf    EECON1,WR    ; Initiate the Write cycle
           nop                 ; Dummy nops
           nop                 ; Processing stops until complete
           bcf    EECON1,WREN  ; Disable any more Writes
           bsf    STATUS,GIE   ; Re-enable interrupts

           bcf    STATUS,RP1   ; Go back to Bank 0
           bcf    STATUS,RP0
           return              ; & return when cycle has finished
```

Special/Test configuration area at h'2007' give protection for all the Program store (00), the top half of the store (01), the top 256 bytes only (10) or no protection (11); the default situation; see Fig. 10.6(c) on page 281. If protection is given to *any* area of memory then the external programmer cannot subsequently write data into *anywhere* in the Program store. However, reading is only inhibited in protected areas. Protection should normally be *disabled during prototyping* where frequent updates to the code is to be expected. However, in Flash devices the complete Program store can be erased and the Configuration reverted back to its default all 1's state by the external Programmer if it is necessary to remove protection. This latter is not normally possible for PIC16CXXX non-Flash devices.

Code protection also affects *internal* Writes into the Program store using code such as in Program 15.6. Internal Writes can be made into unprotected areas of Program memory, provided that the WRT fuse is 1; its default setting. Setting this fuse to 0 (_WRT_ENABLE_OFF) will disable internal Writes to anywhere in the Program store irrespective of the main code protection setting. Internal Reads are not affected by code protection. Program 15.5 shows the __config directive used to disable all code protection in Program memory; which is actually superfluous as this is the default situation.

Fig. 15.6 Configuration word for the PIC16F87XA devices.

In the A-variant of the family the Program store code protection is a little different, as shown in Fig. 15.6. This time a single Code Protection fuse **CP** optionally protects the entire Program store from external Reads or Writes. When Program memory is code protected, internal Reads and Writes are still possible. Two WRiTe fuses **WRT[1:0]** can be used to optionally prevent internal Writes to the listed fractions of the Program store. Internal Reads are not inhibited.

Apart from the progenitor PIC16F84, all devices with a EEprom Data module have a **CPD** fuse (Code Protection Data module) which if cleared to 0 will inhibit all external access to this module.

The original PIC16F87X group can write single words into the Program store. However, the internal organisation of the Flash Program memory has been changed in the PIC16F87XA devices. As a consequence, these newer devices use a 4-word Write cycle. From the perspective of the software this means that Flash Program memory must be written in blocks, defined as a group of four sequential words with a lower boundary address with EEADR[1:0] = 00. For instance, if the programmer wishes to copy a 14-bit word into the Program store at location h'500' a new word will also have to be written into address h'501', h'502' and h'503'. The processor has four internal 14-bit buffer registers, as shown in Fig. 15.7. As each Write process is actioned, the datum is simply copied into the appropriate buffer; as dictated by the bottom two bits of the Program store address. When the datum word is written into the last buffer register, with EEADR[1:0] = 11 (in our example that corresponds with address h'503'), the block of four words in locations h'500:1:2:3' are erased and then all four buffered words blasted into the Program store in one action.

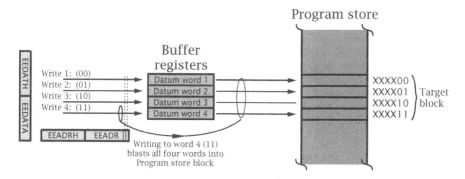

Fig. 15.7 Writing to Flash memory with the PIC16F87XA group.

The internal buffered registers are not directly accessible to the software, using the normal Move instructions. Rather, the same interlock sequence used for the single-word Write process implemented in the FLASH_PUT subroutine of Program 15.6 is used for each word. For intermediate Writes the datum word simply ends up in a buffer register and the nominal 4 ms burn time during which processing halts does not occur. Using the same subroutine with an address at the end boundary of the block triggers a "real" burn with its 4 ms hiatus.

In order to illustrate the block Write procedure, consider an example where four double-byte data are located in GPR Files named DATA_ARRAY through DATA_ARRAY+7, ordered most- to least-significant byte, and are to be burnt into the Program store of a PIC16F877A.

The subroutine described in Program 15.7 assumes that the data is already present in the Data store and that the address of the first target Program store word is in EEADRH:EEADR on entry. The implementation uses a loop to write each word into the buffer register and initiate the burn. The File Select Register is used as a pointer to walk along the array in Data memory, as described in Fig. 5.8 on page 106. Each double-byte is copied into EEDATH:EEDATA and the subroutine FLASH_PUT of Program 15.6 used to activate the pseudo Write cycle. After each loop pass the Program address is incremented in situ in EEADR (EEADRH doesn't to be altered). After the fourth pass, a true Write cycle will be initiated. Unfortunately there is no status flag in EECON1 to allow the software to distinguish between a pseudo and true Write. Instead, the state of the two lowest EEADR bits are checked for roll over back to b'00'. When this occurs, the process is complete.

Although a complete block needs to be burnt, only one, two, or three words can be updated by first of all reading words that are to remain unaltered out of the Program store into Data memory, and then writing them back along with any data that is to be altered.

Program 15.7 Block writing to the PIC16F87XA-family Program store.

```
;         ************************************************************
; FUNCTION: Writes block of 4 14-bit words into Program store*
; ENTRY   : Block start address is in EEADRH:ADDR              *
; ENTRY   : Four words in Data store at DATA_ARRAY:8           *
; EXIT    : Four words blasted into Program store             *
; EXIT    : System in Bank 0                                   *
; RESOURCE: Subroutine FLASH_PUT                               *
;         ************************************************************
FLASH_BLAST
          bsf     STATUS,RP1   ; Change to Bank 2
          bcf     STATUS,RP0

          movlw   DATA_ARRAY+7 ; Point File Select Register to
          movwf   FSR          ; LS byte of Data array in RAM
; Now do four Writes ---------------------------------------
FB_LOOP   movf    INDF,w       ; Get low byte of word and
          movwf   EEDATA       ; put in Low EEPROM Data register
          incf    FSR,f        ; Point to high byte
          movf    INDF,w       ; Get high byte of word and
          movwf   EEDATH       ; put in High EEPROM Data register

          call    FLASH_PUT    ; Put in Buffer register

          bsf     STATUS,RP1   ; Change to Bank 2 again
          bcf     STATUS,RP0
          incf    EEADR,f      ; Increment Program store address
          movf    EEADR,w      ; Check; are the lower bits 00
          andlw   b'00000011'  ; Isolate these bits
          btfss   STATUS,Z     ; IF both zero then done
          goto    FB_LOOP      ; ELSE do next word

          bcf     STATUS,RP1   ; Go back to Bank 0
          return               ; and return when cycle has finished
```

Examples

Example 15.1
The CCS compiler has the following built-in functions dealing with the EEPROM Data module:

read_eeprom(address)
Reads a byte from the specified EEPROM address.

write_eeprom(address, value)
Writes the value to the specified address and returns only when the Write cycle has finished.

   Write a **C** function to duplicate the odometer update implemented at assembler level in Program 15.3.

Solution
Like its assembly-level counterpart of Program 15.3, the function of Program 15.8 is divided into three phases.

1. This phase creates an array of three bytes named **odometer[]** which will act as a temporary store in RAM for the odometer count located in EEPROM. The array is given the current reading by extracting the data one byte at a time from EEPROM locations h'10:11:12' using the **read_eeprom()** function.
2. After the 3-byte data is in the Data store it is incremented using an **if-else** tree:
   (a) Increment low byte and check for zero. IF not zero THEN the overall addition is complete ELSE pass on carry to next byte.

Program 15.8 C-based coding for the odometer.

```
void odometer(void)
{
unsigned int odometer[3];          /* Define the 3-byte count  */
odometer[0] = read_eeprom(0x10);  /* Get the existing count   */
odometer[1] = read_eeprom(0x11);
odometer[2] = read_eeprom(0x12);

/* Increment array                                            */
if(++odometer[0] != 0) break;
else if(++odometer[1] != 0) break;
else odometer[2]++;

/* Now return incremented count to the EEPROM                 */
write_eeprom(0x10, odometer[0]);
write_eeprom(0x11, odometer[1]);
write_eeprom(0x12, odometer[2]);
}
```

    (b) Increment the middle byte and check for zero. IF not zero THEN the overall addition is complete ELSE pass on carry to next byte.

    (c) Increment most significant byte.

3. Finally each byte is written back into the EEPROM Data module using the write_eeprom() function.

Comparing the hand-coded assembly of Program 15.3 against the code generated from Program 15.8 for a PIC16F62X target gives 54 instructions against 105.

Example 15.2

A feasibility study is being undertaken in using a PIC-based Sauna controller. This is to monitor temperature and control heating and cooling units. There is also to be an over-temperature emergency alarm and shut-off.

One possibility is to use an 8-pin PIC, together with an external temperature transducer. Someone has suggested that an efficient approach to this problem would be to use the variation in the internal Watchdog timer's period with temperature as a cost-effective, albeit crude, sensor.

Experimental data was collected using a sample of eight production devices from the same manufacturing lot, with a soak time of 30 minutes at each tested temperature and 500 uncalibrated periods averaged to produce the graph of Fig. 15.8.

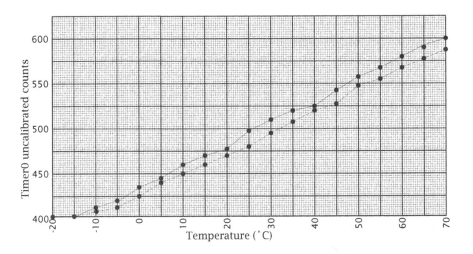

Fig. 15.8 Watchdog timer period versus temperature.

The data presented in Fig. 15.8 is based on Microchip's application note AN720 *Measuring Temperature Using the Watch Dog Timer (WDT)*. The two loci give the maximum and minimum across the range of tested devices. The Watchdog period was measured in Timer 0 overflows counting an internal 4 MHz clock. The Watchdog timer used a 1:8 prescale ratio.

From this data it can be seen that there is a correlation between period and rising temperature. However, although the overall trend is predictable, different devices will have varying offsets and slopes. For example, within the eight devices tested here, the scale factor (scalar) varies from 2.28 to 2.42 counts per degree Celsius. This will necessitate a calibration phase before the system is used. If the count at one temperature T_0 and the scalar are known for any device, then for the specific case where a count $COUNT_n$ is recorded, the offset from T_0 is:

$$\Delta T = (COUNT_n - COUNT_0) \times \text{Scalar}$$

In order to calibrate these devices it has been decided to soak batches in a refrigerator at $0°C$ and save the 2-byte count in the EEPROM Data module of a PIC16F627. The soak test is to be repeated in an oven at $30°C$ with the *difference* between this second count and the original value to be stored in a single EEPROM byte. After this has been done, the device can be reprogrammed with the final running program overwriting the original calibration program. This running program can then calculate temperature using the actual Watchdog period $COUNT_n$:

$$T = (COUNT_0 - COUNT_n) \times (\text{Difference}/30)$$

where $COUNT_0$ is the 2-byte EEPROM datum showing the count at $0°C$ and Difference is the 1-byte EEPROM datum showing the change in count over $30°C$.

Show how you could code the calibrate program to implement this specification.

Solution
Five tasks can be identified which must be undertaken at the end of a Watchdog period.
- Average the current Timer 0 roll-over count with the previous low-temperature (i.e., $0°C$) count unless the first reading.
- Average the current Timer 0 roll-over count with the existing high-temperature (i.e., $30°C$) count unless the first reading.
- Store the low-temperature count in EEPROM.
- Calculate the difference between low- and high-temperature count and store in EEPROM.
- Do nothing.

Port lines can be used to signal which of the first four active tasks are to be actioned. For instance; if RA0 is High then add the current Timer 0 roll-over count to the existing 2-byte low-temperature count. Unless this is the first reading, divide by two to give an average. The complete batch of devices could have their RA0 pin held High for a few minutes after the refrigerator stabilizes at 0°C.

Bringing RA0 to its Low state and then RA2 High for a short time signals an EEPROM storage action. With all port lines Low no action is taken, representing both time before temperature stabilization and after EEPROM programming.

The Timer 0 and Watchdog timers are initialized together with the count values on a once-only basis on Power-on reset. This will typically only occur when the devices are powered up when in the temperature bath. Subsequent resets will normally be due to Watchdog time-outs. The state of the Status register's $\overline{\text{TO}}$ flag can be used to ascertain the source of reset; see page 403.

Program 15.9 shows the routine used to initialize the timers and variables entered if $\overline{\text{TO}}$ is 1 on reset; that is, on Power-on. After this is done, the system enters an endless loop goto $ (the assembler replaces the label $ by the instruction's address) which simply keeps going to itself!

Also shown is the ISR servicing a Timer 0 interrupt. This increments the double-byte variable ROLL_OVER:ROLL_OVER+1 and this is the count value read by the system on a Watchdog reset giving a numerical value for period.

Eventually the Watchdog timer will time-out and reset the processor. Now $\overline{\text{TO}}$ will be 0 and the routine labeled READING in Program 15.10 will be entered. This checks the state of each of the four RA[3:0] pins in turn, executing one of the four listed tasks. If no pin is High, the program simply clears ROLL_OVER:ROLL_OVER+1 and Timer 0 and the Watchdog timer are restarted, and an endless loop entered. This READING_EXIT routine is also entered at the end of the four tasks.

The first two tasks are shown in Program 15.10. Here the 2-byte Timer 0 roll-over count is either added to the existing value LO_TEMP:LO_TEMP+1 or HI_TEMP:HI_TEMP+1 as appropriate and the outcome shifted once right to divide by two to give the average. As the count total is modest, 2-byte arithmetic is sufficient to avoid overflow. If this is repeated over a duration of several minutes an averaged value will result.

If this is the very first time a reading has been made then the divide by two operation is skipped and the flag variable FIRST_LO or FIRST_HI as appropriate is made non-zero.

The core routine with respect to this chapter is given in Program 15.11. If RA2 is 1 then the 2-byte low-temperature count LO_TEMP:LO_TEMP+1 is copied into the bottom two bytes of the EEPROM Data module using the EE_PUT subroutine of Program 15.2.

If RA3 is 1 the difference between the 2-byte high and low temperatures is then calculated. With the data shown in Fig. 15.8 it can be seen that a 30°C difference will not exceed a byte's worth of storage so only the

Program 15.9 The Sauna Power-on reset sequence and ISR.

```
            __config    _WDT_ON & _CP_OFF & _RC_OSC

            cblock h'20'
            _work:1, _status:1
            FIRST_HI:1, FIRST_LO:1
            ROLL_OVER:2, LO_TEMP:2, HI_TEMP:2
            DELTA_TEMP:1
            endc

            org     0
START       goto    MAIN
            org     4
            goto    ISR

MAIN        btfss   STATUS,NOT_TO   ; IF Watchdog timeout
            goto    READING         ; THEN must have a reading

            clrwdt
            movlw   b'11011010'     ; Wdt enabled with a 1:8 prescale
            bsf     STATUS,RP0      ; Change to Bank1
            movwf   OPTION_REG      ; and TMR0 internal clock
            bcf     STATUS,RP0      ; and back to Bank0
            clrf    FIRST_HI
            clrf    FIRST_LO
            bsf     INTCON,T0IE     ; Enable Timer0 interrupt
            clrf    TMR0            ; Zero the Timer
            clrf    ROLL_OVER+1     ; Zero the 2-byte Timer roll-over
            clrf    ROLL_OVER
            bsf     INTCON,GIE      ; Enable all interrupts
            goto    $               ; Endless loop
;   ************************************************************
;   * The ISR to increment the 2-byte COUNT IF TMR0 interrupt    *
;   ************************************************************
; First save context in usual way
ISR         movwf   _work           ; Put away W
            swapf   STATUS,w        ; and the Status register
            movwf   _status

;   ************************************************************
; The core code
            incf    ROLL_OVER+1,f; Record one more roll-over
            btfsc   STATUS,Z        ; Skip if not zero
            incf    ROLL_OVER,f     ; Increment upper byte
            bcf     INTCON,T0IF     ; Clear interrupt flag
;   ************************************************************
            swapf   _status,w       ; Untwist the original Status reg
            movwf   STATUS
            swapf   _work,f         ; Get the original W reg back
            swapf   _work,w         ; leaving STATUS unchanged
            retfie                  ; and return from interrupt
```

Program 15.10 Reading a new period count.

```
READING btfsc    PORTA,0      ; Check; new low temp desired?
        goto     NEW_LO       ; IF yes THEN go to it!
        btfsc    PORTA,1      ; Check; new high temp desired?
        goto     NEW_HI       ; IF yes THEN go to it!
        btfsc    PORTA,2      ; Check; update low temp desired?
        goto     UPDATE_LO    ; IF yes THEN go to it!
        btfsc    PORTA,3      ; Check; update high temp desired?
        goto     UPDATE_HI    ; IF yes THEN go to it!
        goto     READING_EXIT ; ELSE nothing doing

NEW_LO  movf     ROLL_OVER+1,w; ELSE get low byte TMR0 roll-over
        addwf    LO_TEMP+1,f  ; and add it to low byte low temp
        btfsc    STATUS,C     ; Check for Carry
        incf     LO_TEMP,f    ; IF so THEN record it
        movf     ROLL_OVER,w  ; Now get high byte of roll-over
        addwf    LO_TEMP,f    ; and add it to high byte low temp
        movf     FIRST_LO,f   ; Is this the 1st low reading?
        btfsc    STATUS,Z
        goto     FIRST_TIME_LO; IF so THEN go to it!
        rrf      LO_TEMP,f    ; ELSE divide sum by two
        rrf      LO_TEMP+1,f
        goto     READING_EXIT ; and finished
FIRST_TIME_LO                 ; IF first reading simply transfer
        incf     FIRST_LO,f   ; No longer the first reading
        goto     READING_EXIT

NEW_HI  movf     ROLL_OVER+1,w; ELSE get low byte TMR0 roll-over
        addwf    HI_TEMP+1,f  ; and add it to low byte high temp
        btfsc    STATUS,C     ; Check for Carry
        incf     HI_TEMP,f    ; IF so THEN record it
        movf     ROLL_OVER,w  ; Now get high byte of roll-over
        addwf    HI_TEMP,f    ; and add it to high byte high temp
        movf     FIRST_HI,f   ; Is this the 1st high reading?
        btfsc    STATUS,Z
        goto     FIRST_TIME_HI; IF so THEN go to it!
        rrf      HI_TEMP,f    ; ELSE Divide sum by two
        rrf      HI_TEMP+1,f
        goto     READING_EXIT ; and finished
FIRST_TIME_HI                 ; IF first reading simply transfer
        incf     FIRST_HI,f   ; No longer the first reading

READING_EXIT
        clrf     TMR0         ; Zero the timer
        clrwdt                ; Reset the Watchdog timer
        clrf     ROLL_OVER+1  ; Zero the roll-over count
        clrf     ROLL_OVER
        goto     $            ; Wait for another Watchdog reset
```

Program 15.11 Updating the Sauna EEPROM.

```
UPDATE_LO
        movf    LO_TEMP,w       ; Get high byte of low temperature
        bsf     STATUS,RP0      ; To Bank1
        movwf   EEDATA          ; In EEPROM Data register
        clrf    EEADR           ; EEPROM address h'00'
        call    EE_PUT          ; Write datum in
        movf    LO_TEMP+1,w     ; Get low byte of low temperature
        bsf     STATUS,RP0      ; To Bank1
        movwf   EEDATA          ; Is new datum
        incf    EEADR,f         ; EEPROM address h'01'
        call    EE_PUT          ; Write Datum in
        goto    READING_EXIT

UPDATE_HI ; Work out HI_TEMP-LO_TEMP & store at h'02' in EEPROM
; Only need to subtract the lower bytes as diff fits in one byte
        movf    HI_TEMP+1,w     ; Get low byte high temperature
        subwf   LO_TEMP+1,w     ; Subtract low byte low temperature
        movwf   DELTA_TEMP      ; giving the difference

        bsf     STATUS,RP0      ; To Bank1
        movwf   EEDATA          ; Delta temperature in h'02'
        movlw   2
        movwf   EEADR
        call    EE_PUT
        goto    READING_EXIT
```

lower byte of the subtraction is implemented. This single byte difference is then written into EEPROM in the normal way.

Example 15.3

In Example 14.3 the discharge energy of a defibrillator was calculated by calculating the sum of the squared voltage differences of the sampled inputs from a baseline value. In this case we used a baseline value of 2.6 V, from observation of the waveform. This average value may vary from instrument to instrument and over time with usage. It is proposed to enhance the software by introducing a learning feature which could be called up when a switch connected to, say, pin RA4 is closed. This subroutine will sample the quiescent voltage 256 times to give a double-byte total. Taking the upper byte is tantamount to dividing by 256 and thus gives an average value. This datum is to be burnt into location h'00' of the Data EEPROM, and this can subsequentially be used as a learnt baseline value, which if necessary can be updated at regular intervals. Assuming that the GET_ANALOG subroutine of Program 14.1 on page 461 is available, show how a suitable subroutine could be coded.

Solution

From Fig. 14.20 on page 476 we see that the voltage from the defibrillator's Hall effect current sensor is connected to the RA0/AN0 pin. With the assumption that the ADC module has been enabled, as described in Program 14.6 on page 476, our task is to read the digitized Channel 0 byte 256 times inside a loop, accumulating to give a 16-bit total sum. Taking the top byte of this pair effectively gives an average value for this analog input (that is, divides by 256). If the defibrillator is quiescent during this learning run, this average gives the baseline voltage at this time.

When we have a baseline value, this byte can be burnt into the EEPROM Data module in the normal way. This can be subsequently read and treated in the same way as the Defined constant BASELINE in Program 14.6.

Program 15.12 uses the File COUNT to count 256 loop passes. Each pass adds the digitized byte to the double-byte Accumulator File pair. On exit

---

Program 15.12 Learning the baseline.

```
; ***************************************************************
; * FUNCTION: Sums 256 analog samples to find an average byte *
; * FUNCTION: value for the Baseline voltage which is blasted *
; * FUNCTION: into the EEPROM Data module                     *
; * RESOURCE: GET_ANALOG, EE_PUT subroutines                  *
; * ENTRY   : None                                            *
; * EXIT    : Average of Channel 0 in EEPROM location h'00'   *
; ***************************************************************
LEARN clrf    ACCUMULATOR      ; Zero double-byte sum MSB
      clrf    ACCUMULATOR+1    ; Zero LSB
      clrf    COUNT            ; Loop count zero

LEARN_LOOP
      clrw                     ; Start an Analog channel 0
      call    GET_ANALOG       ; Digitize
      addwf   ACCUMULATOR+1,f  ; Add to LSB of total
      btfsc   STATUS,C         ; Was there a Carry
      incf    ACCUMULATOR,f    ; IF yes THEN increment MSByte

      decfsz  COUNT,f          ; Count down one
      goto    LEARN_LOOP

; Burn in datum into EEPROM Data module ----------------------
      movf    ACCUMULATOR,w    ; Get the average value
      bsf     STATUS,RP0       ; Change to Bank 1 and put this
      clrf    EEADR            ; into Data EEPROM @ location h'00'
      movwf   EEDATA           ; This is the datum
      call    EE_PUT           ; Blast it in

      return                   ; All done
```

from the loop, subroutine EE_PUT burns this top Accumulator byte into location h'00' in the EEPROM Data module.

In a real situation a better outcome could be obtained by sampling 65,536 times and accumulating a triple-byte sum. The top byte of this triplet would again represent an average.

Self-Assessment Questions

15.1 Good program practice dictates that the datum written into Data EEPROM should be verified as the value that was intended to be written. Show how you could modify the EE_PUT subroutine of Program 15.2 to return a value of −1 in a File called ERROR if the action is not successful, otherwise zero.

15.2 In Program 15.5 we placed the look-up table at a 256-byte boundary in the Program store (specifically h'300') to simplify the computation of the 1-byte table index. Thus to look up table entry nn we simply place the address h'3nn' in the EEADRH:EEADR register pair.

Placing program segments at user-defined addresses is never a good idea, unless care is taken, as subsequent program alterations can cause code to overlap. Letting the assembler sort out locations of labels is much more reliable. However, in our case we would need to add nn to the address the assembler selects for the label TABLE. Unfortunately Program store addresses are 13-bits wide and PIC MCU arithmetic is only 8-bit. Microchip compatible assemblers have the directives high and low to separate the upper and lower bytes parts of a label; e.g., movlw low TABLE. Using these directives modify the subroutine SQUARE if the directive org h'300' is removed.

15.3 Microchip-compatible assemblers have the directive da (DAta) which can be used to store strings of character codes in Program memory. For example:

```
MESSAGE   da "Hello world\n",0
```

which places the characters inside quotes, coded in 7-bit ASCII code packed two at a time in each 14-bit word, followed by an all zeros word. The \n escape character means New Line; ASCII code h'0A'.

Assuming that this is done in a PIC16F87X device, write a subroutine called PDATA (Print DATA) to fetch each character from Program memory and transmit to a terminal using the subroutine PUTCHAR of Program 12.14 on page 378.

15.4 A certain hotel security system is to use a PIC-based reprogrammable
smart card for electronic guest room locks. On registration the card
is to be charged up with the following details:
1. A 4-digit room number, e.g., 1311.
2. Start data, e.g., 13072005.
3. End date, e.g., 15072005.

Assume that the PIC MCU has an integral EEPROM Data module
and communicates with the receptionist's terminal via a serial input
subroutine, such as described in Program 12.14 on page 378. Data
is coded in ASCII in the order outlined, preceded with the character
STX, terminated by ETX and delimited by SP; see Table 1.1 on page 5.
Design a routine to interpret the data and store them in EEPROM.

CHAPTER 16

Enhancing the Family

In 1994 Microchip released the PIC17C42 25 MHz high-range device based on the same predecessor Harvard RISC structure, but with an extended upwardly compatible 55-instruction set. The architecture potentially supported larger memory sizes and an enhanced Indirect addressing, interrupt, and stack capability.

In 1999 the PIC18CXXX 40 MHz 16-bit instruction word enhanced-range family of devices was introduced. This further extended the instruction set to 75, with features aiding the use of high-level languages and with an even more comprehensive interrupt, Indirect addressing, and stack-handling capability.

In order to complement our mid-range centered discussion, this chapter overviews the enhanced-range family, specifically using the PIC18FXX2 group of devices. These are designed as drop-in replacements for the mid-range PIC16F87X group, used as one of our mid-range exemplars.

After finishing this chapter you will:

- Be equipped to compare and contrast the mid- and enhanced-range architectures.
- Appreciate the byte-organized structure of the 16-bit Program store.
- Understand how the Data store is organized in conjunction with the Bank Select Register to access up to sixteen 256-File banks.
- Know how the three 12-bit FS Registers can be used as pointers, with pre/post increment/decrement and Working register offset options.
- Comprehend how interrupt handling with priority is implemented.
- Identify the major extensions to the standard peripheral module line-up.
- Be aware of how the instruction set has been extended.

The PIC18FXX2 group was introduced in 2001 to provide an easy upgrade path for the PIC16F87X and older PIC16C73/4 devices. The four group devices are:

PIC18F242
This 28-pin device has a 8 Kword Program store and a 788-byte Data store.

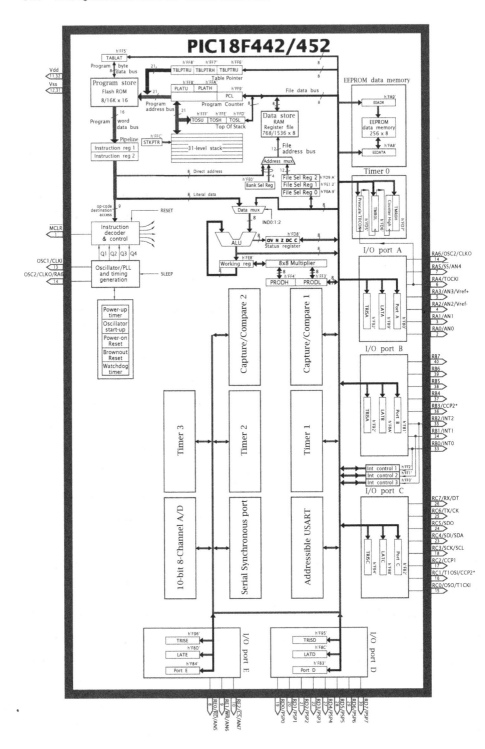

Fig. 16.1 Architecture of the 40-pin PIC18F442/52 devices.

PIC18F252
This is identical to the above but with a 16 Kword Program store and 1536-byte Data store.

PIC18F442
A 40/44-pin version of the PIC18F242.

PIC18F452
A 40/44-pin version of the PIC18F252.

Apart from the cropped Peripheral module line-up in the 28-pin PIC18F2X2 devices, they are virtually identical to the PIC18F4X2 equivalents, whose 40-pin architecture is shown in Fig. 16.1.

In comparing these devices to their PIC16F87X relatives of Fig. 10.1 on page 272 we will look at the fetch unit, execute unit, peripheral services and finally software issues.

Fetch Unit

The Fetch unit, top left of the diagram, uses a Harvard architecture with a 2-deep Pipeline holding two 16-bit instructions, allowing parallel fetch and execution cycles. The Flash EEPROM store holds the instructions stored as 16-bit words. As described in Chapter 15, data can be internally read from or written into the store. In the PIC18XXXX family this internal data is treated as 8-bit bytes, which are funneled in and out via the **TABle LATch** SPR. The actual address in the Program store of the data byte being accessed is held in the three **TaBLe PoinTeR** registers. The instruction **tablrd** is used to copy the addressed byte out to TBLAT. The instruction **tablwt** aids internal writing into the Program store, although this needs to be done as a block of eight bytes at a time, in conjunction with EECON1, in a similar manner to that described in Fig. 15.7 on page 497.

Instructions in the enhanced-range architecture are mainly coded as single 16-bit words; see Fig. 16.4, with a few double-word exceptions. However, to facilitate the access of fixed byte data tables and strings the Program store is organized as bytes. As can be seen from Fig. 16.2, each instruction occupies two bytes and is addressed only at even addresses. For instance, the address of instruction 6 is h'000A' and instruction 7 is h'000C'. This is consequence of the omission of bit 0 of the 21-bit Program Counter and thus the PC effectively increments in steps of two as the program linearly advances. The maximum capacity of this architecture is thus 2^{20} words or 2^{21} bytes. Specifically, for the exemplar devices, the PC is 17 bits wide.

As in the case of the mid-range architecture, the PC is zeroed on reset. The low byte PC[7:0] is directly accessible as the PCL special-purpose register, and PCLATH acts as a buffer, as described in Fig. 4.8 on page 86, allowing the high byte of PC to be altered at the same time as writing a

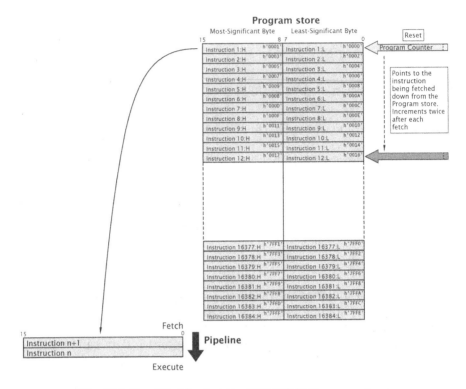

Fig. 16.2 Simplified look at the PIC18FX42 Program store.

new byte to PCL. However, unlike the Mid-range case, reading the state of PCL updates PCLATH with PC[15:8]. The **PCLATU** register works the same way but allows access to the Upper byte PC[21:16]. This allows the software to read-from and write-to the entire 21-bits of the PC.

The stack has now been enlarged from 8 to 31 levels. A stack overflow or underflow will automatically reset the MCU. The Stack Pointer **STKPTR** can be accessed as a SPR in the Data store, and the 21-bit contents at the pointed-to stack location read (pulled) out to the three Top Of Stack registers or written (pushed) into the stack, allowing a great deal of flexibility in manipulating and passing data via the stack.

Execute Unit

Like the low- and mid-range families, the ALU processes data eight bits at a time. Thus the enhanced-range is also categorized as an 8-bit MCU. The ALU has been augmented with a hardware 8-bit × 8-bit multiplier, which supports the instructions **mullw** (MULtiply Literal with W) and **mulwf** (MULtiply W with the File), with the 16-bit outcome being located in the two **PRODH:PRODL** SPRs; see Program 16.1.

The Working register has been enhanced in so far as it now can be accessed as a SPR in the Data store. This means that instructions that operate directly on a File can also be used for W. For instance, `decfsz WREG,f` decrements the contents of the Working register as a SPR, named **WREG**.

The Status register no longer has a role in bank switching and the space is used to implement **N** (Negative) and **OV** (OVerflow) flags, to allow full signed 2's complement addition and subtraction; see page 9.

The enhanced-range Data store shown in Fig. 16.3 holds the source and destination of most of the Data processed by the ALU, as well as the SPR Files. Like all PIC microcontrollers, each location holds one byte, but the File addressing scheme differs radically from the mid-range equivalent of Fig. 4.7 on page 80.

From the diagram we see that the Data store is divided into 16 banks, each holding up to 256 Files, giving a potential 4 Kbyte capacity. Entry to

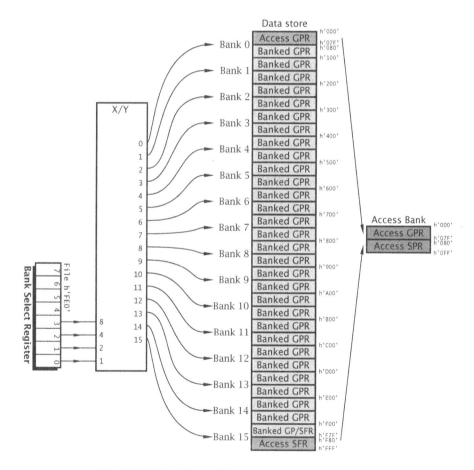

Fig. 16.3 The enhanced-range Data store structure.

any one of these banks is controlled via the **Bank Select Register**. Bits **BSR[3:0]** select any one bank by means of a 4- to 16-line decoder. As the BSR is itself located as a SPR File in Bank 15, and points to Bank 0 on a Power-on reset, the special instruction Load Bank Select Register **lbsr** is used to alter BSR irrespective of which bank is active. For instance, to change to Bank 5 we have **lbr 5**.

The bottom 128 Files in Bank 0 holding general-purpose register (GPR) Files and the top 128 Files in Bank 15 holding SPRs are labeled **Access bank** in Fig. 16.3. The Access bank bypasses the BSR banking scheme and allows access to 128 GPRs and 128 SPRs. This special access to the Data store is controlled with bit 8 of the 16-bit binary instruction word, as shown in Fig. 16.4. Comparing this machine structure with the mid-range equivalent on page 98, we see that the File address field is increased from 7 to 8 bits, giving an expended bank size from $2^7 = 128$ to $2^8 = 256$. When the Access bit is 0 then the Access bank is enabled, otherwise the bank pointed-to by the BSR is targeted by the instruction. If the Power-on default BSR = h'00' is kept, instructions have entry to all 256 Files in Bank 0 and all 128 SPRs in Bank 15 with no change in the contents of the BSR being necessary. For instance, to copy the contents of File h'026' to W we have movf h'026',w,0 and to copy File h'096' we have movf h'096',w,1, the access scheme being indicated in the last comma-delimited digit ,0 or ,1. In practice this explicit access designator is rarely used in the source code, as without it the Microchip assembler will automatically use the Access bank (i.e., ,0) for addresses in the range File h'000' – h'07F' and File h'F80' – h'FFF'.

The PIC18FX52 has general-purpose storage up to the end of Bank 5; that is, up to File h'5FF'. The less generously endowed PIC18FX42 populates Bank 0 through Bank 2, with GPR storage up to File h'2FF'.

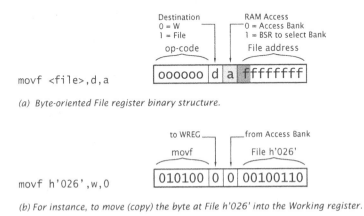

(a) Byte-oriented File register binary structure.

(b) For instance, to move (copy) the byte at File h'026' into the Working register.

Fig. 16.4 Machine-code structure of a typical 16-bit instruction accessing the Data store.

All micros have the capability of using Indirect or Indexed addressing to efficiently deal with arrays and tables of data in RAM. In the low- and mid-range series this is implemented by using the File Select Register (FSR) as a pointer into the Data store and the mechanism invoked by accessing the phantom INDirect register File (INDF). This somewhat roundabout approach is also used by the enhanced-range family, but writ large. Here there are three separate File Select Registers — **FSR0**, **FSR1**, and **FSR2**. Each of these pointers are implemented as two SPRs, as shown in Fig. 16.5(a). This means that a FSR can hold a complete 12-bit code, able to point to *any* File address in the Data store independent of its banking structure. The Load FSR **lfsr** instruction can copy a 12-bit constant into any one of the three FSR registers in a single action. For instance, to point FSR2 to File h'500' we have lfsr 2,h'500'.

Each FSR has five distinct modes of operation, as listed in Fig. 16.5(a). The mode is selected by invoking one of five different virtual trigger registers. These are, where *i* indicates 0, 1, or 2:

INDirect*i*
For instance, clrf INDF2 zeros the File pointed to by the 12-bit FSR2.

POSTDEC*i*
For instance, clrf POSTDEC2 zeros the File pointed to by the 12-bit FSR2 and then decrements FSR2 *after* the action has been executed.

POSTINC*i*
For instance, clrf POSTINC2 zeros the File pointed to by the 12-bit FSR2 and then increments FSR2 *after* the action has been executed.

PREINC*i*
For instance, clrf PREINC2 *first* increments FSR2 and only then zeros the File pointed to by the 12-bit FSR2.

PLUSW*i*
For instance, if the contents of the Working register were h'06' then the instruction clrf PLUSW2 would zero the File pointed to by FSR2 + 06. Neither the contents of FRS2 nor W would be altered. The contents of W are treated as a 2's complemented signed number ranging from −128 through +127.

As an example, consider two arrays of eight unsigned numbers labeled in Fig. 16.5(b) as NUM1 and NUM2. We wish to multiply each element *i* to give an array of eight unsigned 16-bit numbers NUM3. The MSByte of NUM3[i] is displaced from the LSbyte by −8 Files. Program 16.1 uses FSR0 to walk through NUM1[], FSR1 to point to the corresponding NUM2[] array. As NUM3[] is really two byte arrays displaced by eight, FSR2 is used to point to the LS Byte of the Product array. These three pointers are initialized to the location of element 0 of each array, shown shaded in Fig. 16.5(b), using lfsr.

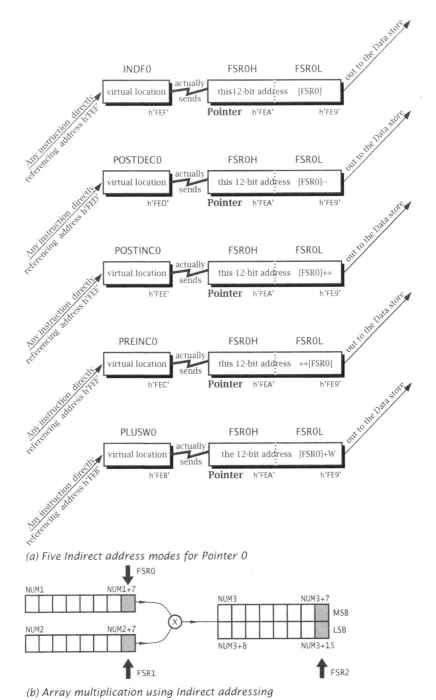

(a) Five Indirect address modes for Pointer 0

(b) Array multiplication using Indirect addressing

Fig. 16.5 Indirect addressing via FRS0.

Program 16.1 Multiplying two byte arrays.

```
;********************************************************************
; FUNCTION: Multiplies NUM1[8] x NUM2[8] = NUM3[16]              *
; ENTRY    : NUM1[8], NUM2[8] global                            *
; EXIT     : NUM3[16] global                                    *
;********************************************************************
ARRAY_MUL lfsr  0,NUM1+7        ; Point to LSByte of Number 1
          lfsr  1,NUM2+7        ; Point to MSByte of Number 2
          lfsr  2,NUM3+d'15'    ; Point to LSWord of Number 3
          movlw 8               ; Set up loop count of 8
          movwf COUNT

M_LOOP    movf  POSTDEC0,w      ; Get NUM1[n]
          mulwf POSTDEC1        ; Multiply by NUM2[n]

          movlw -8              ; Offset for NUM3 LSB:MSB
          movff PRODH,PLUSW2    ; Copy High byte Product into MSB
          movff PRODL,POSTDEC2  ; Copy low byte Product into LSB
          decfsz COUNT,f        ; One more loop pass
           goto M_LOOP

          return
```

The core of the program is the mulwf instruction. This generates a 16-bit product into the SPRs PRODH:PRODL from data in W and the designated File. By using the Postincrement Indirect addressing mode to move the first datum into W and to identify the File, the multiplication operation is walked through the two arrays in a loop of eight.

In order to copy the contents of each byte from PRODL into the LS Byte array NUM3[] and PRODH into the MS Byte array, the program uses the **movff** MOVe from source File to destination File instruction. This 2-word instruction (see page 521) uses two 12-bit File addresses to locate source and destination data anywhere in the Data store without any need to use the banking mechanism. The contents of PRODH are first copied into the MS Byte by using the Plus W Indirect mode. As W has been preset to -8 (h'F8'), effectively this datum is copied to a location eight Files lower than that pointed to by FSR2. Finally, PRODL is copied into the LS Byte array location pointed to by FSR2. By using the Postdecrement Indirect mode for the destination, this pointer too can walk through the Product array.

Peripheral Services

Broadly speaking, the Peripheral modules in our exemplar devices match those in the mid-range family apart from small deviations. Major differences are:

Parallel Ports

Each of the several parallel ports in Fig. 16.1 are shown with three registers, rather than the two described in Chapter 11. The PORTX register is still driving the I/O pins, whose status is controlled by the corresponding TRISX register. In addition, each port has an associated **LATX** register. Although each LATch register has a unique address—e.g. LATB is located at h'F8A' as compared to h'F81' for PORTB—they don't physically exist. The relationship between these SPR Files can be see in Fig. 16.6, which should be compared to Fig. 11.3 on page 300. The only real difference is the addition of the LAT 3-state buffer, shown shaded. When a LATch register is read — for instance, movf LATB,w — then the state of the Data flip flop itself is sensed. The corresponding movf PORTB,w reads the state of the actual Port B pins. These are usually the same, but as we observed on page 303, if the current being sunk or sourced at a pin exceeds its specification or if the loading capacitance is large and the switching

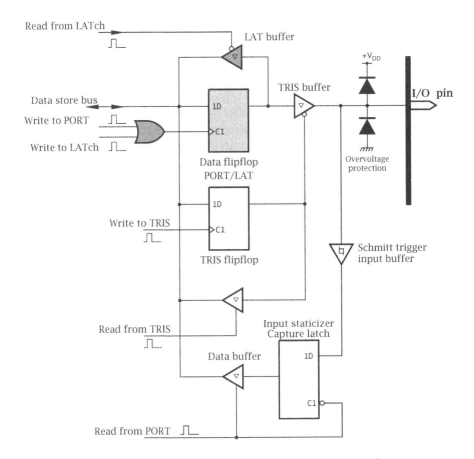

Fig. 16.6 A simplified typical enhanced-range I/O port line.

period is short, then a read–modify–write instruction on a Port SPR can give unpredictable results. Doing the modification on the LATch register rather than the corresponding Port register, makes the process independent of conditions in the outside world. For instance, btfsc LATB,7 will skip over the next instruction if bit 7 of Port B is clear, even if pin RB7 is pulled high due to a large sink current, and is therefore more reliable than the equivalent btfsc PORTB,7.

Writing to either a LATch register or its corresponding PORT register are identical. Thus, e.g., movwf LATB is identical to movwf PORTB in their outcome.

Timer 0

Timer 0, which remained essentially unchanged from the low- to mid-range host has been expanded to a 16-bit counter/timer with its own Prescaler separate from the Watchdog timer Postscaler. Its new Control register **T0CON** can be used to switch Timer 0 to an 8-bit mode for legacy applications.

Timer 3

The enhanced-range family has an additional 16-bit timer, which is similar to Timer 1. **TMR3** can share the TMR1 low-frequency external oscillator, which can also be switched in as the system oscillator if power is to be conserved. Each CCP module can select either TMR1 or TMR3 as a a timebase, allowing two separate timebases.

Some members of the enhanced family have a larger footprint and a greater number of Peripheral modules than our exemplar. For instance, the 80-pin PIC18F8720 has a 128 Kbyte Program store with a 3840 byte RAM, 1024 byte Data EEPROM module and up to 68 I/O port pins. The 10-bit ADC module has a 16-channel input multiplexer. There are two USARTs and five CCP/PWM modules, with an additional 8-bit Timer.

Interrupt Handling

Each Peripheral module can generate an interrupt request in the same way as the mid-range family. As well as the external INT interrupt, now called **INT0**, two additional hardware interrupts are supported. **INT1** shares pin RB1 and **INT2** is at RB2. There are now three corresponding INTCON registers to support these requests.

The largest difference in the interrupt handling capability of the two ranges is the introduction of two levels of **priority**. In the mid-range family once the response starts to an enabled interrupt request, any interrupt from another source is locked out automatically, as GIE in INTCON[7] is automatically cleared. When the interrupt service request (ISR) returns via a retfie instruction, GIE is set again and pending or subsequent interrupt requests can again be honored. While it is important to prevent interference between interrupt requests, this can lead to problems. For instance, consider a biomedical monitor in an ISR handling a slow communication link sending an hour's worth of telemetry data to a central processor, when the patient goes into cardiac arrest. The priority is obvious, but the sensor generating the latter's interrupt is effectively locked out!

With the enhanced-range each source of an interrupt request[1] has three associated bits. For instance, the ADC module has an interrupt flag ADIF to signal a service request in PIR1[6], a mask bit ADIE in PIE1[6] to enable requests from this source and **ADIP** in **IPR1[6]** (**Interrupt Priority Register 1**) to determine if interrupts from this source are to be treated as high priority (ADIP = 1, the Power-up default) or low priority (ADIP = 0).

Any request arriving from a source set to be Low-priority goes to its ISR via the Low-priority Interrupt vector address h'018' and sets the **GIEL** (Global Interrupt Enable Low) bit in INTCON[6]. This locks out any other Low-priority response until the ISR finishes. However, if a Low-priority ISR is being executed when a High-priority request is received, then this ISR will be suspended and the processor will go via the High-priority Interrupt vector at h'008' (which corresponds to the word address h'004' Interrupt vector in the mid-range family). This time the **GIEH** Global Interrupt Enable High-priority bit in INTCON[7] is set, which locks out any further request of whatever priority.

When the processor responds to a request by going to the appropriate vector, the Program counter is copied into the stack as in the mid-range family. However, in addition the state of the Working, Status and Bank Select registers are also saved in three buried shadow locations, sometimes called a **Fast stack**. In order to retrieve this stashed-away context, a single bit in the machine code for the `retfie` instruction if set to 1 will restore this context as well as the Program Counter, i.e., `retfie 1` or more clearly `retfie FAST`. If this **Shadow** bit is 0, which is the default where no suffix is used in the assembly source code, then only the Program Counter is retrieved.

Great care has to be taken with this fast context save/retrieve mechanism, as the Fast stack only stores *one* copy of these system Files. If a High-priority request interrupts a Low-priority ISR, then the context of the former will overwrite that of the latter! Thus, where a mixture of priority levels are to be serviced, the `retfie FAST` instruction should only be used for the High-priority ISR. If no interrupts are set-up, then the `call FAST` variant can be used to call up a subroutine with the context shadowed on entry and `return FAST` to retrieve it on exit.

The three INTCON registers handle interrupt flags, enables and priority for Timer 0, Port B Change, and the three Hardware interrupts. All other interrupts are handled by the PIR1:2 and PIE1:2 and IRP1:2 registers. On Power-up reset, the interrupt handling scheme defaults to the

---

[1] Except INT0 which is always High-priority.

mid-range non-prioritized scheme and the state of the Priority bit settings are ignored. The **IPEN** (Interrupt Priority ENable) bit in the Reset Control register RC[7] must be set to turn on the enhanced-range scheme.

Software

Table 16.1 lists all 75 instructions supported by the enhanced-range family. Comparing this to the instruction set of page 70, we see that all the mid-range instructions have been kept, apart from `clrw` which is now superfluous, as the Working register can be accessed as an addressable SPR File in the Data store; thus `clrf WREG`. Some of these instructions have been enhanced; e.g., activating the Negative and OVerflow flags. Several have extended versions; for instance, **addwfc** ADDs the contents of W to that of the specified File with Carry; see Program 16.2.

To complete this chapter we will list the highlights of the new instruction set, using the same categories as Chapter 5.

Movement Instructions

The most important of the new instructions in this set is `movff`, which we have already used in Program 16.1. This enables the software to copy the contents of *any* source File to *any* destination File, independent of the bank structure of the Data store, without having to go via the Working register. As both the source and destination require a full 12-bit address, `movff` is an example of a 2-word 2-cycle instruction. The binary structure of this instruction is:

1100	f_s	f_s	f_s	f_s	f_s	f_s	f_s	f_s	f_s	f_s	f_s	f_s
1111	f_d	f_d	f_d	f_d	f_d	f_d	f_d	f_d	f_d	f_d	f_d	f_d

The initial four bits of the second word b'1111' mimic the op-code for the `nop` instruction.[2] This is a deliberate feature of all four 2-word instructions, to avoid problems where the program skips into the middle of the instruction; see Program 16.4 for an example.

The `lfsr` instruction, which also requires a double-word coding, loads (moves) the 12-bit constant address into one of three FSR pointers; as shown in Program 16.1. The `lbsr` instruction is coded in a single word.

Arithmetic Instructions

One of the niche markets targeted by the PIC18XXXX series is low-end[3] real-time digital signal processing (DSP). DSP requires considerable arithmetic activity, and this category of instructions arguably has benefited the most in upgrading.

---

[2] There are two `nop` op-codes; b'0000' and b'1111'.
[3] The dsPIC™ family are explicitly designed for heavy-duty number crunching.

Table 16.1: Enhanced-range instruction set.

16-bit Instruction	PIC18XXXX Mnemonic	Dest W	Dest F	N	V	Z	D	C	Operation summary
ADD Literal to W	addlw LL	√		√	√	√	√	√	w <- w + #LL
ADD W and F	addwf f,d,b	√	√	√	√	√	√	√	d <- w + f
ADD W and F with Carry	addwfc f,d,b	√	√	√	√	√	√	√	d <- w + f + C
AND Literal to W	andlw LL	√		√	•	√	•	•	w <- w · #LL
AND W to F	andwf f,d,b	√	√	√	•	√	•	•	d <- w · f
Bit Clear File bit n	bcf f,n,b		√	•	•	•	•	•	f_n <- 0
Bit Set File bit n	bsf f,n,b		√	•	•	•	•	•	f_n <- 1
Bit ToGgle File bit n	btg f,n,b		√	•	•	•	•	•	f_n <- $\overline{f_n}$
BRAnch (relative) to <Label>	bra Offset			•	•	•	•	•	PC <- PC±Offset
Branch (relative) if Carry set to <Label>	bc offset			•	•	•	•	•	PC <- PC±offset IF C==1
Branch (relative) if Not Carry set to <Label>	bnc offset			•	•	•	•	•	PC <- PC±offset IF C==0
Branch (relative) if Zero to <Label>	bz offset			•	•	•	•	•	PC <- PC±offset IF Z==1
Branch (relative) if Not Zero to <Label>	bnz offset			•	•	•	•	•	PC <- PC±offset IF Z==0
Branch (relative) if Negative to <Label>	bn offset			•	•	•	•	•	PC <- PC±offset IF N==1
Branch (relative) if Not Negative to <Label>	bnn offset			•	•	•	•	•	PC <- PC±offset IF N==0
Branch (relative) if OVerflow to <Label>	bov offset			•	•	•	•	•	PC <- PC±offset IF V==1
Branch (relative) if Not OVerflow to <Label>	bnov offset			•	•	•	•	•	PC <- PC±offset IF V==0
Bit Test File bit n & Skip if Clear	btfsc f,n,b			•	•	•	•	•	PC++ IF f_n == 0
Bit Test File bit n & Skip if Set	btfss f,n,b			•	•	•	•	•	PC++ IF f_n == 1
CALL (jump to) subroutine[1]	call aaa			•	•	•	•	•	(TOS) <- PC, SP--, PC <- aaa
CALL and shadow context[1]	call aaa,FAST			•	•	•	•	•	plus save W, STATUS,BSR
CLeaR File	clrf f,b		√	•	•	√	•	•	f <- 00
CLeaR Watch Dog Timer	clrwdt			•	•	•	•	•	WDT <- 00
COMplement File	comf f,d,b	√	√	√	•	√	•	•	d <- $\overline{f}$
ComPare File, Skip if EQual	cpfseq f,b			•	•	•	•	•	PC++ IF f_n == W
ComPare File, Skip if Greater Than	cpfsgt f,b			•	•	•	•	•	PC++ IF f_n (unsigned) > W
ComPare File, Skip if Less Than	cpfslt f,b			•	•	•	•	•	PC++ IF f_n (unsigned) < W
Decimal Adjust W	daw	√		•	•	•	•	√	Correct sum of packed BCD digits
DECrement File	decf f,d,b	√	√	√	√	√	√	√	d <- f--
DECrement File & Skip if Zero	decfsz f,d,b	√	√	•	•	•	•	•	d <- f--, PC++ IF f == 0
DeCrement File & Skip if Not Zero	dcfsnz f,d,b	√	√	•	•	•	•	•	d <- f--, PC++ IF f != 0
GOTO (jump to) aaa[1]	goto aaa			•	•	•	•	•	PC <- aaa
INCrement File	incf f,d,b	√	√	√	√	√	√	√	d <- f++
INCrement File & Skip if Zero	incfsz f,d,b	√	√	•	•	•	•	•	d <- f++, PC++ IF f == 0
INcrement File & Skip if Not Zero	infsnz f,d,b	√	√	•	•	•	•	•	d <- f++, PC++ IF f != 0
Inclusive OR Literal to W	iorlw LL	√		√	•	√	•	•	w <- w + #LL
Inclusive OR W to F	iorwf f,d,b	√	√	√	•	√	•	•	d <- w + f
Load File Select Register n (0-2)[1]	lfsr n,LLL			•	•	•	•	•	FSR_n <- #LLL

√	: Flag operates normally	•	: Not affected	LL	: 8-bit Literal
LLL	: 12-bit Literal	d	: Destination; 0 = w, 1 = f	b	: Banking via the BSR
f_n	: File bit n	PC	: Program Counter	PC++	: Jump over next instruction
w	: Working register	WDT	: Watch Dog Timer	(TOS)	: Contents of Top Of Stack
==	: Equivalent to	!=	: Not equivalent to	++	: Add one
- -	: Subtract one	FSR_n	: File Select Register 0, 1, 2	SP	: Stack Pointer
#	: Constant	offset	: ±128-word offset	Offset	: ±1024-word offset
Note 1 : Two-word instruction					

(continued next page)

Table 16.1: (*continued*).

16-bit Instruction	PIC18XXXX Mnemonic	Dest W	Dest F	N	V	Z	D	C	Operation summary
MOV Literal into Bank SElect Register	movlb L			•	•	•	•	•	BSR <- #L
MOVe in File (load)	movf f,d,b	√	√	√	•	√	•	•	d <- f
MOVe source File to destination File[1]	movff fs,fd			•	•	•	•	•	$f_{destination}$ <- f_{source}
MOVe Literal into W	movlw LL	√		•	•	•	•	•	w <- #LL
MOVe W out to File (store)	movwf f,b		√	•	•	•	•	•	f <- w
MULtiply Literal by W	mullw LL			•	•	•	•	•	PRODH : PRODL <- W × #LL
MULtiply W by File	mulwf f,b			•	•	•	•	•	PRODH : PRODL <- W × f
NEGate (2's complement) File	negf f,b		√	√	√	√	√	√	f <- - f
No OPeration	nop			•	•	•	•	•	Do nothing
POP top of stack	pop			•	•	•	•	•	SP++
PUSH top of stack	push			•	•	•	•	•	(TOS) <- PC+2, SP--
Relative CALL subroutine	rcall Offset			•	•	•	•	•	(TOS) <- PC, SP--, PC <- PC±Offset
RESET software MCLR	reset			•	•	•	•	•	All registers to MCLR Reset state
RETURN from subroutine	return			•	•	•	•	•	PC <- TOS
RETURN with shadowed context	return FAST			√	√	√	√	√	PC <- TOS; context
RETurn from subroutine with L in W	retlw LL	√		•	•	•	•	•	w <- #LL, PC <- TOS
RETurn From IntErrupt	retfie			•	•	•	•	•	GIEL|GIEH <- 1, PC <- TOS
RETurn From IntErrupt with context	retfie FAST			√	√	√	√	√	GIEL|GIEH <- 1, PC <- TOS, context
Rotate Left thru Carry File	rlcf f,d,b	√	√	√	•	√	•	b7	⟵ [C] ⟵ [7 File 0] ⟵
Rotate Left Not thru Carry File	rlncf f,d,b	√	√	√	•	√	•	•	⟵ [7 File 0] ⟵
Rotate Right thru Carry File	rrcf f,d,b	√	√	√	•	√	•	b0	⟶ [7 File 0] ⟶ [C] ⟶
Rotate Right Not thru Carry File	rrncf f,d,b	√	√	√	•	√	•	•	⟶ [7 File 0] ⟶
SET File to all 1s	setf f,b		√	•	•	•	•	•	f <- b'11111111'
SLEEP mode on	sleep			•	•	•	•	•	WDT <- 0, Clock off
SUB W from Literal	sublw LL	√		√	√	√	√	√	w <- #LL - w
SUBtract W from F	subwf f,d,b	√	√	√	√	√	√	√	d <- f - w
SUBtract W from F with Borrow	subwfb f,d,b	√	√	√	√	√	√	√	d <- f - (w + !C)
SUBtract F from W with Borrow	subfwb f,d,b	√	√	√	√	√	√	√	d <- w - (f + !C)
SWAP File nybbles	swapf f,d,b	√	√	•	•	•	•	•	d <- f[7:4] <-> f[3:0]
TABLe ReaD/TABLe WriTe from/to Program store as pointed to by TBLPTR[20:0] into/out of TBLAT									
Read	tablrd *			•	•	•	•	•	TABLAT <- (TBLPTR)
with TABLPTR post incremented	tablrd *+			•	•	•	•	•	TABLAT <- (TBLPTR++)
with TABLPTR post decremented	tablrd *-			•	•	•	•	•	TABLAT <- (TBLPTR- -)
with TABLPTR pre incremented	tablrd +*			•	•	•	•	•	TABLAT <- (++TBLPTR)
Write	tablwt *			•	•	•	•	•	(TBLPTR) <- TABLAT
with TABLPTR post incremented	tablwt *+			•	•	•	•	•	(TBLPTR++) <- TABLAT
with TABLPTR post decremented	tablwt *-			•	•	•	•	•	(TBLPTR- -) <- TABLAT
with TABLPTR pre incremented	tablwt +*			•	•	•	•	•	(++TBLPTR) <- TABLAT
TeST File Skip on Zero	tstfsz f,b			•	•	•	•	•	PC++ IF f == 0
eXclusive OR Literal to W	xorlw LL	√		√	•	√	•	•	w <- w ⊕ #LL
eXclusive OR W to F	xorwf f,d,b	√	√	√	•	√	•	•	d <- w ⊕ f

√ : Flag operates normally	• : Not affected	L : 4-bit Literal (Bank no.)	
LL : 8-bit Literal	d : Destination; 0 = w, 1 = f	b : Banking via the BSR	
f_n : File bit n	PC : Program Counter	w : Working register	
WDT : Watch Dog Timer	TOS : Top Of Stack	PC++ : Jump over next instruction	
== : Equivalent to	context : W, STATUS, BSR	++ : Add one	
- - : Subtract one	FSR_n : File Select Register 0, 1, 2	GIE : Global Interrupt Enable mask	
# : Constant	offset : ±128-word offset	Offset : ±1024-word offset	
Note 1 : Two-word instruction			

One of the most glaring deficiencies in the low- and mid-range instruction set is the absence of an Add with Carry and Subtract with Borrow operation. This leads to rather cumbersome and slow processing when multiple-byte data are added or subtracted. The **addwfc** instruction ADDs the byte contents of the Working register to the specified File plus the state of the Carry flag. The outcome is placed in either W or back in the File in the usual way, with **C** being set as appropriate, ready for the next addition. As an example, a routine to add two 3-byte data giving a 4-byte sum is shown in Program 16.2. Apart from the least-significant byte, the Carry is added in a natural way as the addition moves to the higher digits to whatever precision is desired. Without such an instruction, any addition must be followed by a sequence of possible increments as the carry ripples upwards.

Program 16.2 A triple-precision addition.

```
TP_ADD  clrf   NUM3_V     ; Zero the Number 3 overflow
        movf   NUM1_L,w ; Get Low byte Number 1
        addwf  NUM2_L,w ; Add Low byte Number 2
        movwf  NUM3_L     ; Gives Low byte Number 3

        movf   NUM1_H,w ; Get High byte Number 1
        addwfc NUM2_H,w ; Add High byte Number 2 plus carry
        movwf  NUM3_H     ; Gives High byte Number 3

        movf   NUM1_U,w ; Get Upper byte Number 1
        addwfc NUM2_U,w ; Add Upper byte Number 2 plus carry
        movwf  NUM3_U     ; Gives Upper byte Number 3

        btfsc  STATUS,C ; Skip IF no Carry
         incf  NUM3_V,f ; ELSE add one to byte 4
```

In a similar way the **subwfb** instruction subtracts the contents of W plus any borrow from a previous subtraction from the designated File. Also provided is **subfwb** which commutes the subtraction order to (F + B) − W.

All Addition, Subtraction, together with their Increment and Decrement counterparts, implement full 2's complement signed number arithmetic in conjunction with the **N** and **OV** flags. The incf and decf instructions now actuate all flags, instead of only the **Z** flag and so multiple-precision operations are facilitated. The new **negf** instruction 2's complements the contents of the target File, which if a signed byte effectively changes the sign of the datum.

Multiplication is a critical DSP operation. The two Multiply instructions generate a 16-bit product of two unsigned bytes in a single cycle, as described in Program 16.1. These can be used as the basis of relatively short multiple-precision 2's complement multiplication routines.

The Decimal Adjust Working register (**daw**) instruction facilitates Addition operations on numbers formatted as BCD digits. As discussed on page 93, operations involving such data using normal binary arithmetic requires a post-correction process to eliminate the redundant six binary codes b'1011 – 1111'. daw implements this correction when it directly follows an Addition or Increment of a packed BCD datum; that is, two BCD nybbles packed into a byte. Program 16.3 shows the equivalent of Program 4.1 using this instruction. Note that daw does not convert natural binary data to an equivalent BCD format; it simply provides for a post-addition correction of data already in packed BCD form.

Program 16.3 Packed 2-digit BCD incrementation using the daw instruction.

```
BCD_INC   incf    BCD,w    ; Binary increment BCD datum
          daw              ; Correct it to follow BCD rules
          movwf   BCD      ; and return to memory
```

Also in this category, the ability to clear or set a single bit in the specified File has been augmented to allow a single bit to be toggled using **btg** (Bit ToGgle). For example, Program 16.4 generates a squarewave train of 20 pulses at pin RA0, each of eight machine cycle duration, in the same manner as the code associated with Fig. 5.19 on page 134. Noteworthy is the use of the decfsz instruction instruction directly on the Working register, treated as a SPR named WREG.

Program 16.4 Toggling pin RA0.

```
; Set-up PortA
        movlw   b'0110'  ; Set up PORTA as digital
        movwf   ADCON1
        bcf     LATA,0   ; Start off with pin RA0 Low
        bcf     TRISA,0  ; Make RA0 an output

; Sometime later -------------------------------------------------
        movlw   d'40'    ; Put decimal 40 into W

LOOP    btg     LATA,0   ; Pin RA0 change-over             1~

        decfsz  WREG,f   ; Count down                      1|3~
          goto  LOOP     ; ELSE break from the loop        2~

        .....   .....    ; Next
```

In the enhanced-range core, goto is a double-word instruction. Thus decfsz will skip into the second word of the goto code. In common with all four double-word instructions, the machine code of the second word is such that a direct entry here is treated as a nop. This has the effect of adding one cycle to the execution time, giving a 3˜ exit period.

Finally, the instruction **setf** gives the programmer a direct means to set the specified File to all 1s, as a counterpart to clrf.

Logic and Shifting Instructions

The only change to this category is the Rotate instructions. As we observed in Fig. 5.13 on page 127, the rlf and rrf instructions rotate the datum in the target File through the Carry flag. Rotate instructions which do not shift through **C** have been added to the repertoire, with mnemonics **rlncf** (Rotate Left Not through the Carry Flag) and **rrncf**. To be consistent, the mid-range equivalents have been renamed **rlcf** and **rrcf**. In all cases, as the Working register is now accessible as a SPR in the Data store, the contents of WREG can now be shifted in the same manner as any other File.

Program Counter Instructions

The goto, call and three Return instructions have all been retained. Both the former are double-word instructions, able to jump directly far away to anywhere in a 20 Mword Program store without the page limitation of the mid-range core. The call instruction has a Fast option which can be used to save the context in Shadow stack if interrupts are not being used, together with the Fast return option.

Many goto jumps are relatively short; for instance, in Program 16.4 the goto LOOP only hops backwards over two instructions. The new single-word **bra** instruction BRAnches up to 1024 words forward or 1023 words backward. Thus bra LOOP saves one word of Program code, but still takes two machine cycles to execute. In the same manner the **rcall** (Relative CALL) instruction can jump to a nearby subroutine, albeit without a shadowing option.

The unconditional Branch is complemented by a set of ten **conditional Branches**, which only hop back or forward depending on the state of each of the Status flags; apart from **DC**. Although flags can be checked using the btfss and btfsc instructions, a true outcome only translates to a single skip. A conditional Branch can go 128 words forward or 127 words back based on the specified Status flag being set, e.g., **bc** (Branch if Carry set) or clear, e.g., **bnc** (Branch if Not Carry set). For instance, assuming that a transducer presents a digitized signed 2's complement at Port C representing temperature; the following code fragment clears a File called SIGN if the temperature is positive and sets it to all 1s if negative. The modulus of the value is relocated to a File named TEMPERATURE.

```
        clrf    SIGN                ; Zero the Sign Flag and make
        movff   PORTC,TEMPERATURE   ; a copy of transducer reading

        movf    TEMPERATURE,f       ; Test for zero or negative

        bnn     NEXT                ; IF Not -ve THEN skip ahead

        negf    TEMPERATURE,f       ; ELSE 2's complement to +ve
        setf    SIGN                ; and make Sign File all ones

NEXT    .....   .....               ; Continue
```

The **bnn** instruction branches to the label NEXT if the datum is positive. If this is not true then the datum is negative, so negating it gives the positive equivalent and the setf instruction makes SIGN all 1s, as specified.

We saw on page 119 that in order to compare the magnitude of two unsigned numbers, say, W and the contents of a File, they should be subtracted and the outcome deduced by testing the **C** and **Z** flags. The three new Compare-and-Skip instructions automate this operation, with **cpfseq** skipping over the next instruction if the unsigned datum in the specified File is EQual to the byte in W, **cpfsgt** if it is Greater Than and **cpfslt** if it is Less Than the byte in W.

For an example, consider the code on page 119 which implemented a fuel warning system, lighting a warning lamp if below 20 liters and sounding a buzzer as well if below 5 liters. The code now becomes:

```
ALARM bcf     DISPLAY,BUZZER ; Turn off the Buzzer
      bcf     DISPLAY,LAMP   ; Turn off the Lamp

      movlw   4              ; Test for < 5 liters
      cpfsgt  FUEL           ; Skip IF FUEL is > 4 liters
      bsf     DISPLAY,BUZZER ; ELSE sound Buzzer

      movlw   d'19'          ; Test for < 20 liters
      cpfsgt  FUEL           ; Skip if FUEL is > 19 liters
      bsf     DISPLAY,LAMP   ; ELSE light Lamp

NEXT  .....   .....          ; continue
```

The Compare instructions only work correctly for unsigned numbers. For signed 2's complement numbers the data should be subtracted and the **N, OV**, and **Z** flags tested to determine the appropriate relationship.

Finally, the new **tstfsz** instruction TeSTs the contents of the specified File and Skips if it is Zero but does not change the tested datum. The incfsz and decfsz instructions have been paired with their complements **infsnz** and **dcfsnz** which skip if the outcome is *not* zero.

CHAPTER 17

A Case Study

Up to this point our microcontroller material has been presented piece-meal. To complete our study we are going to put much of what we have learned to good use and design both the hardware and software of an actual widget (gadget). This is not an easy task to do in a single short chapter. However, very little new material needs to be presented at this point, rather a process of coalescence.

We begin with our specification. Students invariably talk too long during their oral presentations. It is proposed that a dedicated embedded microcontroller-based system be designed to act as a time monitor. This monitor should default to a time-out of 10 minutes, but should have the provision to vary the allotted time from 1 to 99 minutes.

Once triggered, the monitor should perform the following sequence of operations:

1. When the RESET switch is closed, a green lamp will illuminate and a dual 7-segment display will show a count-down from the time-out value to 03 at 1-min intervals.
2. After a further minute, an amber lamp only will illuminate, the count of 02 will be displayed and a buzzer will sound for nominally one second.
3. After a further minute, a red lamp only will illuminate together with a display of 01. The buzzer will sound for 2 seconds.
4. Finally, after another minute the display will show 00, the red lamp will continue to be illuminated and the buzzer will sound continuously until the STOP switch is pressed. This will halt the timer and turn off all displays, lamps, and the buzzer. Indeed, closing the STOP switch at any time during the sequence above will cause the system to permanently halt. The system may be restarted from the time-out value by resetting the processor.
5. At any time the sequence can be frozen by toggling the PAUSE switch. When toggled again, the sequence will continue on from where it left off.
6. In order to alter the time-out from the default value of 10, the SETT switch must be closed when the system is reset. The display will then show 99 and will count down slowly. The value showing when the

SETT switch is released will be the new time-out and will be retained indefinitely until another SeT Time process.

The first decision to be made is the choice of microcontroller (MCU). In this case we are constrained by the need to use one of our book's model device, i.e., PIC12F675/29, PIC16F627/8, or PIC16F87X. As we will not need to use the ADC module, any of these devices will be a suitable hardware host for the software. Choosing the 18-pin PIC16F627/8 device, compared to the alternative 40-pin PIC16F874/7, will require additional support functions to expand the port pin budget, but should better illustrate the tradeoffs of more complex systems. The alternative single-chip solution using the latter is illustrated in the book's website. The first edition of this book used the PIC16F84, and as this is pin-compatible with the PIC16F627, then this can be used as a drop-in option with the only software changes being that relating to the non-existent Analog Comparator module in the original device.

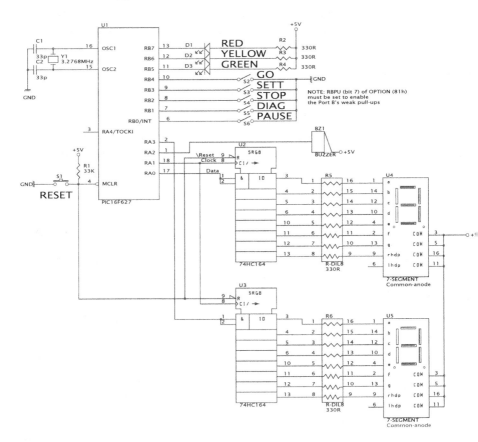

Fig. 17.1 The annunciator hardware.

Based on this decision the final target hardware is shown in Fig. 17.1. The port pin budget is allocated as follows:

Switches

The five switches S2...S6 requesting the functions GO, SET-Time, STOP, DIAG, PAUSE are read from Port B at RB[4:0]. By using this port's internal pull-up resistors (see Fig 11.9 on page 308) no external resistors are required.

S1 with R1 provides a Manual reset in order to restart the count. This $\overline{MCLR}$ signal also provides a Reset signal feed for external circuitry. In the PIC16F627/8 pin 4 can optionally be configured as an additional Port A line. By default this defaults as $\overline{MCLR}$ but in Program 17.3 the MCLRE fuse is explicitly set-up to enable this function.

All six switches can be conveniently implemented as momentary contact keyboard switches.

Lamps

Three suitably colored 10 mm (0.4") high-brightness LEDs D3...D1 driven from RB[7:5] provide the light signals. The 330Ω series resistors limit the current to nominally 10 mA.

Buzzer

The buzzer should be a miniature solid-state device. A typical piezo-electric implementation will operate over a wide dc voltage range of typically 3 to 16 V and require little more than 1 mA at 5 V.[1] The buzzer is driven via RA2.

Numerical Display

Two 7-segment displays give the required 2-digit read-out, facilitating the maximum specified period of 99 minutes. As only four port pins remain, a serial interface is implemented. This is similar to that shown in Fig. 12.2 on page 333 but each 8-bit 74HC164 SIPO shift register has a separate data feed, with RA0 being used for the ten's digit and RA3 for the units digit. Both digits can therefore be simultaneously updated with eight shifts.

The common-anode 7-segment display pinning shown in the diagram is that of the 16-pin Dual In Line (DIL) footprint with both left and right decimal points—lhdp and rhdp. Only the latter is used here, to indicate that the system has paused. Alternative 16- and 14-pinouts are commonly available and even dual-digit packages. However, even the 16-pin footprint pinout is not standardized.

Smaller-sized displays, typically below 20 mm/0.8", use a single LED for each bar, with a conducting voltage drop of around 2 V.[2] The DIL

---

[1]If you want to preclude any possibility of the speaker continuing, a piezo-electric sound bomb producing 110 dB at 1 m distance needs a 12 V d.c. supply at 200 mA.

[2]Larger displays, e.g., 2.24"/56 mm, have typically two or four LEDs in series. In the latter case a separate 12 V supply and current buffering would be needed.

330 Ω series resistors R5 and R6 limit the current to around 10 mA. The common anodes are connected directly back to the normal +5 V power supply to avoid current surges affecting the logic circuits, and should be decoupled by small tantalum capacitors. Although the displays are normally rated for 20 mA, restricting the current to this value gives sufficient illumination and means that the 74HC164 shift registers do not need current buffering.[3]

Crystal
A 3.2768 MHz crystal provides the timing for the MCU's clock oscillator, giving an instruction rate of 819.21 kHz. A typical crystal of this value has a frequency tolerance of ±30 ppm and temperature coefficient of ±50 ppm over the operating range.

This unusual choice is $2^{16} \times 50$ so if we use the 8-bit Timer 0 with a Prescaler value of 1:64 then we can create an interrupt 50 times per second; see page 533. An alternative low-power configuration would be to use a 32.768 kHz crystal and generate an interrupt every 2 seconds using the 16-bit Timer 1. However, compared to the current consumption of the LED display components, the MCU's power dissipation is minor.

The PIC16F627/8 device can use the OSC1 and OSC2 pins as additional Port A lines with a fully internal nominal 4 MHz oscillator; see Table 10.2 on page 278. Selecting the _XT_OSC Configuration word option in Program 17.3 allows the use of an external crystal timing element.

With the hardware environment designed, we can now concentrate on the software.

Figure 17.2 shows the basic modular structure for our system. Here the distinctive double right/left edged box denotes a subroutine or interrupt service routine (ISR). Three distinct processes can be identified together with two major supporting tasks.

Timebase Task
All processes are time related. Timekeeping is implemented in hardware by generating an interrupt 50 times each second. By keeping a Jiffy count, seconds and minutes are updated and are used to sequence the appropriate process.

By monitoring the PAUSE switch this decrementing time chain can be bypassed, hence freezing the countdown for as long as necessary.

Output Display Task
All processes need to output the state of the count or status information to the two 7-segment displays. As this involves parallel-to-serial conversion and shifting, the task is better gathered into one module.

---

[3] Alternatively low-current 7-segment displays are available.

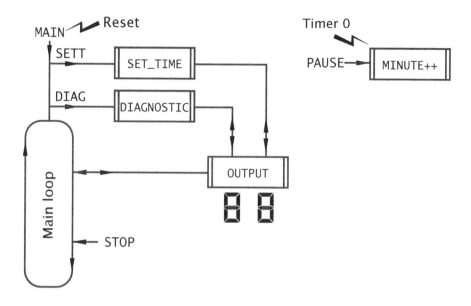

Fig. 17.2 The modular software structure.

Main Process
The Main process is a loop displaying the Minute count until it reaches zero, with a premature break if the STOP switch is closed.

Set-Time Process
If the SETT switch is closed when the PIC MCU is reset, then the SET_TIME subroutine quickly decrements the display count until the switch is released. This displayed value is then written into Data EEPROM and is used by all subsequent Main processes as the starting value for the Minute count.

Diagnostic Process
If the DIAG switch is closed on reset, the system enters a diagnostic subroutine. This essentially exercises each peripheral device in a manner calculated to ease hardware fault finding.

All processes are dependent on the Timebase task to pass basic real-time clock information back. As shown in Program 17.1 this is interrupt driven and is based on the Timer 0:Prescaler dividing down the 3.2763 MHz crystal-driven oscillator to give overflow every $\frac{1}{50}$ s. As can be seen in Program 17.3, the Timer 0 interrupt is enabled and thus the PIC MCU will enter interrupt handler ISR whenever the timer overflows—every 256 outputs from the Prescaler. Remembering that the instruction cycle is $\frac{1}{4}$ of the crystal frequency, a Prescaler ratio of 1:64 will give a timebase rate of 50 per second; that is, $\frac{3.2763 \times 10^6}{4 \times 64 \times 256} = 50$.
The task list for this function is then:

1. IF PAUSE switch open THEN
   (a) Decrement the time chain by one Jiffy.
   (b) IF new second THEN flag it.
2. ELSE
   (a) Toggle the Pause flag.
   (b) IF set THEN tell the world that the system is paused.
   (c) ELSE display time to indicate normal running.
   (d) Wait until SETT switch is released.
3. Return from interrupt.

From Program 17.1 we see that time is kept as a 3-byte count chain using Files MINUTE, SECOND and JIFFY to hold the total. Assuming that the state of bit 0 of File Pause is 0, then one is added to the Jiffy count. Normally the ISR then exits but when Jiffy reaches 50 it is reset to zero and the Seconds count decremented. The File NEW_SEC is also made non-zero to indicate to background software that a second has elapsed. In the situation where the Second count reaches zero then it is reset to 59 and the Minute count is decremented. The procedure is similar to the incrementing count of Example 7.3.

The Timebase task also handles the Pause function. The simplest approach would be to skip over the time decrement code if the PAUSE switch is closed. However, the necessity to keep the switch closed could be irksome if the period is more than a few minutes.

Implementing a push-on push-off scenario is ergonomically superior and can be more economically implemented in software rather than using a different type of switch compared to the others. In Program 17.1 the Pause handling code is located in the separate subroutine FREEZE. It is permissible to call a subroutine from an ISR in the same manner as calling one subroutine from another, that is, nesting. The hardware stack allows nesting up to eight deep. In our situation only two of the eight stack locations are used.

Subroutine FREEZE is only entered if the PAUSE switch is closed. On each entry the value of bit 0 of the File Pause is toggled. This is implemented by simply incrementing the File Pause.

Once Pause[0] is toggled, its state is tested and if it is 1 then the pattern to illuminate only the two decimal points is sent to the SPI_WRITE subroutine. This is an arbitrary indicator display, another possibility would be PR. If Pause[0] is 0 then the state of the Minute count is sent to the OUTPUT subroutine and indicates to the user that the Pause function has ended.

Finally, the subroutine does not exit until the user releases the PAUSE switch. This is important, as on exit the ISR will be re-entered again at the next Timer 0 overflow, and this would cause Pause to be repeatedly retoggled. Some measure of switch debounce is obtained by zeroing Timer 0 and the Prescaler when the switch is released and only then clearing TOIF. This means that the switch will not be retested for a whole $\frac{1}{50}$ second.

Program 17.1 The timebase software.

```
; ****************************************************************
; * The ISR to decrement the real-time clock                    *
; * Adding a 20ms Jiffy on each entry                           *
; * Sets NEW_SEC to a non-zero value each Second update         *
; ****************************************************************
; First save context in usual way
ISR      movwf   _work       ; Put away W
         swapf   STATUS,w    ; and the Status register
         movwf   _status

; ==============================================================
; The core code
         btfss   INTCON,TOIF ; Was it a Timer0 time-out?
         goto    ISR_EXIT    ; IF no THEN false alarm

         btfsc   Pause,0     ; Check the Pause flag
         goto    ISR_EXIT    ; IF closed THEN don't increment

         incf    JIFFY,f     ; Record one more 1/50 second
         movlw   d'50'       ; Has Jiffy count reached 50?
         subwf   JIFFY,w
         btfss   STATUS,Z
         goto    ISR_EXIT    ; IF not THEN finished
         clrf    JIFFY       ; ELSE zero Jiffy count

         movf    SECOND,f    ; Test for Seconds count = 00?
         btfsc   STATUS,Z
         goto    NEW_MIN     ; IF it is THEN a new minute
         decf    SECOND,f    ; ELSE decrement Seconds count and
         incf    NEW_SEC,f   ; tell background prog new second
         goto    ISR_EXIT    ; and exit

NEW_MIN  movlw   d'59'       ; Reset Seconds to 59 seconds
         movwf   SECOND
         movf    MINUTE,f    ; Test for Minutes count = 00?
         btfsc   STATUS,Z
         goto    ISR_EXIT    ; IF it is THEN no more decrement
         decf    MINUTE,f    ; ELSE decrement Minutes

; ==============================================================
ISR_EXIT btfss   PORTB,SETT  ; Check the SETT switch
         call    FREEZE      ; IF closed THEN update Pause flag

         bcf     INTCON,TOIF ; Clear interrupt flag
         swapf   _status,w   ; Untwist the original Status reg
         movwf   STATUS
         swapf   _work,f     ; Get the original W reg back
         swapf   _work,w     ; leaving STATUS unchanged
         retfie              ; and return from interrupt
```

(continued on the next page)

Program 17.1 (*continued*).

```
;*******************************************************
; * FUNCTION: Increments the Pause flag.               *
; * FUNCTION: IF = 1 THEN displays the decimal points  *
; * FUNCTION: IF = 0 THEN displays the normal count     *
; * RESOURCE: Subroutine SPI_WRITE. Var Pause          *
; * ENTRY    : SETT switch closed                       *
; * EXIT     : Pause switch open; appropriate display   *
;*******************************************************
FREEZE  incf    Pause,f    ; Update Pause flag, bit 0
        btfss   Pause,0    ; Check status of Pause flag
        goto    UNFREEZE   ; Change 1 -> 0, unfreeze
; Display freeze
        movlw   b'01111111' ; Code for display decimal point
        movwf   DATA_OUT_L
        movwf   DATA_OUT_H
        call    SPI_WRITE
        goto    FREEZE_EXIT

UNFREEZE ; Land here if Pause 0 -> 1.
        movf    MINUTE,w   ; Display the normal Minute count
        call    OUTPUT

FREEZE_EXIT
        btfss   PORTB,SETT ; Wait til switch is opened again
        goto    FREEZE_EXIT
        clrf    TMR0       ; Reset Timer and Prescaler
        return
```

The task of displaying the contents of the Working register in decimal is handled by the subroutine OUTPUT in Program 17.2. The task list for this function is:

1. Convert the binary datum to 2-digit BCD.
2. Convert both digits to 7-segment.
3. Serially shift out both bytes to the appropriate display.

Subroutine OUTPUT listed in Program 17.2 follows the task list by calling up the following utility subroutines.

Binary to BCD Conversion

Subroutine BIN_2_BCD repetitively subtracts 10 from the binary datum in the manner described in Program 5.7 on page 138. Assuming that this datum is never greater than decimal 99 (h'63') then this count gives the ten's digit. The residue is the unit's digit. The two nybbles are packed together and returned in W.

Binary to 7-Segment Decoder
Subroutine SVN_SEG converts a single datum nybble in W to its 7-segment coded equivalent as described in Program 6.6 on page 162.

SPI Output
Subroutine SPI_WRITE is similar to that described in Program 12.1 on page 334 but transmits two serialized data streams simultaneously. The datum in DATA_OUT_L is sent via RA3 whilst that in DATA_OUT_H is sent out via RA0. A common clock is used.

Before considering the coding for the three processes, we will briefly look at the software configuration of the system when the Program store is set-up, and the initialization code which is performed at run time after reset; see Program 17.3.

Burn-In Time
The __config directive specifies the fuse setting of the Configuration word in the Special/Test area. The Watchdog timer is disabled, an external crystal timing element is employed together with an external $\overline{\text{MCLR}}$ pin. Disabling the Low-Voltage Programming mode releases pin RB3 for use as a normal parallel port pin; see page 280.

Program 17.2 The data display function.

```
; ***************************************************************
; * FUNCTION: Displays datum as a 2-digit decimal output       *
; * RESOURCE: Subroutines BIN_2_BCD, SPI_WRITE, SVN_SEG         *
; * RESOURCE: Vars DATA_OUT_L, DATA_OUT_H, NEW_SEC, NUMBER      *
; * ENTRY    : Datum in W, <100d                                *
; * EXIT     : Data displayed, NEW_SEC zeroed                   *
; ***************************************************************
OUTPUT  bcf    PORTA,SCK        ; Initialize the clock line
        call   BIN_2_BCD        ; Convert to BCD
        movwf  NUMBER           ; Put packed BCD MINUTE in NUMBER

        movf   NUMBER,w         ; Get number count for display
        andlw  b'00001111'      ; Get Units nybble
        call   SVN_SEG          ; Convert to 7-segment code
        movwf  DATA_OUT_L       ; Copy into the serial low register
        swapf  NUMBER,w         ; Put ten's digit into lower nybble
        andlw  b'00001111'      ; Isolate ten's digit
        call   SVN_SEG          ; Convert to 7-segment code
        movwf  DATA_OUT_H       ; Copy into the serial high register
        call   SPI_WRITE        ; Shift both digits out

        clrf   NEW_SEC          ; Reset NEW_SEC flag
```

(continued on the next page)

Program 17.2 (*continued*).

```
; ***********************************************************
; * FUNCTION: Clocks out a two byte in parallel/series     *
; * ENTRY   : Data in DATA_OUT_L and DATA_OUT_H            *
; * ENTRY   : The former to be LSD, the latter MSD         *
; * EXIT    : DATA_OUT_L and DATA_OUT_H altered            *
; ***********************************************************
SPI_WRITE
        bcf    PORTA,SCK      ; Make sure clock starts at Low

        movlw  8              ; Initialize loop counter to 8
        movwf  COUNT
LOOP    bcf    PORTA,SDOH     ; Zero data bit for MSD
        rlf    DATA_OUT_H,f   ; Shift datum left into Carry
        btfsc  STATUS,C       ; Skip if Carry is 0
         bsf   PORTA,SDOH     ; ELSE make data bit 1

        bcf    PORTA,SDOL     ; Zero data bit for LSD
        rlf    DATA_OUT_L,f   ; Shift datum left into Carry
        btfsc  STATUS,C       ; Skip if Carry is 0
         bsf   PORTA,SDOL     ; ELSE make data bit 1

        bsf    PORTA,SCK      ; Pulse clock
        bcf    PORTA,SCK

        decfsz COUNT,f        ; Decrement count
         goto  LOOP           ; and repeat until zero
        return

SVN_SEG addwf PCL,f          ; Add N to PC giving PC + N

        retlw b'11000000'    ; Code for 0
        retlw b'11111001'    ; Code for 1
        retlw b'10100100'    ; Code for 2
        retlw b'10110000'    ; Code for 3
        retlw b'10011001'    ; Code for 4
        retlw b'10010010'    ; Code for 5
        retlw b'10000010'    ; Code for 6
        retlw b'11111000'    ; Code for 7
        retlw b'10000000'    ; Code for 8
        retlw b'10010000'    ; Code for 9
; ***********************************************************
; * FUNCTION: Converts a binary byte to a packed BCD byte  *
; * RESOURCE: TEMP byte                                     *
; * ENTRY   : Binary byte in W range 00 - 63h (0 - 99d)    *
; * EXIT    : Packed BCD byte in W                          *
; ***********************************************************
; Divide by ten
BIN_2_BCD clrf TEMP          ; Zero the loop count
LOOP10    incf TEMP,f        ; Record one ten subtracted
          addlw -d'10'       ; Subtract decimal ten
          btfsc STATUS,C     ; IF a borrow (C==0) THEN exit loop
           goto LOOP10       ; ELSE do another subtract/count

          decf TEMP,f        ; Compensate for one inc too many
          addlw d'10'        ; Add ten to residue to give units
          swapf TEMP,f       ; Put ten's digit in upper nybble
          addwf TEMP,w       ; Add units nybble right justified
          return             ; and return to caller
```

Program 17.3 The initialization code.

```
             include  "p16f627a.inc"
SDOH         equ 0
SCK          equ 1
BUZ          equ 2
SDOL         equ 3
GREEN        equ 5
YELLOW       equ 6
RED          equ 7
SETT         equ 0
DIAG         equ 1
STOP         equ 2
SETT         equ 3
GO           equ 4

             cblock   h'20'
               MINUTE:1, SECOND:1, JIFFY:1, NUMBER:1, NEW_SEC:1
               DATA_OUT_L:1, DATA_OUT_H, COUNT:1, TEMP:1, TIME_OUT:1
               Pause:1, _work:1, _status:1
             endc

             __config _XT_OSC & _WDT_OFF & _PWRTE_ON & _CP_OFF
                             & _LVP_OFF & _MCLRE_ON

             org      h'2100'     ; The EEPROM Data module
             de       d'10'       ; Default value is 10 minutes

RESET        org      0           ; Reset vector
             goto     MAIN
             org      4           ; Interrupt vector
             goto     ISR
% These two lines added cf PIC16F84
MAIN         movlw    b'00000111' ; Make all PortA pins digital by
             movwf    CMCON       ; setting Comparator to Mode b'111'

             bsf      STATUS,RP0  ; Change to Bank 1
             movlw    b'11100000' ; RA[4:0] outputs
             movwf    TRISA
             movlw    b'00011111' ; RB[7:5] outputs; RB4:0 inputs
             movwf    TRISB
             movlw    b'00000101' ; Clock TMR0 internally; assigned PS
             movwf    OPTION_REG  ; Set to 1:64. Enable PORTB pull-ups
             bcf      STATUS,RP0  ; Back to Bank 0

             clrf     Pause       ; The PAUSE switch toggle
             clrf     NEW_SEC     ; Reset NEW_SEC second flag

             clrf     TMR0
             bcf      INTCON,T0IF
             bsf      INTCON,T0IE ; Enable Timer0 interrupts
             bsf      INTCON,GIE  ; Enable all interrupts

             btfss    PORTB,SETT  ; Check the SET-Time switch
              call    SET_TIME    ; IF closed THEN set total time
             btfss    PORTB,DIAG  ; Check the DIAGnostic switch
              call    DIAGNOSTIC  ; IF closed THEN set total time
```

When the code is blasted into the main flash EEPROM Program, store location 0 of the Data EEPROM is set to 10. This means that a freshly programmed PIC MCU will default to a 10-min count down. This value can subsequently be altered using the SET-Time process described in Fig. 17.5.

The address of this cell is in the Special/Test Configuration area at h'2100' and the de (Data EEPROM) directive is used to specify the load time data as described on page 489.

Run Time
The code executed each time a reset is actioned is used to initialize the run-time environment.

Vectors
To initialize the Reset vector at h'000' to point to MAIN and Interrupt vector at h'004' to point to ISR.

Port Setting
To make Port A[4:0] and Port B[7:5] outputs and all other lines inputs.

Timer 0 Setting
To set-up the Prescaler ratio to 1:64 and Timer 0 clock source to internal. The Timer 0 interrupt is also enabled.

Process Select
To check the state of the DIAG and SETT switches to choose either the Diagnostic or SET-Time processes. If neither switch is closed the normal Main process is entered.

If the DIAG switch is closed when the PIC MCU comes out of reset then the code transfers to the subroutine DIAGNOSTIC in Program 17.4. The Diagnostic process aims to exercise the various peripheral devices interfaced to the process in order to verify in a reproducible manner the status of the interconnection and the devices themselves.

Switches
Each of the five switches input via Port B are checked in turn. If a switch is closed, either one of the LEDs or the buzzer is activated. In this manner both the switches and the listed output devices are tested. The DIAG switch is of course verified by moving the system into this Diagnostic process and the RESET switch is tested by initiating the startup process.

If there were more switches than output devices then either combinations of the latter could be activated or else one or more segments in the numerical display could be pushed into service.

LEDs and Buzzer
The static output devices are tested in conjunction with the switch test listed above. Of course the failure of a LED to light or buzzer to sound

Program 17.4 The Diagnostic process.

```
; ********************************************************************
; * FUNCTION: Checks each switch and activates a corresponding*
; * FUNCTION: LED or buzzer. Continually activates a unary    *
; * FUNCTION: pattern to both 7-segment displays              *
; * RESOURCE: Subroutines SPI_WRITE                           *
; * RESOURCE: Vars TEMP, DATA_OUT_H, DATA_OUT_L               *
; * ENTRY   : DIAG switch closed                              *
; * EXIT    : DIAG switch open                                *
; ********************************************************************
DIAGNOSTIC
          movlw  b'11111110'    ; The initial 7-segment pattern
          movwf  TEMP           ; in memory
D_LOOP    movlw  b'11111111'    ; Turn off all LEDs and buzzer
          movwf  PORTB
          bsf    PORTA,BUZ
; Now scan switches
          btfss  PORTB,SETT     ; IF Pause switch closed
          bcf    PORTB,GREEN    ; THEN Green LED
          btfss  PORTB,STOP     ; IF Stop switch closed
          bcf    PORTB,YELLOW   ; THEN Yellow LED
          btfss  PORTB,SETT     ; IF Set switch closed
          bcf    PORTB,RED      ; THEN Red LED
          btfss  PORTB,GO       ; IF Go switch closed
          bcf    PORTA,BUZ      ; THEN Buzzer
; Now turn on each segment in turn of both displays
          movf   TEMP,w         ; Get pattern
          movwf  DATA_OUT_L     ; Put in output Files
          movwf  DATA_OUT_H
          call   SPI_WRITE      ; Display it

          btfsc  PORTB,DIAG     ; IF Diagnostic switch open
          return                ; THEN exit the diag subroutine
          clrf   NEW_SEC        ; Reset the New Second flag
; Now move the display pattern on one and wait for a second
          bcf    STATUS,C       ; Clear Carry
          btfsc  TEMP,7         ; Check MSB of pattern
          bsf    STATUS,C       ; IF 1 THEN Carry = 1
          rlf    TEMP,f         ; Shift it in <<

D_LOOP2   movf   NEW_SEC,f      ; ELSE wait for the new second
          btfsc  STATUS,Z       ; IF non-zero THEN skip
          goto   D_LOOP2        ; ELSE try again
          goto   D_LOOP         ; Repeat routine
```

may be due to either the input or output device circuit. Determining
which has failed is easily accomplished by using a voltmeter or logic
probe. Also remember that all LEDs should be illuminated during the
SET-Time process.

Display

Each of the display devices is tried out by lighting one segment in turn at a 1-s rate, in an endless loop. This is implemented by generating a walking unary pattern b'11111110 → 11111101 → ··· 01111111' sent out to the output subroutine SPI_WRITE once each time the File NEW_SEC is non-zero. NEW_SEC is incremented in the Timer 0 interrupt-handling routine each time the Seconds count is incremented and cleared in the Diagnostic procedure code. This acts as a ratchet, giving only one new display each second.

The SET-Time process is entered when the SETT switch is closed whenever the processor comes out of Power-on reset. Its function is to allow the operator to change the contents of the EEPROM Data module location h'00' to any value up to 99. This location holds the initial count-down value used by the Main process to determine the length of the procedure.

The strategy behind the coding shown in Program 17.5 is to initialize the Second count to 99 and then let it decrement at a 1-s rate as deter-

---

Program 17.5 The SST-Time process.

```
; ***********************************************************
; * FUNCTION: Slowly counts down from 99-00. When Set switch *
; * FUNCTION: released EEPROM is Written with displayed count *
; * RESOURCE: Subroutines DISPLAY, EE_PUT, ISR; Var TIME_OUT  *
; * ENTRY    : Set switch is closed                           *
; * EXIT     : EEPROM Data address 00 is updated              *
; ***********************************************************
SET_TIME movlw  d'99'                ; Start count at 99 seconds
         movwf  SECOND
         movlw  b'00000000'          ; All LEDs on
         movwf  PORTB
SET_LOOP movf   SECOND,w             ; Get Second count
         call   OUTPUT               ; Display it and clear NEW_SEC

         btfsc  PORTB,SETT           ; Check; does user want to stop?
         goto   UPDATE               ; IF yes THEN update EEPROM & exit
         movf   SECOND,w             ; Get displayed count
         movwf  TIME_OUT             ; Make a temporary copy
S_LOOP   movf   NEW_SEC,f            ; Check NEW_SEC status
         btfsc  STATUS,Z             ; IF non-zero THEN skip
         goto   S_LOOP               ; ELSE try again
         goto   SET_LOOP             ; Repeat display

UPDATE   movf   TIME_OUT,w           ; Get the value and set-up EEPROM
         bsf    STATUS,RP0           ; Change to Bank 1
         movwf  EEDATA               ; Set-up EEPROM data
         clrf   EE_ADDRESS           ; and address 00
         call   EE_PUT               ; Program EEPROM & return in Bank 0
         return                      ; and return to main program
```

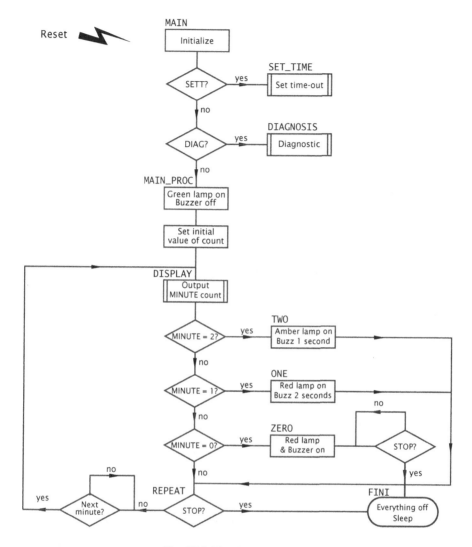

Fig. 17.3 The Main process.

mined by the foreground ISR. The value of SECOND is sent to the Display subroutine each time the ISR sets the flag File NEW_SEC to a non-zero value; that is, once per second. DISPLAY clears NEW_SEC so the net effect is to update the display only once each second. Each second the SETT switch is checked, and when open, the state of the Seconds count is transferred to the EEPROM Data module at UPDATE using the EE_PUT subroutine of Program 15.2 on page 488.

The complete background system flow chart is shown in Fig. 17.3. This shows in outline the decision flow taken after a reset, and in de-

Program 17.6 The Main process.

```
        movlw   b'11000000'  ; Green LED on
        movwf   PORTB
        bsf     PORTA,BUZ    ; Buzzer off

; Get start value from EEPROM

        bsf     STATUS,RP0   ; Change to Bank 1
        clrf    EEADR        ; EEPROM address zero
        call    EE_GET       ; Get the start value
        movwf   MINUTE
        movlw   d'59'        ; Initial value for seconds
        movwf   SECOND       ; is 59
        clrf    JIFFY

DISPLAY movf    MINUTE,w     ; Get Minute count
        call    OUTPUT       ; Output to display

; The 2-minutes-to-go phase --------------------------------
; At a count of two sound the buzzer for one second and turn on
; the amber lamp
TWO         movf    MINUTE,w     ; Minute count = 2?
            addlw   -2
            btfss   STATUS,Z
            goto    ONE          ; IF not THEN try for one minute
            movlw   b'10100000'  ; Amber LED on
            movwf   PORTB
            bcf     PORTA,BUZ    ; Buzzer on
TWO_LOOP    movf    NEW_SEC,f    ; Check NEW_SEC status
            btfsc   STATUS,Z     ; IF non-zero THEN skip
            goto    TWO_LOOP     ; ELSE try again
            bsf     PORTA,BUZ    ; Turn off buzzer after one second
            goto    REPEAT       ; repeat display

; The 1-minute-to-go phase ---------------------------------
; At a count of one sound the buzzer for two seconds and turn on
; the red lamp
ONE         movf    MINUTE,w     ; Minute count = 1?
            addlw   -1
            btfss   STATUS,Z
            goto    ZERO         ; IF not THEN try for zero minutes
            movlw   b'01100000'  ; Red LED on
            movwf   PORTB
            bcf     PORTA,BUZ    ; Buzzer on
ONE_LOOP    movf    NEW_SEC,f    ; Check NEW_SEC status
            btfsc   STATUS,Z     ; IF non-zero THEN skip
            goto    ONE_LOOP     ; ELSE try again
            clrf    NEW_SEC      ; Again clear NEW_SEC flag
UN_LOOP     movf    NEW_SEC,f    ; Again check NEW_SEC status
            btfsc   STATUS,Z     ; IF non-zero THEN skip
            goto    UN_LOOP      ; ELSE try again
            bsf     PORTA,BUZ    ; Turn off buzzer after two seconds
            goto    REPEAT       ; Repeat display
```

(continued on the next page)

Program 17.6 (continued).

```
; The Timed-Out phase *****************************************
; When the Minute count reaches zero, sound the buzzer
; until the Stop switch is closed
ZERO       movf    MINUTE,f      ; Minute count = 0?
           btfss   STATUS,Z
           goto    REPEAT        ; IF not THEN repeat after minute
           bcf     PORTA,BUZ     ; Buzzer on
ZERO_LOOP
           btfsc   PORTB,STOP    ; Check the STOP switch
           goto    ZERO_LOOP     ; and continue until closed
FINI       movlw   b'11100000'   ; Turn lamps off
           movwf   PORTB
           bsf     PORTA,BUZ     ; and buzzer
           movlw   b'11111111'   ; Code for blank
           movwf   DATA_OUT_L
           movwf   DATA_OUT_H
           call    SPI_WRITE     ; Blank both displays
           sleep                 ; and await another reset

REPEAT     btfss   PORTB,STOP    ; Check the STOP switch
           goto    FINI          ; IF closed THEN freeze
           movf    SECOND,f      ; Wait until Second count is again 0
           btfss   STATUS,Z      ; i.e., for the next minute
           goto    REPEAT        ; IF not THEN wait again
           clrf    NEW_SEC       ; ELSE wait one more second
R_LOOP     movf    NEW_SEC,f     ; Check NEW_SEC status
           btfsc   STATUS,Z      ; IF non-zero THEN skip
           goto    R_LOOP        ; ELSE try again
           goto    DISPLAY       ; Repeat display
```

tail, the Main process. Although this looks rather complex, it may be broken down into five phases, with the corresponding coding shown in Program 17.6.

Preamble
On reset if neither SETT or DIAG switches are closed, the Main procedure code is entered at MAIN_PROC. This reads the initial value of the countdown period from EEPROM location h'00' and initializes the count chain. The green lamp is illuminated and other lamps and buzzer are turned off.

Countdown
The Countdown phase continually displays the Minute count—updated behind the scenes by the ISR. The green lamp remains illuminated as long as this display does not drop below ⊡⊒. This phase is complete whenever

the count drops below 3 minutes or else the STOP switch is closed. In the latter case all displays are blanked and the PIC MCU is put into its Sleep state.

In all situations, except where the STOP command is issued, the Minute count is displayed at 1-min intervals. The routine at REPEAT checks the Second count and if zero the loop is repeated; that is, once per minute. The simpler alternative of continually refreshing the 7-segment readouts gives an inferior display as the data being serially shifted at a frequent rate may partially illuminate segments which should otherwise be off. In addition, repeating the loop each minute eases the task of sounding the buzzer once only when the Minute count drops to both two and one.

Two Minutes to Go
When the display is ⏢⏢ the amber lamp is illuminated and the buzzer sounded for one second. The latter is timed using the NEW_SEC variable. Again the loop can be prematurely exited if the STOP switch has been closed.

One Minute to Go
When the display is ⏢⏢ the loop diverts to illuminate the red lamp. The buzzer is sounded for 2 s; implemented in code as two 1-second buzzes.

Timed Out
When the Minute count reaches zero, not only is ⏢⏢ displayed but also the buzzer sounds continually. This cacophony can only be silenced by pressing the STOP switch, or by resetting and starting again. As in previous situations when the STOP switch is closed, all displays are blanked out and the PIC MCU is placed in its Sleep state.

After the source code has been assembled and where possible simulated (see Fig. 8.7 on page 238) it can then be burned into the PIC MCU's Program store. In the first instance only the diagnostic software and associated tasks need to be programmed in order to check the target hardware. The precise details will depend somewhat on the PIC MCU Device programmer being used and its associated software.

The screen shot shown in Fig. 17.4 shows the situation where the Microchip PICSTART Plus® development programmer (see Fig. 17.5)is used in conjunction with the MPLAB IDE. Communication with the host computer is via a RS-232 serial port and contact is made from the Picstart Plus menu. The right-hand window allows the operator to set-up the Configuration Bits (fuses), shown in the left-hand window. When this is set-up, the operator can Blank out, Read from, Program or Verify that the contents of the EPROM or EEPROM Program store is the same as that produced by the current assembly process. This task can only be carried out as long as Code Protect remains off. When the processor is configured to turn on Code Protect, it is irreversible, and neither Program nor Verify tasks can then be carried out.

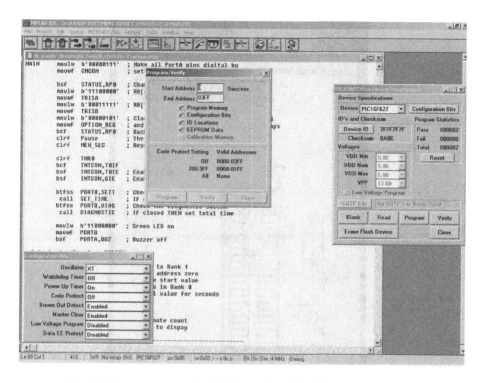

Fig. 17.4 Programming the PIC from the MPLAB Version 5 IDE.

The middle window shows the status of the Program or Verify process. As shown here, it is announcing that it has completed the task up to h'03FF' and is reporting success. The complete process takes less than a minute and the 254 program words that this case study generates.

With F series PIC MCUs, the programming process may be repeated several thousand times without any deterioration of the Flash EEPROM Program store. C series PIC MCUs,[4] such as the PIC16C74, have EPROM Program stores. Where the target has a quartz window (see photograph on page 2) it must be erased using a suitable UV-based EPROM eraser for approximately 20 min before programming, unless code is being added to previously unprogrammed memory. Although quartz windowed devices are necessary for C-part development purposes, they are relatively expensive. Thus cheaper windowless version are used for production purposes, and are called One-Time Programmable (OTP) since they cannot subsequently be erased. Part numbers which are windowed are usually identified by JW postfix; for example, PIC16C74B-20/JW is a 20 MHz ceramic windowed PIC16C74B part and the PIC16C74B-4/P is a 4 MHz OTP

---

[4]The exception being the obsolete PIC16C84 which also has an EEPROM Program store.

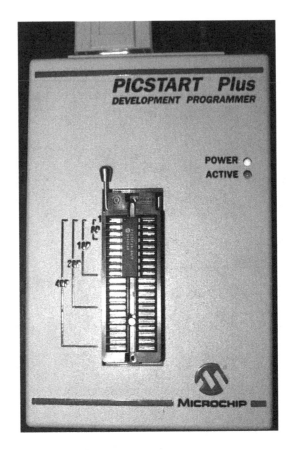

Fig. 17.5 The Microchip PICSTART Plus programmer.

version in a 40-pin plastic DIL package. Ensure that you obtain the correct device!

The hardware and software circuits have been presented here as a simple illustrative case study to integrate many of the techniques described in the body of the text. If you decide to build your own version, files, **C** coding, PCB, comparison with a Motorola 68000 MPU version and other ideas for experimentation, which you are welcome to contribute, are given on the associated website detailed in the Preface. Good luck!

Appendix A

Acronyms and Abbreviations

ADC (A/D)	Analog-to-Digital Converter/Conversion
ADCON0	A/D CONtrol 0
ADCON1	ADC CONtrol 1
ADCS*n*	ADC Clock Select, bits; ADCON0[7:6]
ADDEN	ADDress ENable; RCSTA[3]
ADFM	ADC module outcome ForMat; ADCON1[7]
ADIE	ADC Interrupt Enable mask; PIE1[6]
ADIF	ADC Interrupt Flag; PIR1[6]
ADIP	ADC Interrupt Priority; IPR1[6] (PIC18XXXX)
ADON	ADC module ON; ADCON0[0]
ADRES	ADC RESult
ADRESH	ADC RESult High byte (10+ ADC modules)
ADRESL	ADC RESult Low byte (10+ ADC modules)
ALU	Arithmetic Logic Unit
AN*n*	Analog input pin *n*
ANSI	American National Standards Institution
ALU	Arithmetic Logic Unit
ASCII	American Standard Code for Information Interchange
AUSART	Addressable USART (see also USART)
BSR*n*	Bank Select Register bits; BSR[3:0] (PIC18XXXX)
BCD	Binary Coded Decimal
BF	Buffer Full; SSPSTAT[0]
C	Carry flag in STATUS[0]
C1OUT	Comparator 1 OUTput bit; CMCON[6]
C2OUT	Comparator 2 OUTput bit; CMCON[7]
C1INV	Comparator 1 INVertor; CMCON[4]
C2INV	Comparator 2 INVertor; CMCON[5]
C1OUT	Comparator 1 OUTput pin
C2OUT	Comparator 2 OUTput pin
CCP	Capture/Compare PWM module
CCP1	CCP1 input/output pin
CCPR1H	CCP Register 1 High byte
CCPR1L	CCP Register 1 Low byte
CCP1CON	CCP1 CONtrol register
CCP1IE	CCP1 Interrupt Enable mask; PIE1[2]
CCP1IF	CCP1 Interrupt Flag; PIR1[2]
CCP1M*n*	CCP Mode control bits; CCP1CON[3:0]
CCP2IE	CCP2 Interrupt Enable mask; PIE2[0]
CCP2IF	CCP2 Interrupt Flag; PIR2[0]
CHS*n*	ADC CHannel Select bits; ADCON0[5:3]
CIS	Comparator Input Switch; CMCON[3]

CISC	Complex Instruction Set Computer
CK	USART synchronous ClocK I/O pin
CKE	ClocK Edge; SSPSTAT[6]
CKP	ClocK Polarity; SSPCON[4]
CMIE	CoMparator change Interrupt Enable mask; PIE2[6]
CMIF	CoMparator change Interrupt Flag; PIR2[6]
CMn	Comparator (analog) Mode; CMCON[2:0]
CMOS	Complimentary Metal-Oxide Semiconductor
CMCON	CoMparator (analog) CONtrol register
CPCON	ComParator CONtrol register
CPU	Central Computing Unit
CREN	Continuous Receive ENable; RCSTA[4]
$\overline{\text{CS}}$	Chip Select pin
CTS	Clear To Send, RS-232 handshake signal
CVRn	Comparator Voltage Reference mode bits; CVRCON[3:0]
CVREN	Comparator Voltage Reference ENable; CVRCON[7]
CVRCON	Comparator Voltage Reference CONtrol register
CVROE	Comparator Voltage Reference Output Enable; VRCON[6]
CVRR	Comparator Voltage Reference Range select; CVRCON[5]
D/$\overline{\text{A}}$	Data/$\overline{\text{Address}}$; SSPSTAT[5]
DAC (D/A)	Digital-to-Analog Converter/Conversion module
DC	Digit Carry flag in STATUS[1]
DC1Bn	Duty Cycle 1 Bits; CCP1CON[5:4]
DCE	Data Circuit terminating Equipment
DSP	Digital Signal Processing
DT	USART synchronous DaTa pin
DTE	Data Terminal Equipment
ea	Effective Address
EEADR	EEPROM ADdress
EEADRH	EEPROM ADdress High byte
EECON1	EEPROM CONtrol 1
EECON2	EEPROM CONtrol 2
EEDATA	EEPROM DATA
EEDATH	EEPROM DATa High byte
EEIE	EEPROM Interrupt Enable mask; INTCON[6] or PIE2[4]
EEIF	EEPROM Interrupt Flag; EECON1[4] or PIR2[4]
EEPGD	EEPROM ProGram/$\overline{\text{Data}}$; EECON1[7].
EEPROM	Electrical Erasable PROM
EPROM	Erasable PROM
FERR	Framing ERRor; RCSTA[2]
FSR	File Select Register (FSR2:0 PIC18XXXX)
GIE/GIEH	Global Interrupt Enable mask (High-priority PIC18XXXX); INTCON[7]
GIEL	Global Interrupt Enable Low-priority mask; INTCON[6] (PIC18XXXX)
GO/$\overline{\text{DONE}}$	ADC Start Convert (GO)/End Of Conversion (DONE); ADCON0[2]
GPR	General-Purpose File Register
GPn	General-Purpose Register (I/O port) GPIO pin n; e.g., GP0
GSEN	General SENd receive enable; SSPCON2[7]
IC	Integrated Circuit
ICSP™	In-Circuit Serial Programming
I²C	Inter-Integrated Circuit serial protocol
IDE	Integrated Development Environment
IEC	International Electrotechnical Commission
INDF	INDirect File register

INT	External INTerrupt input pin (INT2:0 PIC18FXXXX)
INTCON	INTerrupt CONtrol Register
INTEDG	External INTerrupt EDGe polarity selection; OPTION_REG[0]
INTE	INTerrupt Enable; INTCON[4]
INTF	INTerrupt Flag; INTCON[1]
I/O	Input/Output
IPR X	Interrupt Priority Interrupt Register X (PIC18XXXX)
IRP	Indirect addressing Register Page; STATUS[7]
ISR	Interrupt Service Routine
LAT X	LATch X (Parallel I/O port LATch register X); e.g., LATA (PIC18XXXX)
LED	Light-Emitting Diode
LSB	Least Significant Bit or Byte
LSI	Large-Scale Integration
LSD	Least-Significant Digit
LVP	Low-Voltage Programming (c.f. High-Voltage Programming)
$\overline{\text{MCLR}}$	Master CLear Reset pin
MCU	MicroController Unit
MPU	MicroProcessor Unit
μs	Microsecond (10^{-6} s)
ms	Millisecond (10^{-3} s)
MSB	Most Significant Bit or Byte
MSD	Most Significant Digit
MSI	Medium-Scale Integration
MSSP	Master Synchronous Serial Port
N	Negative flag in STATUS[4] (PIC18XXXX)
ns	Nanosecond (10^{-9} s)
$\overline{\text{OE}}$	Output Enable pin
OERR	Overflow ERRor; RCSTA[1]
OS	Operating System
OPTION_REG	OPTION REGister
OTP	One-Time Programmable (EPROM)
OSCAL	OScillator CALibrate register
OV	OVerflow flag in STATUS[3] (PIC18XXXX)
P	StoP condition; SSPSTAT[4]
PC	Program Counter
PC	Personal Computer
PCFG n	ADC Port ConFiGuration bits; ADCON1[2:0]
PCL	Program Counter Low byte; File 2
PCLATH	Program Counter LATch High byte
PCLATU	Program Counter LATch Upper byte (PIC18XXXX)
$\overline{\text{PD}}$	Power Down sleep mode in STATUS[3]
PEIE	PEripheral Interrupt Enable mask; INTCON[6]
PIC	Peripheral Interface Controller
PIPO	Parallel-In Parallel-Out register
PIE X	Peripheral Interrupt Enable register X
PIR X	Peripheral Interrupt Register X
PISO	Parallel-In Serial-Out shift register
PORT X	Port X (Parallel I/O port register X); e.g., PORTA
PR2	Period Register for Timer 2
PRNG	Pseudo Random Number Generator
PRODH	PRODuct High byte (PIC18XXXX)
PRODL	PRODuct Low byte (PIC18XXXX)
PROM	Programmable ROM

PSn	Post/Prescale rate Select bits; OPTION_REG[2:0]
PSA	Post/Prescale Scaler Assign; OPTION_REG[3]
PWM	Pulse Width Modulation
RXn	Register (Parallel I/O port register X) pin n; e.g., RA0
RAM	Random Access Memory
RBIE	Register port B Interrupt Enable; INTCON[3]
RBIF	Register port B Interrupt Flag; INTCON[0]
$\overline{\text{RBPU}}$	Register port B Pull-UP; OPTION_REG[7]
RCIE	ReCeive register Interrupt Enable mask; PIE1[5]
RCIF	ReCeive register Interrupt Flag; PIR1[5]
RCREG	ReCeive data REGister
RCSTA	ReCeive STAtus register
RD	ReaD; EECON1[0]
R/$\overline{\text{W}}$	Read/$\overline{\text{Write}}$ packet in SSP module; SSPSTAT[2]
RISC	Reduced Instruction Set Computer (see CISC)
ROM	Read-Only Memory
RPn	Register Page bits; STATUS[6:5]
rtl	Register Transfer Language
RTS	Ready To Send, RS-232 handshake signal
RX	ReCeive pin for USART
RX9	ReCeive 9-bit data control; RCSTA[6]
RTCC	Real Time Counter/Clock (see Timer 0)
S	Start condition; SSPSTAT[3]
SAR	Successive Approximation Register
SCI	Serial Communication Interface module (USART)
SCK	Serial ClocK in SPI protocol
SCL	Serial CLock in I^{2}C protocol
SDA	Serial DAta bidirectional I^{2}C pin.
SDI	Serial Data Input pin in SPI protocol
SDO	Serial Data Output pin in SPI protocol
SEN	Stretch clock ENable; SSPCON2[0]
SIPO	Serial-In Parallel-Out shift register
SISO	Serial-In Serial-Out shift register
SMP	SaMPle incoming data; SSPSTAT[7]
SP	Stack Pointer
SPBRG	Serial Port Baud-Rate Generator
SPEN	Serial Port ENable; RCSTA[7]
SPI	Serial Peripheral Interface protocol
SPR	Special-Purpose File Register
SSP	Synchronous Serial Port
SSPADD	SSP ADDress register
SSPBUF	SSP BUFfer register
SSPCON	SSP CONtrol register
SSPCON2	MSSP CONtrol register 2
SSPEN	SSP Enable; SSPCON[5]
SSPIE	SSP Interrupt Enable mask; PIE1[3]
SSPIF	SSP Interrupt Flag; PIR1[3]
SSPMn	SSP Mode control bits; SSPCON[3:0]
SSPOV	SSP OVerflow; SSPCON[6]
SSPSR	SSP Shift Register
SSPSTAT	SSP STATus register
STATUS	Status Register
STKPTR	STacK PoinTeR (PIC18XXXX)

SYNC	SYNChronous mode in the USART; TXSTA[4]
$\overline{TO}$	Watchdog Time Out in STATUS[4]
T0CKI	Timer 0 ClocK Input pin
T0CS	Timer 0 Clock Select; OPTION_REG[5]
T0IE	Timer 0 Interrupt Enable mask; INTCON[5]
T0IF	Timer 0 Interrupt Flag; INTCON[2]
T1CKI	Timer 1 ClocK Input pin
T2CKSn	Timer 2 ClocK Source Prescale ratio bits; T2CON[1:0]
T0CON	Timer 0 CONtrol register (PIC18XXXX)
T1CON	Timer 1 CONtrol register
T2CON	Timer 2 CONtrol register
$\overline{T1G}$	Timer 1 Gate input pin
T1GPOL	TiMeR 1 Gate input POLarity select; T1CON[7]
$\overline{T1SYNC}$	Timer 1 SYNChronize; T1CON[2]
T1OSCEN	Timer 1 OSCillator ENable; T1CON[3]
TABLAT	TABle LATch (PIC18XXXX)
TMR0	TiMeR 0 RTCC
TMR1CS	TiMeR 1 Clock Select; T1CON[1]
TMR1H	TiMeR 1 High byte Timer 1
TMR1IE	Timer 1 Interrupt Enable mask; PIE1[0]
TMR1IF	Timer 1 Interrupt Flag; PIR1[0]
TMR2IE	Timer 2 Interrupt Enable mask; PIE1[1]
TMR2IF	Timer 2 Interrupt Flag; PIR1[1]
TMR1L	TiMeR 1 Low byte Timer 1
TMR1ON	TiMeR 1 ON; T1CON[0]
TMR2ON	TiMeR 2 ON; T2CON[2]
TMR3H	TiMeR 3 High byte Timer 3 (PIC18XXXX)
TMR3L	TiMeR 3 Low byte Timer 3 (PIC18XXXX)
$\overline{TO}$	Watchdog Time Out in STATUS[4]
TOUTPSn	Timer 2 OUTput Post Scaler bits; T2CON[3:0]
TRISX	TRIState X (Data Direction register X); e.g., TRISA
T0SE	Timer 0 Set Edge; OPTION_REG[4]
TTL	Transistor Transistor Logic family
TTY	TeleTYpewriter
TX	TranSmit pin for USART
TX9	TranSmit 9-bit data in USART; TXSTA[6]
TX9D	Ninth bit for transmission in UART; TXSTA[0]
TXEN	TranSmit register ENable; TXSTA[5]
TXIE	TranSmit register Interrupt Enable mask; PIE1[4]
TXIF	TranSmit register Interrupt Flag; PIR1[4]
TXREG	TranSmit data REGister
TXSTA	TranSmit STAtus register
UA	Update slave 10-bit Address in MSSP; SSPSTAT[1]
UART	Universal Asynchronous Receiver Transmitter
USART	Universal Synchronous-Asynchronous Receiver Transmitter
V_{DD}	Positive (Drain) supply voltage
V_{EE}	Earth (0 V) supply voltage
V_{PP}	Positive Programming voltage
VLSI	Very Large-Scale Integration
VRn	Voltage Reference mode bits; VRCON[3:0]
VREN	Voltage Reference ENable; VRCON[7]
VRCON	Voltage Reference CONtrol register
VROE	Voltage Reference Output Enable; VRCON[6]

VRR	Voltage Reference Range select; VRCON[5]
W	Working register
WCOL	Write COLlision; SSPCON[7]
WR	WRite; EECON1[1]
WREG	Working REGister in Data store (PIC18XXXX)
WREN	WRite ENable; EECON1[2]
WRERR	WRite ERRor; EECON1[3]
Z	Zero flag in STATUS[2]

Appendix B

Special-Purpose Register Structure for the PIC16F87XA

File	Name	7	6	5	4	3	2	1	0	Power Resets	All other Resets
Bank 0											
00	INDF[1]	Uses contents of this to address Data memory (not a physical register)									
01	TMR0	8-bit real-time clock/counter								XXXX XXXX	UUUU UUUU
02	PCL[1,3]	Lower-order 8 bits of the Program Counter								0000 00000	0000 0000
03	STATUS[1]	IRP	RP1	RP0	$\overline{TO}$	$\overline{PD}$	Z	DC	C	0001 1XXX	000? ?UUU
04	FSR[1]	Indirect Data memory address pointer 0								XXXX XXXX	UUUU UUUU
05	PORTA	—	—	RA5	RA4	RA3	RA2	RA1	RA0	—-0X 0000	—-0U 0000
06	PORTB	RB7	RB6	RB5	RB4	RB3	RB2	RB1	RB0	XXXX XXXX	UUUU UUUU
07	PORTC	RC7	RC6	RC5	RC4	RC3	RC2	RC1	RC0	XXXX XXXX	UUUU UUUU
08	PORTD[2]	RD7	RD6	RD5	RD4	RD3	RD2	RD1	RD0	XXXX XXXX	UUUU UUUU
09	PORTE[2]	—	—	—	—	—	RE2	RE1	RE0	—- —UUU	—- —UUU
0A	PCLATH[1]	—	—	—	Write buffer for top 5 PC bits					—-0 0000	—-0 0000
0B	INTCON[1]	GIE	PEIE	T0IE	INTE	RBIE	T0IF	INTF	RBIF	0000 000X	0000 000U
0C	PIR1	PSPIF[2]	PSIF	RCIF	TXIF	SSPIF	CCPIF	TMR2IF	TMR1IF	0000 0000	0000 0000
0D	PIR2	—	CMIF	—	EEIF	BCLIF	—	—	CCP2IF	-R-0 -0-0	-R-0 -0-0
0E	TMR1L	Timer 1 Low Byte								XXXX XXXX	UUUU UUUU
0F	TMR1H	Timer 1 High Byte								XXXX XXXX	UUUU UUUU
10	T1CON	—	—	T1CKPS1	TICKPS0	T1OSCEN	$\overline{T1SYNC}$	TMR1CS	TMR1ON	—-00 0000	—-UU UUUU
11	TMR2	Timer 2								0000 0000	0000 0000
12	T2CON	—	TOUTPS3	TOUTPS2	TOUTPS1	TOUTPS0	TMR2ON	T2CKPS1	T2CKPS0	-000 0000	-000 0000
13	SSPBUF	Synchronous Serial Port Receive Buffer/ Transmit Register								XXXX XXXX	UUUU UUUU
14	SSPCON	WCOL	SSPOV	SSPEN	CKP	SSPM3	SSPM2	SSPM1S1	SSPM0	00000 0000	00000 0000
15	CCPR1L	Capture/Compare/PWM register 1 Low byte								XXXX XXXX	UUUU UUUU
16	CCPR1H	Capture/Compare/PWM register 1 High byte								XXXX XXXX	UUUU UUUU
17	CCP1CON	—	—	CCP1X	CCP1Y	CCP1M3	CCP1M2	CCP1M1	CCP1M0	—-00 0000	—-00 0000
18	RCSTA	SPEN	RX9	SREN	CREN	—	FERR	OERR	RXD9	0000 -00X	0000 -00X
19	TXREG	SCI Transmit Data Register								0000 0000	0000 0000
1A	RCREG	SCI Receive Data Register								0000 0000	0000 0000
1B	CCPR2L	Capture/Compare/PWM register 2 Low byte								XXXX XXXX	UUUU UUUU
1C	CCPR2H	Capture/Compare/PWM register 2 High byte								XXXX XXXX	UUUU UUUU
1D	CCP2CON	—	—	CCP2X	CCP2Y	CCP2M3	CCP2M2	CCP2M1	CCP2M0	—-00 0000	—-00 0000
1E	ADRESH	A/D Result Register High byte								XXXX XXXX	UUUU UUUU
1F	ADCON0	ADCS1	ADCS0	CHS2	CHS1	CHS0	GO	—	ADON	0000 00-0	0000 00-0

X   Not known.
U   Unchanged.
R   Reserved; always maintain these bits clear.
?   Value depends on reset condition.
—   Unimplemented; read as 0.
Note 1: These registers can be addressed from any bank.
Note 2: Not implemented in 28-pin members of the family group.
Note 3: Next instruction address if PIC in Sleep mode.

File	Name	7	6	5	4	3	2	1	0	Power Reset	All other Resets
Bank 1											
80	INDF[1]	Uses contents of this to address Data memory (not a physical register)									
81	OPTION_REG	$\overline{RBPU}$	INTEDG	T0CS	T0SE	PSA	PS2	PS1	PS0	1111 1111	1111 1111
82	PCL[1,3]	Lower-order 8 bits of the Program Counter								0000 00000	0000 0000
83	STATUS[1]	IRP	RP1	RP0	$\overline{TO}$	$\overline{PD}$	Z	DC	C	0001 1XXX	000? ?UUU
84	FSR[1]	Indirect Data memory address pointer 0								XXXX XXXX	UUUU UUUU
85	TRISA	—	—	Port A Direction Register						—-11 1111	—-11 1111
86	TRISB	Port B Data Direction Register								1111 1111	1111 1111
87	TRISC	Port C Data Direction Register								1111 1111	1111 1111
88	TRISD[2]	Port D Data Direction Register								1111 1111	1111 1111
89	TRISE[2]	IBF	OBF	IBOV	PSPM	—	TRISE2	TRISE1	TRISE0	0000 -111	0000 -111
8A	PCLATH[1]	—	—	—	Write buffer for top 5 PC bits					—-0 0000	—-0 0000
8B	INTCON[1]	GIE	PEIE	T0IE	INTE	RBIE	T0IF	INTF	RBIF	0000 000X	0000 000U
8C	PIE1	PSPIE[2]	ADIE	RCIE	TXIE	SSPIE	CCPIE	TMR2IE	TMR1IE	0000 0000	0000 0000
8D	PIE2	—	CMIE	—	EEIE	BCLIE	—	—	CCP2IE	-R-0 0—-0	-R-0 0—-0
8E	PCON	—	—	—	—	—	—	$\overline{POR}$	$\overline{BOR}$	—- —-??	—- —-UU
91	SSPCON2	GCEN	ACKSTAT	ACKDT	ACKEN	RCEN	PEN	RSEN	SEN	0000 0000	0000 0000
92	PR2	Timer 2 Period Register								1111 1111	1111 1111
93	SSPADD	Synchronous Serial Port (I²C mode) Address Register								0000 0000	0000 0000
94	SSPSTAT	SMP	CKE	D/$\overline{A}$	P	S	R/$\overline{W}$	UA	BF	0000 0000	0000 0000
98	TXSTA	CSRC	TX9	TXEN	SYNC	—	BRGH	TRMT	TX9D	0000 -010	0000 -010
99	SPBRG	Baud Rate Generator								0000 0000	0000 0000
9Ch	CMCON	C2OUT	C1OUT	C2INV	C1INV	CIS	CM2	CM1	CM0	0000 0110	UUUU UUUU
9Dh	CVREF	CVREN	CVROE	CVRR	—	CVR3	CVR2	CVR1	CVR0	000- 0000	UUU- UUUU
9E	ADRESL	A/D RESult register Low byte								XXXX XXXX	UUUU UUUU
9F	ADCON1	ADFM	—	—	—	—	PCFG2	PCFG1	PCFG0	0—- —000	0—- —000
Bank 2											
100	INDF[1]	Uses contents of this to address Data memory (not a physical register)									
101	TMR0	8-bit real-time clock/counter								XXXX XXXX	UUUU UUUU
102	PCL[1,3]	Lower-order 8 bits of the Program Counter								0000 00000	0000 0000
103	STATUS[1]	IRP	RP1	RP0	$\overline{TO}$	$\overline{PD}$	Z	DC	C	0001 1XXX	000? ?UUU
104	FSR[1]	Indirect Data memory address pointer 0								XXXX XXXX	UUUU UUUU
106	PORTB	RB7	RB6	RB5	RB4	RB3	RB2	RB1	RB0	XXXX XXXX	UUUU UUUU
10A	PCLATH[1]	—	—	—	Write buffer for top 5 PC bits					—-0 0000	—-0 0000
10B	INTCON[1]	GIE	PEIE	T0IE	INTE	RBIE	T0IF	INTF	RBIF	0000 000X	0000 000U
10C	EEDATA	EEPROM DATA register								XXXX XXXX	UUUU UUUU
10D	EEADR	EEPROM ADdRess register								XXXX XXXX	UUUU UUUU
10E	EEDATH	—	—	EEPROM DATa register High byte						XXXX XXXX	UUUU UUUU
10F	EEADRH	—	—	—	EEPROM ADdRess register High byte					XXXX XXXX	UUUU UUUU
Bank 3											
180	INDF[1]	Uses contents of this to address Data memory (not a physical register)									
181	OPTION_REG	$\overline{RBPU}$	INTEDG	T0CS	T0SE	PSA	PS2	PS1	PS0	1111 1111	1111 1111
182	PCL[1,3]	Lower-order 8 bits of the Program Counter								0000 00000	0000 0000
183	STATUS[1]	IRP	RP1	RP0	$\overline{TO}$	$\overline{PD}$	Z	DC	C	0001 1XXX	000? ?UUU
184	FSR[1]	Indirect Data memory address pointer 0								XXXX XXXX	UUUU UUUU
186	TRISB	Port B Data Direction Register								1111 1111	1111 1111
18A	PCLATH[1]	—	—	—	Write buffer for top 5 PC bits					—-0 0000	—-0 0000
18B	INTCON[1]	GIE	PEIE	T0IE	INTE	RBIE	T0IF	INTF	RBIF	0000 000X	0000 000U
18C	EECON1	EEPGD	—	—	—	WRERR	WREN	WR	RD	X—- X000	X—- U000
18D	EECON2	EEPROM CONtrol register 2 (Not a physical register)									

X   Not known.
U   Unchanged.
R   Reserved; always maintain these bits clear.
?   Value depends on reset condition.
—   Unimplemented; read as 0.
Note 1: These registers can be addressed from any bank.
Note 2: Not implemented in 28-pin members of the family group.
Note 3: Next instruction address if PIC in Sleep mode.

C Instruction Set

C Operators, Their Precedence, and Associativity.

Operator	Operation	Example
Top priority		
Direction (associativity) ⇒		
() [] . ->	Function call Array element Structure element Structure element using a pointer	sqr() x[6] PIA1.CRA
Unary operators		
Direction (associativity) ⇐		
! ~ - + ++ - & * (type) sizeof	Logical NOT Inversion (1's complement) Negative Unary plus Increment Decrement Address of Contents of address Cast Size of object in bytes	!x ~x y=-x y=x- +(y+z) x++ or ++x x- or -x &x *address (long)x sizeof x
Arithmetic		
Direction (associativity) ⇒		
* / %	Multiplication Division Remainder	z=x*y z=x/y z=x%y (Integer types only)
+ -	Addition Subtraction	z=x+y z=x-y
Shift		Integer types only
Direction (associativity) ⇒		
>> <<	Shift left Shift right	z=x>>3 z=x<<3
Relational operators		Boolean objects
Direction (associativity) ⇒		

<
<=
>
>=
	Less than
	Less than or equal
	Greater than
	Greater than or equal
		while (x<3)
		while (x<=3)
		while (x>3)
		while (x>=3)

== !=	Equivalent Not equivalent	while (x==y) while (x!=0)

(continued on the next page)

C operators, Their Precedence, and Associativity (*continued*).

Operator	Operation	Example
Bitwise logic		Integer types only
Direction (associativity) ⇒		
&	AND	x&0xFE (Clear bit 0)
^	Exclusive-OR	x^0x01 (Toggle bit 0)
|	OR	x|0x01 (Set bit 0)
Objectwise logic		Boolean objects
Direction (associativity) ⇒		
&&	Logical AND	x&&y is True if both x and y are True
||	Logical OR	x||y is True if both or either x and y are True
?:	Conditional	x=(y>z)?5:10 x=5 if y>z True else x=10
Assignment		
Direction (associativity) ⇐		
=	Simple	x=3
+=	Compound plus	x+=3 e.g. (x=x+3)
-=	Compound minus	x-=3 e.g. (x=x-3)
*=	Compound multiply	x*=3 e.g. (x=x*3)
/=	Compound divide	x/=3 e.g. (x=x/3)
%=	Compound remainder	x%=3 e.g. (x=x%3)
&=	Compound bit AND	x&=3 e.g. (x=x&3)
^=	Compound bit EX-OR	x^=3 e.g. (x=x^3)
|=	Compound bit OR	x|=3 e.g. (x=x|3)
<<=	Compound shift left	x<<=3 e.g. (x=x<<3)
>>=	Compound shift right	x>>=3 e.g. (x=x>>3)
Direction (associativity) ⇒		
,	Concatenate	if(x=0,y=3;x<10,x++)
Lowest priority		

Index